HANDBOOK OF MATERIALS SELECTION FOR ENGINEERING APPLICATIONS

MECHANICAL ENGINEERING

A Series of Textbooks and Reference Books

Editor

L. L. Faulkner

Columbus Division, Battelle Memorial Institute
and Department of Mechanical Engineering
The Ohio State University
Columbus, Ohio

1. *Spring Designer's Handbook*, Harold Carlson
2. *Computer-Aided Graphics and Design*, Daniel L. Ryan
3. *Lubrication Fundamentals*, J. George Wills
4. *Solar Engineering for Domestic Buildings*, William A. Himmelman
5. *Applied Engineering Mechanics: Statics and Dynamics*, G. Boothroyd and C. Poli
6. *Centrifugal Pump Clinic*, Igor J. Karassik
7. *Computer-Aided Kinetics for Machine Design*, Daniel L. Ryan
8. *Plastics Products Design Handbook, Part A: Materials and Components; Part B: Processes and Design for Processes*, edited by Edward Miller
9. *Turbomachinery: Basic Theory and Applications*, Earl Logan, Jr.
10. *Vibrations of Shells and Plates*, Werner Soedel
11. *Flat and Corrugated Diaphragm Design Handbook*, Mario Di Giovanni
12. *Practical Stress Analysis in Engineering Design*, Alexander Blake
13. *An Introduction to the Design and Behavior of Bolted Joints*, John H. Bickford
14. *Optimal Engineering Design: Principles and Applications*, James N. Siddall
15. *Spring Manufacturing Handbook*, Harold Carlson
16. *Industrial Noise Control: Fundamentals and Applications*, edited by Lewis H. Bell
17. *Gears and Their Vibration: A Basic Approach to Understanding Gear Noise*, J. Derek Smith
18. *Chains for Power Transmission and Material Handling: Design and Applications Handbook*, American Chain Association
19. *Corrosion and Corrosion Protection Handbook*, edited by Philip A. Schweitzer
20. *Gear Drive Systems: Design and Application*, Peter Lynwander
21. *Controlling In-Plant Airborne Contaminants: Systems Design and Calculations*, John D. Constance
22. *CAD/CAM Systems Planning and Implementation*, Charles S. Knox
23. *Probabilistic Engineering Design: Principles and Applications*, James N. Siddall
24. *Traction Drives: Selection and Application*, Frederick W. Heilich III and Eugene E. Shube
25. *Finite Element Methods: An Introduction*, Ronald L. Huston and Chris E. Passerello
26. *Mechanical Fastening of Plastics: An Engineering Handbook*, Brayton Lincoln, Kenneth J. Gomes, and James F. Braden
27. *Lubrication in Practice: Second Edition*, edited by W. S. Robertson
28. *Principles of Automated Drafting*, Daniel L. Ryan
29. *Practical Seal Design*, edited by Leonard J. Martini
30. *Engineering Documentation for CAD/CAM Applications*, Charles S. Knox
31. *Design Dimensioning with Computer Graphics Applications*, Jerome C. Lange
32. *Mechanism Analysis: Simplified Graphical and Analytical Techniques*, Lyndon O. Barton

33. *CAD/CAM Systems: Justification, Implementation, Productivity Measurement,* Edward J. Preston, George W. Crawford, and Mark E. Coticchia
34. *Steam Plant Calculations Manual,* V. Ganapathy
35. *Design Assurance for Engineers and Managers,* John A. Burgess
36. *Heat Transfer Fluids and Systems for Process and Energy Applications,* Jasbir Singh
37. *Potential Flows: Computer Graphic Solutions,* Robert H. Kirchhoff
38. *Computer-Aided Graphics and Design: Second Edition,* Daniel L. Ryan
39. *Electronically Controlled Proportional Valves: Selection and Application,* Michael J. Tonyan, edited by Tobi Goldoftas
40. *Pressure Gauge Handbook,* AMETEK, U.S. Gauge Division, edited by Philip W. Harland
41. *Fabric Filtration for Combustion Sources: Fundamentals and Basic Technology,* R. P. Donovan
42. *Design of Mechanical Joints,* Alexander Blake
43. *CAD/CAM Dictionary,* Edward J. Preston, George W. Crawford, and Mark E. Coticchia
44. *Machinery Adhesives for Locking, Retaining, and Sealing,* Girard S. Haviland
45. *Couplings and Joints: Design, Selection, and Application,* Jon R. Mancuso
46. *Shaft Alignment Handbook,* John Piotrowski
47. *BASIC Programs for Steam Plant Engineers: Boilers, Combustion, Fluid Flow, and Heat Transfer,* V. Ganapathy
48. *Solving Mechanical Design Problems with Computer Graphics,* Jerome C. Lange
49. *Plastics Gearing: Selection and Application,* Clifford E. Adams
50. *Clutches and Brakes: Design and Selection,* William C. Orthwein
51. *Transducers in Mechanical and Electronic Design,* Harry L. Trietley
52. *Metallurgical Applications of Shock-Wave and High-Strain-Rate Phenomena,* edited by Lawrence E. Murr, Karl P. Staudhammer, and Marc A. Meyers
53. *Magnesium Products Design,* Robert S. Busk
54. *How to Integrate CAD/CAM Systems: Management and Technology,* William D. Engelke
55. *Cam Design and Manufacture: Second Edition; with cam design software for the IBM PC and compatibles, disk included,* Preben W. Jensen
56. *Solid-State AC Motor Controls: Selection and Application,* Sylvester Campbell
57. *Fundamentals of Robotics,* David D. Ardayfio
58. *Belt Selection and Application for Engineers,* edited by Wallace D. Erickson
59. *Developing Three-Dimensional CAD Software with the IBM PC,* C. Stan Wei
60. *Organizing Data for CIM Applications,* Charles S. Knox, with contributions by Thomas C. Boos, Ross S. Culverhouse, and Paul F. Muchnicki
61. *Computer-Aided Simulation in Railway Dynamics,* by Rao V. Dukkipati and Joseph R. Amyot
62. *Fiber-Reinforced Composites: Materials, Manufacturing, and Design,* P. K. Mallick
63. *Photoelectric Sensors and Controls: Selection and Application,* Scott M. Juds
64. *Finite Element Analysis with Personal Computers,* Edward R. Champion, Jr., and J. Michael Ensminger
65. *Ultrasonics: Fundamentals, Technology, Applications: Second Edition, Revised and Expanded,* Dale Ensminger
66. *Applied Finite Element Modeling: Practical Problem Solving for Engineers,* Jeffrey M. Steele
67. *Measurement and Instrumentation in Engineering: Principles and Basic Laboratory Experiments,* Francis S. Tse and Ivan E. Morse
68. *Centrifugal Pump Clinic: Second Edition, Revised and Expanded,* Igor J. Karassik
69. *Practical Stress Analysis in Engineering Design: Second Edition, Revised and Expanded,* Alexander Blake
70. *An Introduction to the Design and Behavior of Bolted Joints: Second Edition, Revised and Expanded,* John H. Bickford
71. *High Vacuum Technology: A Practical Guide,* Marsbed H. Hablanian
72. *Pressure Sensors: Selection and Application,* Duane Tandeske

73. *Zinc Handbook: Properties, Processing, and Use in Design,* Frank Porter
74. *Thermal Fatigue of Metals,* Andrzej Weronski and Tadeusz Hejwowski
75. *Classical and Modern Mechanisms for Engineers and Inventors,* Preben W. Jensen
76. *Handbook of Electronic Package Design,* edited by Michael Pecht
77. *Shock-Wave and High-Strain-Rate Phenomena in Materials,* edited by Marc A. Meyers, Lawrence E. Murr, and Karl P. Staudhammer
78. *Industrial Refrigeration: Principles, Design and Applications,* P. C. Koelet
79. *Applied Combustion,* Eugene L. Keating
80. *Engine Oils and Automotive Lubrication,* edited by Wilfried J. Bartz
81. *Mechanism Analysis: Simplified and Graphical Techniques, Second Edition, Revised and Expanded,* Lyndon O. Barton
82. *Fundamental Fluid Mechanics for the Practicing Engineer,* James W. Murdock
83. *Fiber-Reinforced Composites: Materials, Manufacturing, and Design, Second Edition, Revised and Expanded,* P. K. Mallick
84. *Numerical Methods for Engineering Applications,* Edward R. Champion, Jr.
85. *Turbomachinery: Basic Theory and Applications, Second Edition, Revised and Expanded,* Earl Logan, Jr.
86. *Vibrations of Shells and Plates: Second Edition, Revised and Expanded,* Werner Soedel
87. *Steam Plant Calculations Manual: Second Edition, Revised and Expanded,* V. Ganapathy
88. *Industrial Noise Control: Fundamentals and Applications, Second Edition, Revised and Expanded,* Lewis H. Bell and Douglas H. Bell
89. *Finite Elements: Their Design and Performance,* Richard H. MacNeal
90. *Mechanical Properties of Polymers and Composites: Second Edition, Revised and Expanded,* Lawrence E. Nielsen and Robert F. Landel
91. *Mechanical Wear Prediction and Prevention,* Raymond G. Bayer
92. *Mechanical Power Transmission Components,* edited by David W. South and Jon R. Mancuso
93. *Handbook of Turbomachinery,* edited by Earl Logan, Jr.
94. *Engineering Documentation Control Practices and Procedures,* Ray E. Monahan
95. *Refractory Linings: Thermomechanical Design and Applications,* Charles A. Schacht
96. *Geometric Dimensioning and Tolerancing: Applications and Techniques for Use in Design, Manufacturing, and Inspection,* James D. Meadows
97. *An Introduction to the Design and Behavior of Bolted Joints: Third Edition, Revised and Expanded,* John H. Bickford
98. *Shaft Alignment Handbook: Second Edition, Revised and Expanded,* John Piotrowski
99. *Computer-Aided Design of Polymer-Matrix Composite Structures,* edited by S. V. Hoa
100. *Friction Science and Technology,* Peter J. Blau
101. *Introduction to Plastics and Composites: Mechanical Properties and Engineering Applications,* Edward Miller
102. *Practical Fracture Mechanics in Design,* Alexander Blake
103. *Pump Characteristics and Applications,* Michael W. Volk
104. *Optical Principles and Technology for Engineers,* James E. Stewart
105. *Optimizing the Shape of Mechanical Elements and Structures,* A. A. Seireg and Jorge Rodriguez
106. *Kinematics and Dynamics of Machinery,* Vladimír Stejskal and Michael Valášek
107. *Shaft Seals for Dynamic Applications,* Les Horve
108. *Reliability-Based Mechanical Design,* edited by Thomas A. Cruse
109. *Mechanical Fastening, Joining, and Assembly,* James A. Speck
110. *Turbomachinery Fluid Dynamics and Heat Transfer,* edited by Chunill Hah
111. *High-Vacuum Technology: A Practical Guide, Second Edition, Revised and Expanded,* Marsbed H. Hablanian

112. *Geometric Dimensioning and Tolerancing: Workbook and Answerbook*, James D. Meadows
113. *Handbook of Materials Selection for Engineering Applications*, edited by G. T. Murray
114. *Handbook of Thermoplastic Piping System Design*, Thomas Sixsmith and R. Hanselka

Additional Volumes in Preparation

Heat Exchanger Design Handbook, T. Kuppan

Mechanical Engineering Software

Spring Design with an IBM PC, Al Dietrich

Mechanical Design Failure Analysis: With Failure Analysis System Software for the IBM PC, David G. Ullman

HANDBOOK OF MATERIALS SELECTION FOR ENGINEERING APPLICATIONS

EDITED BY

G. T. MURRAY

California Polytechnic State University
San Luis Obispo, California

CRC Press is an imprint of the
Taylor & Francis Group, an **informa** business

CRC Press
Taylor & Francis Group
6000 Broken Sound Parkway NW, Suite 300
Boca Raton, FL 33487-2742

First issued in paperback 2019

ISBN-13: 978-0-8247-9910-6 (hbk)
ISBN-13: 978-0-367-40092-7 (pbk)

Visit the Taylor & Francis Web site at
http://www.taylorandfrancis.com

and the CRC Press Web site at
http://www.crcpress.com

Preface

Engineers involved in design projects, mostly mechanical engineers, are faced with the problem of selecting materials for the subject design. The selection process has been largely empirical. In general, the tendency is to prescribe materials that have been used in similar functions or designs, i.e., to take a conservative approach, which sanctions or promotes "me-tooism." This rather safe approach permits the engineer to avoid criticism for failures that may be material related. But better performance and cost saving can often be realized through the use of more recently developed but well-documented materials.

At present there are approximately 85,000 materials from which the design engineer must make a selection for a specific design. To add to the confusion, new materials are being developed at an almost unbelievable pace. New materials—composites, for example—are now engineered to meet specific design requirements rather than adapting the design to existing materials. It is difficult for the materials engineer, who has been educated in the fundamentals of materials behavior, to keep up with all the new developments, and this problem is even more difficult for design engineers. This book will simplify, to some extent, the task of selecting materials for their designs.

This book was written for design engineers involved with the materials selection process although it can be used as an advanced text for all engineering students. The reader is assumed to have some knowledge of the basic science of materials and materials engineering such as that required for most accredited engineering programs. Hence the fundamentals of materials science have been omitted. An introductory chapter outlines the factors that one must consider in the materials selection process, although not all factors apply to all designs. Each design is unique and the designer will recognize those factors that apply to the design of interest. The introductory chapter is followed by chapters on each of the four materials classes, namely, metals, polymers, ceramics, and composites. (Because of their unique characteristics and related designs, electronic materials have not been included.) The properties of selected representative materials of each class are presented and compared in an easy-to-read form. The effects of temperature and environment have been taken into consideration for each class of materials. Following these chapters, the properties of the various classes of materials are compared. This property data should allow the design engineer to narrow the selection to a certain class of materials and to the subclass of materials within this major class.

Information from many sources has been utilized including data supplied by vendors and data from published articles, the ASM International Handbooks,

Smithells Metals Reference book, and previously published Marcel Dekker, Inc., books. Because of the manner in which the data is presented, it was not possible to cite each source. In many cases average values or a range of values were used. The design engineer should use this data as a guide for preliminary selection only. Vendor-supplied property data and/or quality control test data should be employed for the final material specifications.

Numerous applications of certain materials have been presented; however, the design engineer should be aware that there is no unique correct material. An unbiased approach and some judgment in considering all materials must be employed.

G. T. Murray

Contents

Contributors

Isa Bar-On *Worcester Polytechnic Institute, Worcester, Massachusetts*

Sarit B. Bhaduri *University of Idaho, Moscow, Idaho*

W. E. Burd *Consultant, Bernville, Pennsylvania*

W. H. Hunt, Jr. *Aluminum Consultants Group, Murrysville, Pennsylvania*

R. Nathan Katz *Worcester Polytechnic Institute, Worcester, Massachusetts*

Paul T. Lovejoy *Allegheny Ludlum Corporation, Brackenridge, Pennsylvania*

P. K. Mallick *University of Michigan–Dearborn, Dearborn, Michigan*

Thomas J. McCaffrey *Consultant, Reading, Pennsylvania*

G. T. Murray *California Polytechnic State University, San Luis Obispo, California*

William G. Ovens *Nanyang Technological University of Singapore, Singapore, and Rose-Hulman Institute of Technology, Terre Haute, Indiana*

Thoni V. Philip *Consultant, Reading, Pennsylvania*

Philip A. Schweitzer *Consultant, Fallston, Maryland*

Charles V. White *GMI Engineering and Management Institute, Flint, Michigan*

HANDBOOK OF MATERIALS SELECTION FOR ENGINEERING APPLICATIONS

1
Principles of Selection

William G. Ovens
Nanyang Technological University of Singapore, Singapore, and Rose-Hulman Institute of Technology, Terre Haute, Indiana

1 INTRODUCTION

Material selection is not a process that stands alone. It is part of the larger process of creating new solutions to problems. It is tempting to call this larger process simply "engineering design." However, design itself is part of a larger process of bringing new products to market. One of the names which has been coined for the complete process is "product realization process" (PRP). Design and the PRP will be defined presently.

1.1 The Role of Material Selection in the Design Process

To see where material selection fits in the overall process, consider the definition of "engineering" used by the Accreditation Board for Engineering and Technology (ABET) in the United States. ABET is the agency that determines the strengths and weaknesses of engineering programs, and U.S. trained engineers have almost certainly studied at institutions that build their programs around this definition.

> Engineering is the profession in which knowledge of the mathematical and natural sciences gained by study, experience, and practice is applied with judgment to develop ways to utilize, economically, the materials and forces of nature for the benefit of mankind.

Efficient, sensible utilization of materials is at the heart of engineering, since engineering is the very business of creating new things (from materials) to satisfy needs.

Obviously, selecting the materials that new objects will be made from is part of the process normally called engineering design. The Design Appendix for 1994 ABET visits contains the following definition:

> Engineering design is the process of devising a system, component, or process to meet desired needs. It is a decision-making process (often iterative), in which the basic sciences, mathematics, and engineering sciences are applied to convert resources optimally to meet a stated objective.

In trying to formulate a methodology that will embody that definition and be useful to students and practitioners, various authors have described the design process as a series of steps, each of which has certain characteristics and certain tasks to be performed. The process can be tightly structured with each step well defined. It is possible to give detailed guidance to the designer about what to do, when to do it and how to do it. This is the "systematic" approach which is embodied in the German guidelines VDI 2222 entitled *The Design of Technical Products* (Pahl and Beitz 1988). Alternatively, the process can be described in terms that are less rigid, with more room for adaptation to special circumstances. This latter approach may list, sequentially, certain steps which usually comprise the process. A summary of such an offering might look like this:

Recognizing a need
Defining the problem
  Establishing specifications
Gathering information
Generating concept solutions
Analysis and optimization
  Preliminary layout of design
  Detail design
Evaluation
Presenting results
Approval and release for manufacturing

The initial consideration of material selection usually occurs while generating concept solutions. Many alternatives should be considered from all the material classes. This is where a radical departure from traditional thinking, with its potential for true breakthroughs, will occur if the designer is really open minded. Specific materials for the individual components will be chosen later.

This version of the design process, which focuses on a particular component or system only until the design is approved for manufacturing, can be attacked on several fronts. For instance, it is sequential in nature, implying that one proceeds from step to step in the order given. If the process is altered to include feedback loops (to reflect the "often iterative" requirement in the ABET definition of design) that is an improvement.

Also, this listing would seem to imply that the designer's responsibility is finished at the completion of these steps. After doing these things, the next step on the road to the marketplace is manufacturing. Some designers and their managers would have you believe that if the manufacturing cost is found to be too great, it is the manufacturing department's responsibility to find a less expensive means of production.

A designer's responsibility actually extends beyond the time that manufacturing starts. That is, a designer's involvement in the whole life cycle of the product, not just the portion of the life until the product enters the stream of commerce, is increasingly important. If the designer has to face a dissatisfied, perhaps angry, customer, and recognizes that the dissatisfaction stems from poor functional design or poor design for manufacturability, the lesson is very likely to stick.

There has been a recent wide recognition of the fact that there are issues in the manufacturing, safety, sale and use of a product that are best considered early in the design process which will require the expertise and cooperation of many people because no designer can be sufficiently expert in all of them. This is reflected by the increasing use of "concurrent" or "simultaneous" engineering methods.

The notion that there are several issues in addition to function that should be considered early in the design and material selection process is not new. Forty years ago, one handbook on material selection listed general factors to be considered in material selection, in what seems to be the order of importance (Minor and Seastone 1955):

1. Properties
2. Adaptability to fabricating methods
3. . . .

More than 30 years ago a book devoted entirely to educating engineers about the manufacturability properties of materials was on the market. The preface says, "Both the effects of different processes on materials and the suitability of a material for any given process may be determined from the text, so the suitability of any material for manufacturing a given component may be assessed" (Alexander and Brewer 1963). Since this text was specifically written for students preparing for the Institution of Mechanical Engineers exams in Britain (roughly comparable to the American professional engineering examination), we may infer that the examiners were keen to make sure that neophytes would not become registered engineers without an understanding of the broader issues involved in design. Another author states,

> The object of this volume is to provide a text illustrating how available physical data and knowledge of production methods can be combined at a sufficiently early stage in the design process so as to make a significant contribution towards optimum selection. (S. W. Jones 1970)

Even the word "concurrent" appeared in the texts long before it became a buzzword and started appearing in book titles and continuing education seminars. "Concurrent with mechanical property assessment one must also consider factors such as cost, fabricability and availability" (Hanley 1980).

Concurrent engineering is a name applied to a cooperative relationship among as many of the interested parties as possible in the creation of a new design. This relationship should start at the beginning of the design process. If it does not, there is a real danger that the designer will have created features that are unncessarily difficult to manufacture. The function may be elegant, but the cost and complexity of manufacturing may be prohibitive. It is now widely recognized that most of the cost of a product is fixed at the design stage, before cutting any chips or assembling any parts. When a design is fundamentally flawed from a producibility point of view, attempts to reduce the cost of manufacturing after production has started cannot possibly be as effective as reducing the cost of manufacturing by doing the design correctly in the first place. Efficient manufacturing of a flawed design cannot beat efficient manufacturing of a good design. *And choosing the material that gives the fewest problems during manufacturing is part of creating a good design!* The need for designers to seek advice from the manufacturing department about the manufacturability of a design, including the manufacturability characteristics of materials, has been recommended—and the recommendation has been ignored—for a long time. "A successful designer should work closely with the manufacturing engineer throughout the design process (P. H. Hill 1970).

A few years ago an engineer reviewed failure analysis reports that had appeared in the open literature. He noted that he did not find an instance where the knowledge necessary to avoid the failure had not already been published and was therefore available to, but not used by, the designer. "The general problem is one of making knowledge available to people who do not know that they need it" (Wearne 1979). The quote is in the context of getting designers to utilize failure analysis results to avoid situations that had previously been documented as causing failure. It is just as relevant to the problem of getting designers to consider the manufacturability aspects of design and material selection.

1.2 The Product Realization Process

The word "design" often is used in the limited sense of creating the drawings for some new mechanism or device and then letting someone else figure out how to manufacture the item. To emphasize that there are larger issues in the total life of a product, and the need for cooperation among many people with varying expertise, the term "product realization process" has been suggested as an alternative to the traditional word "design" (Hoover and Jones 1991). The term is beginning to appear in the lterature related to design (Starkey et al. 1994). The PRP starts with a customer's ideas or requirements and ends with the customer's satisfaction with the delivered product.

Between these two lie a host of factors to be considered, among them:

Customer's Ideas or Requirements
Sales and marketing problems and potential
What the competition is doing
Finance department concerns
Configuration design
Available technologies
Codes, standards, patents, legal concerns
Materials
Fabrication
Assembly
Model build and test
Installation
Maintenance
Scrap recovery
Customer's Satisfaction

Concurrent engineering is intended to ensure that as much knowledge of the product realization process is incorporated at the right stages of the design and production process.

1.3 Closing Thoughts on the Design Process

Over the years, several thoughtful and insightful comments have been made by people philosophizing about the difficult nature of the design process. Aphorisms are uncommon in engineering literature, but these descriptive definitions might qualify. They capture the essence without trying to embody the detail. They are useful to mention because receptiveness to the intrinsic vagueness of the design process, and a recognition that there is art as well as science in the process, may be helpful to some who are struggling to come to grips with the fact that, unlike some other aspects of engineering, there is no differential equation which, once it is solved, will ensure either better understanding of, nor success with, the design process. "The design function is the bridge between dreams and drawings" (Bronikowski 1986).

The president of one of America's leading hovercraft manufacturing companies defines design as ". . . an attempt at managing the randomness of events" (Chris Fitzgerald).

J.C. Jones, after reviewing various design methodologies in use prior to 1966 threw up his hands and concluded that design is a "very complicated act of faith" (Gregory 1966, p. 296). Certainly it is no less complicated today.

1.4 Introduction to the Material Selection Process

If design is turning imagination into reality, material selection is putting meat on the bones.

There are numerous schemes and plans proposed to make the design process more systematic. The specific stage of the design process where material selection will occur varies from one practitioner to the next. Some have material selection far down the list of important activities in design, almost as an afterthought. This is based on the assumption that from among all the known materials, one will be suitable. It just needs to be found.

Some forward-looking practitioners envision the day when the design starts with the assumption that the material to be used does not yet exist and part of the design problem is the design of the material (Materials Futures 1988). In such a process, referred to as "materials by design," computer modeling "constructs" new molecules with desired properties, then "tests" and "optimizes" them without the expense of laboratory testing. There are success stories in creating "magic bullet" pharmaceutical products and selective catalysts like ZSM-5 for the conversion of methanol to gasoline. Creating new polymer molecules, i.e., whole new families of polymers, is very expensive when done by experiment in the laboratory. There is promising work in progress to create new molecules by computer modeling at greatly reduced cost.

Other approaches to finding new materials are less fundamental, like the process of looking at the material property data for existing metal alloys and predicting, by simple interpolation, what the properties would be for a nonstandard alloy whose composition lies between standard alloys (Tagaki and Iwata, p. 89 in Doyama et al. 1991). That may be an interesting computer exercise, but it has not created many useful new materials.

On the other hand, a similar method used for polymers has borne fruit. Many of the newest advances in polymer materials comes from judicious blending of existing polymers to overcome the deficiency of one material by adding a desired characteristic from another. Examples are polyphenylene ether modified with high-impact polystyrene to improve processability and ABS modified with polycarbonate to improve burning resistance. A new potentially useful blend is probably created every day in someone's computer.

This book is concerned with a relatively small portion of the product realization process, i.e., that portion in which decisions about materials are needed. Actually decisions about materials are implicit in several steps of the PRP. Whether creating a new material or choosing an existing material, the material will need certain characteristics to perform its intended function. The function to be performed, and therefore the needed characteristics, are determined by what the customer desires. So establishing what the customer wants is a necessary first step. Assuming that the market survey is accurate, the designer will begin to formulate a picture of the features and materials that will meet those requirements. A list of such requirements would typically include, but certainly not be limited to, such items as the desired performance, appearance, expected life, ease of maintenance, and reliability. (A much more complete list is found in Starkey 1992.) Some of the requirements will be absolutely necessary for the product to function, some will merely be desired.

It is important for the designer to understand what the product is intended to do, i.e., what functions the product will be expected to perform. The most formal study of the meaning and use of the concept of function is in value engineering (VE) (Miles 1972, Fowler 1990). The underlying principle of VE is to find the most efficient, lowest cost way to perform the desired functions, without compromising quality, reliability, safety and similar considerations. To perform a function means to do something, to act in a certain prescribed way, to accomplish a desired task. All functions may be described in their simplest terms by a verb and noun which describe what action is to be taken and what object is to be the recipient of the action.

Quality function deployment methods have been used successfully to determine accurately what the customer really wants. One of the problems that arises is the common use of vague and general terms in a customer's responses. "Make it as light as possible," and "Make it easy to assemble/install/use," are typical examples. It is necessary to quantify the words used. "Light" can be expressed exactly as a target weight or as a specific shade of color. "Easy to assemble" can be quantified as the number and duration of the steps required, the number (and perhaps the complexity) of the tools required, the amount of force, power or energy required, the time required to read and understand the instructions, etc.

Each function or characteristic will have a corresponding material property that will be helpful in evaluating each material's ability to perform the desired function. Before choosing the material, it is customary to translate the functional requirements into specifications. There are three common types of specifications: product specifications, process specifications and material specifications. A specification is a statement of the requirements that must be met for a product, system, component, material, etc., to be considered satisfactory. In each of these there will be characteristics, properties, quantities, etc., that can be measured. The measurement method may be specified. It is a way to write down the requirements so that all parties to the operation may agree that they understand what is needed and how they will know that what is needed has been achieved.

A product specification is a statement of what functions the product must perform to be considered satisfactory. Additional, optional performance that might enhance the usefulness of the product must be identified clearly as optional to avoid any possibility of confusion. Clarity, conciseness and the provision for unambiguous, quantitative evaluation of performance are key features of a well-written specification.

A process specification describes that process that must be used in producing the product. It will typically include sections that tell

The name of the process
Purpose
Related documents
Which materials to are be processed
What materials are used in the process
What result is expected
How the result will be measured
The process limitations
Precautions to be observed

Sometimes a process specification is incorporated as a part of a material specification. For instance, in aluminum alloy 7075-T6, the composition (7075) is controlled by one specification and the details of the heat treatment (T6) necessary to achieve the specified properties are contained in another. A separate process specification is usually required for each material used in the product or in each component of the product.

A material specification uniquely identifies the composition, properties or other characteristics of the material to which it refers. It is an attempt to standardize the material so that variations from batch to batch and vendor to vendor are eliminated, or at least restricted to an acceptable range. The information contained in a material specification typically includes name of the material, properties and/or composition (sometimes including tolerances), acceptance criteria and tests, packaging, marking, and various appendices.

Occasionally, materials from different sources may behave differently in service even though they both comply with the same specification. This is usually the result of not correctly identifying the properties which needed to be controlled for consistent performance. For instance, purchasing carbide materials for wear-resistant applications using hardness as the acceptance property may lead to problems. Two sintered carbide inserts meeting the same specification requirements for composition and hardness, but coming from two different sources, may have wear rates that are significantly different. Some specifications, e.g., ASTM, list a range of compositions. More than one alloy (SAE No.) may satisfy a given ASTM specification.

The "specification stage" of the design process refers to the product specification. It is not likely that any material selection decisions are possible at this stage. In some instances, the performance specifications will dictate that, at the current state of the art, a small number of materials, and maybe only one, can meet the requirements. When building transistors, there is a very small number of materials that can ever be considered.

In the more common cases that this book addresses, the performance specifications will lead to the generation of one or more concepts for the design that will meet the requirements. At the beginning of the conceptual stage, all materials must be considered as candidates. For example, a concept may require a spring. Metal, ceramic, polymer and composite springs must all be considered viable options until there is some narrowing of the concepts that precludes one or more of these materials. For each proposed concept, there will be certain classes of materials, and maybe specific materials, that will obviously be suitable. The material selection process proceeds throughout the design process, but the range of materials that might be considered will narrow as the design nears completion. One can very quickly narrow the field to a few promising candidates using experience, expert knowledge or selection aids that cover all classes of materials or give broad ranges of properties.

It is vital to consider the interactions among material, configuration and manufacturing method at this stage. The configuration is likely to be significantly different for a structural component which might be made either from a metal or a composite. For that reason, it will be necessary to carry on parallel design efforts to pursue the ramifications that the different materials will present for the design. With composites in particular, the performance of the component is greatly dependent on the manufacturing method, so the selection of the material and the manufacturing method are almost one process.

Once the winning concept has been established and the details are beginning to be formulated, the range of candidate materials will be narrowed considerably. Certainly the class of material will have been determined. The specific composition and properties may not. Now the need is for specific property data for the candidate materials. The materials are likely to be quite similar in properties. With rather small differences among them, the decision will require more specific and detailed information than was required at the concept stage. Handbooks and data base information will be particularly useful.

In the final detailed design, one material will have been chosen. Complete information about the material and its properties is required. In some cases this means that a supplier needs to be identified because the properties of nominally identical material may actually vary from supplier to supplier. For metals, the standards and specifications are so well established, this is not usually a significant problem. On the other hand, properties of polymers are not standardized. (The tests methods for determining the properties are mainly standardized, but the expected values, say minimum tensile strength, are not.) Indeed, the properties vary significantly from one manufacturer to the next because the competitive edge is often found in purposely offering a material with different properties than the competitors. The manufacturer's data sheets for the actual material should be consulted. These data sheets cannot actually predict accurately the properties that will be achieved in the finished product because the properties are so sensitive to product configuration and manufacturing process parameters.

In certain cases, in-house testing might be needed to verify that the material property values used for the design calculations are actually achievable in the product itself. This might included designs where

1. Suitable factors of safety cannot be decided upon.
2. Reliability is particularly important, but the appropriate statistical information is either not available or not trustworthy.
3. Failure would be catastrophic and it is vital to verify properties for legal defense reasons.

Just as the candidate materials are narrowing as a winning concept emerges, so do the likely manufacturing processes for the various components. The appropriate manufacturing process is dependent on many factors, among which size, shape, quantity, and material are likely to be particularly important. There is an interrelationship between the material, the service conditions and the manufacturing process. Each must be compatible with the others. A material which is suitable for a particular service may not be compatible with the manufacturing process that is appropriate for the part's configuration. There is no point in choosing a material which would be fine in service but which cannot be fabricated into the required shape.

Three examples from actual practice will serve to illustrate this point.

1. If a certain part's configuration and production quantities suggest that injection molding would be appropriate, the material selection must be made among only those plastics which can be injection-molded, even if some important service property has to be compromised. If that important property cannot be compromised, then redesign to a configuration that can be manufactured by a process that is appropriate for that material is likely to be necessary.

2. A designer may choose an aluminum casting alloy with good elevated temperature strength for a high-temperature application. It is a bad choice if the designer fails to note the material has such poor hot tearing resistance that, with the particular configuration of this design, no sound castings can be produced; i.e., all castings crack during solidification and cooling.

3. The allowable stress for welded structures of heat-treated or cold-worked aluminum structural shapes is different from the allowable stress for that same material used in bolted construction. Thus, successful material selection may depend just as much on accurate information about the fabrication characteristics of the material as it does on the mechanical and physical properties of the material as it does on the mechanical and physical properties of the material as received from the supplier.

Eventually, one material will be chosen from among all the candidates. As much precise, reliable, detailed information as possible will be required at this point. Manufacturer's information, in-house test results, history from previous usage of the same or similar materials, and any other available information about this material in this application will be needed. Obviously, much of this may not be in the open literature.

1.5 Reasons for Choosing New Materials

When products are new and have never been produced previously, the need for making material selection decisions is obvious.

With existing products, material changes may occur for the following reasons:

1. *Reduce material cost.* Commodity prices fluctuate for many reasons, so the material price to the producer may vary. When the fluctuations are expected to be cyclic (e.g., seasonal), there is not much reason to search diligently for lower cost substitutes. The fluctuations in market price can be built into the pricing scheme for the product. Alternatively, temporary substitutions may be made until the price comes back down, or the price charged to the consumer may be allowed to fluctuate with the commodity price.

When the upward trend is perceived to be permanent, the search for permanent substitute materials may be initiated. Gold-based brazing alloys were technically superior to other choices for use in gas turbine engines for many years. They were also economical so long as the price of gold was held artificially low because the U.S. government used gold as a fixed price standard for international trade. When the United States abandoned that fixed price for gold, the cost of the alloys increased by a factor of more than 10 in a fairly short time with no indication that the price would ever return to its old level. Technically satisfactory and economically realistic substitutes were found. Copper-manganese-cobalt braze alloys replaced gold-nickel and silver-palladium alloys in gas turbine applications strictly on price (Bradley and Donachie 1977).

On the other hand, when silver prices increased by a factor of ten a few years ago, it was due to manipulation of the market by investors using dubious techniques. That was destined to be a short-lived abberation, and prices returned to normal quite quickly.

2. *Reduce production cost.* Manufacturing methods improve constantly. Wire EDM, for instance, has significantly reduced the cost of making extrusion dies, so that extruded sections and materials are economical now in situations where they would not have been a few years ago. Changing from die cast aluminum or magnesium housings for chain saws to plastic (often glass-filled nylon) is motivated by the reduced cost of the injection molding dies and machines. Numerous other light-duty power-tool housings have been converted from metal to plastic for the same reason. Naturally, there are other benefits that may accompany the change, so the decision to change cannot be said to be production cost exclusively, although that would probably have been sufficient.

3. Accommodate some desired change in function. As a product line matures in the marketplace, derivative products will spin off of the original. The desire to make new variations that are bigger, smaller, differently colored, more luxurious or more economical may dictate a material change.

4. Solve some material processing problem. As material suppliers come and go, new suppliers may not be able to provide material with exactly the same characteristics that the manufacturing process has been tuned for. To get production back on track, the choice may come down to changing the process or changing the material.

5. Take advantage of new materials or processing methods that become available. New materials come onto the market daily. In most companies, the designers

and other engineers are preoccupied with problems of bringing new products on the market or with solving the ***problem de jour***. Reconsideration of material selection for mature products that are not giving immediate problems is a luxury that few companies indulge in. However, engineers experienced in value engineering techniques know that any product that is 10 years old, and many that are 5 years old, are excellent candidates for cost reduction and function improvement using the value engineering methods, which include consideration of new materials.

6. Incorporate failure analysis recommendations. When products fail, engineers often need to determine the cause. There may be a need to establish responsibility in anticipation of litigation or there may be a need to ensure that the failure for the same reason does not recur. Failure analysis methods vary. There has been public discussion of the need for establishing standards for investigations of failures on certain kinds of products (Roberts, in Fong 1981, p 171), but no action has been forthcoming. There is at least one attempt to create a classification system for types and causes of failure (Tallian 1992). Causes of failures are numerous, but among the causes may be improper choice of material for the application. For instance, an alternative material that is more damage tolerant, more corrosion resistant, or perhaps less likely to suffer problems in a weld's heat-affected zone, may be indicated. These are often subtle problems that cannot be anticipated in advance of actual use because some users will have operating conditions that cannot reasonably be foreseen.

2 FACTORS AND CRITERIA IN MATERIALS SELECTION

2.1 Properties

Some engineers must make material selection decisions without the aid of materials experts. They will, by definition, have less knowledge than the expert about the range of materials that might suit a particular purpose. It is common that a relatively small number of materials, the details of which the engineer can master, will satisfy most of the applications. For the remainder, expert help should be sought. There is a real danger that an engineer will chose an "adequate" material from the familiar repertoire when actually a new material would be better. When existing products are modified or updated, generally less thought and effort is put into the material selection for the modification than for the original product. However, the modification may actually change the material requirements enough so that it behaves like a new product.

This problem is likely to occur when engineers get very familiar with a certain material for a particular application. Then the material is inappropriately used for a new, similar product (sometimes called "like parts engineering") because the engineer fails to recognize that the material properties will be altered by the new geometry. For instance, in scaling up heat-treated parts to carry larger loads in tension, it is not appropriate merely to increase the cross-sectional area in proportion to the increased load. Doing so would assume that the strength remains the same as the part size increases. That is not a valid assumption because the response to heat-treating is a function of the part dimensions.

Part of the design and material selection process is predicting the probable modes of failure and then choosing material with properties appropriate to preventing failure. A matrix showing failure modes in metals and the properties related to preventing failure by that mode has been published in more than one place (Smith and Boardman, Metals Handbook, vol. 3, 1980, p. 828; Dieter 1991, p. 235).

Statistical Considerations

Material properties are measured by testing. Many methods are standardized so that data from various sources may be compared without fear that they may not really represent the same conditions. Even when a standard test method is used, attempting to replicate a result (all conditions nominally identical for two or more specimens) will inevitably produce some variability of the results. The variations of nominally similar materials may be caused by differences in chemistry, heat treatment and other factors that will naturally vary from lot to lot. Also, product geometry (e.g., plate thickness), different vendors, different test machines and a host of other factors may cause variations.

Statistical treatment of the data is therefore appropriate. One should expect, at a minimum, to find data for the mean or median and the variance or standard deviation of the distribution. Data compilations, however, usually give a single value for most material properties. This value will usually represent either the mean or some other percentile (with confidence limits) of the distribution of all test values.

For static strength, it is usually safe to assume that the reported value is the one for which the investigator has 95% confidence that 99% of all test values will be at or above the reported value. This is the "A" basis strength defined in MIL-HDBK 5, now widely used by ASTM and most suppliers. It is determined by the statistics of one-side tolerance limits.

The reporting of single values gives no indication of the expected scatter. In most instances that will not be of any practical concern because the assumption is that so long as the minimum value is not exceeded in service there will not be a "failure." Where the variance is important, as in probabilistic design, then additional information about the actual distribution will be required.

Ductility is most likely to be given as a minimum value which does not have any statistical significance associated with it.

For elastic properties, density, thermal properties and many others where failure is not a probable consequence of exceeding the handbook value, the reported value is probably a mean value.

Fatigue behavior of metals shows more scatter than other mechanical properties. Thus, statistical treatment of fatigue data has been common for longer than for other types of mechanical behavior. Confidence limits often appear on published S-N curves. Tolerance limits can be calculated using statistical methods. However, it is very risky to assume that a "fatigue strength" found in a data source has the same percentile and confidence limits as either yield or tensile strength. If the statistical details are not specified, make no assumptions about what they might be.

Mechanical Properties

Stiffness, modulus of elasticity, Young's modulus, modulus of rigidity, and shear modulus all refer to the fact that there is a relationship between the stress and the resulting strain in the elastic materials. The property that expresses this relationship is the constant of proportionality between stress and strain. The common symbol for Young's modulus is *E*, and the units are stress. It is usually determined by measuring the slope of the linear elastic portion of the stress-strain curve. The stress and strain state may be in tension, compression, or shear. When the material is not linear or not elastic, alternative methods are used. One is the secant modulus, i.e., the slope of the line that connects the origin of the stress-strain curve with some arbitrarily chosen point on the curve. The chosen point may be expressed as some fraction of the stress or the strain. In gray cast iron, for instance, that point is at a stress equal to one fourth of the tensile strength of the material. In plastics, the point is sometimes taken at 1% strain. When using the secant modulus to compare one material against another, it is vital that the definition used is the same for all the materials being compared. The other common alternative method for finding elastic modulus is to determine the tangent to the stress-strain curve at the origin. In polymers, where the nonlinearity is considerably more pronounced than in metals, it is important to be sure that the same method is used when comparing data from different sources.

Unless otherwise stated, the modulus values published in data compilations of metals and ceramics may be assumed to represent isotropic, polycrystalline specimens. In single crystals, the modulus may be expected to vary considerably with crystallographic direction. The maximum modulus in iron single crystals is twice the minimum. In directionally solidified metals (whether single crystal or not) and single-crystal ceramics, this may become a significant issue.

For most ceramics and metal alloys of fixed composition, the modulus is essentially constant. Tests of several lots of nominally identical material will have the same mean value and a small standard deviation or variance. This is true regardless of the strength (heat-treat or cold-worked condition) and stress state (tension or compression).

There are exceptions. The modulus of gray cast iron is quite variable. For ASTM class 20, the modulus may vary from 66 to 97 GPa (9.6 to 14.0 Mpsi) (Metals Handbook, 9th ed., vol. 1, p. 19). In other grades the mean value is dependent on the composition, but for each grade the modulus is also quite variable.

Cold-working induces texture (anisotropy) in many materials as the deformation tends to cause certain preferred crystallographic directions to take up specific orientations relative to the principal strain directions. Several mechanical properties, including elastic modulus, may vary with direction. Measuring the elastic modulus in several directions is the underlying principle of one test method to determine the formability of anisotropic sheet metal (Mould and Johnson 1973).

In polymers, flow in a mold is likely to induce anisotropic behavior as well. Liquid crystal polymers are a good example.

Elastic modulus may usually be assumed to be the same in tension and compression. There are exceptions. Ti 6-2-4-2 has 10% greater modulus in compression than in tension at room temperature. The difference rises to 15% at 1000°F.

Designing a material for a certain value of elastic modulus is relatively easy in fiber-reinforced composites where the fiber and matrix have such different elastic properties. By mixing different proportions of high-modulus fiber and low-modulus matrix, the resulting modulus will lie somewhere in between and is predictable by some variation of the "rule of mixtures." Sintered carbides are also specifically designed to have different moduli for different applications.

In direct shear or torsion, if the relationship between stress and strain is linear, the constant of proportionality is the shear modulus, or modulus of rigidity. It is usually denoted by G and has units of stress.

The shear modulus, G, Young's modulus, E, and Poisson's ratio, ν, are related by the equation

$$G = \frac{E}{2(1 + \nu)} \tag{1}$$

If two of the three are known, the third may be estimated.

The strength of a material is a measure of the ability of the material to resist stresses applied to the material without suffering permanent alteration. The stress may be tension, compression, shear or some combination. In material property data compilations, tension is the most likely state of stress to be reported. For brittle materials and wood, compression is commonly reported. Shear strength values are considerably less common.

Many different kinds of strength measurements are defined. The material's inability to resist further application of the stress depends on the stress amplitude and a host of other factors like temperature (cryogenic to elevated temperature), time of application (creep, stress rupture), strain rate (static to dynamic, or impact, loading rates), variability of the load (fatigue), corrosive environment, presence of defects (fracture toughness) and various combinations of these factors. In a test in tension, the effect of the stress might range from merely beginning to deviate from linear behavior (the proportional limit) to fully developed plastic deformation (yield strength) to maximum load-carrying capacity (ultimate tensile strength) to complete fracture. The stress on the specimen or part when each of these effects occurs corresponds to a strength value that might be quoted in a handbook. Each of these strength values has some use in design, quality control, failure analysis and other aspects of engineering.

Proportional limit is the stress at which detectable deviation from linear behavior occurs in a material. Elastic limit is the stress at which some detectable plastic (permanent) deformation occurs. They are difficult to determine accurately, and the sensitivity of the measuring instruments has a large influence on the values of both elastic limit and proportional limit that is recorded. These are of little concern in most instances. Appropriate factors of safety applied to the yield or tensile strength, both of which are much easier to measure, keep the working stresses substantially less than the elastic limit or proportional limit anyway.

Yield strength is the stress at which plastic deformation is fully developed in some portion of the material. A few metallic materials (strain aging types, the most common of which are the annealed or normalized low-carbon steels) show a sudden transition from elastic to plastic behavior as the applied stress reaches a critical value. This give a true yield point, an observable physical phenomenon from which the stress can be determined quite accurately.

In other metals the transition is gradual. The customary definition of yield strength for these materials is the stress necessary to cause a predetermined amount of plastic strain, called the offset. In this case, yielding has clearly occurred at the stress called the yield strength. The amount of strain used to define the strength is arbitrarily chosen. Several different values are used in different countries, for different materials and for different applications. Offset strains as small as 0.01% for some military requirements to as large as 0.5% for brass and soft steel sheet are used: 0.2% is probably the most common. Since greater stress is required to cause the larger offsets, the quoted value of offset yield strength is a function of the offset used. Therefore it is vital before camparing values from different sources to determine that the offset used is the same in all cases. It is not safe to assume which value was used. Well-edited data compilations will always state the offset value.

In brittle metals, yielding does not really occur. Occasionally values of yield strength are quoted for brittle materials like gray cast iron. That is almost certainly the stress at some arbitrary amount of strain. This is a special case because the material has essentially no linear portion to the curve and fracture occurs at very small strains. It does not yield in the conventionally accepted meaning of the term. In other brittle materials, like ceramics, there is no definition for yield strength.

In plastics, the yield strength may be defined as the stress at which the first load peak occurs during the tensile test. It may also be defined as the stress to cause a predetermined amount of total strain, like (1/4)% or 1%, which is not really related to a yielding phenomenon as used in the usual sense for metals.

It is often assumed that the yield strength of metals is the same in tension and compression. This is only approximately true. Statistically significant differences have been noted for some materials, but practical problems seldom arise from neglecting the differences. However, when the compressive stresses are large, as in roller bearings and gear teeth, care should be taken to verify whether or not the reported tensile value is sufficiently accurate for design of compression members.

Yield behavior in pure shear in ductile metals is quite accurately described by the von Mises criterion, which predicts that the shear stress to cause yielding will be about 57% of the axial stress necessary to cause yielding. Most materials of this type actually yield in shear at stresses between 55% and 60% of the axial yield strength.

The tensile strength, also called ultimate strength or ultimate tensile strength, is defined as the nominal stress at the maximum load of the tensile test. Because the nominal stress is calculated using the original cross-sectional area of the specimen, the tensile strength of a material is not the actual, or true, stress on the material when it carries the maximum load. It is fictitious in the sense that the area used to calculate it is not the area that is carrying the load that is used to calculate it. Nevertheless it is useful for several reasons. In design, it permits simple calculations for the maximum load-carrying capacity of tension members prior to final failure. It is generally of no interest what the actual area of the member will be when it reaches its maximum load capacity. The tensile strength is relatively easy to measure, so it can be used as a quality control item to determine the consistency of material from lot to lot. The fatigue strength may be estimated from the tensile strength in some materials. The estimate is subject to considerable uncertainty, so it should not be used if fatigue data are available or can be determined with reasonable effort. Brinell hardness correlates well with tensile strength for non-cold-worked steels.

Modulus of rupture (MOR) is the maximum stress that a specimen can withstand when loaded in three-point bending, i.e., a simply supported beam with central point load. It is calculated by the simple beam bending equation. This kind of test is used when tension testing is difficult or unreliable because of brittleness of the test material. Brittle materials are intolerant of any misalignment of the grips in conventional tension testing that causes bending stresses. Both gray cast iron and ceramics are commonly tested in three-point bending and the results are reported as modulus of rupture. When the modulus of rupture is compared with the tensile strength, the MOR is usually greater. This effect is attributable to the fact that in brittle materials the presence of surface defects greatly influences the fracture behavior. In a brittle material, the fracture strength and tensile strength are the same. In bending, which is used to determine the MOR, a smaller portion of the surface is subjected to the maximum stress in tension. In an axial tension test, which is used to determine the tensile strength, the whole surface is subjected to the maximum stress. Consequently, in the axial test, there is a greater probability of having a critical surface defect simply because a greater proportion of the total surface area is at risk.

In fatigue, the stress on a part varies with time. If the stress is sufficient it may initiate a crack. The crack may grow without any apparent degradation in the performance of the part until the crack reaches a critical size. Then the part fails, without warning. If the final fracture region is a small fraction of the total fracture surface, it is likely to appear to be a "flat" fracture with very little apparent ductility, leading to the conclusion that it is a brittle failure. However, scanning electron microscopic examination of the fracture surface often reveals that the final fracture region has characteristics associated with a ductile overload failure, even though the overall deformation is very small.

The stress fluctuations may be regular, as in automotive valve springs fluctuating between the open and closed positions. The fluctuations might be irregular, or random, as in an automobile suspension where the loads imposed by the road are quite variable. Laboratory specimens may either simulate the actual components and their expected loading, or they may be a standard specimen with regular fluctuating loads. Data from the latter case is what usually appears in handbooks. Relating these data to the complicated geometry and loading patterns encountered in real service is a significant design problem.

Depending on the stress level, the strain resulting from the stress may be mainly elastic or mainly plastic. If the strain is mainly elastic, many more cycles will be required to cause failure than if it is mainly plastic. The former is called high-cycle fatigue, the latter, low-cycle fatigue. The dividing line between the two is not precisely fixed. Some workers use equal elastic and plastic strain as the divider, some say that 10,000 cycles or less of life is indicative of low-cycle fatigue. In fact, they are often nearly the same but one has a physical interpretation and the other is arbitrary. Another physical difference between them is the observation that usually in low-cycle fatigue the strain is sufficient to initiate the crack fairly early in the life of the part. Thus, the majority of the cycles are causing the crack to grow. In high-cycle fatigue, the opposite is true: most of the life is spent trying to initiate the crack.

When the stress on the test specimen varies between two known values, the difference between them is the stress range. Half the range is called the stress amplitude, and the average value of the two extremes is called the mean stress. Mean stress plus stress amplitude = maximum stress; mean stress minus the stress amplitude = minimum stress. The stress ratio is defined as the minimum stress divided by the maximum stress. When the mean stress is zero, the stress ratio is −1. In fact, the two most common fatigue tests are probably sinusoidal variation of stress on rotating specimens loaded in four-point bending (the R. R. Moore test, simulating railroad car axles), and fully reversed bending of a cantilever (ASTM B-593 and ASTM D-671). Both have $R = -1$.

In low-cycle fatigue testing, the strain range is considered the controlling factor instead of the stress range. In general, in low-cycle fatigue a material's ductility is more important than strength. When a material's ductility and strength are inversely related, as is generally true, the lower strength material may be the better choice for certain fatigue applications. When the strain range is fixed, a lower strength material with better ductility will survive more cycles than a higher strength material with poor ductility.

The most common display of fatigue data is the S-N curve where alternating stress (vertical axis) is plotted against number of cycles to failure. Most S-N curves are for zero mean stress. There are analytical methods to estimate the effect of nonzero mean stress (Goodman and Gerber diagrams, etc.). Hence it is not common to see S-N data for nonzero mean stress, except in constant-life diagrams. Positive mean stress will either reduce the life if the alternating stress is unchanged, or the alternating stress must be reduced if the life is to remain the same. Negative mean stress has the opposite effect.

Two distinct types of S-N behavior occur. In many ferrous metals (but not stainless steel), there is some minimum stress required to cause fatigue failure. When operating at stresses less than that endurance limit (fatigue limit), infinite life can be expected. In most other metals, there is no known lower limit to the stress which will ultimately cause failure. However, the stress at which a part can survive for any predetermined number of cycles can be found. This result will be quoted as "the fatigue strength at *X* cycles." Tests to 500 million cycles are fairly common. Reporting the number of cycles for which this fatigue strength is determined is vital for the data to be useful in design.

Fatigue tests are time consuming. A 500-million-cycle test in an 1800 RPM machine requires more than six months. Sometimes an estimate of the fatigue strength is desired for preliminary design purposes. For most metals, the fatigue strength is found to be between 35% and 60% of the tensile strength. High-strength steels may have endurance limits as low as 20% of the tensile strength. For the low- and medium-strength steels, the data tend to cluster around 50%, leading to the rule of thumb that the endurance limit is about half of the tensile strength. One author wisely included the admonition, in italics for emphasis, that this rule of thumb is suitable "*for students only*" (Shigley and Mitchell, 1983, p. 276). Note that the endurance limit is significantly less than the yield strength, also. In fact, the life of fatigue specimens running at a stress equal to the yield strength often exceeds 1000 cycles.

Two other material property data items may be reported. The fatigue notch factor is the ratio of the fatigue strength of an unnotched fatigue specimen to the fatigue strength of a notched fatigue specimen when both fail at the same number of cycles. The symbol is almost universally given as K_f. This is analogous to the static or geometric stress concentration factor, K_t.

The notch sensitivity, q, is a measure of the influence of a notch on the fatigue strength compared to the influence of a notch on the static strength. It is defined as

$$q = \frac{K_f - 1}{K_t - 1}$$

When the value of q is 0, the presence of a notch influences the fatigue behavior only by increasing the local stress over the nominal stress by the factor of the K_t. Gray cast iron is a good example where the graphite flakes already in the microstructure act like stress concentrations, so any notch due to external geometry has little or no additional effect.

A value of 1 means that the presence or absence of a notch has a great influence of the fatigue behavior. Hardened steels are very notch sensitive, annealed steels somewhat less so, and aluminum alloys even less so.

Fatigue loading may occur simultaneously with other conditions that would be damaging even if they occurred alone. There are two common cases. Aggressive environments accelerate fatigue failure (corrosion fatigue), and elevated temperatures may cause an interaction between the creep phenomenon and the fatigue.

Creep is slow continuous deformation under stress. The stress for this plastic deformation is far less than the yield strength that would be obtained at the same temperature in a short-time (conventional) tension test. The design engineer must be aware that such yield stress values cannot be used. In metals, at temperatures less than 40% to 50% of the absolute melting temperature, creep may occur, but it is usually characterized by a maximum creep rate occurring as the load is applied followed by ever-decreasing strain rates. Thus, it may be a problem in some instances because of the initial changes in dimensions, but a steady state is eventually reached. This is called primary creep. At temperatures greater than 40% to 50% of the absolute melting temperature, creep may become a more important problem because the parts will continue to deform after the primary creep rate decreases to some nonzero value. Then the creep proceeds at a constant rate (secondary creep) for some time, after which the rate increases again (tertiary creep) and finally failure occurs. In creep testing, if constant load (or sometimes a constant stress) is maintained on a tension specimen until failure occurs, it is referred to as a creep rupture or stress rupture test. Increasing the stress or the temperature will decrease the time for rupture to occur. The lower melting point metals may creep at room temperature under load. In early applications of aluminum electrical wiring, for instance, creep under the clamping pressure of screw heads caused a reduction of the clamping force and a consequent increase in contact resistance. There are numerous recorded instances of that being the cause of electrical fires. The metallurgy of the wire material has been modified, so that with the correct installation procedures it is no longer a problem. Lead and zinc alloys creep at room temperature, the former being a significant problem when solder is used as a structural element in electronic applications and the latter in structural die castings. Solder joint failure due to creep has also been observed in nonelectronic parts.

Stress relaxation occurs when a part is deformed to a certain strain, and then a creeplike phenomenon reduces the load necessary to maintain that strain. Loss of tension in bolted joints is an example, as is loosening of plastic strapping material around packages. The initial tension in both cases decreases over time even though the dimensions of the part do not change. Springs preloaded by a fixed deformation (automotive valve springs) may suffer stress relaxation and thereby change the force they apply (equivalent to changing their free length).

Data to predict stress relaxation are not available in compact and directly applicable form. Stress relaxation tests are described in ASTM E 328, *Stress Relaxation Tests for Materials and Structures*.

In plastics, creep and stress relaxation at room temperature are normal and must be accounted for in the design of structural parts.

In ceramics, creep is also possible at elevated temperature. Creep has even been observed in diamond under certain conditions.

Many different formats are used to present the data. Graphs of rupture stress (vertical) versus time to rupture at various temperatures are common. Also, the creep curve (strain versus time or log strain versus log time) may be found. The steady-state creep rate may be tabulated. Another format is to report the stress to produce (1/2)%, 1% or 2% strain in a given time at various temperatures.

Some structures subject to creep failure have design lifetimes of many years. It is normally unreasonable to test creep specimens for the full expected lifetime of long-life products, so accelerated tests are performed at higher temperatures to reduce the time to failure. Then design data is extrapolated from them. Extrapolation requires that the behavior can be expressed by some mathematical model that is valid over the temperature and time ranges being considered. Parameter curves are commonly used to condense data of several variables into a single curve. Many expressions have been proposed to represent the relationships among stress, time and temperature in creep behavior (H. R. Vorhees, "Assessment and Use of Creep-Rupture Properties," in Metals Handbook, vol. 8, p. 329, 1985). Of these, the Larson-Miller parameter may be the most widely known because of how long it has been used (Larson and Miller, 1952). It is not necessarily the most accurate. Data for the parameter constants for various materials are published in handbooks that deal specifically with creep. Designing against creep failure is complicated by the amount of scatter in data recorded during testing, the difficulty of obtaining consistent results when specimens from the same lot of material are tested in different labs, and the imperfections in extrapolation models. Consultation with competent metallurgists should be sought.

Hardness is measured by trying to scratch or indent the test material with a standard tool or tools. Harder materials will scratch softer materials and softer materials will have a larger indentation than harder materials. A scale with arbitrary units is established to quantify the relative hardnesses of materials. Many hardness tests have been invented for metals. Brinell, Rockwell, Vickers, Knoop, and Shore are the most common. The results are convertible from one hardness scale to another for some metals using ASTM E 140, *Standard Hardness Conversion Tables for Metals*.

Hardness of nonmetals may also be measured. A durometer is used for plastics and rubbers. A standard indenter is pressed into the surface of the material with a preset force, as described in ASTM D2240. The penetration of the indenter is inversely related to the hardness of the material. There may not be permanent deformation in the polymeric materials when the hardness test is performed. ISO/48 and ASTM D1415 describe another type of penetration under a known load with the result reported as "international rubber hardness degrees." Mohs scratch hardness is used for minerals. In this test, 10 minerals varying from very soft (talc = #1 hardness) to very hard (diamond = #10 hardness) are used to determine which standard material will scratch the unknown sample.

Specialized tests are invented from time to time to address particular problems in particular industries. Because they are not widely recognized and used, or even described in comprehensive, authoritative handbooks, they may add to the confusion that an engineer faces when trying to sort out the relationships among the profusion of hardness tests. Webster and Barcol hardness (ASTM B647 and ASTM B 648) are examples.

In material selection for structural parts, hardness is seldom used. More commonly, once the strength requirements are established, the corresponding hardness becomes the material requirement. This is possible because the hardness correlates closely with strength in many instances. The nondestructive hardness test is much easier to perform in a production and quality control environment than is the tensile test. Consistency of results from the hardness test ensures consistency of strength in the parts, but does it in a much simpler, less costly, way.

Wear rate also correlates closely with hardness in many instances. There are exceptions. Leaded and unleaded 70-30 brass have nearly identical strength and hardness, but their wear rates differ dramatically. Sintered carbide inserts in the wear strip on the bottom of snowplow blades are specified by hardness. However, among all inserts meeting the specification some wear out much sooner than others. Considerable variability in wear resistance may also be shown by carbide cutting-tool inserts from the same manufacturing lot. They all went through all the manufacturing processes together and are as nearly identical as it is possible to make two nominally identical pieces.

Modulus of resilience measures the ability to absorb elastic energy without permanent deformation. This condition occurs in components, like springs, which are designed to absorb and store energy, or components which may be subject to unexpected impact, like a wheelchair wheel accidentally falling off a curb. The area under the elastic portion of the stress-strain curve is the absorbed elastic energy per unit volume and is numerically equal to the modulus of resilience. If σ_y is the yield strength, or for a more accurate determination it is taken as the proportional limit, and ϵ_y is the yield strain, the area, i.e., the modulus of resilience, can be found as

$$\textit{Modulus of resilience} = \frac{\sigma_y \epsilon_y}{2}$$

Since $\sigma_y = E\epsilon_y$, the preceding equation may also be expressed as

$$\textit{Modulus of resilience} = \frac{\sigma_y^{\,2}}{2E}$$

Note that modulus of resilience is not measured directly; it is calculated from other mechanical properties.

Defining a material property that relates to fracture of a material is perhaps more difficult than defining most other material properties. Material properties are useful because they permit us to predict the behavior of things we are designing so that we can ensure successful operation, or prevent failure. Failure by fracture is more complicated to describe than, say, failure by excessive elastic or plastic deformation because the material behavior is more dependent on changes in operating condition. Temperature, strain rate, presence of flaws, etc., have a more profound effect on fracture behavior than on elastic or simple yield behavior.

The ability to absorb energy, or to deform plastically, without fracture are the general principles used to describe the resistance to fracture. The most common measures are ductility, modulus of toughness, Charpy or Izod impact, and fracture toughness.

Ductility, as measured by percent elongation, is a function of the specimen's geometry. For two round specimens of the same diameter, increasing the gauge length will decrease the percent elongation to fracture. This is because after the neck forms, all subsequent deformation takes place in the vicinity of the neck. The behavior immediately around the neck is the same regardless of the length of the specimen. Thus in shorter gauge lengths, a larger fraction of the specimen's length is deforming during the whole test. In longer specimens, the portion away from the neck is not continuing to deform after the onset of necking, so a smaller percentage of the specimen's length is contributing to total deformation.

Flat and circular specimens even with the same cross-sectional area will have different apparent ductilities. An empirical conversion of values from one type of specimen to another is given in ASTM A370.

Since the value of percent elongation is specimen dependent, it is vital that the test conditions be the same before comparing data from different sources. If the gauge length and diameter are not quoted, comparison is of doubtful usefulness.

Ductility, as measured by percent reduction of area, is not specimen dependent for standard round tensile specimens. It is a more reliable measure of how much plastic deformation a material can withstand without fracture. By calculating the true strain at fracture, it is possible to estimate the amount of deformation that can be tolerated in some metal-working operations. The stress state in the neck is complex and triaxial, not the simple uniaxial tension as found in the elastic and subultimate plastic range. The ductility of a material is strongly dependent on the stress state, so using the reduction of area to predict ductility in other situations where the stress state is much different is not generally successful. The strain at fracture in tension may be quite small while the strain to fracture in pure shear may be very large for the same material at the same test conditions (temperature, strain rate, etc.). Shear test data may be useful for studying possible material behavior (workability) in manufacturing operations where plastic strain is expected to be greater than that observed by a standard specimen in tension. If the shear test data can be converted into effective stress and effective strain data, then force and work required in metal-working operations may be estimated for deformations much larger than the fracture strain in tension. The effective stress and effective strain are the equivalent uniaxial tension or compression values that would result in the same amount of work done in deforming the actual part. The conversion involves using either the von Mises or Tresca criterion, so the same torsion data may convert to different effective stress-strain data. But in either case there is usually some difference between the effective values converted from torsion data and the corresponding stress strain curve found directly from axial tension or compression tests. They should be the same, if all the assumptions are correct.

The ductility of a material may be sensitive to alloy additions, heat-treat condition, and other factors. Specifying the ductility as a quality control parameter for acceptance of material is common. For instance, low ductility of tensile specimens cut from across a weld may detect the presence of unacceptably brittle martensite that might not be accurately reflected in the strength measurements alone. Bend tests are also commonly used to verify qualitatively the adequacy of the ductility in weldments, sheet, strip, and plate material.

Ductility also gives a designer some indication of whether the material will yield significantly prior to fracture in service. Plastic deformation prior to fracture is desirable to absorb energy, as in energy-absorbing steering wheels and chassis structures in automobiles. Ductility is also desirable to give warning that serious overload has occurred before fracture makes the situation catastrophic. While the ductility values for the candidate materials are not used directly in design calculations, they do offer a method of comparing potential suitability of the materials, e.g., for fabricability. Otherwise, tabulated values of ductility, while quite common in the literature, are not of much use in the material selection process.

The modulus of toughness is taken as the area under plastic portion of stress-strain curve. This is a crude measure that permits quick comparison for possible relative energy-absorbing ability. It is not reliable as a design tool.

Charpy and Izod tests determine the amount of energy absorbed in deforming and fracturing a standard specimen by impact loading with a hammer. The Charpy specimen is loaded in three-point bending, the Izod in cantilever bending. The procedure is described in ASTM E23 for metals and D256 for plastics. In the Charpy method the specimens are a bit easier to load quickly into the machine. Thus, Charpy is more commonly used, especially when time is important to prevent atmospheric warming of cold metal specimens. Izod is somewhat more commonly used in plastic materials and for metal testing in Europe.

The impact test creates triaxial stresses in a specimen with a stress concentration that is rapidly loaded. These are all conditions known to promote brittle failure. So, the notched Charpy specimen was used to get information about dynamic fracture behavior of materials long before fracture mechanics became well established. The test gives little information that is directly usable to design structures to resist brittle fracture. Fracture mechanics overcomes that objection because it permits determination of flaw sizes that can be tolerated. Many empirical correlations between Charpy and fracture mechanics results have been published (Metals Handbook, 9th ed., vol. 8, p. 265).

The Charpy impact test is quick and inexpensive. It has been in use for a long time and was the only impact test widely used before fracture mechanics test superseded it. Consequently, there are a lot of data reported in the literature. It can be used to see trends in behavior as composition, heat treatment, orientation (as in transverse versus longitudinal direction in rolled material) or other variables are changed.

Also, the effect of the potential changes in service conditions can be evaluated quickly. Charpy and Izod tests give data to compare the relative ability of materials to resist brittle failure as the service temperature decreases. Body-centered metals (ferritic steels, for instance) tend to become embrittled as the operating temperature decreases. The impact tests can detect this transition temperature. Since the transition in behavior is not sharp, it is necessary to define, arbitrarily, what constitutes the transition temperature. Many measures have been proposed or used, such as the temperature at which

1. An arbitrarily chosen energy level, say 30 ft-lb, occurs.
2. The impact energy is the average of the impact energy at high temperature and the impact energy at low temperature.
3. The fracture surface contains a certain percentage of cleavage (brittle) and fibrous (ductile) appearance, say 50-50.
4. Some (arbitrarily) predetermined amount of strain occurs in the lateral direction in the vicinity of the notch.

Fracture toughness is a measure of the ability of a material to resist propagating a crack. It is of particular use in designing against failure when a flaw is known to be likely, as in welded structures or structures that are subject to fatigue failure. The subject of fracture mechanics is complex, partly because of the large number of different possible service conditions, crack geometries, stress states, etc. However, the most common material property is found for a simple crack, in a condition of plane strain, that is being opened, or propagated, by uniaxial applied tension. This is called the plane strain fracture toughness, K_{IC}, that is used in linear elastic fracture mechanics. The subscript I refers to "mode one crack opening," which is for a crack being opened by a tensile stress. Fracture toughness is determined by the methods described in ASTM E339.

Elementary theory relates the stress, σ_f, required to cause fracture, the flaw size, a, and the fracture toughness as

$$K_{IC} = Y\sigma_f\sqrt{\pi a}$$

Here, Y is a factor related to the shape of the flaw which has a value near unity.

For low-strength, tough materials where there is significant plastic deformation, a different property, J_{IC}, is used. Values are determined by the methods of ASTM E813.

2.2 Manufacturability Considerations

The properties considered in the preceding section relate to the ability of a material to perform adequately in service without failing. But, first the part must be capable of being made, i.e., it must be producible. There are additional properties and producibility considerations that should be part of the material selection process. For a design to be producible the configuration should not introduce unnecessarily difficult manufacturing challenges. For example, the use of widely variable wall thicknesses in a casting will invariably lead to complications associated with different rates of cooling in the thick and thin portions. The locations that cool last are most prone to internal voids due to the shrinkage that accompanies the increase in density of the material as it transforms from liquid to solid. Also significant residual stresses may develop when one portion of a component cools faster than another. The stress may be sufficient to cause a variety of problems. Warpage in flexible parts is quite common, and the amount of warpage may be sufficient to-make the parts dimensionally nonconforming. When that occurs, straightening (by plastic deformation) is routinely done by the producer to bring the parts back into conformance with the dimensional requirements of the drawing. The straightening operation usually will introduce a residual stress pattern that invariably accompany any nonuniform plastic deformation. The customer will probably not be informed that mechanical straightening has occurred and consequently will not be alerted to the potential consequences of the residual stress, such as significant changes in fatigue life of the finished parts. In other words, the configuration of the product has induced a condition (residual stress) during the manufacturing process that makes the material property (endurance limit) that was used to design the part no longer a valid predictor of service behavior.

The material choice must be compatible with the manufacturing process and the configuration. Consider two examples. First, if the configuration of a cast part is such that stresses arises as the part shrinks when it cools in the mold, and if the material is susceptible to hot tearing (cracking during or immediately after solidification), the resulting parts are usually scrap. If the configuration cannot be altered to eliminate the stress, then the material selection becomes a critical issue. Some materials are more resistant to hot tearing than others, and if the designer fails to take advantage of that fact, then the design process is deficient.

Second, the radius to which sheet metal may be bent will depend primarily on the ductility of the material. Thus, the decision about bend radius and the decision about material choice cannot be made independently of one another.

So, the material and the manufacturing process cannot be considered independently. For many configurations, there will be only a few processes that are realistic possibilities. Immediately when a tentative process decision is made, a significant number of materials are eliminated from further consideration.

Alternatively, when a product may be successfully produced from a variety of materials, a decision must be made about the material early and other decisions about suitable processes and configurations follow. For example, ashtrays are made from metal, polymer and ceramic materials. The basic function, to provide a place for the temporary storage of ashes from smoking products, is the same regardless of the material choice. (A notable exception is the Wedgwood ashtray which is intended to be displayed and not used for stubbing out cigarettes.) Then, secondary functions become the driving factors behind the material choice. Ceramic ashtrays might be made from either glass or marble, to mention only two out of a large number of possible ceramic material candidates. Glass and marble differ in their secondary function related to apperance (customer appeal). When the designer chooses either of those materials, the choice of manufacturing processes is immediately narrowed. Glass ashtrays may be mass produced in molds from molten glass; marble ashtrays may not. This fact is closely related to the secondary function of cost. Different target markets are involved.

The use of glass provides an example of how the same material can satisfy vastly different secondary functions related to cost, i.e., trying to reach the mass market or the up-market folks. Cut glass, or engraved glass (e.g., Steuben), ashtrays are competing against carved marble.

Suppose the decision is for a plastic ashtray. Here the primary function of not catching fire narrows the choice of material. Phenolic is a good candidate, but it might not adequately satisfy the appearance requirements. Blending it with melamine is a good solution to this problem. Now the designer is almost certainly constrained to a configuration that is suitable for production by compression molding. That means no undercuts and no decorative texture on the surface that would interfere with ejection from the mold. This is not a serious impediment to the design and production of a successful product, but it is a real constraint on design flexibility.

Metal ashtrays may be produced by more processes than either ceramic or plastic ones. Stamped sheet metal (disposable or reusable) and die casting may be the most common. In contrast to the case of plastic where the material choice dictated the appropriate manufacturing process, here the material selection is surely dictated by the choice of process. Any aluminum alloy will satisfy the basic, and probably the secondary, function. However, each candidate alloy will be suitable either for the stamping process or the casting process but can not be suitable for both.

Castability

The mechanical properties of many cast materials are dependent on the cooling rate during and after solidification. This means that the properties will vary with location within the casting according to wall thickness, position relative to gates and risers, use of chills, etc. It also means that test bars used to verify the casting's properties, which are usually cast separately from the cast product itself, will probably have properties different from the casting. Most data presented in handbooks and standards are for separately cast test bars. Thus use of handbook data must be adjusted by appropriate factors of safety or "casting factors" before using the data as a design stress.

The appropriate factor is sometimes specified in the material standard. For instance, ASTM B 26-92a, *Standard Specification for Aluminum-Alloy Sand Castings* says

> [At any location in the finished casting], average tensile strength [and] average yield strength . . . shall be not less than 75% of the tensile and yield strength values . . . specified in [the table which gives the tensile requirements for alloys covered by this standard].

Thus, the designer cannot expect the values of strength that are listed in the standard to be attainable in the actual casting.

Also note the use of "average" strength in the preceding paragraph. Thus, the designer also cannot determine from the standard what the smallest likely strength might be, because there is no information given about the variance to be expected.

Worse still, is the fact that sometimes the appropriate casting factor cannot be found at all. ASTM B85-92a, Appendix X2 might be considered typical. The following quotation refers to Table X2.1, entitled "Typical Mechanical Properties Test Specimens." The table contains the only mechanical property data in the standard, i.e., it is the only possible guide the designer might have to the expected properties of materials covered by this standard. But the data are from separately cast test bars which will "have no correlation with the die casting other than a common chemical composition. . . . [Therefore,] it is considered that the only practical method for mechanical property control is proof testing the whole die casting." Which means that the designer is left without any real guidance about the mechanical properties of the alloys covered in this standard.

Determining the "correct" value for strength is further complicated by the fact that property values for any give alloy as listed in different authoritative sources may not be the same. Consider 356.0 T6 sand castings. ASM International Metals Handbook 9th ed., vol. 2, lists the tensile and yield strengths as 230 and 165 Mpa

(33 and 24 ksi), as does the Aluminum Association Handbook #18, while the ASTM standard B26 requires 205 and 140 Mpa (30 and 20 ksi) respectively.

These strength values are for separately cast test bars, which for sand casting are typically 1/2 inch in diameter. Statements have been published that a 1/2-inch-diameter test bar will be representative of the properties to be expected in 1/2-inch-thick wall of a casting. That is not correct because the cooling rate from the cylindrical surface of the test bar and from the flat surfaces of a casting wall are significantly different. Reported tensile strengths are 34.2 ksi versus 40.5 ksi (236 versus 279 MPa) for a flat 1/2-inch wall and a 1/2-inch-diameter separately cast test bar (Metals Handbook, 9th ed., vol. 2).

The cooling rate of the critical locations in the casting and the cooling rate of a separately cast test bar may be quite different, leading to different properties, as pointed out in the preceding paragraph. It is possible to vary the size of the test bar to try to match the properties of the test bar with those of the casting. This is standard practice in gray cast iron, per ASTM A48, for instance.

Alternatively, a test program to determine the correlation between properties in the casting (which is cut up for testing) and the separately cast test bars may be performed.

The fact that the properties are strongly dependent on cooling rate may be used to intentionally vary the properties within a casting. An example might be the need to strengthen the material in the vicinity of an opening or a screw thread to offset, at least partially, the effect of the stress concentration at that location. The standard MIL-A-21180 (also ASTM B 686-92a) for high-strength aluminum castings is an example where the casting may be required to meet different minimum strength "classes" at different locations. In this case, tensile specimens must be taken from the cut-up casting (on a sampling plan basis) to demonstrate that the casting itself actually meets the requirements. The properties listed in these standards may be taken as properties of the casting. No factor needs to be applied to compensate for the uncertainty of the separately cast test bars that arises in other methods.

The strength of metal castings is also dependent on the number and type of defects in the casting. Internal unsoundness due to gas porosity, shrinkage porosity, segregation, foreign material, cracks, and other defects may be classified as grade A (fewest defects) through grade D. The reference radiographs in ASTM E 155 are the standard against which these grades are judged. There is no universal agreement about how much reduction in design stress must be allowed, except for grade A, which is always taken as full strength. However, foundries usually can be expected to charge a premium for grade A castings, so designing for a lesser grade with a small strength penalty but a large cost savings is normal. As a rule of thumb, expect grade D castings to lose 30% to 40% of their load-carrying capacity, compared to grade A.

The variability in cooling rate from one casting to another in the same lot and the variability in internal unsoundness make the statistical variance of properties in castings generally greater than in wrought products. In comparing the reliability of castings and forgings of the same average strength, the casting would be the more unreliable because of the larger standard deviation of the strength distribution. Thus, for the same reliability, castings require a greater factor of safety to be applied to the design stress than forgings or other wrought products. That would normally result in a greater weight of product. Whether or not the weight penalty is tolerable is a factor to be considered in the material/process selection decision.

Large factors of safety may be used to compensate for the lack of specific data.

In addition, the variability in material properties which is inherent in the casting process may be reduced considerably by hot isostatic pressing (HIPing). It is routinely used to improve the reliability of high-strength, aircraft-quality castings, but it does add to the cost.

Many material standards will give guidance to the designer about the fabricability characteristics of casting alloys. These characteristics are often presented as a semiquantified ranking, say on a scale from 1 to 5, based on expert consensus, not a reproducible test. Common data include measures for the following.

Fluidity. A measure of how well the liquid will flow and fill a mold cavity. Complex, intricate casting cavities require the best fluidity. So also do the casting processes, like the permanent mold process, which uses molds that induce rapid cooling rates. Poor fluidity is of less concern when the metal is to be cast by the slower cooling plaster or investment casting processes.

Resistance to Hot Cracking. Just as a casting solidifies, the hot metal has very little strength, but is nevertheless trying to shrink as it cools. If a stress is produced because of some factor that restrains the free contraction of the metal, the metal may not be able to resist the stress successfully. Cracks, also known as hot tears, will occur. Hot tearing is likely to be more of a problem in metal molds than in sand molds which are weak enough so that they can collapse as the casting shrinks.

Pressure Tightness. The solidification shrinkage in some alloys is sufficient to create a significant number of quite small internal voids. In some cases these voids will permit gases to pass through the wall of the casting. Helium is notorious for being able to leak through a small amount of internal porosity. Castings used in the supply systems for gas-fired furnaces are another example of a product in which pressure tightness is of prime concern. Casting are prohibited by some building codes from being use in that application.

Metallurgically, the preceding three characteristics are related to the difference between the temperature at which solidification begins and the temperature at which solidification is complete. In some alloys that temperature range is quite large, and the problems with fluidity, hot tearing and pressure tightness may be severe. It other alloys—eutectics, in particular, which have a unique melting point—the problems may be almost nonexistent.

Castings often form part of a larger system and need additional processing like machining, welding, brazing, anodizing, etc., to be performed. Information on such fabricability is also often found in material standards and handbooks.

Formability/Workability

In service, it is important not to exceed the yield strength of the component's material. There may be exceptions, for instance, in energy-absorbing structures for crash protection. In forming and forging, however, the applied stress must always exceed the yield strength. Formability and workability both refer to the ability of a material to withstand successfully the plastic deformation that may be required during manufacturing.

Sometimes, fracture of the material is planned as part of the process (blanking, lancing, Mannesmann tube piercing) and sometimes not (bending, deep drawing). Sometimes fracture is not planned but occurs anyway (edge checking in rolling, central burst in extrusion). Thus some measure of ductility is often important. Ductility, as measured by the tension test, is inadequate to describe the maximum strain that may be imposed on the material during a manufacturing process. The stress and strain conditions of the tension test are less complicated than those of most processes, and the ductility is a function of the stress and strain state. So any measure of "workability" of a material must also include consideration of the process involved.

Sheet forming may be as simple as bending or as complex as deep drawing or stretch forming. In all cases, uniformity and isotropy are generally desirable. The principal measure of the amount and type of anisotropy is the "r-value," which compares the plastic strain in the width direction with that in the thickness direction of a sheet specimen at the time of maximum load during a tensile test. (See ASTM E517.) The value can be related to the extent and location of earing in deep drawing, the limit draw ratio (the maximum blank to can diameters in deep drawing) and the resistance to local thinning in stretching.

If the plastic deformation is done cold, strain hardening will occur. The change in the flow stress of the material may be of interest for two reasons. First, the designer may use the work-hardened strength in the design calculations. If there is nonuniform deformation, the strength of the part will vary from place to place. Optimum design can take advantage of the strong locations and minimize the detrimental effects of the weaker locations. Accordingly, estimates must be made of how much strengthening will occur during the forming process. Many constitutive stress-strain relations have been formulated, but the most widely used is the power law.

$$\sigma = K\epsilon^n$$

where σ = true stress
ϵ = true strain
K = constant = (stress at true strain of 1); also called the strength coefficient
n = strain hardening exponent = slope of stress-strain curve on log-log coordinates (see ASTM E 646 *Standard Test Method for Tensile Strain-Hardening Exponents* (n-values) *of Metallic Sheet Materials*)

For two materials with the same yield strength, the one with the larger n-value will have a greater tensile strength. The ratio of tensile to yield strength is sometimes used to make a quick comparison of the likely formability of two candidate materials whose n-values may not be known. A larger ratio indicates better probable formability.

The required press capacity for sheet metal forming will depend in part on the flow stress of the material. Frequently, the size of press needed to accommodate the part's volume will have a force capacity that is adequate.

The flow stress of many materials is dependent on the strain rate. A commonly used power-law relation is

$$\sigma \propto \dot{\epsilon}^m$$

where σ = flow stress
m = strain rate sensitivity
= slope of stress versus strain rate on log-log coordinates
$\dot{\epsilon}$ = strain rate

Most metals have a positive value of m, which implies that as the forming rate or speed increases so will the stress, and force, required to form the part. Steel sheet is likely to have m-values around 0.01. Aluminum alloys are quite insensitive to strain rate, with m-values less than 0.005 or even negative. Superplastic forming sheet will tend to have m-values around 0.5, but the strain rates must be kept quite low. In Newtonian fluids, $m = 1$.

When sheet metal is formed successfully, biaxial strains usually occur in the plane of the sheet, i.e., lying in the plane of the sheet and normal to each other. The forming limit diagram displays the combinations of strain that will be successful and those which will not. The larger of the two biaxial strains is called the major strain, the other is the minor strain. The major strain will never be compressive. The minor strain may be tensile, compressive or zero. A minor strain of zero indicates plane strain, i.e., stretching in only one direction in the plane of the sheet and contracting through the thickness. This condition leads to fracture at the smallest total deformation.

Forming limit diagrams are constructed by etching a grid of circles on the surface of a sheet of material and then subjecting the sheet to various combinations of major and minor strain. The strains are increased until failure occurs. All combinations attempted are plotted on a graph with axes of major and minor strain. The boundary between the combinations of strain that are successful and those that are not is quite sharp.

Presently, forming limit diagrams are mainly used to troubleshoot formability problems, not to provide data for design. It is possible to distinguish between problems caused by improper processing (bad die design, poor lubricant, etc.) and bad material (lot-to-lot variations). However, when analytical design capability becomes adequate to calculate the strains that will occur during forming of sheet metal parts, the designer will be able to use forming limit diagram information as part of the material selection process. For instance, in choosing between two materials which are in all other ways satisfactory, one material may be capable of tolerating the amount of major and minor strain required to form the part and the other may not. Then the selection decision will be clear.

There are many other formability tests that attempt to find a simple correlation between the test result and the behavior of the material in production. They do not measure a material property. Rather they attempt to simulate some condition that is similar to the production operation so that the effects of various variables may be assessed. Olsen, Erichsen, Fukui, Swift and the hydrostatic bulge tests are five examples from among the many specific test procedures that measure some aspect of the formability of sheet material.

Workability in bulk forming processes refers to the ability of the material and the process together to permit a part to be formed successfully. Failures that occur during manufacturing may be due to material deficiencies or to tooling and process problems (incorrect die angles, improper lubrication, inadequate temperature control, etc.). Tests to measures workability generally attempt to simulate material response to a particular stress or strain condition. Thus, there are tests for plane strain compression, partial width indentation, ring compression, cold upsetting (with and without friction) using cylindrical, tapered, or flanged specimens, etc. None of these tests give a quantifiable material property that is commonly tabulated in a material property handbook.

Because these sheet and bulk tests do not measure a material property or give a result that is likely to be useful to a designer trying to make a material selection decision, they will not be described in detail. They are mentioned here because the designer, as part of a concurrent engineering team, should be aware that the team's manufacturing members may evaluate material properties and processing behavior in a different way than usually appears in mechanical design and selection books.

Machinability

Machining always includes removal of material from the workpiece and the creation of a new surface. In conventional machining a tool creates sufficient stress and strain by direct contact of the tool and the workpiece to cause a fracture that separates a chip from the workpiece. That fracture may be quite smooth or it may be rough. It may be tangent to the surface or it may leave cracks penetrating into the material. Also, substantial plastic deformation usually occurs, so the new surfaces will probably contain some residual stress. The plastic deformation also generates heat, as does the friction of the chip sliding against the tool. The heat may be sufficient to cause lasting damage to the workpiece, the extent of which is dependent on the response of the workpiece material to heat. The material's surface may soften, harden, become embrittled or be relatively unaffected.

A candidate material need not be rejected because of the threat of residual stress, heat damage or other surface damage. These effects are controllable by appropriate choice of machining processes, which may include additional steps to remove any harmful effects that result from the principal process. However, the need to take special steps in manufacturing adds cost which must be considered.

In the processes that are lumped under the ill-defined term nontraditional machining, the means of separating the scrap from the workpiece might include chemical, electrochemical, laser, electron beam, and high-speed water stream with or without abrasive, among others. Some are noncontact processes, some have no significant heat generated, etc. Thus, each may have potential for overcoming some problem created by traditional machining. But, each incurs some penalty: the rate of metal removal is slower, the capital cost of the equipment is greater, the energy consumption per unit of material removed is greater, the work must be performed in a vacuum, etc. (Boothroyd and Knight 1989, Benedict 1987).

The term "machinability," which has no single universally accepted definition, refers to the ease with which material can be removed while leaving an acceptable surface finish on the workpiece. The "ease with which material can be removed" incorporates several factors, among them how long the tool will last, how much power is required to remove a given amount of material, and what the resulting surface finish is like.

The rate at which the tool wears out determines the tool life, or the time between tool changes, which is important in process planning. The material's properties alone do not control the tool life. The tool geometry, speed, feed and depth of cut all have an influence, but the most significant is speed.

The rate at which material can be removed from the workpiece significantly influences the piece part production rate. In high-volume production, this may be a critical factor. This, too, is influenced by factors other than the material.

One material characteristic that is used to evaluate materials relative to one another is the machinability index. In determining the machinability index, the tool material, the tool geometry, tool life and cutting conditions except speed are fixed. The speed at which the test material can be cut with a predetermined tool life (usually 60 min for steel tools) is determined. Slower speeds indicate poorer machinability because the rate of metal removal is slower. The standard materials against which other steels are compared are AISI B1112 and 1212 which can be cut at 170 feet per minute (fpm) for 60 min. The machinability index is the speed, expressed as the percentage of the standard 170 fpm at which the test material can be cut.

Machinability index is subject to considerable variation due to microstructural and compositional variations within the tolerances permitted for the nominal composition. The machinability index for the standard B1112 itself has been measured between 80% and 160% of the accepted standard 170 fpm. The 100% value that is taken as the standard for this material represents approximately the median for a large number of tests. Tabulated values of machinability index will always be single values with no indication of the variability. Thus, if machinability is a critical issue in material selection, in-house testing will be required to verify the lot-to-lot variability that must be anticipated.

The test material referred to in the preceding paragraph means not only the composition of the material, but also its microstructural condition. The abrasive materials will obviously be more damaging to a tool than nonabrasive materials. In some materials, the microstructural condition significantly influences the wear rate of the cutting tool. For instance, as the carbon content of annealed or normalized steel increases, so does the amount of pearlite or carbide. In lamellar form, the broken edges of the carbide platelets are quite abrasive when sliding up the face of a tool. At higher carbon contents, it is advantageous to modify the microstructure to spheroidite where the carbide particles do not have sharp abrasive edges to wear the tool.

The power and force requirements for machining are sometime lumped into the machinability concept. Again, the materials, tools and machine settings all can influence power consumption and tool forces. The amount of power to remove a unit volume of material has been measured for many materials. This is called the unit horsepower or specific cutting energy. Materials may be compared using this measure, which constitutes an important factor in the overall concept of machinability.

Many attempts have been made to relate machinability and machinability index to conventional mechanical and physical properties (hardness, thermal conductivity, ductility, strain-hardening exponent, etc.) that are routinely tabulated in handbooks. No method has yet emerged that is of use to the engineer making material selection decisions. In general, the material properties which lead to best machinability are low yield strength and little strain strengthening; high thermal conductivity; easy crack initiation and propagation which might best be characterized by low ductility or fracture toughness; little interaction with the tool material in terms of chemical reactivity or diffusivity, abrasiveness of the microstructure or tendency for cold-welding to the tool.

The strength, ductility and microstructure that promote good machinability are likely to be the opposite of what is desired for best performance in service. Fortunately, many materials respond to attempts to change their properties by thermal treatment. Thus it is not uncommon for a material to have one set of properties at the time of machining and another when the part goes into service. For instance, machining high-carbon steel is easiest with a fully spheroidized microstructure, but the high carbon content is almost certain to have been chosen so that high hardness can be developed during heat treating.

In some instances it is possible to alter the composition in such a way that the machinability is enhanced without significantly altering the service properties. In steels, sulfur and lead additions were particularly effective in many grades. Further modifications with calcium, bismuth, phosphorous, selenium and tellurium can be very effective. For instance, some proprietary grades of AISI 12L14 have a machinability rating nearly three times as good as AISI 1212 (which is used as one of the standard materials against which all other steels are measured). Sulfur and selenium are used in stainless steels (AISI 303Se, 430FSe). Lead is added to brass to create "free machining brass" (CDA 360), the standard against which all other brasses are compared. These additions not only permit faster cutting but they will generally result in a better surface finish. Sometimes the improvement in surface finish, not increased production rate, is the dominant reason for choosing the more expensive materials possessing enhanced machinability.

Free machining grades cost more than plain grades with equivalent service properties. Unless the added material cost can be offset by reduced production cost or justified because of improved surface finish, free machining grades should not be chosen. Complete economic analysis is always desirable, but as a simple rule of thumb the following may suffice: if machining removes less than 10% of the material, free machining grades will probably not be justified; if machining removes more than 20% of the material, free machining grades should be seriously investigated.

The surface finish is strongly dependent on the machining method. For traditional chip cutting, the ability to create a smooth fracture surface tangent to the tool path is critical. Thus, in soft and very ductile materials so much deformation is required to cause fracture that it is difficult to get good surface finish. Annealed materials are particularly prone to this problem. Decreasing the material's ductility by cold-drawing is a common. In aluminum alloys, machining in a heat-treated condition, say T6, will generally improve the surface finish.

Unfortunately, one of the major causes of poor surface finish is worn tools. There is no material property that the designer can consider that will prevent poor surface finish that results from bad tooling.

Some machinability classification systems consider the type of chip that results from the machining. A process and material that produce continuous chips pose the danger that the chip will become entangled with the workpiece, tool holder or operator. Curled and easily broken chips are generally considered desirable because of the ease with which the scrap can be handled. Only in cases where strict accounting is required of the scrap produced, as in machining of radioactive materials, is the continuous chip preferred.

Coatability

The materials discussed in this book are mainly solids used to make components. The properties presented are used to make decisions about the ability of the material to perform adequately in service, and this mainly pertains to the interior—not the surface—of the component. However, we know that the surface is often critical to a part's ability to perform adequately in service. Considerations of surface characteristics are important in many instances, like fatigue behavior, the structural strength of ceramics, and the loss of cross-sectional area due to general corrosion.

A component's surface is the interface between the interior material and the surrounding material. Surfaces of parts may perform one or more of a variety of functions. Those functions usually include resisting the deleterious effects of contact with the surroundings. The surrounding environment may be solid, liquid or gas, or it may be a vacuum through which some radiation impinges on the component's surface. The contact may be dynamic or static, as a journal bearing or a bolted joint, respectively, or an underground pipeline with dynamic fluid on the inside and static soil on the outside.

In addition to the technical requirements that surfaces must meet, frequently the appearance is important. The suitability of the appearance may depend on the roughness, texture, color, reflectivity, corrosion resistance and a host of other factors.

To describe all the different possible surface conditions that might need the attention of the designer, and the possible solutions would take up meters of bookshelf space. Many different processes have been developed to prevent deterioration, repair damage, improve appearance, reduce friction, etc., of surfaces. Many of them are proprietary. They are not chosen by calculation, as yield strength, for instance, would be. Each coating or surface treatment has characteristics that are suited to a limited range of applications. Thus, the decision process for coatings is likely to be more experienced-based than with other material selection decisions. Therefore, the need for consultation with coatings experts is also more likely except when there are obvious solutions with well-known methods.

Painting is the most widely used surface treatment for protection and appearance. In some form, it has been used on most materials. Surface preparation may be necessary, adding to the cost. The initial cost is likely to be low compared to other surface treatments. However, the durability is generally poor because both adhesion and abrasion resistance are relatively poor. Thus, repair or routine maintenance may be anticipated. With most paint systems, repair and repainting are generally straightforward procedures. Uniformity of thickness and completeness of coverage are often significant problems in paint selection. Both depend on the type of application system (air spray, airless spray, electrostatic, brush, roller, dip, etc.) and the part geometry (sharp edges, blind holes, combination of horizontal and vertical surfaces, etc.). These are not material selection problems, so they are more appropriately discussed in texts dealing with design for manufacturability.

Plating of a metal onto the surface of a part performs much the same function as painting, i.e., protective and/or decorative. Metal plating is likely to be harder, more wear and abrasion resistant, and more adherent than paint. It is also more expensive. Plating may be either assisted by electrical current or simply a precipitation from a chemical solution (electroless plating). Some of the metals that can be plated may also be applied by other means. Zinc is commonly applied to steel by hot-dipping to make galvanized sheet. Aluminum is vapor-deposited on steel (Ivadize) for superior corrosion resistance in high-humidity and saltwater environments.

Component surfaces may be hardened, usually for improved wear resistance, by applying a coating material. Various carbides may be flame-sprayed or plasma-sprayed onto the substrate.

Before choosing to apply a coating for increased hardness, it is important to investigate the possibility of achieving satisfactory results by simply altering the structure of the component's surface, thus avoiding the complications of adding another material and process to the part's manufacturing process. Also, by avoiding applying the new material to the component's surface the problems of loss of adhesion and flaking off are sidestepped. This is possible if the base material is susceptible to case hardening either by selective heating or local alterations in chemistry that enhance the response to heat treatment.

Selective hardening of certain areas in steel by local heating (usually flame or induction) so that only the heated portion austenitizes and hardens is one possibility. Alternatively, increasing the carbon content of a steel only at its surface, by carburizing, followed by full austenitize and quenching is another. Since the hardness of martensite is strongly dependent on carbon content, the surface is capable of achieving a high hardness while the interior is not. Other surface treatments that alter the chemistry, like nitriding and carbonitriding, are variations on this basic principle.

In either case, the material selection decision for the base material must include consideration of the fact that the finished product will have different properties at different locations.

There are too many other surface treatments for protection, appearance, or local property modifications to mention all of them. The names of many are well know, even if the details of their chemistry and processing are not. Anodizing, "black oxide," "gun blue" and chromate conversion coatings might be cited as examples. It would be an unusual circumstance if the base metal were chosen because a particular surface treatment is required. Almost always the base material is chosen because of its ability to perform its structural, or other, function. Then a compatible, suitable surface protection system is chosen.

Heat Treatment

Depending on the material and its condition, heat treatment may be used for softening a previously cold-worked or hardened material, or for strengthening those materials whose microstructures are susceptible to hardening by heat treatment. Thermal treatments are also involved in production processes other than the ones intended to alter the material properties. Sintering of metals and ceramic powders, diffusion bonding, heating for homogenizing an inhomogeneous structure, and stress relieving are examples.

Materials may be subjected to thermal cycling either during processing or during use. The problems that arise with thermal cycling are usually related either to the expansion and contraction that most materials undergo as their temperature is changed or to the volumetric changes that accompany phase changes. Rapid heating and cooling are often desirable because it reduces cycle time or because it produces a desired phase change. Rapid heating and cooling are accompanied by temperature gradients in the part, resulting in thermal stresses that may be sufficient to cause distortion, fracture or thermal fatigue. Material selection decisions may be influenced by the need to prevent these problems.

To explore what material properties and processing factors are important in controlling thermal stresses, the following simplified analysis may be performed. Let a material with temperature T_0 be immersed in a quench liquid of temperature T_q. The wall temperature at the interface between the solid and liquid, T_w, will vary with time, beginning at T_0 and ending at T_q. Heat will flow across the interface out of the solid at a rate predicted by the Fourier conduction law. The rate is proportional to the thermal conductivity, H, of the solid, the area, A, across which the heat is flowing, and the temperature gradient at the wall.

$$\dot{Q} = HA \frac{\Delta T}{\Delta X}$$

All the heat from the solid will be absorbed by the liquid, and the rate of heat transfer will be described by the Newtonian convection law. The rate is proportional to the convective heat transfer coefficient, h, the area, A (which is the same area found in the conduction equation), and the difference between the wall temperature and the quenchant temperature.

$$\dot{Q} = hA\,(T_w - T_q)$$

Because the two rates of heat transfer must be equal if there is no heat buildup in the boundary layer, the resulting equation is

$$HA \frac{\Delta T}{\Delta X} = hA\,(T_w - T_q)$$

This can be rearranged to give

$$\frac{\Delta T}{\Delta x} = \frac{h}{H}(T_w - T_q)$$

Also, the stress, σ, in an elastic solid may be expressed as the product of the elastic modulus, E, and the strain, ϵ:

$$\sigma = E\epsilon$$

However, the strain arising from thermal expansion is the product of the coefficient of thermal expansion and the temperature change:

$$\sigma = E\alpha\,\Delta T = E\alpha \frac{\Delta T}{\Delta X}\Delta X$$

Finally, the mechanical and thermal portions of the analysis may be coupled through the temperature gradient term to give

$$\sigma = E\alpha \frac{h}{H}(T_w - T_q)\,\Delta X$$

The last term represents the size of the part, giving rise to a mass effect in thermal stress problems, i.e., large parts are more susceptible to cracking and distortion than smaller, geometrically similar, parts.

The elastic modulus, thermal conductivity, and coefficient of thermal expansion are properties of the component's material, while the convective heat transfer coefficient and the quenchant temperature are process variables. Since a material for a component is almost always chosen because it has the right properties for the expected service, the elastic modulus and thermal conductivity will normally be predetermined and probably not negotiable as a means of reducing thermal stress, distortion or cracking during heat treatment or thermal cycling.

In instances where resistance to thermal stress or distortion is a main service condition, it is possible to alter the coefficient of thermal expansion to accomplish that objective. An example is the change in glass chemistry to make heat-resistant (Pyrex) coffee pots. The only significant material property that is different between an ordinary drinking glass, which would crack if put directly on the stove to heat water, and Pyrex, which successfully withstands the direct application of heat, is the coefficient of thermal expansion. For the two materials, the coefficients of thermal expansion differ by a factor of approximately 10. This means that for the same conditions, the thermal stress in the Pyrex is only about one tenth of the stress in the drinking glass. That is, the Pyrex is suited to this application not because it is stronger than other types of glass but because it generates less stress. The stresses in laboratory quartz glassware are even smaller because its coefficient of thermal expansion is only about one fifth that of Pyrex. Thus, it is possible to heat quartz to white heat and plunge it into ice water without cracks forming because the stresses generated even by these extreme temperature gradients are so low.

Usually, it is the process variables that are altered to control the thermal stress. The convective heat transfer coefficient varies over several orders of magnitude as the quench medium is changed from still, warm air through oil to water and finally to an ethylene glycol/water mixture or brine spray. The likelihood of distortion or cracking increases as one proceeds from furnace cooling through brine quenching. Thus, if excessive stress is produced at one rate, one possible solution is to slow the quench rate to reduce the stress. Unfortunately, this also reduces the effectiveness of the quench process and usually reduces the final hardness of the part. However, a cracked part is totally unacceptable. A noncracked but distorted part may be acceptable if the distortion does not make the part dimensionally nonconforming. However, there will be a residual stress, the effect of which may be adverse. If the distortion causes dimensional nonconformities, straightening may be possible, but will add cost and may also create an adverse residual stress. Consequently, the engineer is stuck with accepting a reduced strength level or finding a material that will harden adequately at the reduced quench rate. This latter approach is precisely why alloy steels were developed. There is a broad range of alloy steels that harden adequately even at quench rates as slow as air cooling. However, before changing to a more expensive alloy, be sure to investigate the possibility of redesigning the part. By reducing the stress concentrations associated with sharp corners, fillets, etc., cracking may often be eliminated. Another common design error that results in unnecessarily large thermal stress is the joining of thick and thin sections. The cooling rates will obviously be different because of the mass/area ratio differences, so there will be a stress that arises at the interface because of the difference in shrinkage.

This problem of cracking due to differential heating and cooling rates in thick and thin sections occurs in ceramic and powder metal parts also as the parts are heated for sintering or cooled afterwards. It is most economical to heat as rapidly as possible because this contributes to minimizing the cycle time. However, the heating and cooling must be slow enough to prevent cracking. With large variations in wall thickness, the time may be unacceptably slow, so that redesign to make the wall thickness more uniform throughout the part may be a viable alterative. This is one more example of the rule of thumb in the design of most parts that it is best to have the wall thickness as nearly uniform as possible.

When parts are quenched, their rejected heat builds up in the quench tank, raising the temperature of the quenchant. This naturally reduces the quench rate, which reduces the severity of the quenching stress problem. It is possible for the quenchant temperature to become so high that it has an adverse effect on the part's properties. Cooling of the quenchant may therefore be necessary to hold the quench temperature in an acceptable range so that the part's properties do not have unacceptable variability.

In steels the quenching stresses are the result of a combination of stresses from differential cooling and stresses that arise from the volumetric expansion accompanying the phase change when austenite transforms into the final product. The distortion and cracking problem is sometime overcome by raising the quench temperature very high using a salt bath. For the steels that have a "bay" between the pearlite nose and the martensite start temperature, neither the slower quenching nor the time spent in the bay have a deleterious effect on the final properties. However, both effects reduce the quenching stresses. Austempering and Martempering are both based on this principle.

Hardenability is defined as the measure of the ease of producing martensite as a function of cooling rate. The most hardenable steels transform into the most martensite at the slowest cooling rate. There are numerous aids for the selection of steels for suitable hardenability, principal among them being the isothermal transformation diagram (also called the time-temperature-transformation, or TTT diagram), the continuous cooling diagram and the end-quench hardenability (Jominy) curves. They all relate the alloy content and the cooling rate to the expected microstructure and properties.

The cooling rate is mainly a function of the quench medium, the location inside the part, and the thermal diffusivity of the metal. Air cooling and furnace cooling are slow, water and brine quenching are fast. The surface always cools faster than the interior, and small parts cool faster than large parts. In low-alloy structural steels, the alloy content is too small to have a significant effect on the thermal properties of the metal, i.e., plain carbon and low-alloy steels do not differ much in thermal diffusivity. All of these effects can be accounted for to determine the cooling rate, either analytically or experimentally. Once the cooling rate is known, the various charts and diagrams will tell what structure and properties can be expected.

After the quenching is complete, the parts must be tempered because freshly quenched martensite is structurally unreliable due to poor ductility and unacceptably low fracture toughness. The final hardness is a function of both tempering temperature and time. It is possible to get almost any desired properties between fully hard and fully soft by proper control of the tempering conditions. The ease with which one can control such a wide possible range of properties is an important reason why steel is so useful as a structural material.

If the cooling rate is fast enough, the transformation results in an amorphous structure that has properties quite different from the crystalline counterpart obtained at normal, commercial cooling rates. Such fast cooling rates might be achieved by flinging liquid droplets onto a cool, thermally conductive surface, or by quenching very thin sheets, like the cutting edge of a heat-treated razor blade.

When the alloy content exceeds about 5% total (which is often cited as the arbitrary upper limit for "low-alloy" steels), the thermal properties of the alloy may begin to differ from plain carbon steel. In the most common stainless steels, for instance, where the alloy content exceeds 20% of the composition, the thermal diffusivity is significantly less than that of plain carbon and low-alloy wheels.

Many metals other than alloy steels may be heat-treated, but there are no selection aids corresponding to the TTT or Jominy curves. The heat-treating cycle is defined and the resulting properties are tabulated as minimums or ranges of expected properties.

Failure and Repairability

Failures are unavoidable. The failure may by any of several types: wearout, corrosion, fracture, etc. When the failure occurs, a decision will need to be made regarding repair or replacement. A designer should consider the anticipated failure mode and, if repair is warranted, should choose a material that can be repaired. Wearouts can be repaired by building up the worn surface. Replacing worn metal by depositing weld metal, plasma spray, electroplating, etc., may be possible. Obviously, if this is anticipated, the initial selection must be of a material that will permit such repair procedures.

Fractures in many metals may be repaired by welding if the designer chose a metal that is weldable. The welding process usually involves rapid input of heat which alters the properties of the material in the heat-affected zone (HAZ) compared to the base metal adjacent to the HAZ. When the welding is expected to occur in a low-stress region, the welding may do no harm. Higher stress regions must be post-weld-treated to restore adequate strength. This may be difficult or impossible. For instance, weld repair of aluminum structures made of heat-treated or cold-worked aluminum will reduce the strength of the material because of the annealing effects of the heat input of the welding. There is no realistic way to restore the strength that was imparted to the material by the original heat treatment or cold work that was done prior to the initial fabrication of the structure. If the weld repair is done in either an area that is under low stress in service, i.e., a location that had a large safety factor initially, or in an area that was designed to be welded, a weld repair is likely to be successful. However, if the repair is done in a high-stress area, the reduction in strength that accompanies the repair may be catastrophic.

If the original strength cannot be restored, the designer must account for the property change that accompanies the fabrication. Allowable stresses for HAZ are tabulated in some books.

Fractures in stone may be partially repaired by staples that hold the broken pieces together. However, if the metal of the staples corrodes and produces a corrosion product that expands in the hole holding the staple, the expansion itself may further crack the stone at the staple hole. This was the fate of some stones in ancient structures in Greece that were "repaired" in modern times by well-meaning, but insufficiently diligent, curators.

Failures sometimes occur before a part is even finished in production. Cracks and holes in castings are an example of defects that must be anticipated in production. These are often repaired by welding or filling with epoxy as part of the normal production operation. There are instances where this procedure is harmful and must be prohibited. When the softening or the residual stress that inevitably accompanies welding will have a deleterious effect on the fatigue life of the part, welding must either be prohibited or carefully controlled. Also, particularly in aluminum alloys, the properties of the material in the vicinity of the weld repair will not be the same as the properties of the bulk material even after heat treatment. This nonuniformity of properties may be unacceptable in high-strength regions of critical castings. Thus, weld repair of casting defects is prohibited in certain aircraft parts.

2.3 Cost Analysis

The cost of a finished product depends on the cost of purchasing the material before manufacturing starts and the costs associated with the manufacturing processes used to make the finished product ready for the final customer. Selection obviously affects the material cost, but it is not always so obvious how the material choice affects the manufacturing cost. In this discussion, the material cost to the manufacturer is assumed to be the price paid to the warehouse or wholesealer for material in a form appropriate to begin the manufacturing process. To a foundry, that would be the ingot price; to an airframe manufacturer, that would be the price of sheet metal and extruded sections; to an injection molder, that would be the price of pelletized polymer.

Any tabulation of the actual price of materials at the time of this writing would be foolish because the data would be out of date before the book is published. The price of many raw materials fluctuates markedly. Fluctuations exceeding 1000% over a relatively short period have occurred for several reasons: political decisions (let the price of gold float), commodity speculations (a failed attempt to corner the silver market), formation of cartels (oil, which affected some plastics), strikes and other supply interruptions (copper).

The price of materials depends on the price of the base material, the alloy additions, any blending, polymerization and other processes required to alter the chemistry and any costs associated with special preparation. Some changes in chemistry do not affect the price. In plain carbon steel, the carbon content does not affect the price. Alloy additions usually do add to the cost, especially for high concentrations of expensive elements. Price fluctuations in the alloying elements may dictate material change. Nickel in austenitic stainless steel is an example. When nickel prices are high, ferritic stainless (no nickel) may be a viable substitute. In spite of its slightly poorer corrosion resistance and formability, the more attractive price may make it the better choice until the price of the technically superior alternative becomes attractive again. The change in alloy content may affect the cost of the additional preparation needed to put the material in to the standard shapes that the customer wants. Extruded aluminum sections of 7075 and 6063 differ in price partly because the 7075 base material is somewhat more expensive and partly because 7075 is more difficult to extrude. Numerous other "extras" might add to the cost, like grain size control, vacuum melting, and custom chemistry.

In polymers, differences in price are more the result of differences in the difficulty of processing the raw material into the final product than in differences in the price of the raw material (oil, natural gas, etc.) itself.

The choice of material may affect the cost of manufacturing as well. Any change in the price of the finished product associated with a change in material will be more complicated than simply adjusting for the change in the incoming material cost. For instance, there is not much difference in the per-pound cost of the cast iron as the class (strength) changes. However, as the strength increases, the solidification characteristics are such that more risers and more complicated feeding are needed. Hence, the amount of metal that has to be poured to make finished castings of low- and high-strength gray cast iron is different. The extra risers of the higher strength castings must be paid for, even though the finished castings have the same geometry and weight.

Similar concerns might be true for alternative material choices where machining is a significant part of the total cost of the product. Changing the carbon content of steel has virtually no effect on the per-pound price of bar stock, but it has a significant effect on the machinability of the material. Thus, changing the carbon content to improve the strength of a product (without changing the geometry) may significantly increase the product's cost, even though the warehouse charges the same price for both materials.

Availability

Obviously there is no point in choosing a material that is not available. A material may be unavailable either because it is not produced at all or because it is not produced in a form that is suitable for the prospective application. Sometimes a standard alloy becomes obsolete and simply goes out of production. Some reference sources actually list "former" or "obsolete" alloy compositions to aid in find a suitable substitute (Metals Handbook, Ross 1992, Woldman 1990).

The shape of material that a manufacturer can purchase significantly affects the amount of work that will be needed to turn the incoming stock into the finished product. Thus, availability in a suitable form or shape can be a critical factor. Numerous examples have been published over the years. A gas turbine manufacturer substituted 410 stainless steel for the 4130 and 8630 low-alloy steel that had been used to make certain brackets. The low-alloy materials were no longer available in suitable sizes as incoming stock for the brackets (Bradley and Donachie 1977).

Most engineers who make material selection decisions can cite other examples which have not been published previously where availability in the appropriate form was a significant factor. A cannon-launched, laser-guided projectile (a form of smart bomb) for destroying tanks is made of seamless tubing. The material choice was determined, in part, by the fact that one material under consideration was already being produced in seamless form of the appropriate size for large hydraulic cylinders. Other materials that would also have met, or exceeded, all the technical requirements were not already being produced in this form. The cost of special ordering could not be justified. In the end, availability became the deciding factor. Critical questions to be answered about availability might be: Is the material available for immediate delivery, in the required quantities, from stock? If not, immediately, what delivery time must be expected? There may be long lead times on items with special processing. What is the minimum quantity that must be ordered? Are there alternative suppliers? Is there an alternative material to fall back on in event of delivery problem with the preferred material?

Most materials are priced by the unit of weight. Even when the price is per unit of length, as with structural tubing, the price is actually derived from the weight: the thicker the wall, the greater the price, approximately in proportion to the weight. Only timber, along with concrete, aggregates, earth fill and the like are usually priced by volume. The material price is not likely to be an important factor in making low-volume, or one-off, products. It may be the critical factor in high-volume production.

It is worth stating the obvious: not every manufacturing process can be used for every material. When changes in material are being considered as a way to reduce cost, it is vital to remember that a new material can alter the manufacturing characteristics of the process enough to offset the material savings. With commodity plastics, the prices are similar but they fluctuate some. Accordingly, the choice may be made strictly on price. However, there are small variations in mold skrinkage, warpage, cycle time, etc., that must be considered along with the price. That is part of the reason why it is so important that price be only *part* of the selection decision, regardless of what the bean counters say.

Materials Life and Maintenance

When total life-cycle cost is part of the selection criteria, high initial cost may well be justified if it results in low maintenance cost or long life. Initial cost, not life-cycle cost, is the purchasing criterion for many customers. Therefore, it is desirable to know the customer's thinking before making the selection decision. Since this is often not possible, the decision is made with uncertainty. Then it becomes important that the sales people be made aware of the decision criteria so that they can communicate successfully with the customer why the selling price might seem a little high. Concurrent engineering design makes this communication easy.

When different materials are being considered for a particular application, they are likely to have both different properties and different cost per weight. Assume that the criterion for adequate performance is the same for all materials being considered; e.g., all candidate materials must carry the same load or torque without yielding, or all materials must have the same elastic deformation. If the properties are different, the part configuration will be different also. For geometrically similar cross sections, weaker materials require a larger cross section than stronger materials. If both the density and the price per unit weight are the same, as might be the case in comparing various plain carbon steels, components made from the weaker material will be heavier and more expensive. However, both the price and the density may vary. To illustrate the method of making comparisons in the general case, consider a simple circular beam with diameter, d, subjected to a bending moment, M. The bending stress, σ, is

$$\sigma = \frac{My}{I} = \frac{32M}{\pi d^3}$$

where

$$y = \frac{d}{2} \qquad \text{and} \qquad I = \frac{\pi d^4}{64}$$

therefore,

$$d = \sqrt[3]{\frac{32M}{\pi\sigma}}$$

The volume, V, of a unit length of beam is

$$V = \frac{\pi}{4} d^2 = \frac{\pi}{4} \left[\sqrt[3]{\frac{32M}{\pi\sigma}}\right]^2$$

Suppose a component may be made from two different materials that have different strengths, S_1 and S_2. Assuming that the same factor of safety, N, is to be used with both materials, the allowable stress, σ, is either S_1/N or S_2/N. Thus, the ratio of the required volumes is

$$\frac{V_1}{V_2} = \left[\sqrt[3]{\frac{S_2}{S_1}}\right]^2 = \left[\frac{S_2}{S_1}\right]^{2/3}$$

The cost of the component is the product of the unit material cost, c (say, dollars per pound), the specific weight, w (say, pounds per cubic inch), and the volume, V. The ratio of the costs of the two components is then

$$\frac{C_1}{C_2} = \frac{c_1}{c_2}\frac{w_1}{w_2}\frac{V_1}{V_2} = \frac{c_1 w_1}{c_2 w_2}\left[\frac{S_2}{S_1}\right]^{2/3}$$

A similar analysis may be made for other loading schemes, like axial loading and torsion. The failure criterion used above was exceeding the yield strength. Other failure criteria might include excessive elastic deformation (including column buckling) and exceeding the fracture strength when a crack is present and the fracture toughness is known.

Numerous other conditions have been analyzed using an approach similar to this. See the discussion on the Ashby method in Section 3.6.

2.4 Other Considerations

Ergonomic and Safety Considerations

When the user is expected to have direct interaction with the item being designed, ergonomic and safety considerations take on increased importance. For a very large number of components, the user has no direct (visual or tactile) contact with the item. Therefore the appearance, feel, texture and other such factors are not important to ensuring the customer's level of satisfaction. For instance, with gears in an automobile transmission, the functional and manufacturability considerations will be most important. There is no reason to be concerned with whether the aesthetics or the shape of the gears is compatible with the user. A possible exception could be related to assembly considerations of the gears, but this is not likely to be a very large concern compared to the function of the item. On the other hand, consider a drinking cup that is expected to be used with boiling hot beverages. Many possible material choices exist among the three main classes of materials. The liquid contained in the cup is capable of causing discomfort and actual damage to the user, but only if its heat reaches the lips or tongue of the user at a sufficiently high temperature. The cup's material should not exacerbate the problem by readily conducting the heat to the lower lip before the liquid is actually drawn into the mouth. Traditional ceramic teacups and coffee mugs quite successfully protect the lower lips from the heat of the cup's contents because the ceramics that are used for this item are relatively poor thermal conductors. (Not all ceramics are poor thermal conductors. While diamond has significantly better thermal conductivity than any metal, it would not be considered for this application.) However, for certain conditions of use, the ceramic is unsuitable. Camping and hiking, for instance, require a cup that is light weight but will withstand the stress of being stuffed into a knapsack or knocked against a rock. While plastic is possible and would have thermal properties similar, perhaps even superior, to ceramic, metal is more resistant to damage in this harsh environment and can have thinner walls. Since corrosion resistance and formability are of paramount concern, aluminum and stainless steel might be suitable candidates. Here the stainless gets the nod for thermal protection for two reasons:

1. The thermal conductivity of aluminum alloys is about 10 times greater than stainless steels.
2. Thinner walls are possible with stainless because of its greater strength. This offsets the greater density of the stainless, thus keeping the weight down, but more importantly the thinner walls cuts down the heat flow to the lip of the cup above the liquid level.

Both stainless and aluminum cups are available. Those customers who opt for the aluminum will, if they are quick learners, only burn their lips once.

Recycling

Recycling of scrap and used material occurs routinely in industrial production. Steel and cast iron have been produced using recycled material for more than 100 years. Some of that scrap is the waste from production processes and some is from discards after use by consumers. The plastics industry has recycled sprues and runners of thermoplastic materials since the earliest days of injection molding. The "regrind" is mixed in with virgin material. Engineers do not usually need to factor these considerations into their material selection decisions.

The recycling of plastic packaging material and certain consumer products, notably automobiles and electronic goods, is rapidly becoming required by law in many jurisdictions. In the design of such products, material selection decisions must include consideration of the ease of recovery and recycling. Recovery includes ease of disassembly so that the different materials may be separated appropriately. This is an example of why the concurrent engineering approach to design is so important. Assembly and disassembly methods may include the use of snap-fit components. The material must be strong enough to withstand the assembly stresses and the stresses encountered in operation, but not so strong that the disassembly procedure is unnecessarily difficult. Thus the material selection decision must satisfy the needs of assembly, use, and disassembly of the product as well as the recyclability of the material itself. In addition, some means must be found for the recycler to identify the type of material.

An arbitrary division is commonly made, based partly on price and partly on performance, between commodity plastics, engineering plastics and high-performance plastics. Almost all packaging materials are chosen from the group of six commodity plastics that are identified by the recycling symbols created by The Society of the Plastics Industry, Inc.

Symbol	Polymer
1 = PETE	polyethylene terephthalate
2 = HDPE	high-density polyethylene
3 = V	vinyl, mainly polyvinylchloride
4 = LDPE	low-density polyethylene
5 = PP	polypropylene
6 = PS	polystyrene
7 = other	other and/or mixed polymers

The abbreviations are standardized by ASTM D4000, D1600, and D1972-91.

Almost all of the postconsumer recycling currently done is with these six plastics. Usually, the HDPE (milk and water containers) and PET (beverage containers) are separated and reprocessed into new products of essentially the same material as the original. The remainder is a mixture of uncertain and variable composition which is generally sent to a landfill or incinerator. There are processes for making plastic lumber, pallets, and other heavy-sectioned items from the mixed tailings of the plastics waste steam.

Some engineering plastics, notably ABS and nylon, are recycled alone from postconsumer products. British Telecoms and ATT/Western Electric have both had programs to recycle the ABS from discarded telephones.

Recycling directly into new products that have essentially the same characteristics as the former ones is one way of reusing the material. The material, because of degradation that may take place during the recycling, may also be used to make products that are used in less demanding applications. Processes have also been developed to depolymerize the plastic and recover the starting material or some other useful chemical or fuel. Finally, burning to recover the heat of combustion of the scrap is a way to avoid landfilling while recovering some useful energy. These four recycling schemes are referred to as primary, secondary, tertiary and quaternary. The recovery value for each of these processes might be calculated and used as part of the material selection decision when determining the life-cycle cost of various options.

One impediment to successful use of recycled material is the insistence by designers that more than one type of plastic be used in the product. Separation is possible, but expensive, so it is preferable to minimize the use of mixed materials. Some products have been skillfully redesigned to cut the number of different plastics by more than half. That is, an important aspect of the material selection decision may be to reduce the number of different types of materials in the product, specifically to enhance the recyclability of the product. Even with relatively simple packaging products, the labels are likely to be either paper or a different plastic than the main container. This either complicates the sorting process prior to regrinding or contaminates the regrind and degrades the resulting properties. Designing labels that are integral with, and of the same material as, the main component is a potential method for improving the recyclability of products.

Any plating or coating on the base material also complicates the recycling process. Thus, if recyclability is an issue in the material selection, designs using materials that do not need the foreign material should be considered. This principle applies regardless of the base material. Plastics are often favored in such situations because they can be colored, instead of painted or plated, for appropriate appearance. They are also more corrosion resistant than metals in many environments, which again avoids the contamination of paint or plating during recycling.

U.S. automobiles now have about 300 pounds of plastic per vehicle. Current production is around 10 million units per year, most of which will eventually be scrapped. Most of the ferrous metal and some of the nonferrous metal are presently recovered. Most of the nonmetal is not, but is sent to the landfill after shredding. Thus the potential for avoiding landfill is large. As the pressure mounts for disassembling and recycling more material from junk automobiles, the need to identify the various plastics becomes more acute. The Society of Automotive Engineers has responded by adopting a recommended practice of marking of plastic parts so that they may be easily identified at the time of disassembly. The ASTM abbreviations are used inside a sharp-ended oval with proportions about like the longitudinal cross section of an American football.

There are technically difficult problems in reusing mixed engineering plastics because the different species are not compatible. Research is ongoing to find modifiers that will overcome this problem by linking the incompatible molecules together. When this is successful, there will be a new type of material available for selection and use. It will also affect the selection decision for the original product because there will be pressure to use only those plastics which can subsequently be reprocessed most easily.

Thermosets are widely perceived as unsuitable for recycling. That is not correct. Practical uses are found for thermosets that have been reground or depolymerized. Regrind may either be mixed with adhesive to hold the particulate together (carpet underlayment) or used as a filler in new thermoplastic or thermoset products.

3 DECISION MAKING IN MATERIAL SELECTION

3.1 Decisions

Selection from among alternatives requires decisions. Sometimes experience and judgment are sufficient to permit a sound decision without a formal process. This is especially true when a candidate material has some characteristic that is outstanding or all but one have some serious deficiency. When there is no obvious choice, confidence in the decision can be enhanced by using a formal process. This eliminates vague and unreliable techniques based on who is the boss, who holds the strongest opinions or who argues most persuasively.

There will be a set of requirements which the material must satisfy. Each candidate material must at least minimally satisfy all the requirements. It is likely that each material will possess some characteristics that exceed the minimum requirements by varying amounts. Thus, we need a method for rating how well each material meets the requirements. Some requirements may be more important than others. Greater weight should be given the more important requirements.

Many methods have been developed to formalize the decision process. Mainly, they attempt to

1. Quantify how important each desired characteristic is by determining a weighting factor
2. Quantify how well a candidate material satisfies each requirement by determining a rating factor
3. Combine factors 1 and 2 to determine which material offers the best compromise

There are methods to make each of these three steps as objective as possible. Judgment is required to choose the most satisfactory method, but once that is done the decision mechanism leads inexorably to an answer. There is danger in accepting the answer just because it is the result of a formal, numerate process. If the engineer's wisdom and experience cause doubts about the correctness of the answer thus obtained, reexamination of the assumption and previous judgments about methodology is appropriate. Sometimes the methodology is wrong or leads to an inappropriate conclusion because the input is deficient. But, sometimes it is not, and it leads to an appropriate solution that is different from that obtained by "gut feeling."

3.2 Ranking of Attributes

Attributes are the characteristics that can be described to differentiate one item from another. That may be size, color, weight, reliability, etc. In choosing materials to satisfy product requirements, a property can normally be found which describes how well that material satisfies each of the requirements. Quality function deployment formalizes this process of finding engineering terms, usually quantifiable properties, to describe the list of attributes that constitutes the customer's requirement for a product.

Sometimes an attribute cannot be characterized by a quantifiable engineering property. The feeling of softness of the liner of disposable diapers is important to customers (moms, not babies; nobody really knows what the babies prefer). The perception of softness is known to depend in part on engineering properties like modulus of elasticity and coefficient of friction. However, it is also known that the pattern of holes, bumps and other geometric features that is embossed on the sheet during manufacture is critical. Quantifying this aspect of the problem is done by customer preference surveys: What fraction of customers prefer pattern A? What fraction prefer pattern B? etc.

Some attributes will be more important than others. Determining the relative importance of the various properties assigned to these attributes is necessary if one is to use the weighted rating decision matrix. Establishing this relative importance may be easy or very difficult, depending on the circumstances. Cost, for instance, may either be very important, as with low-margin consumer products, or quite unimportant, as with military aircraft and Rolls Royces. In the cases where it is not obvious, there are methods that may help to arrange attributes in order of importance.

There are two useful schemes for recording the results. The first simply ranks the attributes in order of their importance, with no account taken of how much more important one attribute may be than another. The second assigns a weight to the importance of each attribute. It will be described in the next section.

In simple rank ordering, pairs of attributes are compared one after another. The more important attribute of the pair under consideration is give a score of 1, the less important attribute is scored as 0. Sometimes, the two numbers of the pair seem to be equally important. Take some extra time to see if it is possible to find even some small differences between them. There is no method for choosing the better of two items that are equal. Finding something to distinguish between candidates will lead to an unequivocal decision.

Tallying the result is simplified by numbering each attribute and using a matrix to keep score of the number of times one attribute is considered more important than another. The matrix can be constructed by assigning numbers to the attributes and creating one row and one column for each attribute. The intersection of a row and a column represents a pair of attributes that are to be compared. The intersections on the diagonal have no meaning because they compare an attribute with itself. Leave blank spaces on the diagonal.

To illustrate, suppose the five attributes in Table 1 are considered important for a particular application. The first row and column represent strength, the second row and column represent weight, etc. Reading across the first row, strength, ignore the first intersection. At the second column, ask the question, "Is strength more important than weight?" If yes, put a mark in the intersection under column 2

Table 1 Unweighted Comparison of Attributes

Attributes		1	2	3	4	5	# marks
Strength	1		1	1	1	0	3
Weight/density	2	0		0	0	0	0
Color	3	0	1		0	0	1
Texture	4	0	1	1		0	2
Cost	5	1	1	1	1		4

(which corresponds to weight). If not, either put 0 or no mark. Continuing across, if strength is more important than color, put a mark at row 1, column 3. Continue this until all intersections have been considered. Note that the result of the comparison of the first row/second column (strength versus weight) will be the opposite of the result of the second row/first column (weight versus strength). After filling in a row, cover it. This forces more careful consideration of each succeeding row because it prevents simply filling in the complementary question with the opposite answer without really thinking about the question. Then, double-check the accuracy of the recording by checking for antisymmetry across the diagonal. This increases the likelihood of finding discrepancies. Determine the number of marks in each row. The attribute receiving the largest number of marks will be the most important. The attribute receiving the smallest number of marks will be the least important. All the others will fall in descending order of importance between the most and the least important. By breaking the process into a large number of relatively easy decisions, it is more likely that the correct order is achieved than if one is forced to chose the most and least important attributes in one step.

3.3 Weighting Factors

The method in the preceding section gives no clue about how much more important one attribute is than another. It may be desirable to quantify the relative importance of the attributes. One attribute may be very much more important than another, while others are quite similar in importance. It is simple to show the relative importance using a point scale that totals 100 points. If, for example, one attribute is four times as important as another, this would be represented by an 80/20 division. Equally important attributes are split 50/50. This division is a judgment that must be made. There is no objective scheme that can take the place of expert opinion at this step.

In the method of the preceding section, every possible combination of pairs was considered. In quantifying the relative importance, it is not necessary to compare every possible combination. If attribute 1 is compared first with attribute 2, then 3, then 4 etc., the relative importance of 3 and 4 can be found simply by dividing the ratio for attribute pair 1 and 3 by the ratio for attribute pair 1 and 4. Thus, it is not necessary to actually survey and record the opinion of experts on pair 3 and 4.

Table 2 shows a hypothetical breakdown for a new application. Weight is four times as important as strength, strength is four times as important as cost, color is

Table 2 Weighting of Attributes

		1/2	1/3	1/4	1/5	Ratio	Weight
Strength	1	20	60	50	80	1.0	.14
Weight/density	2	80				4.0	.58
Color	3		40			.66	.10
Texture	4			50		1.0	.14
Cost	5				20	.25	.04
Total						6.91	1.00

two thirds the importance of strength, etc. The simple ratio for each property relative to strength is given. Strength relative to itself has the value of 1. The weight, which shows the relative importance, is the last column. This is the decimal equivalent of how big each individual ratio is relative to the sum (total) of all the ratio numbers. The strength rating is 1.0/6.91 = 0.14.

This method uses attribute 1 as a baseline, or datum, for comparison of all other attributes. Choosing the baseline should be done so as not to prejudice the result. Listing the attributes in alphabetical order is one possible way. Listing in random order (put all the names of the attributes on slips of paper in a hat and draw them one at a time) is another.

How many attributes should be listed? Five to 10 will adequately reflect most needs. Using many more than 10 will require such a fine distinction between attributes that the result may become muddy because distinctions that are really not truly significantly different will have numerically different scores. Thus, if a large number of attributes seem important to the designer, some judicious culling of the list is probably prudent. If fewer than five attributes are important, either the designer has overlooked something or it is a special case for which mathematical methods like this decision matrix may not be appropriate.

The methods just described show the order of importance and fix a numerical value of the relative importance. In some circumstances the relative importance assigned to any attribute may still be negotiable, especially if a potential material has some characteristic that is overwhelmingly desirable. This shifting of relative importance may be handled by the more advanced methods of fuzzy mathematics (Thurston in Kusiak 1993, p. 437). Essentially this kind of mathematics can quantify the statement, "I might be willing to consider property A a little less important than the matrix shows if some material has a really good value for property B."

There are other methods for determining the importance of various factors that should influence a decision. The analytic hierarchy process (Saaty 1980) has been applied successfully by Thurston (in Kusiak 1993) to material selection problems. She shows clearly how the material choice changes as the design requirements (product lifetime, recyclability, etc.) change.

Many other decision-making methods have been published (Porter et al. 1990, Starkey 1992, Hill 1979, Kepner and Tregoe 1981, Ackoff 1978, Rubinstein 1980, 1986). Published examples of their application to material selection decisions are rare. It should not be inferred from the shortage of published examples that the methods are not applicable.

3.4 Rating Factors for Material Properties

Each property will have different quantitative values for each of the candidate materials. There are several methods that are used to list the candidate materials in order of suitability for a particular property, i.e., to give a numerical score corresponding to how they rate as a candidate material. The simplest one is to rank the materials in order for each property that is being considered. The figure of merit for each property will simply be the number corresponding to the material's place in the ordering. The best material will have a rating of 1, the second best material will have a rating of 2, etc. The material with the poorest relative ranking for that property will have a rating number equal to the number of materials being considered. There is no way to tell how close number one is to number two or how far it is from number "last." But, it is quick and it gives a satisfactory overview for preliminary investigations. This method has been used for more than 25 years by LNP plastics selector guide described later.

More discrimination about the relative merit of each property of the materials being considered may be incorporated by assigning the value of 100 (percent?) to the best material in that property category. Merit ratings for the remaining materials may be determined simply as the percentage that the new material's property is of the best material's property. This linearizes the relative ratings. It implies that a material which has a tensile strength half as great as the highest strength material should be awarded a strength merit figure half as good as the strongest material.

Note that the "best" material may either have the largest value of the given property or the smallest. One would expect that increasing strength is good and the strongest candidate would be assigned a value of 100% For some situations smaller values of the property are "best." For lightweight structures, low density is good, so the lowest density material would be rated 100% and all others would be less than 100% A material with twice the density of the best candidate would have a scaled property of 50% In this case it is the reciprocals of the material properties that are being linearized.

While this is a relatively simple calculation, it may be inadequate or misleading for several reasons. First, as the strength decreases, the amount of material required increases, affecting the weight, cost, placement of other components and potentially a host of other secondary factors. Accordingly, some nonlinear functional relationship for the merit ratings may be more appropriate.

Second, there may be a lower limit below which the material would either be unsatisfactory (useless), or its usefulness would diminish very rapidly. In that case, it may be possible to construct a graph of usefulness or utility versus the material property. The scaled value for the candidate materials may be taken from that. For instance, consider a plating material that is intended to give satisfactory surface appearance to a component for a given life. Suppose the cost is roughly proportional to the plating thickness, which in turn is proportional to the expected life. An automobile bumper with a target lifetime of 10 years might serve as an example. If a candidate material will last only five years, it is not worth considering. For $9\frac{1}{2}$ years there would be considerable penalty, but the material might still be considered. At the other extreme, suppose there is a candidate material that might last for 20 years. That would not be twice as useful as the 10-year candidate and therefore would not be worth twice the cost, because the car would likely be unusable for other reasons. There is no rationale for designing components or choosing materials which will last significantly longer than the assembly to which they belong.

Third, some property values are themselves either nonlinear or have values that are derived from an arbitrarily chosen scale, which may include both zero and negative values. Hardness is such a property. In the Rockwell scales, for instance, it does not make sense to say that a part with Rockwell C 60 is twice as hard as RC 30. First of all, all Rockwell scales have an arbitrarily chosen datum of zero hardness. Second, when one scale is converted to another, the same ratio of hardness values in not necessarily retained. Third, two materials which have the same value of hardness when measured on one scale may have different values when measured by another scale because of differing shapes of indentors and, consequently, of different deformation patterns under the indentor.

Fourth, some properties may be quantified by two or more different methods which are attempting to measure the same property but which would give different merit rating for the same materials. For instance, relative hardnesses, which are easy to measure, are sometimes used to represent the relative values of wear resistance, which is more difficult to quantify. Great care has to be exercised in this instance because the hardness may not really be proportional to the other property. For instance, leaded and unleaded 70-30 brass may have nearly identical hardnesses but differ in wear rate by a factor of 10.

Some property values, especially ones related to the ability of materials to undergo certain kinds of manufacturing processing, are not normally quantified as a property that is measured in conventional units. Rather, an arbitrary numerical value is assigned that quantifies a descriptive relative rating for the property. Weldability and resistance to hot tearing (in castings) are examples. These are assigned values of, say, 1 to 5 representing the best to worst. They may be assigned ratings of A to E (as the Aluminum Association does to rate the processing characteristics of aluminum casting alloys), also representing best to worst. These values represent the judgment of experts in the field about the relative merits of each material for the characteristic that is being measured. It is not the result of a test that the reader can verify.

Some processing characteristics may appear in the literature either as an apparent property, complete with units of measure, or as a relative or comparative value. Fluidity of casting alloys, for instance, as measured by the distance that a stream of hot liquid will flow before freezing in a "fluidity spiral," is given in inches or centimeters of flow around the spiral. The dimensions of the spiral are not standardized, so comparison of one data set to another is dangerous. One scheme for reporting machinability of metals may appear as "cutting speed for 60 min of tool life" in feet per minute. Machinability may also be given as relative machinability rating, which compares the cutting speed for 60 min of tool life of the candidate material with the cutting speed for 60 min of tool life of a standard material (AISI 1112 or SAE 1212 for steel, AA7075-T6 for aluminum and CDA 360 for brass). Machinability may also be given on the A to E scale mentioned above, and when it is, the rating takes into account not only the cutting speed for a given tool life but also other factors as well. Such factors might include the nature of the chip (broken versus continuous) and the surface finish (excellent finish versus satisfactory finish, whatever those words may mean). When translating the A-to-E letter system into corresponding numerical values, some people use 0 to 4 and some use 1 to 5 for a 5-point scale. Some use 1 for best and some use 5 for best.

Translating words into numbers for the rating schemes is possible. The words corresponding to the five numerical values might be unacceptable, poor, satisfactory, good, and excellent. A more complete list that might be used on a scale with more points could be useless, almost useless, very poor, poor, barely adequate, satisfactory, a little better than just OK, good, very good, excellent, ideal.

Needless to say, the assignment of values to weighting and rating factors is of utmost importance in the accuracy of data on which the ultimate decision will be based. Therefore one must base these values on the greatest possible amount of information, a full understanding of the problem, and judgment of the highest degree. (P. H. Hill 1970).

Professor Hill does not say how to ensure that only engineers with good judgment and full understanding of the problem be permitted to assign rating and weighting factors.

3.5 Decision Matrices

Decision matrices can be used for making decisions in many engineering situations. Evaluation of competing concepts in design is one commonly used application outside the area of material selection. Purchasing decisions among competing products are another. Matrices to determine the relative importance of properties were illustrated above. Decision matrices are also used to evaluate the relative merit of candidate materials. Simple matrices do not weight the importance of different factors. That may make the matrix very easy to construct and use, but it ignores the fact that it is unlikely that all properties or characteristics of that candidate material are equally important. This shortcoming is surmounted by using a weighted matrix.

Unweighted Matrices

The materials are usually listed in the first column on the left and the properties are listed on the top row. In each row/column intersection, the appropriate relative value for the property (rating factor from Section 3.4) of that material is shown.

The LNP plastics selection guide is an example of a simple matrix that has been used for many years (Fig. 1). Six principal families of plastics are listed down the left column. Each family has several individual plastic types as a subgroup. This is not intended to be an exhaustive survey of all available plastics, only a preliminary and rough guide to the most important characteristics of the most widely used plastics. The principal properties that are likely to be used for material selection are listed in the top row. For each property, the materials are ranked ordered, with the order shown as numbers 1 to 6. 1 indicates most desirable, 6 indicates least desirable. Within each family, the subgroups are also rank-ordered starting with 1 as the most desirable and continuing for as many numbers as there are members of the subgroup. At each intersection the rank number is shown.

The properties shown here have broad definitions that are shown at the bottom of the chart. Purists would argue that strength and stiffness are quite different properties and that they should not be confused by being lumped together in the same column. Similarly, environmental resistance, which includes resistance to solvents, could be considered overly broad because such resistance is highly dependent on the specific solvent. However, this kind of chart is used mainly to eliminate most of the materials that would be unsuitable for one reason or another, leaving behind for further consideration the plastics that warrant more detailed investigation of their usefulness. It is not intended to be the vehicle for the final decision about which one specific material to use. Thus, rather broad, loose, all-encompassing definitions are justified.

This matrix is used by crossing out the columns that represent properties that are not important to the application for which a new material is sought. Then simple

G/R Resin Groups \ Design Criteria	Strength & Stiffness	Toughness	Short Term Heat Resistance	Long Term Heat Resistance	Environmental Resistance	Dimensional Accuracy In Molding	Dimensional Stability	Wear & Frictional Properties	Cost
Styrenics	**3**	**6**	**6**	**6**	**6**	**1**	**5**	**6**	**2**
ABS	2	1	1	1	1	3	2	3	3
SAN	1	2	2	2	2	1	1	1	2
Polystyrene	3	3	3	3	3	2	3	2	1
Olefins	**5**	**4**	**4**	**5**	**3**	**5**	**5**	**3**	**1**
Polyethylene	2	2	2	2	2	1	1	2	1
Polypropylene	1	1	1	1	1	1	1	1	2
Other Crystalline Resins	**1**	**1**	**2**	**4**	**4**	**4**	**4**	**2**	**3**
Nylons 6	2	2	2	2	5	1	4	3	1
6/6	1	3	1	1	4	2	3	2	2
6/10, 6/12	3	1	3	3	3	2	2	3	4
Polyester	4	4	2	1	2	2	1	4	1
Polyacetal	5	5	5	2	1	3	2	1	1
Arylates	**3**	**2**	**3**	**3**	**5**	**1**	**2**	**4**	**4**
Modified PPO	4	3	4	4	3	4	4	4	1
Polycarbonate	2	1	3	3	4	1	3	3	2
Polysulfone	2	2	2	2	2	2	2	1	3
Polyethersulfone	1	3	1	1	1	3	1	2	4
High Temp. Resins	**2**	**4**	**1**	**1**	**2**	**4**	**1**	**4**	**5**
PPS	1	2	2	2	1	1	2	2	1
Polyamide-imide	2	1	1	1	2	2	1	1	2
Fluoropolymers	**6**	**2**	**2**	**1**	**1**	**6**	**6**	**1**	**6**
FEP	2	1	2	1	1	2	2	1	2
ETFE	1	2	1	2	2	1	1	2	1

Ratings: 1—most desirable 6—least desirable. Large numbers indicate group classification, small numbers are for the specific resins within that group.

Strength & stiffness:—The ability to resist instantaneous applications of load while exhibiting a low level of strain. Materials which demonstrate a proportionality between stress and strain have been assigned better relative ratings.

Toughness:—The ability to withstand impacting at high strain rates.

Short-term heat resistance:—The ability to withstand exposure to elevated temperatures for a limited period of time without distortion.

Long-term heat resistance:—The ability to retain a high level of room-temperature mechanical properties after exposure to elevated temperature for a sustained period.

Environmental resistance:—The ability to withstand exposure to solvents and chemicals.

Dimensional accuracy in molding:—The ability to produce warp-free, high tolerance molded parts.

Dimensional stability:—The ability to maintain the molded dimensions after exposure to a broad range of temperatures and environments.

Wear and frictional properties:—The ability of the plastic to resist removal of material when run against a mating metal surface. The lower the frictional values, the better the relative rating.

Cost:—The relative cost per cubic inch.

Figure 1 An unweighted matrix with data for material selection. (Reproduced with permission of LNP Plastics, Inc.)

addition of each row will give a score that indicates the rank order of the most likely candidates. The lowest total represents the most desirable choice.

As an example of selecting a candidate material with this kind of chart, consider potential materials for chain saw housings with an eye toward changing from die-cast aluminum or magnesium to plastic (Fig. 2). This example, which was published before most chain saw housings had been converted to plastic, suggests that glass-filled nylon is the most likely candidate material. Today, glass-filled nylon is the material of choice for a significant number of commercially available chain saw housings.

Weighted Matrices

When it is clear that some properties are significantly more important than others, the unweighted matrix will be inadequate. One possible configuration for a weighted decision matrix is shown in Table 3a. An alternative layout for the decision matrix is shown in Table 3b.

As with the unweighted matrix, the candidate materials are listed in the left column and the properties of interest are listed in the top row. The weighting factor for each property appears with the property indicating how important that property is relative to the other properties to be considered. In each material/property intersection two figures will appear. First, the rating factor indicates how good the property for that material is relative to other materials listed. Second, the product of the weighting factor and the rating factor will be recorded. This is a measure of the weighted score of that material for that property, or a weighted rating factor.

On the right-hand side immediately after the last property column, enter the sum of the weighted rating factors for each material. The order of preference of the materials will be evident in that column. Some people prefer to normalize the total weighted rating by dividing by the total of the weighting factors. This is shown in the last column. It will not change the order of the result, and there is no compelling reason to do this.

Also shown in the table is a preliminary screening device that has been used in design, problem-solving and value engineering decision-making schemes. It is a list of the features or characteristics that are mandatory. Check if the characteristic is present in the candidate. If it is not, that line is eliminated immediately from further consideration.

A similar scheme may be useful in material selection decisions also. List property value minimums or other characteristics that are mandatory. Any material that does not meet all of these requirements is eliminated from further consideration right there.

However, it often happens that a "minimum requirement" may actually be negotiable, if the price is right. Material review boards routinely order rejected material to be used "as is." This is possible because failure to meet the specification does not in fact make the material unusable. Tolerance bands and minimum values imply that very small changes—from just inside the tolerance bands to just outside the tolerance band—make the difference between completely acceptable and completely unacceptable material. In fact there is not such a step change in the usefulness of material. The Taguchi loss function, for instance, accounts for this anomaly in the use of tolerance bands and minimum values simply by assigning ever larger penalties as material deviates from the target value. A similar concept, i.e.,

Material Characteristics	Strength & Stiffness	Toughness	Short Term Heat Resistance	Long Term Heat Resistance	Environ-mental Resistance	Dimensional Accuracy In Molding	Dimensional Stability	Wear & Frictional Properties	Point Sub Total	Cost	Point Total
G/R Resin Groups / Design Criteria	X	X	X		X						
Styrenics ABS SAN Polystyrene	3	6	6		6				21	2	23
Olefins Polyethylene Polypropylene	5	4	4		3				16	1	17
Other Crystalline Resins Nylons 6 6/6 6/10, 6/12 Polyester Polyacetal	1 2 1 3 4 5	1 2 3 1 4 5	2 2 1 3 2 5		4 5 4 3 2 1				8 11 9 10 12 16	3 1 2 4 1 1	11 12 11 14 13 17
Arylates Modified PPO Polycarbonate Polysulfone Polyethersulfone	3	2	3		5				13	4	17
High Temp. Resins PPS Polyamide-imide	2	4	1		2				9	5	14
Fluoropolymers FEP ETFE	6	2	2		1				11	6	17

Figure 2 Material selection using an unweighted matrix.

Table 3a Weighted Matrix

Material	Preliminary screening	Properties					
	(1)	(3) A	B	C	D	(6)	
Material 1	(2)	(4) (5)				(7)	(8)
Material 2						(7)	(8)
Material 3						(7)	(8)
Material 4						(7)	(8)

(1) Enter the mandatory properties.
(2) Enter X if satisfactory. Any material without all boxed X'ed is eliminated.
(3) Enter weighting factor, WF.
(4) Enter rating factor, RF.
(5) Enter the product WF × RF. This is the weighted rating of this property material for this material.
(6) Enter the summation of the weighting factors, Σ (WF). If the method of Table 2 is used, this number will be 1, but other totals are possible if other methods are used.
(7) Enter the sum of the weighted ratings for this row, Σ (WF × RF). There will be one sum for each material.
(8) If the total of the weighting factors is not 1, the weighted ratings may be rationalized by dividing by the total weighting factor. Enter {Σ (WF × RF}/{Σ WF}. This does not change the order of preference shown in the preceding column.

increasing the penalty as the property goes outside the most desirable range, was discussed in connection with nonlinear rating factors.

3.6 Alternatives to Decision Matrices

Polygons

A graphical method (Hanley 1980, also illustrated in Cornish 1987) requires that the designer make a judgment about the set of *n* properties that would constitute the ideal material. Usually, *n* is between 5 and 10 for the same reasons described in Section 3.3.

When asked how strong the candidate material should be, a designer might say, "as strong as possible." That might be a satisfactory answer in normal conversation, but in the polygon method, the designer is forced to make a sensible, quantitative choice. Then the properties of the candidate materials are compared, one by one, to the ideal properties by forming the ratio, *r*, of the candidate's property to the ideal material's property value. Each property may be greater, smaller or the same as the ideal property. These ratios could easily be displayed in matrix form and be manipulated mathematically to determine the preferred material. However, the graphical representation as a polygon is more striking. See Fig. 3.

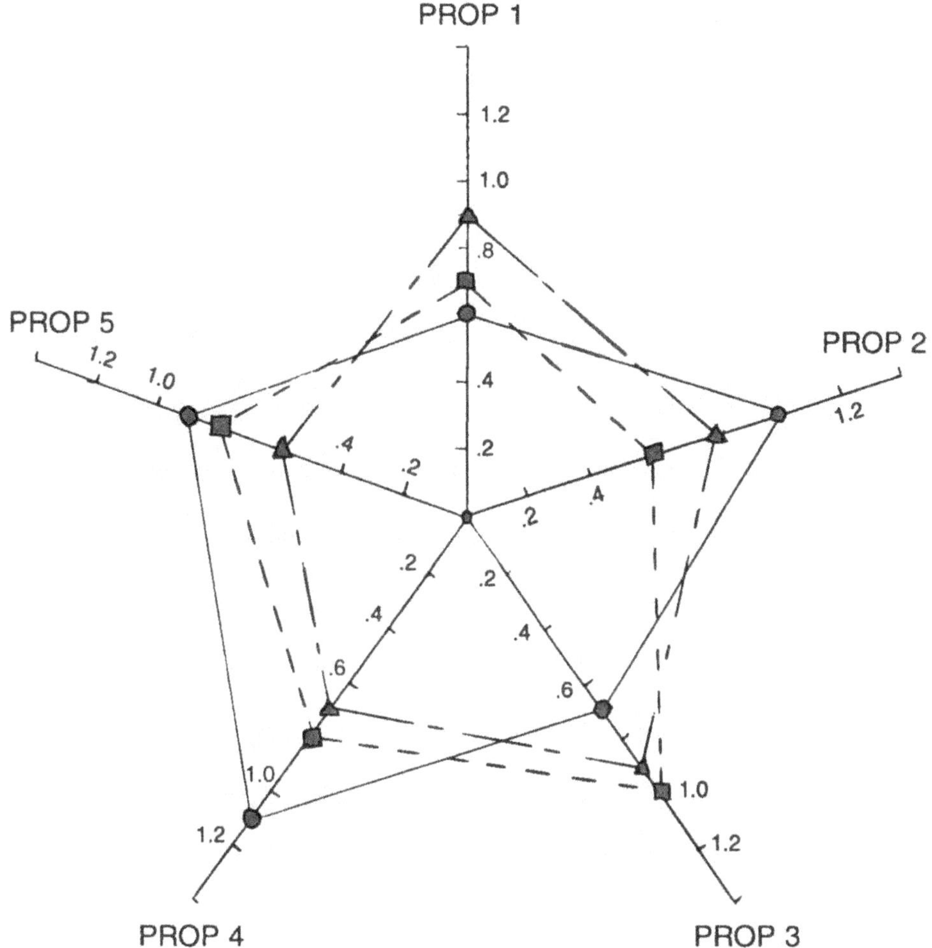

Figure 3 Polygon method using data from Table 4.

The ideal material has property ratios, r, that are all unity. If plotted in cylindrical coordinates as n lines of equal length, equally spaced about the origin, they will connect the centroid with the apexes of a regular n-sided polygon.

When the r-values for each real material are plotted as lines with the same angular spacing as the ideal material, the polygon composed of lines that connect the outer ends of the radial (property ratio) lines will form a polygon representative of that material. If the polygon is congruent with the ideal polygon, an unlikely situation, the candidate material is ideal. The polygon may be regular, but either larger or smaller than the ideal, indicating a good balance of properties but wrong magnitude. That may still prove to be the best choice, even though it clearly is not ideal. Other nonideal candidate materials will form irregular polygons that will deviate either in size or shape from the ideal. Then the candidate materials may be compared and some judgment applied as to how significant the deviation from the ideal actually is. Incidentally, the graphics functions of some spreadsheet programs will create the appropriate graph from tabular data.

Weighting factors, *w*, may be applied to each of the property values. Assuming that the sum of all weighting factors is unity, as described in Section 3.3, the weighted rating, *R*, for any candidate material may be determined by

$$R = \sum_{i=1}^{n} r_i w_i$$

This is the same mathematical procedure for determining the weighted rating of materials listed in the weighted matrix of Section 3.5, Table 3, except the procedure for determining the rating numbers is different. The weighted rating for the ideal material is unity. The candidate material whose weighted rating is closest to unity might seem to be the best choice. However, if one property is very large compared to the ideal and another property is smaller enough to compensate when calculating the mean, that is less likely to be better than one with the same mean but less deviation from the ideal shape of the polygon. This is analogous to statistical distributions that have the same mean but different standard deviations. The mean value alone does not adequately describe how similar two different distributions actually are. Similarly, in this material selection method, the root-mean-square deviation, *D*, of the candidate material from the ideal is calculated by

$$D = \sum_{i=1}^{n} \sqrt{(r_i - R)^2}$$

The best candidate material is likely to be the one with the minimum value of the root-mean-square combination of the deviation of the mean weighted characteristic from the ideal and the balance factor, as follows:

$$C = \sqrt{(1 - R)^2 + D^2}$$

The first term under the radical shows how far from ideal the weighted average is, and the second term shows how much spread there is in the property values.

A less complicated alternative method is to simply minimize the sum of the deviations of the properties from the ideal values of the specified properties. The best material will then be the one with minimum value of *Z*, where

$$Z = \sum_{i=1}^{n} |w_i r_i - 1|$$

Note use of the absolute value signs. That eliminates the possibility that a value for one property that is significantly greater than ideal can be compensated for by another property that is significantly less than ideal. It penalizes deviation either over or under the target value. This does not address the condition where being on opposite sides of the target may not be equally acceptable (or unacceptable). For instance, fusible plugs have a specific melting point that is desired. Is it really equally acceptable to be 2% over the target as to be 2% under the target? That is a question for the judgment of the engineer, indicating the potential pitfall of relying on a purely mathematical model for the decision.

Table 3b Alternative Layout for Weighted Rating Decision Matrix

	Preliminary screening	Properties				
Weighted rating matrix	(1)	WF_1	. . .	. . .	WF_n	$\Sigma\ WF_i$
		Property 1	Property . . .	Property . . .	Property n	
Material 1	(2)	RF_1 $WF_1 RF_1$	. . . . . .	. . . . . .	RF_n $WF_n RF_n$	$\Sigma\ (WF_i \times RF_i)$ or $\frac{\sum_{i=1}^{n} WF_i\ RF_i}{\sum_{i=1}^{n} WF_i}$
Material 2						
etc.						

(1) Enter mandatory properties.
(2) Enter √ if all mandatory properties are satisfied. Any material that does not meet all mandatory requirements is eliminated.
WF = weighting factor = How important is this property relative to other properties?
RF = rating factor = How good is this material relative to other materials?

Table 4 Matrix of Material Properties Used to Illustrate the Polygon Method in Fig. 3

	●	■	▲
Property	Material 1	Material 2	Material 3
1	0.6	0.7	0.9
2	1.0	0.6	0.8
3	0.7	1.0	0.9
4	1.1	0.8	0.7
5	0.9	0.8	0.6

Probability Methods

When choosing a material for a new product, there is risk simply because you are trying something that has not previously been tried or been proven to be feasible. Some choices are likely to be less risky than others. Situations which are quite similar to ones which have been used previously are probably better understood (and all the potential problems can be anticipated and planned for) than where there is a totally new configuration, new material, new manufacturing method, etc.

There are two independent aspects of this situation which need to be considered: the technical success and the commercial success of the product. One does not ensure the other. When alternative approaches to design or material selection are being considered, each approach will have certain unique technical difficulties which make commercialization of the product more or less likely. Also, each approach will have a certain customer appeal that will contribute differently to the commercial success of the product. Estimating the relative customer appeal of the various proposed alternatives may be difficult, but market surveys attempt to do that quite often.

Estimating the likelihood of surmounting the technical difficulties is a matter for the technical experts to decide. There are known methods for assessing technical risk (Porter et al. 1990).

The most attractive alternative will be the one that maximizes the profitability of the product. A figure of merit, F, can be expressed as

$$F = p\{t\}p\{m\}\frac{P}{C}$$

where $p\{t\}$ = probability of successfully overcoming the technical problems
$p\{m\}$ = probability of market success
P = profit
D = development cost

Cost estimating methods are widely used and notoriously unreliable. "Cost overrun" is part of the common vocabulary. Not only are the methods unavoidably inexact, but one must be alert to the possibility of a hidden agenda that purposely underestimates cost and overestimates profits to enhance the attractiveness of some person's favorite alternative.

Material Performance Index

When there is more than one material that meets all the minimum requirements, a decision must be made to select the final candidate. Weighted matrix decision methods of the preceding sections may be useful in trying to balance competing or conflicting needs in choosing the "best" material. Other methods have been used which combine the most important properties into a single parameter which is then either maximized or minimized. Common examples would be: Find the material which gives maximum strength at minimum cost, or find the material which gives maximum stiffness at minimum weight. Section 2.3 gave a very restricted example of a method for combining parameters to compare two materials whose cross sections are geometrically similar.

The more general case would arise when the engineer has not yet chosen materials to compare or determined the probable shape of the cross section (which may vary with the material). The reason to used different cross sections usually originates with limitations imposed either by the manufacturing process or extremes in the anisotropic nature of the properties. For example, when wood and metal compete for use as beams, the wooden beams are likely to have rectangular cross sections and the metal ones an I shape. For circular sections, metal may have a standard hollow cross section, while wood is almost certain to be solid.

Ignoring for the moment the complications of shape, components are designed to perform some function using some material which is formed into the proper size. The function statement usually includes a requirement that the material should not fail. That is, a load should be carried without permanent deformation, a bearing should operate for the specified life without excessive wear, etc. (In some cases failure may be the objective. Frangible discs fracture to protect pressure vessels, fusible links melt to protect circuits from damaging currents or start water sprinkler systems in case of fire, and energy-absorbing steering columns in automobiles buckle and collapse.)

Suppose the function is to carry a load without excessive elastic deformation. The load may cause axial tension, axial compression, or bending, among others, depending on the situation. The deformation will be dependent on both the shape and size of the cross sections. For geometrically similar cross sections, however, changing the dimensions of the cross section of the part will not change the deformation by the same amount in the three cases. The deformation may be dependent on the cross-sectional area itself, as in the case of axial tension and compression of short columns. Or the deformation may be dependent on the second moment of the cross-sectional area, as in the case of bending. In the case of the compressive loading of slender columns, we do not usually care how much deformation takes place, but we only want to know how much load will cause the column to become unstable.

Even though the deformation is calculated in very different ways for these three loading modes, in all cases the deformation equations will contain factors to express the load, the size and shape of the member, and relevant material properties.

In trying to select a material that will perform a certain function while minimizing or maximizing some aspect of the design, we can usually find a parameter that permits several materials to be compared for that aspect of their performance. Minimizing the weight of a structure is a common goal. Texts on aircraft structures, lightweight structures and optimal design will give details on the common techniques. A short example here will suffice to illustrate the point and introduce the most comprehensive set of data available on comparing the performance of different materials.

Suppose we want to minimize the weight (mass) of a rod that is resisting the elastic deformation caused by a load in tension, subject to the constraint that the length of the rod and the amount of load to be carried are predetermined by a design requirement. Thus we are looking for

1. High modulus to reduce the elastic deformation under load.
2. Minimum cross-sectional area to minimize the volume of material. This must also be subject to the constraint that the area is sufficient to prevent plastic deformation or fractures.
3. Low density to minimize the weight of a given volume of material.

However, we do not know the relationship among these variables. Some simple calculations will reveal the relationships.

$$m = Al\rho \qquad \text{or} \qquad A = \frac{m}{l\rho}$$

$$\delta = \frac{Fl}{AE} \qquad \text{or} \qquad A = \frac{Fl}{\delta E}$$

$$\frac{m}{l\rho} = \frac{Fl}{\delta E}$$

$$m = \frac{Fl^2\rho}{\delta E}$$

$$m = \left[\frac{F}{\delta}\right][l^2]\left[\frac{\rho}{E}\right]$$

where m = mass
A = cross-sectional area
l = length
ρ = density
δ = deflection
F = load
E = modulus of elasticity

In the last expression, the first group tells the function that the object has to perform, namely carry a load, F, and not deform by more than δ. The second group, in this case simply l^2, tells about the geometry that is required. The load, deformation and length are fixed by the statement of the problem. (Some authors [e.g., Dieter] lump these two groups together into a single "performance factor" for load and geometry.) The last group is the material property function or material performance index that is to be maximized or minimized. In this case, where we want minimum weight, this material performance index is to be minimized, or its reciprocal is to be maximized. All materials with the same ratio of density and elastic modulus are equally good, or bad, at satisfying the desired condition.

In other cases, one or more of the material properties may be raised to some power. The material performance index for column buckling, for instance, is the ratio of the density and the square root of the elastic modulus. A slightly different approach was used in Section 2.3 to show that the appropriate exponent in the material performance index for equal strength in bending is 2/3.

One author (Ashby 1992) has summarized the conditions for 57 cases in designing for minimum cost, minimum weight, damage-tolerant design, thermal design, and assorted other conditions. His data are presented graphically with lines of equal material performance indices clearly shown for all the major types of materials. Thus, is it possible to see, when stiffness-to-weight ratio is important (e.g., for aircraft structures, both model and full scale), why balsa wood and spruce were widely used before aluminum and fiber-reinforced composites were available in appropriate cross-sectional shapes. It is also clear that the reason why balsa is still used for model airplanes is more complicated than just its "light weight." See Fig. 4.

The information is also available in computerized form as the Cambridge materials selector. This kind of aid is most valuable in the early stages of material selection when all materials are possible candidates. In the later stages of detail design, handbooks like this one with compilations of properties for individual materials are necessary.

4 INFORMATION SOURCES

4.1 Applications of Materials and Case Histories

Much of the progress of engineering is made by correcting the errors made by our predecessors. Henry Petroski's book *To Engineer Is Human: The Role of Failure in Successful Design* makes this point at considerable length. Engineers should push the limits of the possible, else there can not be progress. Only by testing a design until it fails can we really verify what the limits are. Test programs are often specifically designed to accomplish just this. Companies do not want the customer to be the guinea pig who has to pick up the pieces of the failed product. When failures and successes are documented in the open literature, all engineers may benefit from the knowledge.

The *Engineering News Record* magazine has been documenting failure investigations, many of which involve material selection decisions, for nearly a century. Several collections of failure investigations are available, some of which cover a broad range of materials and component types. Others are specialized in one type of failure or component. Sometimes, case studies are included in texts as illustrations of the principles the text is mainly concerned with. Seventeen such books are listed in the bibliography.

Many texts on material selection also include examples of successful material selection decisions.

4.2 Manufacturer's Literature

Manufacturers need to get information about their products to potential customers. Specific technical information is usually available on request. There is no coordi-

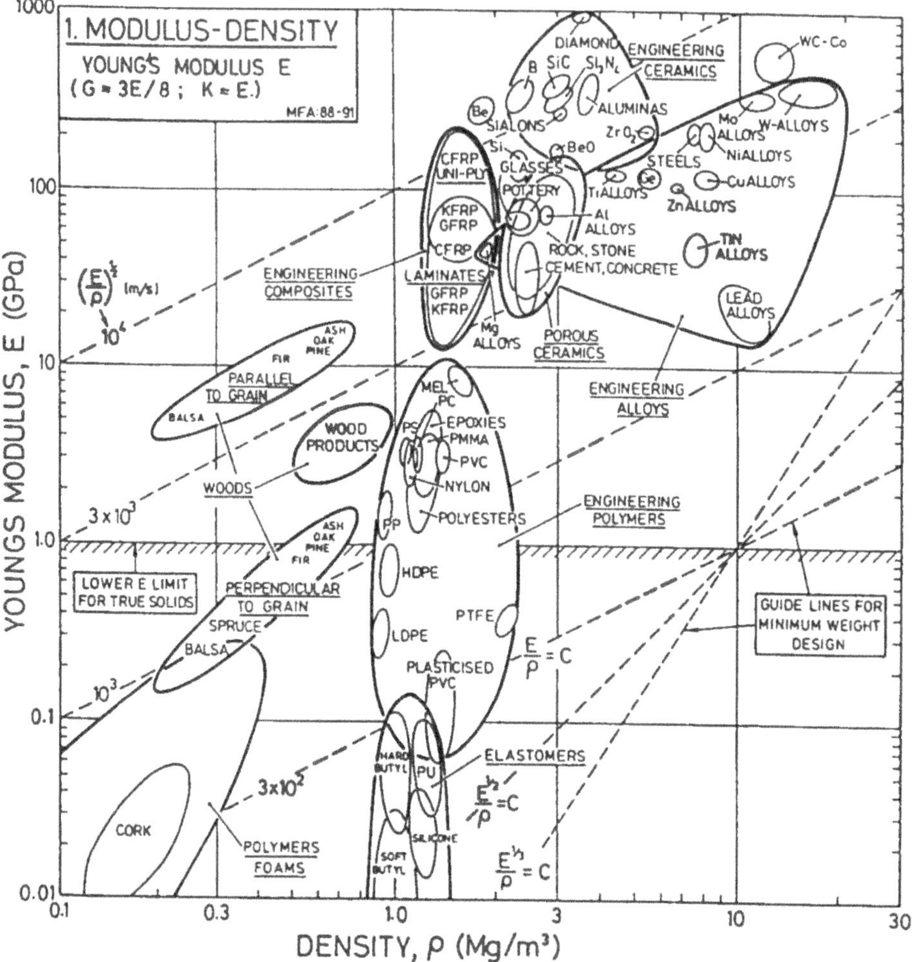

Figure 4 Ashby's material selection chart. (Reproduced with permission of Butterworth-Heinemann, Ltd. and M. F. Ashby.)

nation among manufacturers about format, test methods, etc. It is wise to check carefully when comparing information from different manufacturers that the test methods are the same or at least will give comparable results. Plastics manufacturers normally cite the ASTM test method used. The properties of the material in the final product may be different from those given in the manufacturer's literature because the processing required to produce the final product may be significantly different from that used to produce the test specimens. Plastic materials are more sensitive to property variations due to processing variations than metals are.

Metal manufacturers do not usually cite the test method. It may safely be assumed that a standard ASTM test method has been used for determining the common properties like yield strength. For some properties in metals, even when the test method is not specified, it is necessary to provide information about the particular method used. For instance, various offsets may be used to determine offset yield strength and must be reported. Ductility as determined by percent elongation is a function of the specimen's gauge length, which must be reported.

4.3 Edited Literature

There are hundreds of sources of information related to material properties, selection, processing, etc. These are in the form of textbooks and handbooks, periodicals, CD-ROMs, abstracts, on-line data bases, patents, professional society publications, and government reports. It would be impossible to present a comprehensive list in the space available. Names of many of the most prominent sources are included in lists published in Dieter (1991, Chaps. 6, 14) and Cornish (1987, Chap. 13).

BIBLIOGRAPHY

This bibliography is not intended to represent either a complete list of relevant information on materials selection or the best available information. It gives bibliographic data for the works referred to in the text and adds some references that are known to be useful. Considerable additional information is available from material manufacturers, suppliers, and professional societies. Also, specialized property data, or bibliographies that will lead the reader to such data, can frequently be found in textbooks dealing with specialized fields, e.g., acoustics, corrosion, solid state diffusion, etc.

References

Metals Handbook 8th, 9th, and 10th eds., American Society for Metals, (now ASM International), Metals Park, OH, 1961 to present. The Handbook series contain more than 30 volumes related to metals. New works are being published related to other materials. There are hundreds of publications available dealing with all aspects of material selection and processing. ASM International publications are the most comprehensive single source of materials and processing information.

Robert B. Ross, *Metallic Materials Specification Handbook*, 4th ed., Chapman and Hall, London, 1992. Lists worldwide materials by properties and material specification.

N. A. Waterman and M. F. Ashby, *Elsevier Materials Selector*, 3 vols., Elsevier Applied Science, London, 1991.

Hansjurgen Saechtling, *International Plastics Handbook* Hanser, Munich, 1987.

G. S. Brady and H. R. Clauser, *Materials Handbook*, 13th ed., McGraw Hill, New York, 1991.

N. E. Woldman, *Woldman's Engineering Alloys*, 7th ed., ASM International, Metals Park, OH (1990)

M. B. Bever, ed., *Encyclopedia of Material Science and Engineering*, Pergamon, Oxford 1986.

ASHRAE Handbook of Fundamentals, 1967 and later (thermal properties).

Heat Transfer Databook, Genium, Schenectady, New York, 1987 (thermal properties).

Bryan D. Tapley, ed., *Eschbach's Handbook of Engineering Fundamentals*, 4th ed., Wiley-Interscience, New York, 1990. Chapter 16 is 294 pages of "Properties of Materials" by G. E. Dieter and J. H. Westbrook. Data are reproduced from standard sources like AISI, Metal Progress Databooks, ASM handbooks, Alcoa products data book, Aluminum Association, and ASTM standards, among others.

K. W. Reynard, ed., *Materials Information Directory 1988*, The Design Council, London SW1Y 4SU.

J. Binner, P. Hogg, and J. Sweeney, eds., *Advanced Materials Source Book*, Elsevier Advanced Technology, Mayfield House, Oxford OX2 7DH UK.

E. F. Bradley and M. J. Donachie, *Changes and Evolution of Aircraft Engine Materials, Source Book on Material Selection*, ASM; 1977, p. 1. This is one of a series of "Source Books" from ASM International that give practical information in the form of compilations on a particular topic of work previously published in diverse sources.

VDI 2222, *The Design of Technical Products*, Verein Duetscher Ingenieure, VDI-Verlag, Dusseldorf, 1973.

C. W. Hoover and J. B. Jones, *Improving Engineering Design: Designing for Competetive Advantage*, National Research Council, National Academy Press, Washington, DC, 1991.

Texts

G. E. Dieter, *Engineering Design: A Materials and Processing Approach*, McGraw-Hill, New York, 1991.

J. A. Charles and F. A. A. Crane, *Selection and Use of Engineering Materials*, 2nd ed., Butterworths, London, 1989. Good discussion of the various conditions that materials must withstand. Several detailed case studies.

M. F. Ashby, *Materials Selection in Mechanical Design*, Pergamon Press, Oxford, 1992. Compares all classes of materials against each other. Diagrams let the engineer find comparable materials for many different loading schemes and cross sectional shapes. Many short case studies.

G. Lewis, *Selection of Engineering Materials*, Prentice Hall, Englewood Cliffs, NJ, 1990. Many case studies.

E. H. Cornish, *Materials and the Designer*, Cambridge University Press, Cambridge, 1987.

J. F. Young and R. S. Shane, *Materials and Processes*, 3rd ed., Marcel Dekker, New York, 1985.

K. G. Budinski, *Engineering Materials: Properties and Selection*, 3rd ed., Prentice-Hall, Englewood Cliffs, NJ, 1989.

M. M. Farag, *Selection of Materials and Manufacturing Processes for Engineering Design*, Prentice-Hall, Englewood Cliffs, NJ, 1989.

Philip Sargent, *Materials Information for CAD-CAM*, Butterworth-Heinemann, Oxford, 1991. Eight pages of references, good info on decision making.

R. F. Kern and M. E. Suess, *Steel Selection*, Wiley-Interscience, New York, 1979. Much practical information that would be impossible to find in any other single source.

T. C. Fowler, *Value Analysis in Design*, Van Nostrand Reinhold, New York, 1990.

L. D. Miles, *Techniques of Value Analysis and Engineering*, McGraw-Hill, New York, 1972. Value engineering techniques help the engineer concentrate on fulfilling the basic required functions and avoiding unnecessary cost.

A. L. Porter and A. T. Roper, T. W. Mason, F. A. Rossini, and J. Banks, *Forecasting and Management of Technology*, Wiley, New York, 1990.

C. V. Starkey, *Engineering Design Decisions*, Edward Arnold, London, 1992.

T. Saaty, *The Analytic Hierarchy Process*, McGraw-Hill, New York, 1980.

J. E. Shigley and L. D. Mitchell, *Mechanical Engineering Design*, 4th ed., McGraw-Hill, New York, 1983.

Henry Petroski, *To Engineer Is Human*, St. Martin's Press, New York, 1985.

P. H. Hill, *The Science of Engineering Design*, Holt, Rinehart, Winston, 1970.

P. H. Hill, *Making Decisions: A Multidisciplinary Approach*, Addison-Wesley, Reading, MA, 1979.

C. H. Kepner and B. B. Tregoe, *The New Rational Manager*, Princeton Research Press, Princeton, NJ, 1981.

R. L. Ackoff, *The Art of Problem Solving*, Wiley-Interscience, New York, 1978.

M. F. Rubinstein, *Tools for Thinking and Problem Solving*, Prentice Hall, Englewood Cliffs, NJ, 1986.
M. F. Rubinstein, *Concepts in Problem Solving*, Prentice Hall, Englewood Cliffs, NJ, 1980.
K. Pohlandt, *Materials Testing for the Metal Forming Industry*, Springer-Verlag, Berlin, 1989.
B. Mills and A. H. Redford, *Machinability of Engineering Materials*, Applied Science, London, 1983.
G. Boothroyd and W. A. Knight, *Fundamentals of Machining and Machine Tools*, 2nd ed., Marcel Dekker, New York, 1989.
G. F. Benedict, *Nontraditional Manufacturing Processes*, Marcel Dekker, New York, 1987.
G. Boothroyd, P. Dewhurst, and W. Knight, *Product Design for Manufacture and Assembly* Marcel Dekker, New York, 1994, Chap. 2.
G. Pahl and W. Beitz, *Engineering Design*, Springer-Verlag, Berlin, 1988.
James V. Jones, *Engineering Design*, TAB Professional Books, Blue Ridge Summit, PA, 1988. Contains a lot of information on MIL Specs.
J. N. Siddall, *Probabalistic Engineering Design*, Marcel Dekker, NY, 1983.
Morris Asimow, *Introduction to Design*, Prentice-Hall, Englewood Cliffs, NJ, 1962.
J. M. Alexander, R. C. Brewer, and G. W. Rowe, *Manufacturing Technology: Engineering Materials*, Halsted Press, New York, 1987. Concise discussion of many interactions between material and processes.
Josef C. Bicerano, *Prediction of Polymer Properties*, Marcel Dekker, New York, 1993.
R. J. Bronikowski, *Managing the Engineering Design Function*, Van Nostrand Reinhold, London, 1986.

Periodical Articles

D. H. Breen and G. H. Walter, and J. T. Sponzilli, A methodology for selection of steels, *Metal Progr. 103*, 83–88 (1973).
J. Gabrovic, J. V. Hackworth, and H. M. Lampert, The physics of failure applied to material selection, Paper 64 MD 44, ASME June 1964.
S. H. Wearne, A review of reports of failures, *Proc. IMechE 193*, No. 20 (1979).
E. D. Larson, M. H. Ross, and R. H. Willisams, Beyond the era of materials, *Sci. Am. 254*, no. 5, 34–41 (1986).
T. M. Trainer and J. S. Glasgow, Material selection in the design method. *Mech. Eng. 87*, 41–43 (1965).
H. R. Clauser, R. J. Fabian, and J. A. Mock, How materials are selected, *Design Eng. July*, 109–128 (1985).
N. Swindells and R. J. Swindells, "System for engineering materials selection, *Metals Mater. 1*, 301–304 (1985).
P. R. Mould and T. E. Johnson, Radio assessment of drawability on cold rolled low carbon steel sheets, *Sheet Metal Industry, 50*, 328–348, June 1975.
F. R. Larson and J. Miller, A Time-Temperature Relationship for Rupture and Creep Stress. *Trans. ASME, 74*, 765, 1952.

Failure

R. E. Dolby and K. G. Kent, eds. *Repair and Reclamation*, The Welding Institute, Cambridge, 1986.
D. N. French, *Metallurgical Failures in Fossil Fired Boilers*, 2nd ed., Wiley Interscience, New York, 1993.
H. P. Bloch and F. K. Geitner, *Machinery Failure Analysis and Troubleshooting*, vol. 2 of *Practical Machinery Management for Process Plants*, Gulf, Houston, 1983.

R. D. Port and H. M. Herro, *The Nalco Guide to Boiler Failure Analysis*, McGraw-Hill, New York, 1991.
H. Carlson, *Springs: Troubleshooting and Failure Analysis*, Marcel Dekker, New York, 1980.
J. A. Collins, *Failure of Materials in Mechanical Design*, 2nd ed., Wiley-Interscience, New York, 1993. Many references
V. J. Colangelo and F. A. Heiser, *Analysis of Metallurgical Failures*, 2nd ed., Wiley, New York, 1987.
D. Kaminetzky, *Design and Construction Failures: Lessons from Forensic Investigations*, McGraw-Hill, New York, 1991.
C. E. Witherell, *Mechanical Failure Avoidance*, McGraw-Hill, New York, 1994 355 references.
Failure Analysis: Techniques and Applications, Proc. 1st Int. Conf. on Failure Analysis, ASM International, 1992.
Source Book in Failure Analysis, American Society for Metals, 1974.
Case Histories in Failure Analysis, American Society for Metals, 1979.
Handbook of Case Histories in Failure Analysis, Vol. 1, ASM International, 1992.
E. R. Parker, *Fracture of Brittle Materials*, Wiley, New York, 1963.
S. S. Ross, *Construction Disasters*, McGraw-Hill, New York, 1984.
Failure Analysis and Prevention, Metals Handbook, 9th ed., 1986.
C. M. Hudson and T. P. Rich, eds., *Case Histories Involving Fatigue and Fracture Mechanics*, ASTM STP 918, 1986.
F. R. Hutchings and P. M. Unterweiser, *Failure Analysis: The British Engine Technical Reports*, American Society for Metals, 1981.
T. E. Tallian, *Failure Atlas for Hertz Contact Machine Elements*, ASME, New York, 1992.
D. R. H. Jones, *Engineering Materials: Materials Failure Analysis, Case Studies and Design Implications*, Pergamon Press, New York, 1993.

Older Books

There are many ideas, methods, examples, etc., that appear in older texts that do not get repeated later. The data may be dated, but the thinking and experience that are related by the authors often are not.

E. R. Parker, *Materials Data Book*, McGraw-Hill, New York, 1967.
H. J. Sharp, ed., *Engineering Materials and Value Analysis*, American Elsevier, New York, 1966.
E. D. Verink, ed., *Methods of Material Selection*, Gordon and Breach, New York, 1968.
An Approach for Systematic Evaluation of Materials for Structural Applications, NMAB publication 246, Feb. 1970.
J. M. Alexander and R. L. Brewer, *Manufacturing Properties of Materials*, Van Nostrand, London, 1963.
D. F. Minor and J. B. Seastone, eds., *Handbook of Engineering Materials*, Wiley, New York, 1955.
S. W. Jones, *Materials Science–Selection of Materials*, Butterworths, London, 1970.
D. P. Hanley, *Introduction to the Selection of Engineering Materials*, Van Nostrand Reinhold, New York, 1980.
H. R. Clauser, *Industrial and Engineering Materials*, McGraw-Hill, New York, 1975.
J. R. Dixon, *Engineering Design*, McGraw-Hill, New York, 1961, Chaps. 11, 16.
S. A. Gregory, *The Design Method*, Butterworths, 1966. Chap. 23 by Al Davies "Selection of Materials," p. 211.
J. N. Siddall, *Analytical Decision Making in Engineering Design*, Prentice Hall, 1972.

N. A. Waterman, *The Selection of Materials*, Engineering Design Guides #29, Oxford University Press, 1979.
J. P. Sholes, *The Selection and Use of Cast Irons*, Engineering Design Guides #31, Oxford University Press, 1979.
M. F. Day, *Materials for High Temperature Use*, Engineering Design Guides #28, Oxford University Press, 1979.
P. C. Powell, *The Selection and Use of Thermoplastics*, Engineering Design Guides #19, Oxford University Press, 1977.
Fulmer Research Institute, *Fulmer Materials Optimizer*, Slough, 1974.
N. A. Waterman and A. M. Pye, *Guide to the Selection of Materials for Marine Applications*, Fulmer Research Institute, 1980.
A. E. Javitz, ed., *Material Science and Technology for Design Engineers*, Hayden, New York, 1972.
C. T. Lynch, ed., *Handbook of Material Science*, CRC Press, Boca Raton, FL, 1974.
C. A. Harper, ed., *Handbook of Materials and Processes for Electronics*, McGraw Hill, New York, 1970.
Arnold M. Ruskin, *Materials Considerations in Design*, Prentice-Hall, 1967.
Arnold M. Ruskin, *Selection of Materials and Design*, American Elsevier, 1968.

Recycling

Thomas J. Cichonski and Karen Hill, eds., *Recycling Sourcebook* Gale Research, Detroit, 1993. Full of data, sources of information, etc.
Herbert F. Lund, ed., *The McGraw-Hill Recycling Handbook* McGraw-Hill, New York, 1993. Comprehensive, authoritative. Lund is one of the biggest names in the business.
R. J. Ehrig, *Plastics Recycling*, Hanser, Munich, 1992. TP1122.P715p. Best technical coverage of problems, techniques, and prospects.
J. Leidner, *Plastic Waste*, Marcel Dekker, New York, 1981.
J. H. L. van Linden, D. L. Stewart, and Y. Sahai, Second Int. Symp. on Recycling of Metals and Engineered Materials, Minerals, Metals and Materials Society, Warrendale, PA, 1990. Very interesting collection of papers. Lots of anecdotal information about many materials and products. Aluminum beverage cans, for instance, is good article.
T. R. Curlee, U.S. Environmental Protection Agency, *Plastic Wastes: Management, Control, Recycling and Disposal*, Noyes Data Corp, Park Ridge, NJ, 1991.
N. Mustafa, ed., *Plastics Wastes Management: Disposal, Recycling and Reuse*, Marcel Dekker, New York, 1993.

Conference Proceedings and other books composed mainly of chapters by various authors

G. E. Dieter ed., *Workability Testing Techniques*, American Society for Metals, Metals Park, OH, 1984.
B. A. Niemeier, A. K. Schmieder, and J. R. Newby, eds., *Formability Topics—Metallic Materials*, ASTM STP 647, 1978.
J. R. Newby and B. A. Niemeier, eds., *Formability of Metallic Materials*—2000 AD, ASTM STP 753, 1982.
R. H. Waggoner, *Novel Techniques in Metal Deformation Testing*, The Metallurgical Society of AIME, Warrendale, PA, 1983.
B. Golden, E. Wasil, and P. Harker, eds., *The Analytic Hierarchy Process*, Springer-Verlag, Berlin, 1989.

V. A. Tipnis, ed., *Influence of Metallurgy on Machinability*, American Society for Metals, 1975.

E. H. Nielsen, J. R. Dixon, and M. K. Simmons, *GERES: A Knowledge Based Material Selection Program for Injection Molded Resins*, Proc. ASME 1986 Computers in Engineering Conf., Chicago, IL, vol. I, pp. 255–262.

Andrew Kusiak, ed., *Concurrent Engineering: Automation, Tools and Techniques*, Wiley-Interscience, New York, 1993. Chapter 17, p. 437, D. L. Thurston and J. V. Carnahan, Intelligent evaluation of designs for manufacturing cost.

Materials Science and Engineering for the 1990's, National Research Council, National Academy Press, Washington, DC, 1989.

E. Haug, ed., *Concurrent Engineering of Mechanical Systems*, ASME Design Automation Conference, 1990.

Shigeyuki Somiya, Masao Doyana, Masaki Hasegawa, and Yoshitaka Agata, *Transactions of the Materials Research Society of Japan*, Elsevier Applied Science, London, 1990. Contains Hobart G. Rammrath, The future of engineering plastics, pp. 60–69.

Nader Ghafoori, ed., *Utilization of Industrial By-products for Construction Materials*, American Society of Civil Engineers, New York, 1993.

J. Fong, ed., *Critical Issues in Materials and Mechanical Engineering*, American Society of Mechanical Engineers, PVP, vol. 47, 1981.

Materials Futures: Strategies and Opportunities, Proc. US-Sweden Joint Symp., Philadelphia, PA, Materials Research Society, Pittsburg, PA, 1988.

E. J. Haug, ed., *Concurrent Engineering*, NATO ASI Series F, vol. 108. Contains world class concurrent engineering, by D. P. Clausing.

D. Holland, ed., *New Materials and Their Applications*, Institute of Physics, Bristol, 1990.

Evolution of Advanced Materials, Proc. Int. Conf., Associazione Italiana Di Metallurgia, Milano, 1989.

M. Doyama, T. Suzuki, J. Kihara, and R. Yamamoto, eds., *Computer Aided Innovation of New Materials*, North-Holland, Amsterdam, 1991.

B. F. Dyson and D. R. Dayhurst, eds., *Materials and Engineering Design: The Next Decade*, The Institute of Metals, London, 1989. Containing: J. A. Charles, The interaction of design, manufacturing method and material selection, p. 3; M. F. Ashby, Materials selection in conceptual design, p. 13; D. R. Hayhurst, Computer aided engineering, p. 113; K. W. Reynard, Computerized materials data and information, p. 118.

2
Introduction to the Selection of Metallic Materials

G. T. Murray
California Polytechnic State University
San Luis Obispo, California

Metals have been traditionally classified as ferrous and nonferrous, i.e., those whose base metals are iron and those whose base metals are iron free. This distinction arose for two reasons. First, the iron-base alloys have been used in quantities exceeding all others combined, and thus to some extent warrant a separate category. Production of iron and steel on a weight basis is about seven times the total quantity of all other metals combined. Second, the process of extracting iron from its ore is a complex process and somewhat different from the processes used, for example for copper, nickel and aluminum extraction. It was these extractive metallurgists that first separated metals into the ferrous and nonferrous categories. The physical metallurgy of alloys, which includes the composition and processing of such alloys, is basically the same in principal for all metals.

Metals have been widely used for several thousand years, beginning with the Bronze Age, which historians date to about 3000 to 1100 B.C. The Bronze Age was supposedly replaced by the Iron Age, which we are currently experiencing. Actually, we still use a considerable quantity of bronze, but steel use is many times that of the bronzes.

Metals have been defined over the years according to their characteristics. They are hard; reflect light, giving a metallic luster; and have good thermal and electrical conductivities. However, with the advent of semiconductors, which are also hard and possess a metallic luster, metals can best be described by comparing their electrical conductivities with these materials and with insulating materials, e.g., ceramics and polymers. Another characteristic of metals (and most semiconductors) is that in their solid state they are crystalline in form; that is the atoms within the body occupy well-defined positions, in contrast to noncrystalline amorphous solids such as glasses, wood and many polymers.

We currently have available for potential use in engineering applications over 45,000 different metallic alloys. Although the steels and cast irons comprise the largest usage on a weight basis, the number of different nonferrous alloys exceed the number of the ferrous alloys.

The iron-base alloys include the cast irons, a number of steels, and a few iron-base alloys that are not called steels. Some wrought iron, which consist of iron silicate fibers in an iron matrix, is used today in the form of pipe grill and decorative objects. Tools made of iron appeared around 1300 B.C. in Palestine where an iron furnace has been found. Both steels and cast irons are alloys of iron and carbon with the former containing up to about 2 wt% carbon and the latter about 2 to 4 wt% carbon. The cast irons are too brittle to be mechanically formed and must be cast to shape. They are widely used in large, intricately shaped structures that cannot be machined or forged to shape. Cast irons comprise about 95% of the weight of a typical automobile engine in the form of the engine block, head, camshaft, piston rings, lifters, and manifolds. There are several types of cast irons from which to choose, referred to as either white, gray, ductile or malleable iron (see Chapter 8). Steels can be broadly classified as plain carbon (which are alloys of iron and carbon containing small quantities of manganese and silicon); the high-strength low-alloy (HSLA) steels that contain less than 0.2 wt% carbon and minute quantities of niobium, titanium, vanadium and molybdenum; the heat-treatable steels, which contain 2 to 5 wt% alloying elements in addition to carbon; the high-strength steels such as the maraging steels; the Cr-Mo type steels used to about 650°C (1200°F) temperature; and the speciality steels of which the stainless, tool and bearing steels are the major types.

The principal nonferrous alloys are those in which the base metal consists of either aluminum, copper, nickel, magnesium, titanium or zinc. Alloys of these six metals account for over 90% of nonferrous alloy use. Nickel- and cobalt-base alloys that have been especially designed for elevated temperature applications and are the alloys employed in the hottest regions of jet engines (i.e., the compressor blades and turbine parts) have been termed "superalloys" (see Chapter 9).

Metals have the distinct advantage over some of the other materials, particularly polymers, in that they not only can be recycled but in fact it is usually economically wise to do so. A very large percentage of metals are currently recycled. The recycling rate for steel and aluminum cans in 1994 was 53% and 65.4% respectively. There is increased pressure being placed on design engineers to consider recycling issues when determining how a product is designed and what materials are used in its manufacture.

1 MECHANICAL PROPERTIES

The room-temperature tensile yield strengths of the more common alloys have been divided into four groups as depicted in Table 1. The highest strength alloys are the ultrahigh strength steels, while on the other end of the strength scale we find the low-strength tin and magnesium alloys. The nickel- and cobalt-base superalloys have room-temperature strengths approaching those of the ultrahigh strength steels. However, these alloys are noted for their high-temperature strength and corrosion resistance and, because of their high cost, not often employed in room-temperature applications. The aluminum, copper, nickel and titanium alloys fall into more than one strength category with the heat-treated compositions possessing higher

Table 1 Room-Temperature Tensile Yield Strengths of Metals

	MPa (ksi)	Approx. specific strength (10^6 in.)
Ultrahigh strength		
Maraging steels	1380–2415 (200–350)	1
Ultrahigh strength steels	1380–1860 (200–270)	0.85
PH stainless steels	1035–1520 (150–220)	0.7
Martensitic stainless steels	1000–1380 (145–200)	0.6
High strength		
Nickel-base superalloys	800–1365 (116–198)	0.5
Nickel, K-500 Monel	790 (115)	0.4
Low-alloy steels, heat-treated	720–1030 (105–150)	0.45
Titanium high-strength alloys	700–1175 (100–170)	0.85
Cobalt-base superalloys	500–1825 (72–265)	0.3
Copper-berylium alloys, aged	900–1345 (130–194)	0.54
Medium strength		
Copper alloys-bronzes	345–552 (50–80)	0.23
Commercially pure titanium	172–483 (25–70)	0.3
Copper alloys, C. W. brasses	276–518 (40–75)	0.2
Aluminum alloys, aged	138–552 (20–80)	0.5
Tantalum alloys, annealed	138–552 (20–80)	0.08
Niobium alloys, annealed	276–518 (40–75)	0.2
Low to medium strength		
Brasses, annealed to 1/4 hard	69–276 (10–40)	0.08
Phosphor bronzes, to 1/4 hard	69–276 (10–40)	0.08
Copper, cold-worked OFHC	69–207 (10–30)	0.07
Strain-hardened aluminum alloys	69–242 (10–35)	0.02
Nickel, commercial purity	104–138 (15–20)	0.05
Magnesium, cast	76–173 (11–25)	0.27
Tin alloys	14–41 (2–6)	0.01

strengths than that for the strain-hardened alloys, which in turn are stronger than the same alloys in the annealed state. The specific strength, i.e., the strength/density ratio for these alloys, is also listed in Table 1. These approximate values are of use in applications where weight of the component is important. The values listed were obtained by dividing the mid-range of the strength values in psi by the typical density for these alloys in lb/in^3.

The strength of metals decreases with increasing temperature. The short-time strength of certain groups of metals as a function of temperature is illustrated in Fig. 1. This data can be used for screening purposes only. For long periods of use one must consider the creep of metals, i.e., the slow plastic deformation that occurs over long periods of time and at stresses far below the yield strength measured in short time tests at the same temperature. Table 2 lists some estimated upper temperature limits for some common alloys based on creep and stress-rupture data. Furthermore, many metals cannot be used above certain temperatures because of reactions with oxygen and nitrogen. Titanium is limited to about 500°C (932°F) for

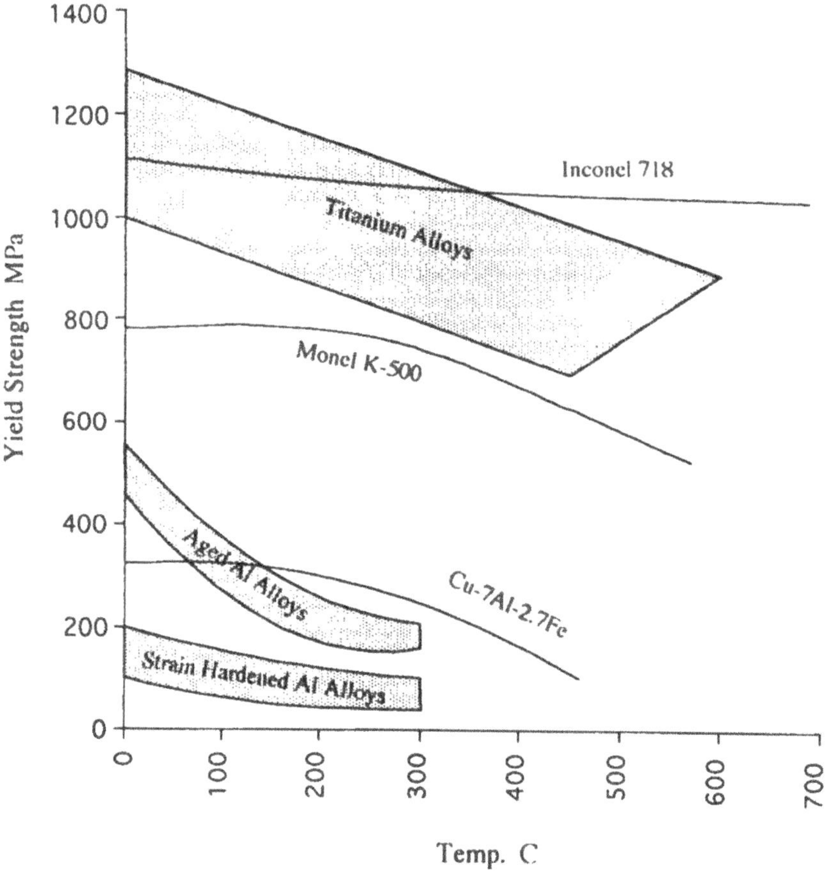

Figure 1 Yield strength versus temperature for selected alloys.

Table 2 Estimated Maximum Use Temperatures for Selected Alloys

Alloy type	Estimated maximum use temperature
Nickel-base superalloys	1200°C (2192°F)
Autenitic stainless steels	800°C (1472°F)
2.25 Cr–1 Mo	600°C (1112°F)
Low-alloy steels	500°C (932°F)
Titanium alloys	500°C (932°F)
Carbon steels	400°C (752°F)
Copper alloys (Al bronze)	300°C (572°F)
Aluminum alloys (6xxx series)	275°C (528°F)

this reason. The refractory metals tungsten, molybdenum, niobium and tantalum are also limited in air to temperatures of a few hundred degrees centigrade unless some type of protective coating is provided.

The tensile moduli of some metals and alloys are also presented in groups of high to low moduli in Table 3. The modulus, sometimes referred to as the elastic strength, is a very important material property in design since it allows one to compute the elastic (nonpermanent) deflection of a member subjected to a given load. Clearance must be provided for these deflections.

The ductility of metals is also an important design parameter since it indicates the degree of plastic formability and often the brittleness. In many cases high strength is compromised in order to obtain a more ductile alloy. Figure 2 depicts the yield strength-ductility relationship for certain classes of metals. The wrought aluminum and copper alloys are preferred for applications where ductility is necessary and low to medium strengths are ample.

PHYSICAL PROPERTIES

The physical properties of most interest for design purposes include the electrical and thermal conductivities and the thermal expansion coefficients. Where good electrical conductivity is required copper and aluminum are the alloys of choice. The electrical conductivity varies considerably with alloy content. Values for specific alloys are given in Chapter 3. The electrical resistivities for selected metals are presented in Chapter 16 on Comparative Properties. Typical values for the thermal properties of the common metals are listed in Table 4.

MACHINABILITY

Machinability is an important factor in consideration of the metal to be selected for a particular application. Because of the variety of ways in which machinability is rated a clear comparison is not always available. Often it is necessary to contact

Table 3 Tensile Moduli of Metals

	GPa (10^6 psi)
High moduli	
Tungsten	405 (58.7)
Molydenum	324 (47)
Medium moduli	
Steels	190–215 (27.5–31.1)
Nickel alloys	180–210 (26.1–30.4)
Titanium alloys	105–125 (15.2–18.1)
Copper alloys	105–130 (15.2–18.8)
Low moduli	
Aluminum alloys	68–73 (9.9–10.6)
Tin alloys	42–50 (6.1–7.2)
Magnesium alloys	44–45 (6.4–6.5)

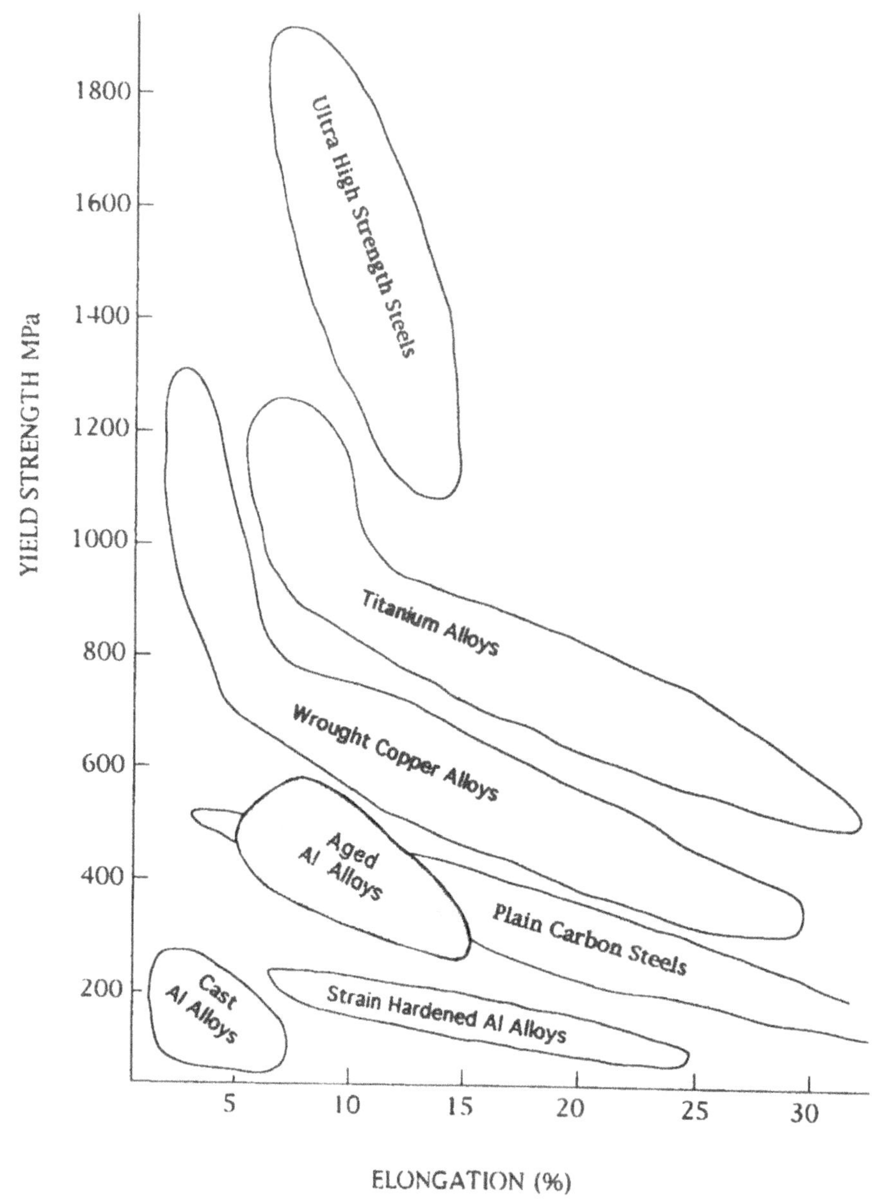

Figure 2 Yield strength versus ductility for alloy groups.

vendors for specific information. Within alloy groups, copper, e.g., the machinability of each alloy is ranked in a comparative fashion. When comparing different groups the problem becomes a more difficult one. ASTM E 618 describes a ranking method based on the effect of production rate on tool wear. Using free-machining brass as 100 a typical ranking appears as follows:

Table 4 Thermal Properties of Alloy Types

Alloy type	Thermal conductivity W/m·K at 20°C	Thermal expansion coeff. 10^{-6}/°C 20–300°C
Aluminum	100–220	20–21
Copper	30–400	16–21
Magnesium	70–150	25–27
Nickel	20–45	13–17
Carbon & low-alloy steels	50–60	11–13
Stainless steels	12–30	11–18
Titanium	5–20	8–11
Zinc	25–30	27–40

Alloy C36000 free-cutting brass	100
Alloy C46400 naval brass	50
201-T3 aluminum alloy	50
12L14 leaded free-machining steel	21

A relative comparison of some selected alloys based on metal removal rates with free-cutting mild steel as 100 give the following approximate order of machinability:

Free-cutting brass	500
Aluminum, cold-worked	350
Malleable cast iron	120
Free-cutting mild steel	100
Cr-Mo steel	65
Structural steel	60
Austenitic stainless steel	45

2 COSTS

For comparison of relative costs of metals it is convenient to use mild steel as unity. The relative costs per unit volume are compared to that of mild steel in Table 5. The numbers listed are cost per unit weight of alloy divided by that for mild steel times the density of the alloy. These relative costs will vary with form and quantity and are useful only as ball park relative figures. These cost values are based on simple forms, e.g., ingot or rod, and thus do not include processing costs such as casting and machining. Often the processing costs will exceed the metal cost and must be considered separately for each item to be manufactured. Specific costs of metals and other materials are given in Appendix C.

Table 5 Relative Cost/Unit Volume[a] for Selected Alloys

Alloy	Relative cost/unit volume
Mild steel	7.87
Low-alloy steel	10
Aluminum alloy 380	7
Aluminum alloy 2024	17
Stainless steel 304	35
Stainless steel 316	45
Stainless steel 17-4 PH	50
Copper rod	33
Copper-beryllium strip	247
Cartridge brass Cu–30%Zn	65
Copper alloy, aluminum bronze	52
Nickel alloy Monel 400	172
Nickel alloy Inconel	198
Magnesium aloy AZ91	10
Zinc alloy ZA 8	15
Molybdenum	255
Titanium alloy 6Al-4V	

[a]Cost per unit weight/cost per unit weight of mild steel) times density of alloy.

3
Nonferrous Metals

G. T. Murray
California Polytechnic State University
San Luis Obispo, California

1 ALUMINUM ALLOYS—INTRODUCTION

Aluminum is produced by the electrolytic reduction of aluminum oxide. Approximately 18 million tons of primary metal is produced each year. About 1.5 million tons are used per year for beverage cans alone, of which about 75% is comprised of recycled metal. The energy required for recycling is about 5% of that required for electrolytic reduction from aluminum oxide.

The attractiveness of aluminum as a structural metal resides in its light weight (2.7 g/cm^3 density compared to 7.83 g/cm^3 for iron), corrosion resistance, ease of fabrication, and appearance. On a strength-to-density ratio (i.e., the specific strength) aluminum alloys are about three times that of structural steel and roughly equivalent to the higher strength steels used in gears, shafts, axles, and the like. Aluminum alloys are manufactured in the usual flat rolled sheet and plate, and in wire, rod, tube and bar forms as well as castings, stampings, and extruded and forged shapes.

2 WROUGHT ALLOYS

2.1 Classifications and Designations

Aluminum alloys produced in wrought form are classified according to their major alloying element via a four-digit numbering system as listed in Table 1. The alloy number so designated and its respective temper is covered by the American National Standards Institute (ANSI) standard H 35.1. The Aluminum Association (900 19th St. NW, Washington, DC 20006) is the registrar from which numbers for various compositions and tempers can be obtained. In the 1xxx group the second digit indicates the purity of the aluminum used in manufacturing this particular grade. The zero in the 10xx group is used to indicate aluminum of essentially

Table 1 Wrought Aluminum and Aluminum Alloy Designation System

Aluminum ≥99.0%	1xxx
Aluminum alloys grouped by major alloying elements	
Copper	2xxx
Manganese	3xxx
Silicon	4xxx
Magnesium	5xxx
Magnesium and silicon	6xxx
Zinc	7xxx
Other elements	8xxx
Unused series	9xxx

commercial purity while second digits of 1 through 9 indicate special control of one or more individual impurity elements. In the 2xxx through 8xxx alloy groups, the second digit indicates an alloy modification. If the second is zero, the alloy is the original alloy, while the numbers 1 through 9 are assigned consecutively as the original alloy becomes modified. The last two of the four digits serve only to identify the different alloys in the group and have no other numerical significance.

A unified numbering system (UNS) has been developed that correlates all alloys, aluminum included, with many national and trade association systems. In addition the International Organization for Standardization (ISO) has developed its own alphanumeric designation system for wrought aluminum alloys based on the systems that have been used for certain European countries. For wrought alloys only, compositions have been registered with the Aluminum Association by a number of foreign organizations. These organizations are signatories of a Declaration of Accord for an International Designation System for Wrought Aluminum and Wrought Aluminum Alloys. A partial listing for some common wrought alloys showing the Aluminum Association designations along with the UNS, ISO and corresponding foreign designations, are shown in Table 2. A complete listing of UNS and ISO numbers can be found in the ASM International Metals Handbook, vol. 2, 10th ed.

Temper Designations

The temper designation system adopted by the Aluminum Association and used in the United States is used for all product forms excluding ingots. In most handbooks the temper designations follow the alloy designation. A listing of these combined alloy and temper designations, date registered and source, product form, thickness, strength and ductility may be obtained from the Aluminum Association.

Aluminum alloys are hardened and strengthened by either deformation at room temperature, referred to as strain hardening and designated by the letter H, or by an aging heat treatment designated by the letter T. If a wrought alloy has been annealed to attain its softest condition the letter O is used in the temper designation. If the product has been shaped without any attempt to control the amount of hard-

Table 2 Related International Alloy Designations Wrought Alloys

USA UNS	ISO	Al Assoc. UK BS/DTD	Germany DIN No.	Japan	France
A91050	Al99.5	1050	Al 99.5	A1X1	A5
A91080	Al99.8	1080	Al 99.8		A8
A91100	Al999.0Cu	1100	—	—	—
A91200	Al99.0	1200	Al 99.0	A1X3	A4
A92014	AlCu4SiMg	2014	AlCuSiMn	A3X1	AU4SG
A92024	AlCu4Mg	2024	AlCuMg2	—	—
A93003	AlMn1Cu	3003	AlMn	A1Mn	AM1
A93004	AlMn1Mg1	3004	AlMn1Mg1	—	Am1G
A94043	AlSi5	4043	AlSi5	—	A-S5
A94047	AlSi12	4047	AlSi12	—	—
A95005	AlMg1	5005	AlMg1	A2X8	A-G06
A95052	AlMg2.5	5052	AlMg2	A2X1	A-G2
A95454	AlMg3Mn	5454	Al2.7Mn	—	A-G3
A96061	AlMg1SiCu	6061	E-AlMgSi0.5	—	A-GS/L
A96063	AlMg0.5Si	6063	AlMgSi0.5	—	A-GS
A96463	AlMg0.7Si	6463	E-AlMgSi	—	—
A97005	AlZn4.5Mg1	7005	AlZnMg1	—	—
A97050	AlZn6CuMgZr	7050	—	—	—
A97075	AlZn5.5MgCu	7075[a]	AlZnMgCu1.5	—	A-Z5GU

[a]U.K. number is DTD5074A.

ening the letter F (as-fabricated) is used for the temper designation. The letter W is used for the solution treated (as-quenched) condition. The strain-hardened and heat-treated conditions are further subdivided according to the degree of strain hardening and the type of heat treating (called aging). These designations are listed below.

Strain-Hardened Subdivisions

H1x—Strain-Hardened Only. The second digit is used to indicate the degree of strain hardening. H18 represents full strain hardening which is produced by about 75% cold reduction in area. H14 would designate a product strained to about a half of maximum strain hardening which is attained by approximately 35% cold reduction in area. "Cold" in this sense means room-temperature reduction.

H2x—Strain-Hardened and Partially Annealed. Tempers ranging from quarter-hard to full-hard obtained by partial annealing of a cold-worked material is indicated by a second digit. Typical tempers are H22, H24, H26, and H28 with the highest number representing the highest hardness and strength just as for the H1x series.

H3x—Strain-Hardened and Stabilized. The H3x series is used to designate alloys whose mechanical properties are stabilized by a low-temperature heat treatment or as a result of heat introduced during fabrication. This designation applies only to those alloys that, unless stabilized, gradually age-soften at room temperature. Again a second digit is used as before to indicate the degree of strain hardening remaining after stabilization, e.g., H32, H34, H36, and H38, the highest number relating to the highest hardness and strength.

For two-digit H numbers whose second digits are odd, the standard limits for strength are the arithmetic mean of the standard limits for the adjacent two-digit H numbers whose second digits are even. When it is desirable to identify a variation of a two-digit H number, a third digit is used when the degree of control of the temper or mechanical properties are different but close to those for the two-digit designation to which it is added.

Heat-Treated Subdivisions

T1 Cooled from an elevated temperature process and aged at ambient temperatures (natural aging) to a substantially stable condition.

T2 Cooled from an elevated temperature shaping process, cold-worked and naturally aged to a substantially stable condition.

T3 Solution heat-treated, cold-worked and naturally aged to a substantially stable condition. The solution heat treatment consists of heating the product to an elevated temperature, often in the 500–550°C (932–1022°F) range, where the alloy becomes a single-phase solid solution, and subsequently cooling at a fast rate (quenching) such that a second phase does not form. This is designated as the W condition, but, since it is an unstable condition, the properties for this condition will not be found in the handbooks.

T4 Solution heat-treated and naturally aged to a substantially stable condition.

T5 Cooled from an elevated temperature shaping process and artificially aged, i.e., holding for a specific time at temperatures above the ambient temperature.

T6 Solution heat-treated and artificially aged. The aging process usually continues until maximum or near maximum strength is achieved.

T7 Solution heat-treated and overaged. This designation applies to products that have been aged beyond the point of maximum strength to provide some special characteristics, e.g., enhanced corrosion resistance or ease of forming.

T8 Solution heat-treated, cold worked, and artificially aged.

T9 Solution heat-treated, artificially aged, and cold-worked.

T10 Cooled from an elevated temperature shaping process, cold-worked, and artificially aged.

There are some additional T temper designations that vary from the above, are fairly complex, and may occasionally be encountered for very specific standards and applications.

2.2 Mechanical and Physical Properties

For strength in tension, the yield strength rather than the ultimate tensile strength will be reported. The 0.2% offset yield strength, i.e., the stress at which measurable plastic deformation commences in a tension test, is the strength commonly used for design purposes. Aluminum alloys are strengthened by either cold working (room-temperature deformation) or by a precipitation hardening heat treatment (aging), although for the strain-hardened alloys some additional strengthening is achieved by the addition of small amounts of alloying elements such as magnesium, manganese and chromium. The aged alloys attain strengths considerably higher than those achieved by the strain-hardening process. Nevertheless, strain-hardened produces are widely used in many applications where higher strengths are not required. Table 3 lists some typical applications for the 1xxx, 3xxx, 4xxx, and 5xxx strain-hardened series of alloys.

The 2xxx, 6xxx, and 7xxx series alloys are hardened by heat treatment and/ or by natural aging. The 4xxx series are most often used for welding and brazing wire and strip and may not be aged, although alloy 4032 has a low coefficient of thermal expansion and high wear resistance and is used for forged engine parts in the T6 aged temper condition. Table 4 lists some typical applications for these aged alloys.

Aluminum alloys do not show a ductile to brittle transition as temperature decreases and consequently neither ASTM nor ASME specifications require Charpy or Izod tests. The fracture toughness of the 1xxx, 3xxx, 4xxx, 5xxx, and most 6xxx alloys usually exceed the minimum value requirements and hence their toughness values are not normally reported in handbooks. Certain wrought alloys have been identified as control-toughness high strength products and include 2024-T8, 2124-T3, T8, 2419-T8, 7049-T7, 7050-T7, 7150-T6, 7175-T6, T7, and 7475-T6, T7. The fracture toughness of these alloys exceed 20 $MPa\sqrt{m}$ (18 $ksi\sqrt{in}$). Table 5 lists some typical room-temperature values of toughness of some high-strength alloys.

Table 3 Applications of Strain-Hardened Aluminum Alloys

Series	Applications
1xxx: Fe and Si of less than 1% are the major alloying elements. Excellent corrosion resistance, high electrical and thermal conductivities, and excellent workability.	Electrical conductors, chemical equipment, reflectors, heat exchangers, capacitors, foil, architectural and decorative trim, railroad tank cars and sheet metal work.
3xxx: Mn to 1.5% is major alloying element. Cu at less than 0.5% is also present. Good workability. Some alloys are clad with high-purity Al for better corrosion resistance.	3003, 3004, and 3105 are widely used as general purpose alloys where moderate strength and good workability are required, such as beverage cans, cooking utensils, tanks, signs, heat exchangers, furniture siding, food and chemical handling equipment, hardware, lamp parts, clad roofing, gutters and general sheet metal work.
4xxx: Si to 12% is major alloying element, which is used to lower melting point and increase fluidity.	Welding wire, brazing alloys, and in anodized condition, for architectural trim. Aged 4032 is used for pistons.
5xxx: Mg to 4% is major alloying element, sometimes with Mn to less than 1%. Alloys possess good welding behavior and good corrosion resistance in marine environments.	Architectural and decorative trim, cans and can ends, appliances, boats, pressure vessels, cryogenic tanks, hardware, armor plate, automotive parts; in clad form as nails, rivets, screens, cable sheathing, and for weldable applications requiring moderate strength.

Table 4 Applications of Aged Aluminum Alloys

Series	Applications
2xxx—Cu to 6.8% is the principal alloying element, often with Mg to 1.8%. These alloys have less corrosion resistance than other Al alloys and are frequently clad with high-purity Al. Mechanical properties are comparable to those of low-carbon steel.	Aircraft parts, e.g., wheels, fuselage skins and structures; truck frame and suspension parts, screw machine products. Generally useful to 150°C (300°F). Alloys, except for 2219, have limited weldability. Also used for rivets, jet engine impellers, and compression rings.
6xxx—Mg to 1.5% and Si to 1.0% are major alloying elements. These alloys do not attain the strength levels of the 2xxx and 7xxx series but good weldability, formability, and corrosion resistance promote their usage. Alloys are often formed in the lower strength state and subsequently aged to the higher T6 condition.	Architectural applications, pipe railings, furniture, bicycle frames, forgings and extrusions for welded railroad cars, forged and extruded auto parts, marine products, screw products, ladders, TV antennas, auto body sheet; in alcad form where strength, corrosion resistance and weldability are needed; also as fuses, die forgings, tubing and pipe for transporting water, oil or gasoline.
7xxx—Zinc to 8%, Mg to 4.7% and Cu to 2.5% results in alloys heat-treatable to moderate and to high-strength levels.	Airframe structures, mobile equipment, trucks, trailers, sports equipment, cargo containers, armor plate, and landing gear cylinders.

Table 5 Fracture Toughness Values of Selected Aluminum Wrought Alloys

	Fracture toughness	
Alloy (plate; T-L direction)	$MPa\sqrt{m}$	$(ksi\sqrt{in.})$
7050-T7651	31	(28)
7075-T651	25	(23)
7475-T7351	45	(41)
2014-T651	23	(21)
2024-T851	27	(24)
2219-T87	31	(28)
5083-O	27	(25)
6061-T651	29	(27)
2024-T351	32	(29)
7475-T651	41	(37)
2097-T8	36	(33)

All values are for plate specimens where the crack plane is in the transverse-longitudinal (T-L) direction. The fracture toughness for alloy 5083-0 has been converted from J_{1c} values. The fracture toughness of this alloy actually increases with decreasing temperature and hence is widely used for cryogenic applications.

Aluminum alloys generally do not exhibit sharply defined fatigue limits as are found in steels in S-N tests. The fatigue resistance in alloys where lifetime is governed by crack initiation is expressed as a stress (fatigue strength) for a given number of cycles. Where crack growth rate is important it is now common practice in aerospace applications to use the crack growth rate (da/dn) as a function of stress intensity range (ΔK).

Tables 6 and 7 list typical mechanical and physical properties of selected wrought and cast alloys respectively. Much of this data was extracted from manufacturer's data sheets. ASM Metals Handbook vol 2, 10th ed., gives the values of most Aluminum Association registered alloys.

2.3 Formability

One of the many attractive features of aluminum alloys is the ease with which it can be formed in the annealed condition. In addition to the common rolling and stamping processes these alloys can be formed by blanking and piercing, press-brake forming, contour roll forming, deep drawing, spinning and stretch forming. The workability of the non-heat-treatable alloys in general are better than that for the heat-treatable ones, although the latter can be readily formed in the fresh solution-annealed condition. In general, the higher the strength of the alloy the more difficult it is to form. Forming characteristics, including bend radii may be found in *Aluminum Standards and Data*, published by the Aluminum Association.

Aluminum and its alloys are easily forged and extruded at relatively low temperatures, being in the 375–475°C (705–887°F) range for forging and in the 340–510°C (650–950°F) range for extrusion. Alloys that are commonly forged include 1100, 6061, 2025, 4032, 2014, 2618, 7050, 7075 and 5083 and are listed here in order of increasing difficulty of forging. Forgings are most often produced in closed dies resulting in parts with good surface finish, dimensional control and soundness. Forging also permits the production of intricate shapes, although for large runs of small parts die casting is a less expensive process. Forgings are more common for larger parts and is widely used for aircraft structural elements. In general, alloys containing the most alloy content are also the most difficult to forge.

Aluminum alloys also differ somewhat in their extrudability and as in the case for forging, the higher the alloy content the more difficult it is to extrude the alloy. A wide variety of shapes can be obtained by the extrusion process. Designs that are symmetrical around one axis are especially suitable to fabrication by extrusion. Alloys that are frequently selected for extrusion include 1100, 3003, 6063, 6061, 2011, 5086, 2014, 5083, 2024 and 7075 and are listed here in the order of increasing difficulty of extrusion. Other factors being equal, the more unsymmetrical an extruded shape cross section, the more difficult it is to extrude. Since the relative extrudability is based on an identical cross section, the design engineer should select where possible high-extrudability alloys for products of complex cross sections.

Table 6 Mechanical Properties of Selected Wrought Aluminum Alloys

Alloy & temper Al Assoc. no.	Alloy type	Y. S. MPa (ksi)	Elong. %[a] in 50 mm	BH No. 500 kg 10 mm ball	F. S. 5×10^7 cycles MPa (ksi)	Shear strength MPa (ksi)	Modulus GPa (10^6 psi)	Stress rupture[c] MPa (ksi)	Mach. index
1100-O	99.0Cu	35 (5)	45	23	35 (5)	60 (9)	69 (10)	—	E
1100-H14	99.OCu	115 (17)	20	32	50 (7)	75 (11)	69 (10)	—	D
2014-T6	Cu4SiMg	415 (60)	13	135	125 (18)	290 (42)	73 (10.6)	—	B
2024-T4	Cu4Mg1	325 (47)	19	120	140 (20)	285 (41)	73 (10.6)	—	—
2048-T851	Cu3.3Mg1.5Mn	416 (60)	8	—	221 (32)	271 (39)	70 (10)	221 (32); 175°C	B
3003-H14	Mn1Cu	145 (21)	16	40	60 (9)	95 (14)	69 (10)	—	D
3004-H4	Mn1Mg1	259 (36)	6	77	110 (16)	145 (21)	69 (10)	—	C
4032-T6	Si12	315 (46)	9	120	110 (16)	260 (38)	79 (11.4)	290 (42); 149°C	B
5005-O	Mg1	40 (6)	25[b]	28	—	75 (11)	69 (10)	—	E
5005-H18	Mg1	195 (28)	24[b]	—	—	110 (16)	69 (10)	—	D
5050-H34	Mg1.5	165 (24)	8[b]	53	90 (13)	125 (18)	69 (10)	—	D
5052-H34	Mg2.5	215 (31)	14	68	125 (18)	145 (21)	70 (10.2)	—	C
5154-H32	Mg3.5	205 (30)	15[b]	67	125 (18)	150 (22)	70 (10.2)	—	D
6061-O	Mg1SiCu	55 (8)	30	30	60 (9)	85 (12)	69 (10)	—	D
6061-T6	Mg1SiCu	275 (40)	17	95	95 (14)	205 (30)	69 (10)	—	C
6063-T6	Mg0.5Si	215 (31)	12[b]	73	70 (10)	150 (22)	69 (10)	—	C
7050-T7651	Zn6CuMgZr	490 (71)	11	—	—	325 (47)	72 (10.4)	179 (26); 149°C	B
7075-T6	Zn5.5MgCu	505 (73)	11	150	160 (23)	330 (48)	72 (10.4)	150 (22); 150°C	B
2090-T83	Cu2.7Li-2.3	483 (70)	3	—	160 (23)	—	79 (11.4)	—	—
AA2097-T8	Cu2.8Li-1.5	400 (58)	7	—	—	—	—	—	—
7150-T7751	Cu2.2Zn6.4Mg	538 (78)	8	—	—	—	—	—	—
7055-T7751	Cu2.3Zn8Mg2	614 (89)	10	—	—	—	—	—	—

[a]0.5 in. dia. spec.
[b]1.6 mm thick spec.
[c]Stress to produce rupture after 1000 h at temp. shown.

Table 7 Typical Physical Properties of Selected Wrought Aluminum Alloys

Alloy no. Al Assoc. & temper	Alloy type	Density (g/cm^3)	Elect. cond. (% IACS)	Ther. cond. (W/ m·K at 25°C)	CTE (10^{-6}/°C)
1100-O	Al99.0Cu	2.70	61	222	24
2014-T6	AlCu4SiMg	2.80	34	154	22
2024-T4	AlCu4Mg1	2.70	30	121	23
3003-H14	Al Mn1Cu	2.73	41	159	23
3004-H38	AlMn1Mg1	2.72	42	163	24
4032-T6	AlSi12	2.68	35	138	19
5005-O	AlMg1	2.69	52	200	24
5050-H34	AlMg1.5	2.69	50	193	24
5052-H34	AlMg2.5	2.68	30	138	24
5154-H32	AlMg3.5	2.67	33	130	23
6061-T6	AlMg1SiCu	2.69	43	165	24
6063-T6	AlMg0.5Si	2.70	53	200	20
7075-T6	AlZn5.5MgCu	2.81	33	130	24

Machinability

In general the machinability of aluminum alloys are considered good when compared to other metals. The softer 1xxx series are probably the most difficult to machine. The machinability rating of selected alloys are presented in Table 6, where an A rating would indicate the most easily machined aluminum alloys compared to that of other aluminum alloys.

Joinability

Aluminum alloys are amenable to joining by a variety of processes including riveting, welding, brazing,, soldering and adhesive bonding. For welding operations the aluminum oxide must first be removed and prevented from reforming by shielding the joint with the inert gases argon and helium. Chemical fluxes are of limited use because of lesser efficacy for oxide prevention plus the possibility of residues that may promote corrosion. Aluminum has a high electrical conductivity hence high electrical currents are required to produce sufficient heat to attain the melting points of the metal. Nevertheless aluminum alloys are resistance welded. Resistance spot and seam welding processes are used in the manufacture of cooking utensils, containers, flooring, and many aircraft and automotive components.

The absence of precipitate particles make the strain-hardening alloys favorable for welding although any benefits of straining will be lost during the welding operation. The design engineer thus cannot use strain-hardened strength values for design purposes. Likewise the age-hardenable alloys will lose strength in the heat-affected zone (HAZ) as a result of welding and this loss must be accounted for in the design process. Post-weld heat treatments can be employed to restore strength to some extent.

In recent years advances in powder metallurgy and rapid solidification technology have enabled the development of dispersion strengthened alloys for elevated temperature use. Fast cooling by electron beam or pulsed laser welding methods help to reduce loss in strength.

The relative weldability of a large number of wrought alloys are listed as follows: *readily weldable wrought alloys* (GMAW/GTAW): 1350, 1060, 1100, 2219, 3003, 3004, 5005, 5050, 5052, 5083, 5086, 5154, 5254, 5454, 5456, 5652, 6010, 6061, 6063, 6101, 6151, 7005, 7039; *weldable in most applications (GMAW/GTAW)*: wrought alloys 2014, 2036, 2038, 4032, 308.0, 319.0, 333.0, 355.0, 551.0, 512.0, 710.0, 711.0, 712.0; *limited weldability*: wrought alloy 2024; *not recommended for welding*: wrought alloys 7021, 7029, 7050, 7075, 7079, 7150, 7178, 7475.

Brazing: Brazing involves the joining of base metal parts by using fused filler metals of lower melting points than that of the base metal. For aluminum alloys the recommended brazing temperature is approximately 21°C (70°F) below the temperature at which the base metal commences to melt. Most brazing is done at temperatures between 560 and 615°C (1040 and 1140°F). Filler metals are usually the low melting Al-Si alloys containing 7% to 12% Si. Conventional brazing performed in air requires the use of chemical fluxes. The non-heat-treatable wrought alloys of the 1xxx, 3xxx, and low-magnesium 5xxx series including 1350, 1100, 3003, 3004, 5005, 5050 and 5052 are readily brazed. The commonly brazed heat-treatable alloys include 6053, 6061, 6063, 6951 and 7005. The 2xxx and most 7xxx series alloys are not normally joined by brazing processes.

3 CAST ALLOYS

3.1 Classifications

The Aluminum Association four-digit numbering system for cast alloys incorporates a decimal point to separate the third and fourth digits. The first digit identifies the alloy group listed as follows:

1xx.x Controlled unalloyed compositions
2xx.x Copper is major addition
3xx.x Al-Si alloys also containing magnesium and/or copper
4xx.x Binary Al-Si alloys
5xx.x Magnesium is major addition
6xx.x Currently unused
7xx.x Zinc is major addition but also containing Cu, Mg, Cr, or Mn
8xx.x Tin is major addition
9xx.x Currently unused

For the 2xx.x through 8xx.x alloys, the alloy group is determined by the alloying element present in the largest percentage, except in cases where the composition has been a modification of a previously registered alloy. In such cases the alloy group is determined by the element that comes first in the sequence. The second two digits identifies the aluminum alloy or, for the 1xx.x series, indicates the aluminum purity. The last digit indicates the product form, whether a casting or an ingot, being 0 for a shaped casting and a number 1 for an ingot having limits for alloying elements the same as those for the alloy in the form of shaped castings, and a 2 if the limits are different. A modification of the original alloy or impurity limits is indicated by a serial letter preceding the numerical designation. Temper designations are the same as those for wrought alloys, with the exception that there are no strain-hardened H tempers for cast alloys. There is no Declaration of Accord for an international system of tempers to be registered with the Aluminum Association. Foreign designations are now being grouped under systems of the American National Standards Institute, the ISO system and the European Committee for Standardization. Foreign designations for some selected cast alloys are listed in Table 8.

3.2 Casting Processes

Aluminum alloy castings are produced by pressure-die, permanent-mold, green- and dry-sand, investment and plaster casting. Aluminum alloys are also readily cast with vacuum, low-pressure, centrifugal, and pattern-related processes. Aluminum casting alloys contain sufficient quantities of eutectic forming elements, most often silicon, in order to provide adequate fluidity to fill the shrinkage regions with molten metal. Alloying elements added for strengthening purposes are similar to those used for wrought alloys. Hence, both heat-treatable and non-heat-treatable alloys are cast. Since these alloys are cast to final or near-final shape, the strain-hardening category does not apply. The casting process often dictates the alloy(s) that one would select. The casting process must thus be established prior to property considerations. The casting process is determined by feasibility, quality, and cost factors. When two or more casting processes for a certain part are feasible, and with acceptable quality, the cost factors will usually dictate the process to be used. For a small number of parts the tooling costs will often be the major cost factor. The pros and cons of the more prominent casting methods are summarized below.

Die Casting

Die cast products are, in terms of product tonnage, more than twice that of all other aluminum cast products combined. Die cast alloys are not usually heat treated. The die casting process consists of rapid injection of molten metal into metal molds

Table 8 Related International Alloy Designations Cast Aluminum Alloys

USA UNS	ISO	USA Al Assoc.	UK	Germany DIN no.	France
A01001	Al 99.0	100.1	—	—	—
A01501	Al 99.5	150.1	LMO	—	—
A03190	AlSi5Cu3	319.0	LM4, LM22	GD-AlSi6Cu3	A-S5U3
—	AlMg5	314.0	LM5	G-AlMg5	A-G3T
A03551	AlSi5Cu1Mg	355.1	LM16	G-AlSi5Cu1	—
A03562	AlSi12Mg	A360.2	LM9	G-AlSi10Mg	—
A13800	AlSi8Cu3Fe	A380	LM24	G-AlSi8.5Cu	—
A03900	AlSi7Cu4Mg	390	LM30	—	—
A04430	AlSi5	443	LM18	—	—
A05200	AlMg10	520	LM10	G-AlMg10	AG10Y4

under pressures of 2 MPa (300 psi) or higher. This process is favored for large quantities of relatively small parts. Die casting equipment is relatively expensive compared to other casting processes. However, on large runs the cost of tooling per part becomes minimal and coupled with low labor costs per part makes die casting an attractive process for large production runs. Typical products include gear cases, instrument cases, lawn mower and street light housings, aircraft and marine hardware and small motor pistons, connecting rods and housings. Linear tolerances of ±4 mm/m (±0.004 in./in.) can be achieved with surface finishes as fine as 1.3 μm (50 μin.). Casting weights are usually less than 5 kg (11 lb) and in many cases less than 1 kg (2.2 lb). Alloy 380.0 is the most frequently used alloy for die casting except where good corrosion resistance is required. In the latter situation alloys 360.0, 413.0, A413.0 and 518.0 may be selected. Alloy 390.0 has also found many useful applications in recent years.

Permanent Mold Casting

Permanent mold casting consists of gravity feeding of molten metal at a relatively low rate into metal molds. As in die casting rapid solidification is achieved making this process also suitable for high-volume production. Production rates are less than that for die casting, thereby increasing labor costs. Linear tolerances of about ±10 mm/m (±0.10 in./in.) with finishes in the 3.8 to 10 μm (150 to 400 μin.) range are obtained. Casting weights are usually less than 10 kg (22 lb). This process has also been modified by the use of low pressures of 170 kPa (25 psi) or less. Common alloys for permanent mold casting include 356, 355.0, 444.0, A444, and B443.0. Typical parts include automotive pistons, gears, impellers, machine tool parts, wheels, pump parts, and carburetor bodies.

Sand Casting

This casting process involves the forming of a mold by ramming sand with an appropriate bonding agent around a removable pattern. Its main advantages are versatility and low equipment costs. The casting rate is low and labor intensive, making this process much more suitable for a small number of relatively large castings. Linear tolerances of about ±30 mm/m (± 0.030 in./in.) and surface finishes of the order of 10 μm (394 μin.) are achieved. Typical alloys used include C355.0, 356.0, A356.0, 357.0, 443.0, 520.0 and 713.0. Automobile transmission cases, oil pans, rear axle housings and aircraft fittings, levers and brackets are a few of the typical parts cast by this process.

Lost Foam Pattern Casting

This process also uses a sand mold (unbonded) around an expendable polystyrene pattern that vaporizes during the pouring of the molten metal. Labor costs are considerably more than that for the conventional sand casting method. Tolerances are improved, however, and the dimensional variation associated with pattern removal are eliminated. Cleaning of the casting is also greatly reduced. The major concern is shrinkage of the foam pattern. Alloys similar to those employed in the conventional sand casting process are usually suitable to the lost foam process.

Shell Mold Casting

In shell mold casting the molten metal is poured into a thin shell of resin bonded sand. Its chief advantages are excellent surface finish and dimensional accuracy.

Production rates are higher than for sand casting processes, however, equipment costs are higher.

Investment Casting

This process employs plaster molds coupled with expendable wax patterns, the latter being melted out as the plaster is baked. Christmas-tree gating systems are often used thus producing many patterns per mold. The investment casting process is frequently used for large numbers of intricately shaped parts. This process has been referred to as precision casting since walls can be as thin as 0.4 mm (0.015 in.) with linear tolerances off ±5 mm/m (±0.05 in./in.) and surface finishes of the order 0.2 μm (80 μin.). The cost, however, is high and can only be justified when accurate dimensional control is justified. Typical alloys used are 308.0, 355.0, 356.0, 443.0 and 712.0.

Plaster Casting

In plaster casting the plaster slurry is formed around a removable pattern and the plaster is baked before the molten metal is poured. Fine details and good linear tolerances, of the order of ±5 mm/m (±0.05 in./in.), are obtained. Alloys may be selected from 295.0, and/or the 355.0 and 356.0 varieties.

Other Casting Processes

Modifications of the above processes do exist but they do not significantly alter the alloys that can be used. The molten metal can be forced into steel, sand, plaster, or graphite molds by centrifugal forces during casting. One of the advantages of centrifugal casting is that inclusions and gases tend to be forced into the gates. Costs are high, however, and shapes and sizes are limited. Hot isostatic pressing of aluminum castings reduce porosity and improve properties. Squeeze casting of molten metal under pressure achieves similar results and is easily automated to produce near-net shape products. Semisolid processing, which is a hybrid method that incorporates both elements of casting and forging is also a near-net shape-forming process.

3.3 Mechanical and Physical Properties

The properties of cast products will vary considerably depending on whether the alloys will be used in the as-cast or the cast and heat-treated condition. The physical properties of most interest to the design engineer are the thermal and electrical conductivities and the coefficients of thermal expansion. For aircraft design the densities are of significance. The electrical and thermal conductivities vary considerably with treatment and the design engineer is cautioned to examine these properties in terms of both the alloy number and its temper condition. The mechanical and physical properties of interest for selected cast alloys are listed in Table 9. It is evident that the ductility of cast aluminum is somewhat limited.

Machinability

Casting alloys that are readily machinable include 201.0, 222.0, 511.0, 513.0, 514.0, 518.0, 520.0, 535.0, A535.0, B535.0, 705.0, 707.0, 710.0, 711.0, 712.0, 713.0, 771.0, 772.0, 850.0, 851.0, and 852.0. Alloys 213.0, 242.0, A242.0, 295.0, 512.0, 213.0, 238.0, 512.0, and 390.0 are considered to have good machining qual-

Table 9 Typical Properties of Selected Aluminum Casting Alloys

Alloy no.	Comp.	Y. S. [MPa (ksi)]	Elong. % in 50 mm	Elect. cond. (% IACS)	Ther. cond. (W/mk at 25°C)	CTE [10^{-6}/°C (10^{-6}/°F) 20–300°C]
208.0- F(S)	Cu4Si3	97 (14)	2.5	31	121	23.9 (13.3)
295.0 -T4(S)	Cu4.5Si1.1	110 (16)	8.5	35	138	24.8 (13.8)
A206.0-T4(S)	Cu4.5MnMg	250 (36)	6.0	30	121	24.7 (13.7)
A346.0-T6(S)	Si7Mg0.3	207 (30)	6.0	40	151	23.4 (13.0)
360.0-F(D)	Si9Mg0.5	172 (25)	3.0	37	142	22.9 (12.7)
380.0-F(D)	Si8.5Cu3.5	165 (24)	3.0	27	96	22.5 (12.5)
390.0-F(D)	Si17Cu4.5Mg	241 (35)	1.0	25	134	18.5 (10.3)
390.0-T5(D)	Si17Cu4.5Mg	265 (39)	1.0	24	134	18.0 (10.0)
296.0-T4(P)	Cu4.5Si2.5	131 (19)	9.0	33	130	23.9 (13.3)
354.0-T6(P)	Si9Cu1.8Mg	283 (41)	6.0	32	122	22.9 (12.7)
355.0-T6(S)	Si5Cu1.3Mg	172 (25)	3.0	36	142	24.7 (13.7)
356.0-T6(S)	Si7Mg0.3	164 (24)	3.5	39	151	23.4 (13.0)
443.0-F(D)	Si5.2	110 (16)	9.0	37	142	23.8 (13.2)
443.0-F(S)	Si5.2	55 (8)	8.0	37	142	24.1 (13.4)
413.0-F(D)	Si12	145 (21)	2.5	39	121	22.5 (12.5)
A444.0-T4(S)	Si 5	62 (9)	12.0	35	138	25.9 (14.4)
514.0-F(S)	Mg4	83 (12)	9.0	35	146	25.9 (14.4)
518.0-F(D)	Mg8	186 (27)	8.0	24	96	26.1 (14.5)
520.0-T4(S)	Mg10	330 (48)	16.0	21	88	25.0 (13.9)
713.0-T5(S)	Zn7.5CuMg	205 (30)	4.0	35	140	24.1 (13.4)

S = sand; D = die; P = permanent mold

ities. Some of the more difficult to machine alloys include 443.0, A443.0, 444.0, and 332.0.

Joinability

The general statements previously made about the joinability of the wrought alloys also apply to the cast alloys. Cast alloys readily weldable by the shielded arc processes include 356.0, 443.0, 413.0, 514.0 and A514.0. Alloys 208.0, 308.0, 319.0, 333.0, 355.0, C355.0, 511.0, 512.0, 710.0, 711.0, and 712.0 are weldable in most applications.

4 TEMPERATURE LIMITATIONS

Metals generally lose strength as the temperature is increased. The heat-treatable alloys may overage (soften) with time at temperature due to the coalescence of the precipitate particles which, when present in a smaller dispersed size, enhance room-temperature strength. For long-time usage one must examine the strength as a function of time at temperature at a fixed stress. The design engineer must examine the detailed creep and stress-rupture data that can be obtained from vendor literature. Creep occurs due to slow plastic deformation under stress at elevated temperatures. Pure aluminum and a few off the non-heat-treatable alloys will creep at room temperature under sufficient stress. Pure aluminum electrical cables have been known to creep at very low rates under their own weight. For preliminary screening purposes the maximum recommended use temperature for certain alloys are listed in Table 10. Both creep of non-heat-treatable alloys and overaging of heat treatable alloys have been considered in compiling this data.

Low-Temperature Limits

Most aluminum alloys can be used below room temperature. The wrought alloys that are frequently used for low-temperature service include 1100, 2024, 2014, 3003, 5083, 5456, 6061, 7005, 7039 and 7076. Alloy 5083-O and 2519-T7 are recommended for cryogenic applications. Alloy 5083-O experiences about a 10% increase in yield strength along with a 60% increase in ductility (elongation) at the boiling point of liquid nitrogen ⟨−196°C (−320°F)⟩ compared to room-temperature values. It also exhibits an increase in fracture toughness having a K_{Ic} (J) of 43.4 M Pa$\sqrt{m}$ at −196°C.

References and Acknowledgements

1. Alcoa Technical Data Sheets
2. ASM, Int. Metals Handbook, Vol. 2, 10th ed.
3. Smithells Metals Reference Book, 6th ed.
4. Aluminum Association

5 COPPER ALLOYS

5.1 Wrought Alloys

Introduction

Copper and copper alloys are widely used because of their high electrical and thermal conductivities, good strength and ductility, high fracture toughness, ease of fabrication, joining and polishing, attractive colors and good corrosion resistance. Copper itself is inherently corrosion resistant. In addition many alloys have been

Table 10 Approximate Maximum Service Temperature for Selected Wrought Aluminum Alloys

Alloy	Temperature (°C)
3103 - O, H4, H8	100
5005 - O, H3, H6	100
2024 - T6	130
2124 - T851	175
2519 - T87	175
6061 - T6	120
6063 - T6	200
6463 - T4, T6	300

developed that have excellent corrosion resistance. Certain bronzes and copper-nickel alloys are traditionally employed in marine environments. There exist nearly 300 "standard" wrought alloys. Copper alloys are used in the building construction industry for plumbing, hardware, heating and air conditioning, and electrical wiring. Other major industries using copper alloys include electrical products, marine equipment, heat exchangers and telecommunications. Table 11 lists typical applications for each wrought alloy category.

Classifications and Designations

Wrought copper and its alloys can be separated into the following groups:

Coppers	Metal which has a minimum copper content of 99.3%.
High-copper alloys	Those which contain 96.0 to 99.3% copper.
Brasses	Alloys with zinc as the major alloying element. This group consist of the binary copper-zinc alloys; copper-zinc-lead alloys (leaded brass); and copper-zinc-tin alloys (tin brasses).
Bronzes	Copper alloys in which the major alloying element is not zinc or nickel. There are four major families of bronzes: copper-tin-phosphorous (phosphor bronzes); copper-tin-lead-phosphorous (leaded phosphor bronzes); copper-tin-nickel alloys (nickel-tin bronzes); and copper-aluminum alloys (aluminum bronzes).
Miscellaneous copper-zinc alloys	This family contains alloys formerly known as manganese or nickel bronzes. But since zinc is the principal alloying element they are really brasses. The two major subgroups in this category are the copper-nickel alloys often called cupronickels and the copper-zinc-nickel alloys commonly called nickel-silvers because of their silvery color, although they do not contain silver. Other alloys in this group are the brasses that contain small quantities of manganese or silicon.

The Unified Numbering System (UNS) for Metals and Alloys, which is jointly managed by the American Society for Testing and Materials and the Society of Automotive Engineers, is the accepted alloy designation system in North America for both wrought and cast copper and copper alloy products. This designation system is also being used in Australia and Brazil. It is administered by the Copper Development Association, Inc. (CDA) at 260 Madison Ave., New York, NY 10016. Standards handbooks for both wrought and cast products can be obtained from CDA. The UNS copper alloy designations consist of a five-digit number preceded by the prefix letter C. Numbers from C10000 through C79999 denote wrought alloys. Numbers from C80000 through C99999 are used for cast alloys. They are not specification numbers. Each number refers to a specific alloy composition. UNS numbers for specific groups of wrought alloys are given in Table 11 and for cast alloys in Table 18, along with applications for each specific group.

Table 11 Typical Applications of Wrought Copper Alloys

Group and UNS numbers	Applications
Coppers: C10100-C15999	General for group; high electrical conductivity requirements
C10100 to C10700	Highest conductivity coppers
C11000	Electrical wire and cable
C12200	Household water tube
C10800	Refrigerators, air conditioners and gas lines
C14200, C14300	Contacts, terminals; resists softening when soldered
C15715 to C15760	Electronic components, lead frames, integrated circuits
High-copper alloys: C16200 to C19199	General for group; electrical and electronic connectors and contacts
C17000 to C17300	Highest strength copper-beryllium alloys; bellows, diaphragms, fasteners, relay parts
C18100	Switches, circuit breakers, contacts
C18200 to C18300	Cable connectors, electrodes, arcing and bridging parts
C19400	Terminals, flexible hose, gaskets, fuse clips; excellent hot and cold workability
Copper-zinc brasses: C21000 to C28000	General: sheet for stampings, springs, electrical switches and sockets, plumbing
C23000	Red brass; condenser and heat exchanger tubes, plumbing, architectural trim, good formability
C26000 to C26200	Cartridge brass; radiator cores, hardware, ammunition, plumbing accessories
C26800	Muntz metal; architectural sheet and trim, large nuts and bolts, brazing rod
Copper-zinc-lead brasses: C31200 to C38500	General for group: leaded brasses; -high machinability requirements
C34500	Clock parts, gears, wheels
C36000	Free-cutting brass; machinability 100; screw machine materials, gears, pinions
C37700	Forging brass; forgings and pressings of all kinds
Copper-zinc-tin brasses: C40400 to C48600	General: corrosion resistance and higher strength requirements
C42500	Supplied as strip for fabricating electrical connectors, springs, terminals
C46400 to C46700	Naval brass; marine hardware, propeller shafts, structural uses
C48200, C48500	Free machining; marine hardware, valve stems, screw machine products
Copper-tin-phosphorous bronzes (C50100 to C54200)	General: superb spring qualities, good formability and solderability, high fatigue resistance, high corrosion resistance
C50500	Flexible hose, electrical contacts
C51100	Fasteners, bellows, fuse clips, switch parts
Copper-tin-lead-phosphrous (C53400 to C53500)	Leaded phosphor bronzes; general: combines high strength and fatigue resistance with good machinability and wear resistance

Table 11 Continued

Group and UNS numbers	Applications
Copper-aluminum bronzes (C60800 to C64210)	General: combination of high strength and excellent corrosion resistance
C61000	Marine hardware, pumps, valves, nuts, bolts, shafts tie rods, machine parts, condenser tubing
C63000	Nuts, bolts, marine shafts, aircraft parts, forgings (usually hot forged)
Copper-silicon (C64700 to C66100)	Silicon bronzes. General: properties similar to aluminum bronzes; excellent weldability; hydraulic fluid lines, high-strength fasteners, wear plates, marine hardware
Copper-nickel (C70100 to C72950)	General: excellent corrosion resistance, strength retention at high temperatures, condenser tubes, marine products
Copper-zinc-nickel (C73500 to C79800)	Nickel silvers; good corrosion resistance with moderately high strength, silver luster, food and beverage handling equipment, decorative hardware, hollow-ware

Temper Designations. ASTM B601, *Standard Practice for Temper Designations for Copper and Copper Alloys—Wrought and Cast*, describes the more than 100 temper designations for copper alloys. The more common temper designations are listed as follows:

Tempers	Description
HOO to H14	Degree of cold work—1/8 hard to superspring
H50 to H90	Cold-work tempers based on manufacturing process
HR01 to HR50	Cold-worked and stress-relieved tempers
HT04 to HT08	Cold-rolled and thermal-strengthened tempers
M01 to M45	As-manufactured tempers; cast, extruded, forged, etc.
O10 to O82	Annealed tempers to meet specific mechanical properties
OS005 to OS200	Annealed tempers to meet prescribed grain sizes
TD00 to TD04	Solution-treated and cold-worked tempers
TH01 to TH04	Cold-worked and precipitation-hardened tempers

Mechanical and Physical Properties

Copper alloys to a large extent attain their mechanical properties by the introduction of different amounts of cold work. The chief exceptions to this generality are the copper-beryllium alloys, the highest strength copper alloys, the aluminum bronzes and some copper-chromium alloys. (Beryllium poses a health hazard: consult OSHA regulations.) The aluminum bronzes are the second highest strength group.

These alloys attain their high strengths by heat treatments that introduce precipitate or second phase particles.

Typical mechanical properties of representative alloys of each compositional group are listed in Table 12. This data was extracted from the *Copper Alloys for Machined Products Handbook*, published in 1992 by the Copper Development Association Inc.

The electrical and thermal conductivities are the physical properties of most interest. The International Annealed Copper Standard, or IACS, is the recognized standard for metal electrical conductivity. High-purity and oxygen-free coppers have conductivities of 100%. IACS (1.7241 microhm-cm at 20°C). Alloying reduces the electrical conductivity. The high nickel content (30%) of alloy C71500 copper-nickel, for example, reduces the conductivity to only 4% IACS at 20°C (68°F). The high thermal conductivity of copper and its alloys are also exploited in many applications such as heat exchangers, condensers and other heat-transfer devices. And unlike most metals, the thermal conductivity of many copper alloys increase with increasing temperature. Alloying reduces the thermal conductivity. The physical properties of representative wrought alloys are listed in Table 13.

Formability

Copper and its alloys are noted for their ease of fabrication. Most alloys can be shaped by the common forming processes. They are cold-rolled, stamped, drawn, bent and headed. They are also extruded, forged and rolled at elevated temperatures. Some copper alloys work-harden during cold-working more rapidly than others thus requiring annealing (softening) treatments intermittently in order to be cold-reduced extensively. Annealing is not normally required for bending and stamping operations. The age-hardenable coppery-beryllium alloys must be formed in the unhardened (solution annealed) condition and subsequently aged to the desired strength level. The leaded copper alloys, particularly the free-cutting brasses, have low ductility, making forming somewhat more difficult. Machined parts requiring deep knurls or high-pitched roll threads should be made from alloys with reduced lead contents. Alloys C34000, C34500, and C35300 are good choices.

Table 14 categorizes alloys according to their cold and hot formabilities. Generally the design engineer is concerned only with the cold formability since the process metallurgist has reduced the material to selected sizes by hot- and/or cold-working operations. The design engineer should be familiar with the various tempers, however. In general the higher the temper numbers the harder the metal. For cold-worked metal the higher the temper number the more prior cold work the material has received. This cold-worked metal resists further deformation making it more difficult to shape. Extensive additional forming could lead to fracture of the part.

Machinability. Relative machinability of different materials is often based on the ASTM E 618 method that evaluates the material by determining the effect of production rate on tool wear. When rated by this test brass is about five times more machinable than leaded steel (lead is added to enhance the machinability). Copper alloys are compared within their own group by an index based on the allowable machining speed, tool wear, finish, accuracy and power requirements. Alloy C36000, an extra-high-leaded brass, is assigned an index rating of 100, and all other copper alloys are compared to this free-cutting brass. The machinability of

Table 12 Typical Mechanical Properties of Representative Wrought Copper Alloys

Group & UNS no.	Condition	Y.S. 0.5% ext. MPa (ksi)	Elong. in 51 mm (2 in.)	Hard. HRB[a]	F.S. 10^8 cycles MPa (ksi)	Shear strength MPa (ksi)	Modulus GPa (10^6 psi)
Coppers (C10100–C15999)							
C11000	Spring H08	345(500)	4	60	96.5(14)	—	117(17)
C10800	Hard rod H04	303(44)	16	47	—	186(27)	115(17)
C10800	Tube H55	220(32)	25	35	—	180(26)	115(17)
C15760	C. W. 74%	600(87)[b]	14	86	248(36)	—	—
High-copper alloys (C16200–C19199)							
C17200	C.W. & aged	1345(195)[b]	5	44HRC	380(55)	795(115)	128(18.5)
C18100	C.W. & aged	455(66)[b]	10	—	—	—	125(18.2)
C19400	Strip H04	380(55)[b]	7	73	145(21)	—	121(17.5)
Copper-zinc brasses (C21000–C28000)							
C23000	Rod H04	359(52)	10	65	—	296(43)	296(43)
C26000	Strip H02	359(52)	25	70	152(22)	290(42)	110(16)
C26800	Rod H00	276(40)	48	55	—	248(36)	—
Copper-zinc-lead brasses (C31200–C38500)							
C34500	Rod H02	400(58)	25	80	—	290(42)	—
C36000	Rod H02	310(45)	25	75	—	262(38)	103(15)
C37700	Extruded rod	138(20)	45	78HRF	—	207(30)	—

Copper-zinc-tin brasses (C40400–C48600)							
C42500	Strip H08	517(75)	4	92	—	—	110(16)
C46400	Rod H02	365(53)	20	82	—	303(44)	103(15)
C48200	Rod H02	365(53)	15	82	—	283(41)	100(15)
Copper-tin-phosphorous bronzes (C50100–C54200)							
C50500	Strip H04	345(50)	8	75	—	—	117(17)
C51100	Strip H08	552(80)	3	93	—	—	110(16)
Copper-aluminum bronzes (C60800–C64210)							
C61300	Rod H04	331(48)	35	80	—	276(40)	117(17)
C63000	Rod H02	414(60)	15	96	262(38)	—	117(17)
Copper-silicon bronzes (C64700–C66100)							
C65500	Rod H04	379(55)	22	95	—	400(58)	103(15)
Copper-nickel (C70100–C72950)							
C70600	Tube H55	393(57)	10	72	—	—	124(18)
C71500	Rod H02	483(70)	15	80	—	290(42)	152(22)
Copper-zinc-silver "nickel-silver" (C73500–C79800)							
C74500	Strip H06	524(76)	3	92	—	—	121(17.5)
C75200	Rod H02	414(60)	20	78	—	—	125(18)

[a]Except as noted.
[b]0.2% offset.

Table 13 Physical Properties of Representative Wrought Copper Alloys

UNS number	Density 20°C [gm/cm^3 (lb/in.^{3})]	CTE[a] [10^{-6}/°C (10^{-6}/°F)]	Ther. cond. [W/m · K (English)[b] at 20°C]	Specific heat	Elec. cond. cond. (%IACS)
C11000	8.94 (0.323)	17.7 (9.8)	391 (226)	0.092	101
C10800	8.94 (0.323)	17.7 (9.8)	349 (202)	0.092	92
C15760	8.80 (0.318)	16.6 (9.2)	322 (186)	0.092	85
C17200	8.25 (0.298)	17.8 (9.9)	107 (62)	0.10	22
C18100	8.83 (0.319)	19.4 (10.7)	324 (187)	0.094	80
C19400	8.91 (0.322)	17.9 (9.2)	262 (150)	0.092	65
C23000	8.74 (0.316)	18.7 (10.4)	159 (92)	0.09	37
C26000	8.52 (0.308)	20.1 (11.1)	121 (70)	0.09	28
C26800	8.47 (0.306)	20.4 (11.3)	116 (67)	0.09	27
C34500	8.49 (0.307)	20.3 (11.3)	116 (67)	0.09	26
C36000	8.50 (0.307)	20.6 (11.4)	116 (67)	0.09	26
C37700	8.44 (0.305)	20.8 (11.5)	119 (69)	0.09	27
C42500	8.78 (0.317)	20.2 (11.2)	109 (64)	0.09	28
C46400	8.41 (0.034)	21.3 (11.8)	116 (67)	0.09	26
C48200	8.44 (0.305)	21.2 (11.8)	116 (67)	0.09	26
C50500	8.89 (0.321)	17.8 (9.9)	87 (50)	0.09	48
C51100	8.86 (0.320)	17.8 (9.9)	84 (48.4)	0.09	20
C61300	7.94 (0.287)	16.3 (9.0)	55 (32)	0.09	12
C63000	7.58 (0.274)	16.3 (9.0)	39 (22)	0.09	7
C65500	8.52 (0.308)	18.0 (10.0)	36 (21)	0.09	7
C70600	8.94 (0.323)	17.1 (19.5)	45 (26)	0.09	9
C71500	8.94 (0.323)	16.2 (9.0)	29 (17)	0.09	4
C74500	8.69 (0.314)	16.4 (9.1)	45 (26)	0.09	9
C75200	8.75 (0.316)	16.3 (9.0)	32 (19)	0.09	6

[a]20–300°C
[b]BTU/ft^2/ft/h/°F

representative wrought copper alloys based on such an index are given in Table 15.

All copper alloys are machinable in that they can be cut with standard tooling. High-speed tool steels suffice for all but the hardest alloys. For screw machine production use of the free-cutting copper alloys is advised. Most of these alloys have a machinability rating of 60 or higher when compared to alloy C36000.

Joinability. Copper alloys can be joined by a wide variety of processes. Due to their good formability, mechanical means of joining, such as crimping, riveting, and bolting are frequently used. However, soldering, brazing and welding are the most frequently employed joining processes. The high thermal conductivities of copper alloys must be taken into account during welding and brazing operations. Preheating and high heat input are required for most welding operations. Their high thermal expansion coefficient must also be considered when fitting pieces together.

Table 14 Formability of Wrought Copper Alloys

Alloys with both excellent cold and hot workability: C10100 through C14310; C15000 through C15500; C17410, C18100, C18900, C19000, C19200 through C19500; C63800, C65100, C65400, C65500, C6900.
Alloys with excellent cold workability and good hot workability: C16200 through C17300; C71500 through C18090; C18135 through C18500; C21000 through C23000; C40500, C41100, C41300, C42200, C43000, C50500, C70400, C72600 through C72900.
Alloys with excellent cold workability and fair hot workability: C24000 through C26200; C27400, C40800, C41500, C42500, C43400 through C44500.
Alloys with fair to good cold workability but good to excellent hot workability: C14700 through C14720; C28000, C46400 through 46700.
Alloys with both good cold and hot workability: C61000 through C61500; C62300, C64400 through C72200.
Alloys that should be cold worked only: C15715, C15720, C18700, C24000, C27000, C31400 through C35300; C51000 through C54400; C74500 through C78200.
Alloys that should be hot worked only: C35600, through C38500; C48200, C48500, C62400 through C63000; C64200, C67400, C67400, C67500, C69400.

All copper alloys can be joined by soldering; however, their protective film must be removed. Mechanical cleaning is beneficial, but generally fluxes are needed to remove and prevent reformation off tenacious films. Coppery-beryllium and aluminum bronzes tend to resist fluxing and may require cleaning with a 15–20% aqueous solution of nitric acid followed by a water rinse. The possible change in properties of these age-hardenable alloys during application of heat must also be recognized. Small amounts of elements such as silver, cadmium, iron, cobalt, and zirconium are often added to deoxidized copper to impart resistance to softening at times and temperatures encountered in soldering operations. The common joining processes used for representative wrought copper alloys are given in Table 16.

Temperature Limitations

By far the majority of copper alloys are used for room temperature applications, heat-transfer apparatus being the major exception to this generalization. The rather limited low-temperature applications are usually not a problem. Most copper alloys can be used to −200°C (−328°F). Stress relaxation in mechanical components and for electrical connectors is of considerable concern and certain alloys are better suited for these requirements. C16200 (Cu-1Cd) has been used to 95°C (203°F). Alloys C19000 (high phosphorous) and C17500 (Cu–2.5Co–0.6Be) have been used to 165°C (329°F). Both of these alloys must be age-hardened after forming. For electronic components used in the increasingly severe automotive underhood conditions, high stress-relaxation resistance between 135 and 200°C (275 to 390°F) are required and may be found in the C16200 to C19199 group. C15100 has good stress relaxation resistance and high conductivity and thus is found in lead frames, connectors and switches. For mechanical components requiring good stress relaxation resistance the Cu-Be alloys such as C17200 are favored.

Long time creep and/or creep-rupture data are needed in order to select alloys for design purposes at elevated temperatures. Once the field of possible alloys has

Table 15 Machinability of Wrought Copper Alloys

UNS no.	Descriptive name	Rating	UNS no.	Descriptive name	Rating
Type 1 Free cutting alloys for screw machine products					
C36000	Free-cutting brass	100	C54400	Phosphor bronze B-2	90
C35600	Extra-high-leaded brass	100	C19100	Nickel-copper with Te	75
C34200	High-leaded brass	90	C34000	Medium-leaded brass	70
C38500	Architectural bronze	90	C48500	Naval brass, high leaded	70
C14700	Sulfur-bearing copper	85	C17300	Leaded beryllium copper	60
C18700	Leaded copper	85	C33500	Low-leaded brass	60
C31400	Leaded bronze	80	C67600	Leaded Mn bronze	60
C34500	Leaded brass	80	C66100	Leaded silicon bronze	50
C37700	Forging brass	80	C79200	Leaded nickel silver	50
Type 2 Short-chip alloys					
C62300	Aluminum bronze 9%	50	C46200	Naval brass	30
C17410	Beryllium copper	40	C67400	Si-Mn-Al brass	30
C17500	Beryllium copper	40	C67500	Mn bronze A	30
C17510	Beryllium copper	40	C69400	Silicon red brass	30
C28000	Muntz metal 60%	40	C69430	Arsenical-Si red brass	30
C61800	Aluminum bronze 10%	40	C17000	Beryllium copper	20
C63000	Ni-Al bronze 10%	40	C17200	Beryllium copper	20
Type 3 Long-chip alloys; not for screw machine work					
C23000	Red brass 85%	30	C26000	Catridge brass 70%	30
C26800	Yellow brass 66%	30	C61000	Aluminum bronze 7%	30
C65500	High-silicon bronze	30	C65600	Silicon bronze	30
C67000	Manganese bronze	30	C10200	Oxygen-free copper	20
C10800	Oxygen-free low-P Cu	20	C11000	Electrolytic copper	20
C12000	P-deoxididized Cu	20	C16200	Cadmium copper	20
C18135	Cr-Cd copper	20	C22000	Commercial bronze	20
C50700	Tin bronze	20	C51000	Phosphor bronze 5%	20
C61000	Aluminum bronze 7%	20	C70600	Copper-nickel 10%	20
C74500	Nickel-silver 65-10	20	C71500	Copper-nickel 30%	20
C75400	Nickel-silver 65-15	20	C64700	Silicon bronze	20

Table 16 Joining Processes for Wrought Copper Alloys

UNS number	GTAW/GMAW	Oxyacetylene	Brazing	Soldering
C10100-10200	Good	Fair	Excellent	Excellent
C11000-11600	Fair	Not Rec.	Good	Excellent
C12000-12100	Excellent	Fair	Excellent	Excellent
C14500-14520	Fair	Fair	Good	Excellent
C17000-17200	Good	Fair	Good	Excellent
C17410-17510	Good	Fair	Good	Excellent
C18200-17400	Good	Not Rec.	Good	Good
C22000-22600	Good	Good	Excellent	Excellent
C26130-28000	Fair	Good	Excellent	Excellent
C31400-38500	Not Rec.	Not Rec.	Good	Excellent
C51000-52100	Good	Fair	Excellent	Excellent
C61300	Excellent	Not Rec.	Fair	Not Rec.
C62300-63000	Good	Not Rec.	Fair	Not Rec.
C65100-65500	Excellent	Good	Excellent	Excellent
C70600-71500	Excellent	Fair to Good	Excellent	Excellent
C74500-75200	Fair	Good	Excellent	Excellent
C75400-75700	Fair	Good	Excellent	Excellent
C63600-64200	Fair	Not Rec.	Poor-Fair	Not Rec.

been narrowed to a select few it is advisable to obtain creep or creep-rupture data from vendor literature. Approximate upper temperature limits for a number of alloys are listed in Table 17. These limits were approximated using data from a number of sources. (The ASM, Int. Metals Handbook, vol. 2, 10th ed., reports creep data on alloys C10100, C10200, C11300 through C11600; C15000, C19400, C23000, C26000, and C44300 through C44500, and strength versus temperature data for the aluminum bronzes C61300, C61400, C62300, C63000 and C63800.)

5.2 Cast Alloys

Copper alloy castings are selected for a wide variety of products because they possess desirable mechanical properties combined with good corrosion resistance. Their favorable tribological properties also contribute to their rather extensive usage for sleeve bearings, wear plates, gears and other wear-prone components.

Classifications and Designations

In North America copper casting alloys are classified according to the UNS. Numbers from C80000 through C99999 are reserved for casting alloys. The temper designations, which define metallurgical conditions, heat treatment, and/or casting method, further describe the alloy. As in the case for the wrought alloys the terminology associated with tempers can be found in ASTM B601. A summary of the temper numbers follows:

Table 17 Approximate Maximum Service Temperatures for Wrought Copper Alloys

Alloy type	Maximum service temperature (°C)
Coppers	80
Phosphous deoxidized copper	120
Copper-beryllium	250
Copper-chromium	350
Silicon bronze	200
Aluminum bronze (7–9% Al)	300
Cartridge red brass	200
Naval brass	150
Muntz metal	180
Leaded brass	100
Manganese brass	180
Phosphor bronze (9% Sn)	160
Cupro-nickel (5–10% Ni)	150
Cupro-nickel (20–30% Ni)	200
Nickel silvers	200
Cu-Cd-bearing alloys	200

Temper Designations	Description
010	Cast and annealed (homogenized)
011	As cast and precipitation heat-treated
M01 through M07	As cast; includes casting method
TQ00	Quench-hardened
TQ30	Quenched and tempered
TQ50	Quench-hardened and temper-annealed
TX00	Spinodal-hardened
TB00	Solution heat-treated
TF00	Solution heat-treated and precipitation-hardened

The various alloy categories and their numbers are listed along with significant applications in Table 18. However, the selection of the proper alloy for a given application is inherently tied into the casting process to be used which in turn is often dictated by cost considerations.

Casting Processes

The casting process to be selected is dependent on a number of factors the more important ones being listed as follows:

(a) number, size and weight of castings
(b) intricacy of shape
(c) quality of finish
(d) internal soundness required
(e) dimensional accuracy
(f) casting characteristics of alloy selected

The common casting processes include sand, die, investment, centrifugal, plaster and permanent mold methods. These processes were discussed in more detail under aluminum alloys. It is advisable to utilize the knowledge of a skilled foundryman in selecting the casting process to be used since the foundryman's experience can be a source of cost-saving ideas.

Sand castings account for about 75% of U.S. copper alloy foundry production. This method offers the greatest flexibility in casting size and shape and is the most

Table 18 Applications of Cast Copper Alloys

Alloy class	Applications
Coppers C80100–C81200	Electrical connectors, water-cooled apparatus, oxidation-resistant applications
High Copper Alloys C81400–C82800	Electromechanical hardware; high-strength beryllium-copper alloys are used in heavy-duty electromechanical equipment, resistance welding machine components, inlet guide vanes, golf club heads and components of undersea cable repeater housing
Brasses C83300–C87900	Red brasses: water valves, pipe fittings and plumbing hardware Leaded yellow brasses: gears, machine components, bolts, nuts Silicon bronzes/brasses: bearings, gears, and intricately shaped pump and valve components, plumbing goods
Bronzes C90200–C95900	Unleaded tin bronzes: bearings, pump impellers, valve fittings, piston rings Leaded tin bronzes: most popular bearing alloy, corrosion-resistant valves and fittings, pressure-retaining parts Nickel-tin bronzes: bearings, pistons, nozzles, feed mechanisms
Copper-nickel C96200–C96900	Pump components, impellers, valves, pipes, marine products, offshore platforms, desalination plants
Copper-nickel-zinc C97300–C97800	Ornamental, architectural and decorative trim, low-pressure valves and fittings, food dairy and beverage industries
Copper-lead (leaded coppers) C98200–C98840	Special-purpose bearings, alloys have relatively low strength and poor impact properties and generally require reinforcement

economical method if only a few castings are made. All copper alloys can be successfully cast in sand. Dimensional control ranges from about ±0.8 to 3.2 mm with a surface finish between 7.7 to 12.9 μm rms. Permanent metallic mold castings are noted for very good surface finishes (about 1.8 μm) and good soundness. This method is best suited for tin, silicon, aluminum and manganese bronzes and for the yellow brasses. Die costs are relatively high so this method is often favored for medium to large production runs. Dimensional accuracy and part-to-part consistency are unsurpassed in both small and large castings. Copper alloys can be cast into plaster molds which gives surface finishes as good as 32 μin. rms and dimensional tolerances as close as ±0.005 in. (±0.13 mm). Plaster mold casting accounts for only a small fraction of cast copper alloys and should not be used with leaded alloys since the lead reacts with the plaster. Die casting involves the injection of molten metal into a multipart die under high pressure. This method is best known for its ability to produce high-quality products at very low unit costs and is well suited for the yellow brasses. In centrifugal casting molten metal is poured into a spinning mold. Centrifugal force causes impurities to concentrate at the casting's inner surface which can be machined away. Centrifugal castings are made in sizes from approximately 50 mm to 3.7 m (2 in. to 12 ft) and from several centimeters to many meters in length. Virtually all copper alloys can be cast by this method. In investment casting a plastic or a low-melting alloy is cast into an intricate metal die and then removed and carefully finished. This pattern, or clusters of them, are dipped into a plaster slurry, which is allowed to harden as a shell or monolithic mold. The mold is then heated to volatilize the plastic (or remove the low-melting alloy) and to vitrify the plaster. Metal is then introduced into the mold and allowed to cool. Investment casting is capable of maintaining a very high dimensional accuracy with a good surface finish. Because of the high tooling cost this process is usually reserved for large production runs of precision parts.

Many of the copper alloys can be cast by most of the above-mentioned processes. Applicable casting processes for each of the approximately 100 casting alloys may be obtained from the Copper Development Association. A products' shape, size, and physical characteristics may limit the choice of casting method to a single casting process.

Mechanical and Physical Properties

The properties of most interest for copper casting alloys are presented in Table 19. The heat-treated copper-beryllium alloys, as in the case for the wrought alloys, have the highest strengths, followed closely by the aluminum bronzes. The copper-beryllium alloys are relatively expensive and thus these cast alloys have been limited to special applications, e.g., in aircraft parts and more recently in golf club heads. It should be realized that beryllium poses a health hazard and OSHA regulations must be consulted before processing and thus selecting copper-beryllium alloys. The aluminum bronzes, on the other hand, have a wide range of attractive properties including high hardness, wear resistance, excellent corrosion resistance and good castability. Many aluminum bronzes are strengthened by heat treatment. Aluminum bronze castings are widely used in marine equipment, pump and valve components and wear rings.

The manganese bronzes also exhibit high strengths, hardness and wear resistance and compete with the aluminum bronzes in many of the same applications.

Table 19 Properties of Representative Copper Casting Alloys

UNS no.	Descrip.	Y. S. MPa (ksi)	Hardness	Electrical cond. (% IACS)	Thermal cond. (W/m·K)	Mach. index
C80100	99.95 Cu	62 (9)	44Brinell	100	391	10
C81100	99.70	62 (9)	44Brinell	92	346	10
C81500[a]	Cu–1Cr	276 (40)	105Brinell	82	315	20
C82500[a]	2Be–0.5Co	1034 (150)	43 HRC	20	130	20
C82700	2.4Be1.3Ni	896 (130)	39 HRC	20	130	20
C83600	5Sn5Pb5Zn	117 (17)	60Brinell	15	72	84
C85400	Zn29Pb3	83 (12)	50Brinell	20	87.9	80
C86500	Zn40	200 (29)	100Brinell	22	85.5	26
C87400	Zn14–Si3	165 (24)	70Brinell	7	27.7	50
C90300	Sn8–Zn4	45 (21)	70Brinell	12	74.8	30
C92200	Zn4.5Sn6	138 (20)	65Brinell	14	69.6	42
C93200	Zn7Pb7Sn3	124 (18)	65Brinell	12	58.8	70
C94700	Ni5Sn2Zn2	159 (23)	85Brinell	12	31.2	30
C95400	Al11–Fe4	241 (35)	170Brinell	13	58.7	60
C95800	Ni5Fe4Mn1	262 (38)	159Brinell	7	36	50
C96200	Ni10–Fe2.4	172 (25)	—	11	45.2	10
C96400	Ni30–Fe1	255 (37)	140Brinell	5	28.5	20
C97300	Ni12–Zn20	117 (17)	80Brinell	6	28.6	70
C97600	Ni20–Zn8	165 (24)	80Brinell	5	31.4	70
C97800	Ni25Sn5	207 (30)	130Brinell	4.5	25.4	60

[a]Heat-treated

Lead is added to some of these bronzes for improved machinability. Although their corrosion resistance is less favorable than the aluminum bronzes they are frequently specified for marine propellers and fittings. They have been largely replaced by the aluminum bronzes for bearing applications.

Tin bronzes have also found usage in applications requiring corrosion and wear resistance. Brasses are the most common copper casting alloys having a favorable combination of strength, corrosion resistance and cost. They are good general-purpose alloys. The copper-nickel alloys are noted for exceptional corrosion resistance in seawater and find use aboard ships and off-shore platforms. The corrosion resistance and cost increases with increasing nickel content and hence their use must be justified. Applications for each group of copper casting alloys are summarized in Table 18.

Joinability. The general joining characteristics of the copper cast alloys are similar to those presented for the wrought alloys. The various joining processes and their degree of success for selected cast alloys are presented in Table 20.

Machinability. The machinability of selected copper cast alloys are presented in Table 19. The highest machinability alloys are those containing lead. The machinability of the aluminum bronzes are inferior to that for most brasses and tin bronzes.

Table 20 Joining Processes for Copper Casting Alloys

UNS number	Alloy type	GTAW/GMAW	Brazing	Soldering
C81300–82200	High copper	Fair	Good	Good
C82400–82800	Cu–Be	Fair	Fair	Fair
C83800–85200	Leaded brasses	Not Rec.	Good	Excellent
C85300	70–30 brass	Good	Excellent	Excellent
C85400–85800	Yellow brass	Not Rec.	Good	Good
C87200–87600	Silicon brass	Fair	Fair	Not Rec.
C90200–91600	Tin bronze	Fair	Good	Excellent
C92300–94500	Leaded Sn bro.	Not Rec.	Fair	Excellent
C95200–95700	Al bronze.	Good–Excellent	Good	Good
C96200–96400	Copper–nickel	Fair–Good	Excellent	Excellent
C97300–97800	Nickel silvers	Not Rec.	Excellent	Excellent

Note: Oxyacetylene welding is generally not recommended. Exceptions are some yellow brasses, 87200 Si bronze, 87600 Si brass, and tin bronze (fair).

Temperature Limitation. As with the wrought alloys the temperature limitations are usually defined by use of creep data, e.g., the amount of creep strain in a certain number of hours at a constant stress and temperature, or in terms of the stress that produces rupture in a given time at temperature. These stresses are lower than the short-term tensile strength at the same temperature. The coppers and leaded alloys are mostly used at room temperature. Alloys 92200 and 92300 are used as pressure-retaining products to 290°C (550°F) and 260°C (500°F) respectively. The unleaded brasses and bronzes, chromium-copper and copper-beryllium alloys have intermediate to high strengths above room temperature. The useful design range for copper-beryllium alloys is limited to about 220°C (425°F). The aluminum bronzes and copper-nickel alloys have the best elevated temperature strengths. The 97600 copper-nickel alloy will show a creep strain of 0.1% in 10,000 h at 288°C (550°F) and a stress of 153 MPa (22.2 ksi). Aluminum bronze alloy 95200 shows the same creep properties and stress at a temperature of 230°C (450°F).

Low-temperature embrittlement of copper alloys is generally not a serious problem even for the high-strength alloys. The 95200 aluminum bronze has a charpy V-notch impact strength of 34 J (25 ft-lb) at −188°C (−320°F).

References and Acknowledgments

1. Copper Development Association, Inc.
2. ASM, Int. Metals Handbook, Vol. 2, 10th ed.

6 NICKEL ALLOYS

6.1 Introduction

Nickel and its alloys are primarily used in applications where heat and/or corrosion resistance are of concern. Consequently, it is convenient to consider the room or ambient temperature usage separately from the elevated temperature applications. The latter will be presented in Chapter 9.

There are some nickel alloys that are used at both room and at elevated temperatures. However, those alloys that have been developed especially for use at high temperatures are generally more expensive than those designed for room-temperature usage. Hence the expensive alloys would most likely be reserved for the more demanding circumstances. Nickel, as a base metal, is also more expensive than copper and aluminum. Thus to be selected for room temperature applications the nickel alloys must be able to perform in a fashion superior to the less expensive alloys. Some conditions do demand better corrosion resistance, and in some instances more strength, than that available in the aluminum and copper alloys. One must justify the higher cost of the nickel alloys.

6.2 Classification

Nickel-base alloys have for many years been referred to by trade names, most of these being registered by the INCO Alloys International, Inc. or the Inco family of companies. UNS numbers for nickel alloys are also available and will be listed in the property tables along with the major alloying elements for each composition.

The nickel base alloys used for room temperature applications can be classified as follows:

(a) Commercially pure and low-alloy nickels containing more than 93% nickel
(b) Nickel-copper alloys
(c) Nickel-chromium and nickel-chromium-iron alloys that have exceptional corrosion resistance in certain environments (note: most Ni-Cr and Ni-Cr-Fe alloys were developed for elevated temperature applications but find some usage at room temperature because of their good corrosion resistance)
(d) Special electrical and magnetic application alloys

The principal wrought alloys for room temperature usage include the commercially pure nickels, the low-alloy nickels and the nickel-copper alloys.

6.3 Properties and Applications

The properties of selected wrought nickel alloys for room-temperature applications are listed in Table 21 and their corresponding applications in Table 22. Alloys that are frequently cast will be covered in Chapter 9 under high-temperature applications. There are a number of nickel alloys with special electrical and magnetic properties. These include the nickel-chromium alloys used for heating elements and the high-permeability magnetic alloys. However, these materials will not be discussed herein since electrical and electronic materials are beyond the scope of this book.

Acknowledgment

1. Inco Alloys International, Inc.

7 MAGNESIUM ALLOYS

7.1 Introduction

Magnesium is a lightweight metal having a density of 1.74 g/cm^3, which is 36% less than that of aluminum. Hence, magnesium and aluminum alloys compete to

Table 21 Room Temperature Properties of Nickel Alloys

UNS no.	Trade name	Comp.	Y.S. 0.2% MPa (ksi)	% Elong. in 50 mm	Modulus MPa (ksi)	Hardness	Elect. resist. ($\mu\Omega$m)	Therm. cond. [W/m·K (BTU in/ft^2·h·F)]
N02200	Nickel 200	99.6Ni	148 (21.5)	47	204 (29.6)	109 HB	0.096	70 (487)
NO2201	Nickel 201	99.6Ni low C	103 (15)	50	207 (30)	129 HB	0.085	79 (550)
	Duranickel 301	93Ni 0.1Si0.6Fe	862 (125)	25	207 (30)	35 HRC	0.424	23.8 (165)
N04400	Monel 400	63Ni28-34Cu2Fe	240 (35)	40	180 (26)	110–150 HB	0.547	21.8 (151)
NO4405	Monel R-405	Monel 400 +S	240 (35)	40	180 (26)	110–140 HB	0.510	21.8 (151)
C71500	Monel 450	29–33Cu 1Mn1Zn	165 (24)	46	—	—	0.412	29.4 (204)
N05500	Monel K-500	Ni63Cu30 Al 13Fe2	790 (115)	20	180 (26)	300HB	0.615	17.5 (121)
N06600	Inconel 600	Ni72Cr14-17Fe6–9	310 (45)	40	207 (30)	75 HRB	1.03	14.9 (103)
N06625	Inconel 625	Ni58Cr20 Mo9Fe5Nb 3–4	517 (75)	42	207 (30)	190 HRB	1.29	9.8 (68)
N06985	INCO G-3	Cr22Fe20 Mo7	320 (47)	50	199 (29)	79 HRB	—	10 (69)
N10276	INCO C-276	NiMo16 Cr15Fe5	415 (60)	50	205 (30)	90 HRB	1.23	9.8 (68)

Table 22 Room-Temperature Applications of Nickel Alloys

Alloy	Applications
UNS N02200–Nickel 200	Commercially pure wrought nickel with good mechanical properties and resistance to a range of corrosive media. Used for a variety of processing equipment, particularly to maintain product purity in handling foods, synthetic fibers, and alkalies.
UNS N02201–Nickel 201	Essentially the same as Nickel 200 but with low carbon content to prevent embrittlement at temperatures above 315°C. Also has a lower hardness making it more suitable for cold-formed items.
Duranickel 301	A precipation hardened alloy for applications requiring the corrosion resistance of Nickel 200 but with more strength or spring properties. Applications include diaphragms, springs, clips, and press components for extrusion of plastics.
UNS N04400–Monel 400	A nickel-copper alloy with high strength and excellent corrosion resistance in a range of media including sewater, hydrofluoric acid, sulfuric acid, and alkalies. Used for marine engineering, chemical and hydrocarbon processing equipment, valves, pumps, shafts, fittings, and fasteners.
UNS N04405–Monel R-405	The free-machining version of Monel 400. A controlled amount of sulfur is added to provide chip breakers during machining. Used for meters and valve parts, fasteners and screw machine products.
C71500–Monel 450	Actually a copper base alloy but included here because of the name. It is resistant to corrosion and biofouling in seawater, has good fatigue strength, and relatively high thermal conductivity. Used for seawater piping.
UNS N05500–Monel K-500	A precipitation-hardenable nickel-copper alloy that combines the corrosion resistance of Monel alloy 400 with greater strength and hardness. It also has low permeability and is nonmagnetic to under −101°C (−150°F). Used for pump shafts, oil-well tools and instruments, doctor blades and scrapers, springs, valve trim, and fasteners.
UNS N06600–Inconel 600	A nickel-chromium alloy with good oxidation resistance at high temperatures and resistance to chloride-ion stress corrosion cracking, corrosion by high-purity water, and caustic corrosion. Used in furnace components, in chemical and food processing, and in nuclear engineering.

Table 22 Continued

Alloy	Applications
UNS N06625–Inconel 625	A nickel-chromium alloy with an addition of niobium that acts with molybdenum to stiffen the alloy's matrix and provide higher strength without a strengthening heat treatment. The alloy resists a wide range of severely corrosive environments and is especially resistant to pitting and crevice corrosion. Used in chemical processing, aerospace and marine engineering, pollution-control equipment, and nuclear reactors.
UNS N06985–INCO G-3	A nickel-chromium-iron alloy with additions of molybdenum and copper. It has good weldability and resistance to intergranular corrosion in the welded condition. Used for flue-gas scrubbers and for handling phosphoric and sulfuric acids.
UNS N10276–INCO C-276	A nickel-molybdenum-chromium alloy with an addition of tungsten having excellent corrosion resistance in a wide range of severe environments and maintains corrosion resistance in as-welded structures. Used in pollution control, chemical processing, pulp and paper production, and waste treatment.

some extent for the applications where the strength-to-weight ratio is an important consideration. Compared to aluminum however, magnesium has two distinct disadvantages. It is more expensive and possesses less corrosion resistance. Magnesium is very anodic in the electromotive series of metals and can be severely attacked unless galvanic couples are avoided by proper design or surface protection. Although, when exposed to air a film is formed, it only provides limited protection, unlike the adherent protective oxide film that makes aluminum inherently more corrosion resistant.

Magnesium alloys do possess some advantages particularly for die casting processes and applications of the die cast metals. They have the best strength-to-weight ratio for the commonly die cast metals, in general have better machinability, and often a higher production rate. In fact the major usage of magnesium is for die cast products. Magnesium alloys have high electrical and thermal conductivities, high impact resistance, good damping capacity and low inertia. Magnesium alloys are used in a wide range of applications in the aerospace, automotive, optical, computer and office equipment industries.

7.2 Designations

Magnesium alloys are designated by a series of letters and numbers. The first two letters indicate the two principal alloying elements, which are followed by two numbers that state the weight percentages of each element. The next letter in the sequence denotes the alloy developed; the letter C, e.g., indicates the third alloy of the series. AZ91C thus describes the third alloy standardized that contains nominally 9% aluminum and 1% zinc. The heat treatments are designated in a fashion similar to that used for the aluminum alloys: i.e., H10, slightly strained hardened; H23 to H26, strain hardened and partially annealed; and T6, the solution heat-treated and artificially aged condition. It should be noted that some letters are different from the chemical symbols: e.g., E, rare earths; H, thorium; K, zirconium; and W, yttrium.

7.3 Properties and Applications

The mechanical properties of selected magnesium alloys are presented in Table 23. The physical properties do not vary a lot with composition. For screening purposes one may use the following values: tensile modulus, 45 GPa; thermal conductivity, 30 W/m·K; thermal coefficient of expansion, 26×10^{-6}/°C.

Magnesium alloys have been used for automotive engine crankcases, experimental engine blocks, wheel housings, racing pistons, helicopter transmissions, gearbox covers, generator housings, impellers, and in missiles and a host of such applications where weight reduction is important.

References and Acknowledgments

1. Magnesium Elektron Inc.
2. Dow Chemical Company

8 TITANIUM ALLOYS

8.1 Introduction

Titanium and its alloys are noted for their high strength-to-weight ratio and excellent corrosion resistance. Although the needs of the aerospace industry for better

Table 23 Mechanical Properties of Selected Magnesium Alloys

UNS no.	Mfg. number	Composition	Condition	Tensile Y.S. MPa (kSi)	Elong % in 50 mm	Hardness Brinell	F.S. 10^6 cys. unnotched MPA (kSi)
M16710	ZC71	Zn6.5–Cu1.2	Extruded	200 (29)	5	—	—
M16710	ZC 71	Zn6.5–Cu1.2	Heat treated	297 (43)	3	—	—
—	ZW3	Zn3–Zr6	Extruded	228 (33)	8	65–75	124 (18)
M11312	AZ 31	A13–Zn1	Extruded	159 (23)	10	50–65	—
M18410	WE54A	Y5–R.E.2	Cast-heat tr.	207 (30)	4	80–90	102 (14.8)
M16631	ZC 63 T6	Zn6–Cu3	Cast-heat tr.	159 (23)	4	55–65	100 (14.5)
M18430	WE 43A	Y4–R.E.3	Cast-heat tr.	193 (28)	7	75–95	90 (13.1)
—	AZM	Al6–Zn1	Forged	159 (23)	7	60–70	90 (13.1)

Source: Data supplied by Magnesium Elektron Ltd.

strength-to-weight ratio structural materials was the impetus for their early development, increasing usage is found in automotive applications, chemical processing equipment, pulp and paper industry, marine vehicles, medical prostheses and sporting goods. Applications for some of the more popular alloys are listed in Table 24. The high cost of titanium alloy components may limit their use to applications where the lower cost aluminum and stainless steels cannot be used.

Unlike some of the other nonferrous alloys, the titanium alloys are not separated into wrought and cast categories. To date most of the widely used casting alloys are based on the traditional wrought compositions. Because the high cost of titanium alloys resides in processing as well as the raw materials cost, a substantial effort has been devoted to the development of net shape and near-net shape manufacturing processes. Casting, and in particular precision casting, is by far the most fully developed and widely used net shape process. Powder processing, especially the hot isostatic pressing method, has emerged as a viable method to ensure pore closure and reduce the scatterband of fatigue property test results.

Metallurgists have traditionally separated titanium alloys into categories according to the phases present, namely (a) commercially pure or modified titanium, (b) alpha and near-alpha alloys, (c) alpha-beta alloys, and (d) beta alloys. Although these terms may not mean much to the nonmetallurgist, they will be included here for sake of completeness.

8.2 Properties

The room-temperature properties of some of the more widely used wrought titanium alloys are listed in Table 25. One should be aware that the mechanical properties of the alpha-beta alloys are strongly dependent on processing history. When these alloys are processed well below the beta transus (the alpha-beta transition temperature) a fine equiaxed alpha grain structure is attained which is associated with high strength, good ductility and resistance to fatigue crack initiation. The acicular alpha structure is obtained by heating above the beta transus and allowing the beta to transform to alpha during cooling and aging. This microstructure is associated with excellent creep strength, good ductility, high fracture toughness and resistance to fatigue crack propagation. The relationship of processing to strength and fracture toughness is presented in Table 26 for some alpha-beta alloys.

The reaction of titanium with oxygen and nitrogen, together with its elevated temperature and limited creep resistance, generally restricts the use of titanium alloys to temperatures below about 540°C (1000°F). A recently developed alloy (Timet 21S) has been reported to possess sufficient heat and corrosion resistance to permit use at temperatures to 595°C (1100°F). The elevated yield strengths of selected titanium alloys are listed in Table 27 and some creep data in Table 28.

Titanium alloys are finding use in the chemical and petroleum industries as new alloys are developed that are corrosion resistant to reducing brines at high temperatures. These alloys often contain small additions (0.05%) of palladium that adds little to the cost.

References and Acknowledgments

1. RMI Titanium Company
2. Teledyne Wah Chang Albany
3. ASM, Int. Metals Handbook, vol. 2, 10th ed.

Table 24 Applications of Titanium Alloys

Alloy	Applications
0.3Mo–0.8 Ni ASTM grade 12 Modified titanium	Chemical process industries; used in hot brines, heat exchangers and chlorine cells
3Al–2.5V UNS R56320 ASTM grade 9 Alpha alloy	Chemical processing and handling equipment; has high degree of immunity to attack by most mineral acids and chlorides, boiling seawater and organic compounds (specific corrosion data available from Teleydne Wah Chang Albany); normally used in the annealed condition; available as foil, seamless tubing, pipe, forgings, and rolled products
5Al–2.5Sn (ELI) UNS 5421 Alpha alloy	Cryogenic applications
6Al–1Nb–1Ta–0.8Mo UNS R56210 Ti-6211 Near alpha alloy	Marine vehicle hulls; has high fracture toughness
6Al–4V and 6Al–4V (ELI) UNS R56400 Ti 6-4 ASTM grade 5 Alpha-beta alloy	In addition to the aerospace industry this popular alloy has been used for medical prostheses, marine equipment, chemical pumps and in high-performance automative components. ELI = extra low interstital content.
4.5Al–3V–2Mo–2Fe Alpha-beta alloy	Superplastically formed into golf club heads, auto parts, working tools and metal balloons
6Al–7Nb IMI 367 Alpha-beta alloy	Surgical implants
5Al–2.5Fe DIN 3.7110 Alpha-beta alloy	Surgical implants, bone nails, screws and plates
11.5Mo–6Zr–4.5Sn UNS R58030 Beta alloy	Aircraft fasteners, springs, and orrthodontic appliances
3Al–8V–6Cr–4Mo–4Zr UNS R58640 Beta alloy	Fasteners, springs, and casings in oil, gas and geothermal wells
15Mo–3Nb–3Al–2Sn–0.2Si Timetal 21S Metastable beta alloy	Lightly loaded aircraft structures to 595°C
8Al–11Mo–1V UNS R54810 Ti811 alpha-beta alloy	Airframe and turbine use
5.8Al–0.4Sn– 3.5Zr–0.7Nb–0.5Mo–0.35Si IMI 417 Near alpha	Cast or wrought parts for turbine and internal combustion engines
6Al–2Sn–4Zr–6M0 UNS R56260 Ti-6246 alpha-beta alloy	Forgings in turbine and for seals in airframe components

Table 25 Average Room-Temperature Properties of Wrought Titanium Alloys

Alloy	Y. S. MPa (ksi)	% Elong. in 50 mm	Hardness	Modulus [GPa (10^6 psi)]	Charpy impact [J (ft·lb)]	Ther. cond. (W/m·K)	CTE (10^{-6}/°C)
0.3Mo0.8Ni	448 (65)	25	—	103 (15)	—	19	8.6
alpha							
3Al–2.5V	586 (85)	20	—	107 (15.6)	—	8.3	9.5
alpha-beta							
5Al–2.5Sn	700 (101)	16	35 HRC	110 (16)	—	7.6	9.5
ELI							
alpha							
6Al–2Nb	725 (105)	13	30 HRC	113.8	31 (23)	7.3	9.0
1TalMo				(16.5)			
alpha							
6Al–4V	900 (130)	14	36 HRC	113.8	19 (14)	6.7	10.8
nonaged				—			
aged	1103 (160)	10	41 HRC	(16.5)	—	6.7	10.8
alpha				—			
beta							
3Al8V6Cr	1379 (200)	7	—	105.5 (15.3)	10 (7.5)	—	8.7
4Mo4Zr							
aged							
beta C							
11.5Mo6Zr	1317 (191)	11	—	103 (15)	—	—	—
4.5Sn							
Beta III							
35V15Cr	938 (136)	22	—	116 (16.8)	—	10	9.2

Table 26 Mechanical Property Variation with Heat Treatment for Some Alpha-Beta Titanium Alloys

Alloy	Microstructure	Y. S. MPa (kSi)	Fracture toughness MPa$\sqrt{m}$
Ti–6Al–4V	Equiaxed	910 (130)	44–66
Uns R56400	acicular	875 (125)	
Ti–6Al–	Equiaxed	1155 (165)	22–23
2Sn–4Zr–6Mo	acicular	1120 (160)	33–55
UNS R56260			
Ti–6Al–6V–2Sn	Equiaxed	1085 (155)	33–55
UNS R56620	acicular	980 (140)	55–77

Table 27 Elevated Temperature Yield Strength of Selected Titanium Alloys

	Temperature		Yield strength			
			MPA (ksi)		MPa (ksi)	
	°C	°F				
Alloy	315	(600)	425	(800)	540	(1000)
5Al–2.5Sn	448	(65)	280	(40)	—	
6Al–2Sn–4Zr–2MO–0.1Si	586	(85)	586	(85)	489	(71)
8Al–1Mo–1V	621	(90)	565	(82)	517	(75)
5Al–5Sn–2Zr–2Mo–0.25Si	655	(95)	531	(77)	503	(73)
6Al–4V annealed	655	(95)	572	(83)	427	(62)
6Al–4V aged	703	(102)	621	(90)	483	(70)
6Al–2Sn–4Zr–6Mo aged	841	(122)	758	(110)	655	(95)
13V–11Cr–3Al	793	(115)	827	(120)	—	
10V–2Fe–3Al aged	974	(142)	—		—	
6Al–2V–2Sn aged	896	(130)	—		—	
15Mo–3Nb–3Al–0.2Si aged	1034	(150)	863	(125)	—	

9 ZINC ALLOYS

9.1 Introduction

The zinc alloy family, which includes pressure die cast, gravity cast and wrought alloys, provide materials for many engineered components and structures. Zinc alloy castings have been used for over 60 years and offer the advantage that they can be cast to near net or net shape. Zinc castings compete with many of the aluminum, magnesium, bronze and iron casting alloys. The vast majority of zinc alloy castings are produced by pressure die casting following two basic processes, hot and cold chamber methods.

Table 28 Creep Properties of Selected Titanium Alloys

Alloy	Temperature °C	(°F)	Stress to produce 0.1% creep 100 h MPa (ksi)	1000 (h) MPa (ksi)
5Al–2.5Sn	427	(800)	331 (48)	—
6Al–4V	316	(600)	—	483 (70)
"	427	(800)	—	221 (32)
"	500	(932)	221 (32)	—
6Al–2Sn–4Zr–2Mo	427	(800)	—	400 (58)
"	538	(1000)	—	81 (12)
4Al–4Mo–2Sn–0.5Si	400	(752)	551 (80)	471 (68)
"	450	(842)	254 (37)	101 (15)
"	500	(932)	82 (12)	31 (4.5)
4Al–4Mo–4Sn–0.5Si	400	(752)	621 (90)	501 (73)
4Al–4Mo–4Sn–0.5Si	400	(752)	621 (90)	501 (73)
6Al–5Zr–0.5Mo–0.2Si	400	(752)	510 (74)	462 (67)
"	450	(842)	465 (67)	462 (62)

There are two basic groups of zinc casting alloys the more traditional being known as Zamak 3, 5, and 7, which are used for pressure die casting and contain about 4% by weight of aluminum as the major alloying element. The second group, developed in the 1970s, consist of the ZA (zinc-aluminum) series with ZA-8, ZA-12 and ZA-27 being the most popular compositions. The numbers indicate the nominal weight percent of aluminum content, which is also the major alloying element in this series of alloys. These alloys are used in a variety of gravity casting procedures, sand, plaster and permanent molds, as well as for die casting processes. Applications of these zinc casting alloys are presented in Table 29.

Wrought alloys fall into three groups, namely zinc-copper, zinc-titanium and zinc-lead-cadmium alloy compositions. Most wrought alloys are used as sheet products, although a few are forged and even extruded. Wrought products include fuses, eyelets, plumbing fixtures, gutters, flashing, roofing and, perhaps, the best-known canning jar lid. Forms include sheet, wire, forgings and extrusions.

9.2 Processing

The casting methods employed are similar to those previously described for the aluminum and copper alloys. As a general rule die casting is preferred when production quantities exceed 50,000 pieces per year. However, quantities as small as 10,000 per year have been die cast in order to eliminate or minimize machining operations. High-pressure die casting produces high-volume components of intricate detail and excellent surface finish at fast production rates. Cost per part is usually very low. Sand casting is generally used in low- to medium-volume requirements. The chief advantage of sand casting lies in its configurational flexibility along with no size or shape limitations. Gravity permanent mold casting is a common near-net-shape casting method with surface finishes superior to sand casting.

Table 29 Applications of Zinc Casting Alloys

Alloy	Applications
Zamak 3 Al 4Cu0.25Fe0.1 UNS Z33521 ASTM B86; B240 ingot	Most widely used zinc die casting alloy. Has excellent plating and finishing characteristics. Locks, screwdrivers, seat belt buckles, carburetors, faucets, gears, fan housings, computer parts.
Zamak 5 Al 4Cu1Fe0.1 UNS 35530 ASTM B86; B240 ingot	Has greater hardness and strength than Zamak 3. Good creep resistance. Automotive parts, household appliances and fixtures, office and building hardware.
Zamak 7 Al 4Cu0.25 UNS Z33522 ASTM B86; B240 ingot	A high-purity form of Zamak 3 but with better ductility and fluidity and more expensive. Similar uses to Zamak 3, especially those requiring more extreme forming operations.
ZA-8 Al 8.4Cu1Fe0.1 UNS Z35630 ASTM B791; B669 ingot	For pressure die castings, sand and permanent mold. First choice when considering die casting the ZA family. Agricultural equipment, electrical fittings, hand tools, plated parts.
ZA-12 Al 11Cu0.8 UNS 35630 ASTM B791; B699 ingot	Preferred for permanent mold processing. Also popular for sand castings. Used for higher strength applications. Journal bearings, agricultural equipment, pump components, plated parts.
ZA-27 Al 26Cu2Fe0.1 UNS Z35840 ASTM B791: B699 ingot	Components requiring optimum strength, hardness and light weight. Excellent bearing properties and damping characteristics. Engine mounts and drive trains, general hardware, rachet wrench, winch componeent, sprockets, gear housings. Not normally electroplated. Die casting must be by cold chamber method.

Most often steel or cast iron molds are employed. The lower tooling costs for these molds permit economical production of parts for the low and medium volume quantities. High-density, fine-grained graphite permanent molds offer significantly lower tool costs with better dimensional accuracy and surface finish. It is an ideal process for relatively simple medium-sized castings requiring cored holes, good dimensional tolerances and volumes up to 20,000 pieces annually. Shell and plaster mold casting procedures are used occasionally but are part-size-limited and are the more expensive casting methods.

Machinability

Zinc alloys possess excellent machinability. High-speed steel cutting tools are normally suitable for machining the ZA alloys. Moderate to high feed rates and cutting speeds can be employed. The ZA alloys tend to form broken chips. Most of the zinc alloys can be machined by the common procedures. The use of cutting fluids is strongly recommended.

Joinability

The common mechanical methods may be employed for joining zinc alloys including threaded fasteners, crimping, riveting, staking, and force fits. Zinc alloys are also joined by welding and soldering operations. Welding is best accomplished by using the arc-shielded inert gas process. Sheet alloys can be spot- and seam-welded. Several solder formulations are available. Since zinc is anodic to most other metals so the possibility of galvanic corrosion must be considered. Galvanic corrosion can be prevented by insulating the dissimilar metals from contacting each other and by coating the components. Exposure trials showed no significant galvanic corrosion attack after five years of atmospheric exposure on ZA alloys coupled to copper, lead and mild steel.

Finishing

Zinc alloys are buffed, painted, electroplated, anodized, and chromate and phosphate treated. Electroplated coatings include chromium, silver, gold and copper. Anodizing produces a matt olive green color.

9.3 Properties

The mechanical properties of most interest for the zinc casting alloys are listed in Table 30. The densities of these alloys vary from 5.0 to 6.6 g/cm^3 (0.18 to 0.24 $lb/in.^3$) compared to about 2.7 and 1.8 g/cm^3 for aluminum and magnesium casting alloys respectively.

References and Acknowledgments

1. Noranda Sales Corporation Ltd.
2. International Lead Zinc Research Organization, Inc.

10 ZIRCONIUM ALLOYS

10.1 Introduction

Zirconium alloys were initially developed for use in nuclear reactors because zirconium is essentially transparent to neutrons. However, zirconium's natural twin, hafnium, absorbs neutrons at 500–600 times the rate of zirconium. Therefore, the hafnium is removed from zirconium for nuclear applications. For other applications, zirconium alloys will usually contain about 4% by weight of hafnium since the natural occurring minerals contain hafnium of 1% to 4%. Many other applications for zirconium have now been found, due primarily to zirconium's resistance to many corrosive media and its good mechanical and heat transfer properties. Zirconium exhibits excellent resistance to corrosive attack in most organic and mineral acids and strong alkalis. (Detailed corrosion data may be obtained from Teledyne Wah Chang Albany.) This metal is now found in heat exchangers, stripper and drying columns, pumps, chemical reactor vessels and various piping systems. Alloys are available in sheet, strip, plate, clad plate, foil, bar, tube and pipe fitting forms.

Table 30 Typical Properties of Zinc Casting Alloys

Alloy	Y. S. 0.2% offset MPa (kSi)	Elong. % in 50 mm	Modulus MPa (psi × 10⁶)	Hardness Brinell	Impact J (ft-lb)	F.S. MPA (psi × 10³) 5 × 10⁶ cyc.	Elec cond. (% IACS)	Ther. cond (W/m·K)	CTE [10⁻⁶/°C (10⁻⁶/°F)]
Zanak 3[a]	221 (32)	10	85.5 (12.4)	82	58 (43)	47.6 (6.9)	27	113.0	27.4 (15.2)
Zamak 5[a]	228 (33)	7	85.5 (12.4)	91	65 (48)	56.5 (8.2)	26	108.9	27.4 (15.2)
Zamak 7[a]	—	13	85.5 (12.4)	80	58 (43)	46.9 (6.8)	27	113.0	27.4 (15.2)
ZA-8									
Sand cast	198 (29)	1.7	85.5 (12.4)	85	20 (15)	—	27.7	114.7	23.3 (12.9)
Perm M.	208 (30)	1.3	85.8 (12.4)	87	—	51.7 (7.5)	27.7	114.7	23.3 (12.9)
Die cast	290 (42)	8	85.5 (12.4)	103	42 (31)	103 (15)	27.7	114.7	23.3 (12.9)
ZA-12									
Sand cast	211 (31)	1.5	82.7 (12.0)	94	25 (19)	103 (15)	28.3	116.1	24.2 (13.4)
Perm M.	268 (39)	2.2	82.7 (12.0)	—	—	—	28.3	116.1	24.2 (13.4)
Die cast	320 (46)	5	82.7 (12.0)	100	29 (21)	117 (17)	28.3	116.1	24.2 (13.4)
ZA-27									
Sand cast	371 (54)	4.6	77.9 (11.3)	113	48 (35)	172 (25)	29.7	125.5	26.0 (14.4)
Perm M.	376 (55)	2.5	77.9 (11.3)	114	—	—	29.7	125.5	26.0 (14.4)
Die cast	371 (54)	2.5	77.9 (11.3)	119	13 (9)	145 (21)	29.7	125.5	26.0 (14.4)

[a]Die cast.

10.2 Properties

The properties of reactor grade zirconium, commercially pure zirconium plus hafnium and two alloys are listed in Table 31. These properties are for the cold-worked and annealed condition, the condition most often used for fabricating products. For severe forming applications, Zr-706 is available, which shows an elongation of 20% but also a lower yield strength of 50 ksi. The corrosion resistance of 706 is identical to 705. Alloy 705 shows about a 120 MPa (17 ksi) stress to rupture after 10^5 h at 425°C. Its unnotched room-temperature fatigue limit is 290 MPa (42 ksi).

Zirconium alloys can be machined by conventional methods. Slow speeds, heavy feeds and a flood coolant system using a water-soluble oil lubricant are recommended. Satisfactory results can be obtained with both cemented carbide and high-speed tools. Fine chips should not be allowed to accumulate as they can be easily ignited. Zirconium does have a tendency to gall and work harden.

Zirconium alloys are ductile and workable and can be easily bent and formed using standard shop equipment. The same techniques used to bend stainless steel tube can be applied for tube bending. It is necessary to maintain sharp dies for punching operations. Zirconium will gall and seize under sliding contact.

Zirconium is a very reactive metal and must be properly shielded for all welding operations. It is most commonly welded by the gas tungsten arc welding (GTAW) technique.

References and Acknowledgments

1. Teledyne Wah Chang Albany
2. ASM, Int. Metals Handbook Vol. 2, Tenth Edition

Table 31 Typical Properties of Zirconium Alloys

Alloy	Y. S. 0.2% offset MPa (kSi)	% Elong. in 50 mm	Tensile modulus [GPa (10^6 psi)]	Thermal conductivity (W/m·K)	Electrical resistivity (μΩm 20°C)	CTE [10^6/°C (10^{-6}/°F)]
Reactor grade Zr UNS R60001	138 (20)	25	99.3 (14.4)	22	39.7	5.89 (10.6)
Zr–702 Zr–4.5Hf UNS R60702	207 (30)	16	99.3 (14.4)	22	39.7	5.89 (10.6)
Zircaloy 2 Zr–1.4Sn UNS R60802	241 (35)	20	99.3 (14.4)	21.5	74.0	6.0 (10.8)
Zr–705 Zr–2.5Nb UNS R60705	379 (55)	16	96.5 (14.0)	22	55	6.3 (11.3)

Table 32 Typical Properties of Refractory Metals

Alloy	Y. S. 0.2% offset	Elong % in 50 mm	Tens. modulus [GPa (10^6 psi)]	Density [g/cm^3 (lb/in^3)]	Ther. Cond. (W/m·K)	CTE [10^{-6}/°C (10^{-6}/°F)]
Niobium						
recrystallized	138 (20)	25	103 (14.3)	8.57 (0.31)	52.3	7.21 (4.06)
Nb–1Zr recry.	138 (20)	20	68.9 (10)	8.59 (0.31)	41.9	7.54 (4.19)
FS-85 recry.						
Nb28Ta10W1Zr	475 (69)	22	140 (20)	10.61 (0.38)	—	9.0 (5.0)
WC-103 recry.						
Nb10Hf1Ti	262 (38)	20	90.4 (13.1)	8.87 (0.32)	—	8.10 (4.5)
Tantalum						
recrystallized	138 (20)	25	185 (26.8)	16.6 (0.60)	54.4	6.6 (3.7)
Ta–2.5W						
recrystallized	193 (28)	20	179 (26)	16.6 (0.60)	—	—
Ta–10W plate						
recrystallized	379 (55)	20	207 (30)	16.9 (0.61)	—	—
Ta–40Nb						
recrystallized	193 (28)	25	155 (22.4)	12.1 (0.04)	—	—
Molybdenum						
stress rel.	690 (100)	12	324 (47)	10.22 (0.37)	135	4.9 (2.7)
Mo–0.5Ti–0.1Zr						
TZM recry.	380 (55)	20	315 (46)	10.16 (0.37)	—	4.9 (2.7)
stress rel.	8660 (125)	10	—	—	—	—
Mo-1Ti–0.3 Zr						
TZC stress rel.	725 (105)	22	—	—	—	—
Tungsten	—	—	405 (28)	19.3 (0.70)	160	4.5 (2.5)

11 REFRACTORY METALS

The refractory metals (high-melting-point metals) include niobium, tantalum, molybdenum, tungsten and rhenium. Because of the high cost of rhenium it really does not compete with the other metals for structural components and hence will not be covered in the listing herein. The electronic applications of tantalum and tungsten will also be excluded since in this handbook we have elected to omit electronic materials. Tungsten has limited applications as a structural material. Tungsten wire has been used as a fiber reinforcement in composite materials and some tungsten crucibles have been used for recovering uranium and plutonium from spent reactor fuels. Some usage for tungsten has been found for furnace parts, heat sinks, counterweights, and for components in the aerospace industry.

Molybendum and its alloys, in addition to some electronic applications, have been used as tooling materials for forging of superalloys, glassmaking applications, pumps and vessels for handling molten zinc, vacuum furnace parts and structural aerospace components. More recently the metal injection molding process has been used to produce molybdenum gears, dies, and parts that previously could not be economically produced from the hard to machine molybdenum metal.

Tungsten and molybdenum usually require forming by powder metallurgy techniques and/or high-temperature forging and extrusion processes. Tantalum and niobium, on the other hand, differ from the other refractory metals in that they are ductile in the recrystallized condition and can thus be fabricated into many useful shapes at room temperature by the more conventional metal working processes. Tantalum has been widely used in the electronic industry as electrolytic capacitors for many years, but it is now realized that this metal has many other applications because of its corrosion and heat resistance characteristics. Tantalum and its alloys are used in heat exchangers as coils, condensers and coolers, and as tubing and linings for chemical tanks and pipes. Tantalum clad steel, copper and aluminum provide economical construction materials for chemical process equipment. Tantalum crucibles are used in high temperature vacuum operations for fusion and distillations of other materials and for vacuum furnace components. Tantalum has also been found in ballistic applications. Ta-W alloys have been used in aerospace applications where high-temperature strength is required.

Niobium metal usage includes heat exchangers, cathodic protection systems, and electronic and nuclear components. Some alloys have been used, and proposed for use, in missile and space propulsion systems.

Properties of some refractory metals are listed in Table 32.

References and Acknowledgments

1. Cabot Corporation
2. Teledyne Wah Chang Albany
3. ASM, Int. Metals Handbook, vol. 2, 10th ed.

4
Carbon and Low-Alloy Steels

G. T. Murray
California Polytechnic State University
San Luis Obispo, California

1 CLASSIFICATIONS AND SPECIFICATIONS

Steels may be classified in many ways: by composition, such as by carbon content or alloying element in case of the low-alloy steels; finishing methods, such as hot- or cold-rolled; form, such as plate, sheet, strip, rod, tubing, etc.; deoxidation process, such as killed, semikilled, capped, or rimmed steel. The most common manner of classification is according to composition with the carbon steels being grouped according to carbon content and the alloy steels, which are generally heat-treated to obtain their desired properties, according to the major alloying elements (i.e., nickel, chromium, molybdenum, and combinations of these elements).

Carbon steels are frequently subdivided into low-, medium-, or high-carbon steels as follows;

(a) low carbon: to 0.3% C and to 0.4% Mn
(b) medium carbon: 0.3% to 0.6% C and 0.6% to 1.65% Mn
(c) high carbon: 0.6% to 1.0% C and 0.3% to 0.9% Mn

Low-alloy steels are generally considered to be those containing about 2–8% total alloying element content and Mn, Si, and Cu content greater than that for the carbon steels (i.e., 1.65% Mn, 0.6% Si, and 0.6% Cu). These steels can be subdivided as follows:

(a) Low-carbon quenched and tempered steels having yield strengths in the 350–1035 MPa (50–150 ksi) range and total Ni–Cr–Mo contents to about 4%.
(b) Medium-carbon ultrahigh-strength steels having strengths of the order of 1380 MPa (200 ksi). AISI steels of the 4130 and 4340 are typical examples.
(c) Bearing steels, such as 8620 and 52100; see Chapter 6.

(d) Cr–Mo heat-resistant steels containing up to 1% Mo and 9% Cr in the AISI 9xxx series; see Chapter 9.

The high-strength low-alloy structural steels (HSLA), sometimes referred to as microalloyed steels, are a special category of low-carbon steels and must be treated as a distinct group. These are plain carbon steels with less than 0.3% C and 0.7% Mn but with microalloying additions (i.e., a fraction of a percent) of vanadium, niobium, or titanium. Although they may be quenched and tempered, they are seldom used in this condition. Yield strengths are greater than 275 MPa (40 ksi) with many as high as 690 MPA (100 ksi). The weldability of these steels is comparable or superior to the mild steels (<0.2% C). They are primarily hot-rolled and available in sheet, strip, bar, plate, and structural sections (channels, I-beams, wide-flanged beams, and special shapes). They are also available as cold-rolled sheet and as forgings. Processing methods are often proprietary but known to include controlled rolling to obtain fine austenite grains, which transform to fine ferrite grains by using accelerated cooling rates, and often by quenching or accelerated air or water cooling to cause transformation to an acicular ferrite structure. These acicular structures have high yield strengths with excellent weldability, formability, and toughness characteristics. These steels have also been normalized to refine the grain size. Another useful processing method is to anneal in the intercritical range (austenite-ferrite region of the phase diagram) and subsequently to quench to obtain a dual-phase microstructure of martensite islands dispersed in a ferrite matrix.

These steels have been grouped into the following categories:

a. As-rolled pearlitic structural steels which contain carbon and manganese and with minimum yield strengths of 275–345 MPa (40–50 ksi).
b. Microalloyed ferrite-pearlite steels
c. Weathering steels containing small amounts of copper and/or microalloying elements
d. Acicular ferrite steels
e. Dual-phase steels
f. Inclusion-shaped controlled steels
g. Hydrogen-induced cracking resistant steels with low carbon and sulfur

The AISI-SAE designations are the most commonly used alloy numbers in the United States. These two separate designations are nearly identical and will be treated herein as the same alloy numbers. Alloy numbers for this system and a brief description of each type of alloy are presented in Table 1. In addition, the Unified Numbering System (UNS), a system devised and administered by professional societies, government agencies, and trade associations, is frequently used. Foreign alloy designations are listed along with the comparable AISI numbers for selected alloys in Table 2.

Specifications are really a separate subject and are not alloy numbers or designations. A specification is a much more detailed account of the materials that will meet certain requirements. Specifications are written into a document that is used to control procurement. A specification may include many alloy numbers and manufacturing processes, chemical composition ranges, property ranges, tolerances, quality, form, surface finish, etc. Many professional groups write specifications, including government agencies (MIL and JAN), the Society of Automotive Engi-

Table 1 AISI-SAE System of Designations

Alloy number[a]	Type of steel and description
Carbon steels	
10xx	Plain carbon (Mn 1.00% max.)
11xx	Resulfurized
12xx	Resulfurized and rephosphorized
15xx	Plain carbon (max. Mn range, 1.00–1.65%)
Manganese steels	
13xx	Mn 1.75
Nickel steels	
23xx	Ni 3.50
25xx	Ni 5.00
Nickel-chromium steels	
31xx	Ni 1.25; Cr 0.65; 0.80
32xx	Ni 1.75; Cr 1.07
33xx	Ni 3.50; Cr 1.50; 1.57
34xx	Ni 3.00; Cr 0.77
Molybdenum steels	
40xx	Mo 0.20; 0.25
44xx	Mo 0.40; 0.52
Chromium-molybdenum steels	
41xx	Cr 0.50, 0.80, 0.95; Mo 0.12, 0.20, 0.25, 0.30
Nickel-chromium-molybdenum steels	
43xx	Ni 1.82; Cr 0.50, 0.80; Mo 0.25
43BVxx	Ni 1.82; Cr 0.50; Mo 0.12, 0.25; V 0.03 min.
47xx	Ni 1.05; Cr 0.45; Mo 0.20, 0.35
81xx	Ni 0.30; Cr 0.40; Mo 0.12
86xx	Ni 0.55; Cr 0.50; Mo 0.20
87xx	Ni 0.55; Cr 0.50; Mo 0.25
88xx	Ni 0.55; Cr 0.50; Mo 0.35
93xx	Ni 3.25; Cr 1.20; Mo 0.12
94xx	Ni 0.45; Cr 0.40; Mo 0.12
97xx	Ni 1.00; Cr 0.20; Mo 0.20
98xx	Ni 1.00; Cr 0.80; Mo 0.25
Nickel-molybdenum steels	
46xx	Ni 0.85, 1.82; Mo 0.20, 0.25
48xx	Ni 3.50; Mo 0.25
Chromium steels	
50xx	Cr 0.27, 0.40, 0.50, 0.65
51xx	Cr 0.80, 0.87, 0.92, 0.95, 1.00, 1.05
50xxx	Cr 0.50; C 1.00 min.
51xxx	Cr 1.02; C 1.00 min.
52xxx	Cr 1.45; C 1.00 min.
Chromium-vanadium steels	
61xx	Cr 0.60, 0.80, 0.95; V 0.10, 0.15
Tungsten-chromium steel	
72xx	W 1.75; Cr 0.75
Silicon-manganese steels	
92xx	Si 1.40, 2.00; Mn 0.65, 0.82, 0.85, Cr 0.00, 0.65
High-strength low-alloy steels	
9xx	Various SAE grades
Boron steels	
xxBxx	B denotes boron steels
Leaded steels	
xxLxx	L denotes leaded steel

[a]"xx" in the last two (or three) digits indicates the carbon content in hundredths of a percent.

Table 2 Approximate Equivalent Alloys

AISI no. USA	United Kingdom B S	Japan JIS	Germany DIN	France AFNOR NF
1020	970 050A20	G4051 S20C	1.0402 C22	A35-552 XC18
1040	970 060A40	G4051 S40C	1.0511 C40	A33-101 AF60
1060	970 060A57	G4051 S58C	1.0601 C60	A35-553 XC60
1080	970 060A78	—	1.1259 80MN4	XC80
1095	970 060A99	G4801 SUP3	1.0618 D95-2	A35-553 XC100
1137	970 212M36	4804 SUM41	—	A35-562 35MF6
1340	—	—	1.5223 42MNV7	—
4130	970 708A30	G4105 SCM2	—	A35-557 30CD4
4140	970 708A40	G4105 SCM44	1.3563 43CrMo4	A35-556 42CD4
4340	4670 818M40	G4108 SCMB23	1.6565 40NiCrMo6	—
5140	970 530A40	G4104 SCr4	1.7035 14Cr4	A35-552 42C4
5150	3100 BW2	—	1.804 60MnCrTi4	—
6150	970 735A50	G4801 SUP10	1.8159 GS-50CrV4	A35-553 50CV4
8620	970 805A20	G4052 SNCM21H	1.6522 20NiCrMo2	AF35-552 19NCDB2
8640	970 945A40	—	1.6546 40NiCrMo2 2	40NCD2
9260	970 250A58	G4801 SUP7	—	60S7
9310	970 832H13	—	1.6657 14CrNiMo13-4	16NCD13

neers (SAE), Aerospace Material Specifications (AMS), the American Society of Mechanical Engineers (ASME), and the American Society for Testing and Materials (ASTM). In the United States the latter is the most frequently used system for steels. Some of the more common ASTM specifications that apply to the carbon and low-alloy steels are presented in Table 3.

For the design engineer not interested in all of the above details Table 4 was devised to separate the carbon and alloy steels into several major categories according to processing, strength ranges, advantages and disadvantages, and some

Table 3 Selected ASTM Specifications for Carbon and Low-Alloy Steels

ASTM no.	Description
A6	Rolled steel structural plate, shapes, sheet, and bars
A20	Plate for pressure vessels
A29	Carbon and alloy bars, hot rolled and cold finished
A36	Carbon steel plates, bars, and shapes
A108	Standard quality cold-finished carbon bars
A131	Carbon & HSLA steel plates, bars, and shapes
A238	Carbon steel plates of low or intermediate strength
A242	HSLA steel, plates, bars, and shapes
A284	Carbon-Si plates for machine parts and general construction
A304	Alloy bars having hardenability requirements
A322	Hot-rolled alloy bars
A331	Cold-finished alloy bars
A434	Hot-rolled or cold-finished quenched-and-tempered bars
A440	Carbon steel plates, bars, and shapes of high strength
A441	Mn-V HSLA steel plates, bars, and shapes
A505	Hot-rolled and cold-rolled alloy steel sheet and strip
A506	Regular quality hot- and cold-rolled alloy sheet and strip
A507	Drawing quality hot- and cold-rolled alloy sheet and strip
A534	Carburizing steels for antifriction bearings
A535	Speciality quality ball and rolling bearing steel
A510	Carbon wire rods and coarse round wire
A514	Quenched & tempered alloy steel plates of high strength
A546	Cold heading quality medium carbon for hexagonal bolts
A545	Cold heading quality carbon wire for machine screws
A547	Cold heading quality alloy wire for hexagonal head bolts
A548	Cold heading quality carbon wire for tapping or sheet screws
A572	Nb-V HSLA steel plates, bars, and shapes
A549	Cold heading quality carbon steel for wood screws
A573	Carbon steel plates requiring toughness at ambient temp.
A575	Merchant quality hot-rolled carbon steel bars
A576	Special quality hot-rolled carbon steel bars
A568	Carbon and HSLA hot- and cold-rolled sheet and strip
A588	HSLA steel plates with 345 MPa (50 ksi) yield strength
A633	HSLA normalized steel plates, bars, and shapes
A656	V-Al-N and Ti-Al HSLA steel plates
A659	Commercial quality hot-rolled carbon sheet and strip
A689	Carbon and alloy steel bars for springs

Table 4 Characteristics of the Carbon and Low Alloy Steels

Alloy type	Yield strength MPa (ksi)	Advantages	Disadvantages
Hot-rolled 10xx 11xx			
As rolled	400–965 (58–140)	Low cost, good formability, weldability, availability	Low strength
Normalized	400–700 (58–101)	Low cost, good formability, weldability, availability	Low strength
Annealed	385–650 (56–94)	Low cost, good formability, weldability, availability	Low strength
Quenched & tempered	585–1600 (85–232)	Wide range of strengths and toughness, high-temp. strength	High cost, poor weldability
Cold finished			
10xx, 11xx	400–775 (58–112)	Good surface finish, machinability	Low toughness
HSLA			
As rolled	275–345 (40–50)	High strength & toughness	Usually plate and sheet
Microalloyed controlled rolled	325–550 (47–80)	High strength & toughness	High cost
Normalized	300–550 (43–80)	High strength & toughness	
Hardened & tempered	400–650 (58–94)	High strength & toughness	High cost, poor weldability
Low alloy			
Hardened	400–1850 (58–268)	Wide range of strengths, high-temp. strength	High cost, poor weldability

typical alloy numbers of each category. The design engineer is usually not interested in how the steel is manufactured, the deoxidation practice nor the microstructure, but is more involved with the required strength level, such as those specified in ASTM standards.

2 PROCESSING

Certain processing terms may help the design engineer in his or her materials selection procedure. In addition, the design engineer must be aware that the properties can vary considerably, depending on the process to which the steel has been subjected. The most pertinent processing parameters will be summarized in the following.

1. **Heat Treatment:** The properties of all carbon and alloy steels depend on the type of heat treatment given the steel. These treatments involve air cooling and quenching in water or oil, usually followed by tempering (heating) procedures. The tempering temperature is of most importance, as illustrated in Fig. 1, which shows the wide variation in hardness obtained at different tempering temperatures and carbon content. Although the plain carbon steels are widely used in the non-heat-treated hardened state (i.e., annealed and normalized), they also find considerable use in the heat-treated state. Cutlery is a good example since hard (and thus sharp) conditions can be achieved in thin sections by quenching from a suitable elevated temperature. Toughness (lack of brittleness), if needed, can be achieved by subsequent tempering operations.

2. **Hardenability:** The hardenability of a steel is a measure of the depth to which a steel can be hardened. Good hardenability is achieved by adding alloying elements such as nickel, chromium, molybdenum, and, sometimes, vanadium and tungsten. Thus, the alloy steels have a high quenched hardness and strength throughout thick sections (up to a few inches) and correspondingly are used for shafts, thick disks, load supporting rods, etc. Such improved properties are obtained at a higher cost, however, both due to the more expensive alloying elements and the additional heat treatment operations. Figure 2 illustrates the hardenability concept. The 1050 steel is a plain carbon steel containing 0.4% carbon; the 4150 steel contains the same amount of carbon but also has about 1% Cr and 0.2% Mo. These bars were cooled such that a specified velocity of water from a fountain-type orifice impinged on a single-end surface. The quenched end experiences a very high cooling rate, while at the opposite end the cooling rate approaches that of air cooling; i.e., the water has little or no effect on the cooling rate. These data show that a 4150 alloy shaft of 4 in. in diameter would have a quenched core hardness of Rockwell C 50 compared to 15 for a similar 1050 alloy shaft. Figure 3 illustrates the difference in hardenability for several alloy steels, the 4340 possessing the highest hardenability of the group and the 5140 alloy the least.

3. **Case Hardness:** Sometimes it is desirable to have a high surface hardness for good wear and fatigue resistance, but yet retain a softer, tougher, and more ductile core, all at a lower cost. A high surface hardness is realized by a number of surface treatments, which include the addition of carbon in the surface layers by gas carburizing, and by induction hardening. The former is more commonly employed for the plain carbon steels, and the latter, which involves induction heat-

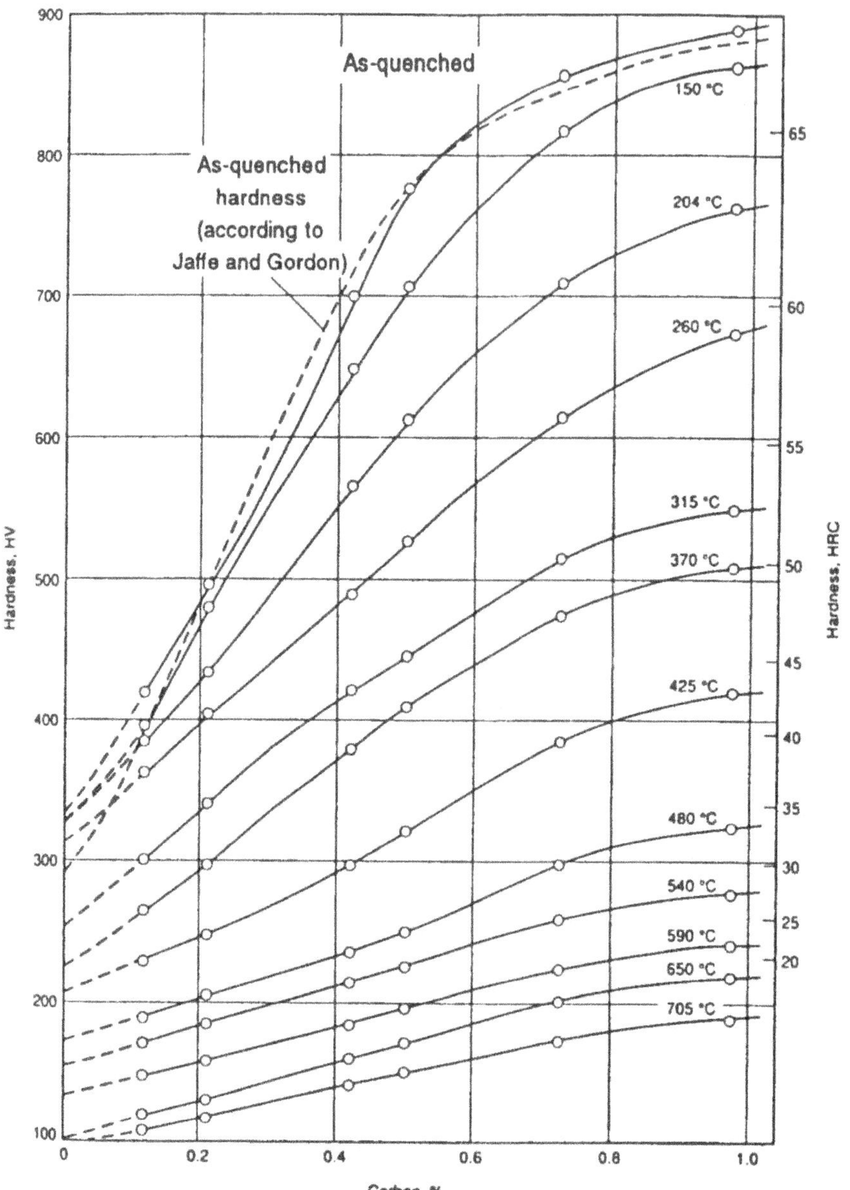

Figure 1 Tempered hardness values of plain carbon steels versus carbon content and tempering temperature. (From R. A. Grange, C. R. Hibral, and L. F. Porter, *Met. Trans. A*, 1977, p. 1775.)

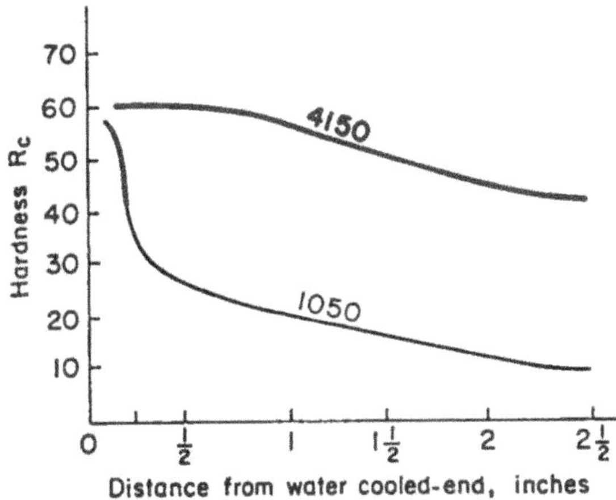

Figure 2 Hardenability curve comparisons for 1050 and 4150 steels.

ing of the surface layers only, is employed for alloy steels. In the induction hardening process the heat and quench hardening is restricted to the surface layers.

4. **Cold Finishing:** Cold finishing is frequently used to improve the surface finish, often by machining and/or cold drawing. A 12% reduction in area by cold

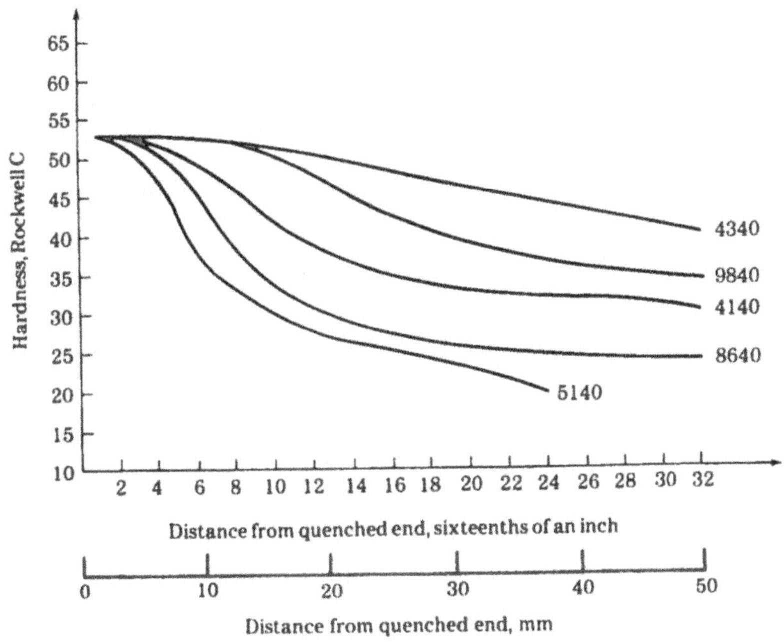

Figure 3 Comparative hardenability curves for 0.4% C alloy steels. (After H. E. McGovern, ed., *The Making and Shaping of Steels*, United States Steel Corporation, 1971, p. 1139.)

drawing can increase the yield strength by as much as 50%. Cold-finished bars fall into four classifications: cold-drawn bars, turned and polished bars, cold-drawn ground and polished bars, and turned, ground, and polished bars. Cold-drawn bars represent the largest tonnage production and are widely used in mass production of machined parts. The most commonly available steels used for cold-finished bars are

Carbon steels	1018, 1020, 1045, 1050, 1117, 1137, 1141, 1212, 1213, 12L14, 1215
Alloy steels	1335, 4037, 4130, 4140, 4340, 4615, 5120, 5140, 8620, 8640

5. **Quality Description:** Various terms are applied to steel products to imply that certain products possess characteristics that make them well suited for specific applications. This fact was evident in Table 3 in the ASTM specifications. A large number of these quality descriptions have been developed for use by process metallurgists which enables them to select a certain quality for rolling, forging, extrusion, etc. For the interested reader detailed quality descriptions appear in appropriate sections of the AISI *Steel Products Manual.* For making a "first cut" in materials selection for a particular design such detailed information is not required. Once the design engineer has decided on a certain steel category, and subsequently narrowed to a small number of potential alloys from this category, then quality becomes of importance. At this time the ASTM specifications should be utilized. For example, the quality descriptions for alloy steel plate and sheet include regular quality, drawing quality, and aircraft quality and are covered by ASTM A505. Low-carbon sheet and strip qualities are covered by ASTM A568.

3 WELDABILITY

Microstructural changes occur in the heat-affected zone (HAZ) during welding, and since microstructure and properties are related, the properties of the HAZ will differ from that of the bulk material. In general, low-carbon steels are less affected by welding since there is less chance for the brittle martensite structure to form in the HAZ.

The weldability of a steel is also related to its hardenability (i.e., alloy content). Hardenability can thus be used as a guide for weldability. Steels with high hardenability have higher hardness in the HAZ for the same cooling rate and, thus, poorer weldability. Steels with higher hardness often contain a higher fraction of martensite, which is extremely sensitive to cracking. Empirical equations have been developed to express the weldability, the carbon equivalent (CE) being a frequently employed equation. The International Institute of Welding uses the equation

$$CE = C + Mn/6 + Ni/15 + Cu/15 + Cr/5 + Mo/5 + V/5$$

where the concentration of alloying element is expressed in weight percent—thus, the higher the concentration of the alloying elements the higher the hardenability and the higher the CE. Carbon, however, is the element that most affects weldability. Steels with low CE values generally exhibit good weldability.

Part thickness affects cooling rate and, hence, weldability, since faster cooling favors martensite formation. The CE has been modified (or compensated) for thickness to produce a compensated carbon equivalent (CCE) that is related to thickness as follows:

$$\text{CCE} = \text{CE} + 0.00254t$$

where t is the thickness in millimeters.

Table 5 lists several families of steels, their CE, and relative weldabilities. These data can be used as a guide only. Strength values are averaged from selected steels in each composition group. Part thickness has not been considered here.

In many cases preheating and postweld heat treatments are required, the former to affect the temperature gradient and hence the cooling rate (faster cooling rates favor martensite formation), and the latter to restore the steel to a less brittle state. If martensite has formed, the heat treatment will temper the martensite to form a less brittle structure. In general, plain carbon steels with carbon content less than 0.22% and with Mn contents to 1.6% are readily weldable and will not require pre nor post heat treatments. A low-alloy filler metal should be used. Pre and post heating treatments are recommended for carbon steels containing more than 0.5% C. Higher-carbon-content steels require such treatments.

For HSLA steels weldability decreases as strength increases. In general, these steels have about the same weldability as plain carbon steels for the same carbon content. Microalloy contents of a few tenths of a percent can be tolerated. Quenched and tempered structural steels with carbon contents less than 0.22% have good weldability. Heat-treated low-alloy steels with CE less than 0.45 may be weldable without heat treatments. Above this value of CE heat treatment is required. These CE values include common steels such as 4130, 4140, 4340, 5140, 8640, and 300M, which should be welded in the annealed condition followed by a postweld heat treatment. Fillers of low carbon and alloy content are recommended.

The hydrogen content of all steels and the filler metals must be maintained at an extremely low level (parts per million) to avoid hydrogen-induced cracking. Vacuum-melted steels are preferred.

4 FORMABILITY

4.1 Sheet Formability

Sheet formability of carbon steels is quite good and accounts for the wide use of sheet steels. Formability is usually approximated from an analysis of the mechanical properties of the steel. However, different shapes require different forming characteristics. Almost all shapes require good ductility for good formability. The plastic strain ratio r is related to the resistance of the steel sheet to thinning during forming operations and is expressed as

$$r = \frac{\varepsilon_W}{\varepsilon_t}$$

where ε_W and ε_t are the true strains in the width and thickness directions, respectively. A deep-drawn steel can requires a high value of r, whereas stretch forming requires a low r value.

The other mechanical properties of interest to formability are yield strength, ultimate tensile strength, total elongation, yield point elongation, and strain-

Table 5 Weldability of Several Families of Steels

Steel type	Strength level MPa (ksi)	Carbon equivalent	Weldability
C − Mn	325–490 (47–71)	0.2–0.4	Good
C − Mn + Cr + Ni or V	580–600 (84–87)	0.50	Fair to good
C − Mn + Cr + Ni + Mo + Si	600–750 (87–109)	0.60	Fair with preheating
C − Mn + Cr + Mo	785–880 (114–126)	0.70	Poor

hardening exponent. The latter, which may be less familiar to the reader, is express as

$$\sigma = k\varepsilon^n$$

where σ is the stress, ε is the strain, k is a proportionality constant, and n is the strain-hardening exponent, all values being determined from the true stress–true strain curve in tension. As a rough approximation, sheet steels possessing yield strengths less than 225 MPa (33 ksi), a total elongation of 35%, a uniform elongation of 20% to 30%, and a value of n to 0.25 will be readily formable for most shapes. However, such strengths may be lower than desired. Higher-strength sheet steels (yield strengths of 345–690 MPa (50–100 ksi)) have been developed that possess moderate formability. Forming limit diagrams have been developed that show failure and safe zones on a major strain–minor strain plot. The interested reader should see the references listed at the end of this chapter.

Formability and steel composition are obviously related since the mechanical properties are a function of the composition. Low-carbon (e.g., 0.1%) steels and steels with less than 1% total alloying element content are preferred for sheet forming. Mn contents usually range from 0.15% to 0.35%. Phosphorous and sulfur are detrimental, and silicon content should be less than 0.1%.

4.1 Bulk Formability

Bulk formability (workability) is defined as the ease of forming shapes by forging, extrusion, and rolling. Bulk formability is qualitatively related to sheet formability, ductility again being the most significant parameter. Most carbon and low-alloy steels have similar bulk formabilities except for the free-machining steels that possess additions such as the sulfides. The design engineer usually purchases steel after the bulk forming operations and usually is not interested in bulk formability. Forging may be an exception since it is a convenient manufacturing process for fabricating a variety of shapes, and the design engineer may wish to specify a forgeable steel that has the ability to readily flow and fill the forging die recess. Hot forging of the carbon and low-alloy steels is not usually limited by forgeability per se. Rather, section shapes and sizes are limited by the cooling rate when the heated workpiece comes into contact with the forging dies. There are a number of forging tests that will not be discussed here (see references). The actual test is not of concern to the design engineer, who only desires to specify a forgeable steel. Figure 4 shows data that compares the forgeability of the plain carbon steels, while Fig. 5 compares the forgeability of plain carbon with low-alloy steels.

5 MACHINABILITY

Machinability depends on many factors, of which strength, hardness, and microstructure are the most important. These factors in turn depend on composition and heat treatment. In addition, certain elements may be added to enhance machinability, such as sulfur, phosphorous, selenium, and tellurium. Cold working (e.g., drawing) generally improves machinability. Plain carbon steels usually have better machinability than alloy steels of comparable carbon content. Through-hardened alloy steels can be difficult to machine. All alloying elements that increase hardenability decrease machinability. The difference in tool life for an alloy steel in the hardened and tempered condition compared to the annealed condition can be quite significant, by factors of 4 to 10 at relatively high cutting speeds. In general, alloy steels with strengths up to about 1300 MPa (194 ksi) can be machined. Not far above this level special techniques may be required.

Machinability is measured in a number of ways and often expressed in terms of a machinability rating or index. These terms are rather ambiguous and have been

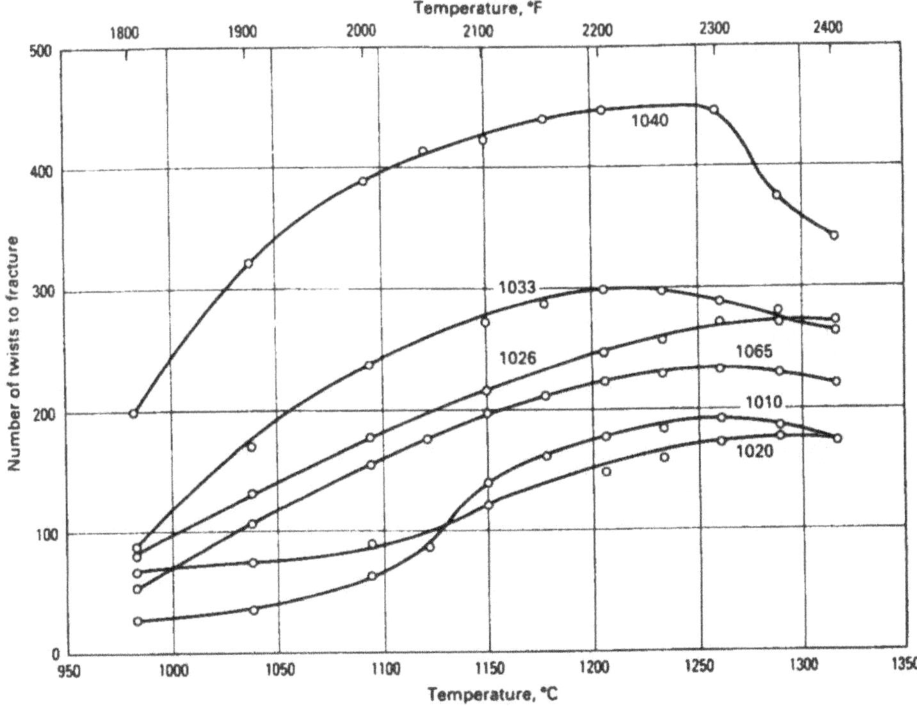

Figure 4 Forgeabilities of various carbon steels as determined using hot twist testing. (From *Evaluating the Forgeability of Steel*, 4th ed., The Timken Company, 1974.)

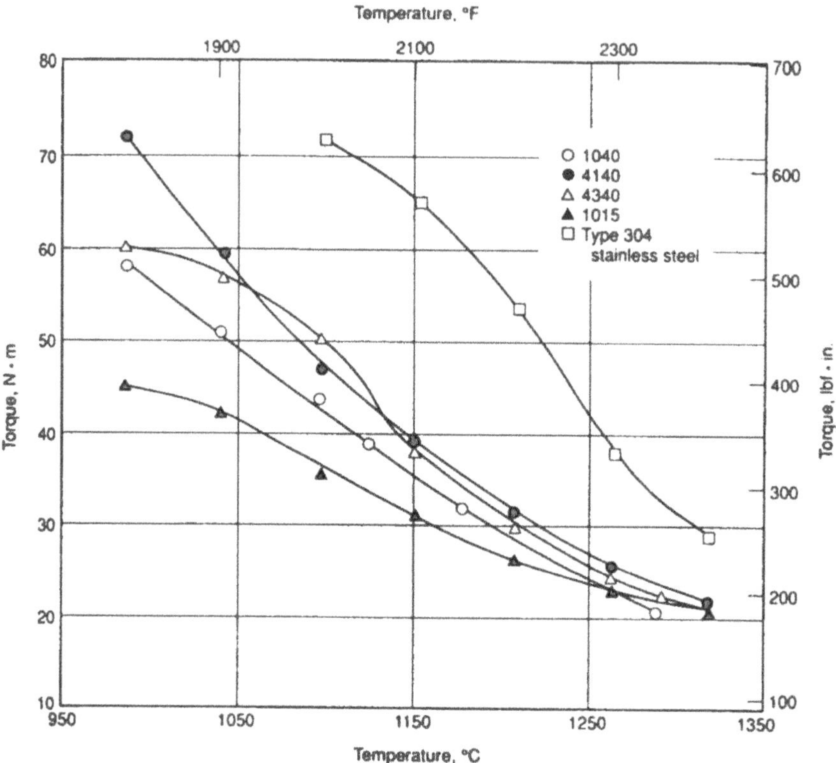

Figure 5 Deformation resistance versus temperature for various carbon and alloy steels. (From C. T. Anderson, R. W. Kimball, and F. R. Cattoir, *J. Met.*, *5*(4), 1953, pp. 525–529.)

determined differently by different investigators primarily because of the wide number of variables involved, such as cutting speed, tool shape, tool material, and type of machining operation. Yet machinability ratings are useful numbers to use as guides in the materials selection process. Tool life and cutting speed are parameters frequently used for measuring machinability. It is more practical to measure machinability as the cutting speed necessary to cause unacceptable tool wear in a specified time. The choice of tool material and tool shape then become pertinent parameters and should be standardized (see references). The type of machining operation also must be specified, the most frequently used being that of simple lathe turning. It should now become apparent that machinability is a difficult and confusing specification for the design engineer to use in the materials selection process. Nevertheless, some comparative ratings for carbon and low-alloy steels are presented in Table 6. The values given are the percentage of cutting speed of a 1212 steel at a given tool life. This steel is a free-machining resulfurized and rephosphorized carbon steel. Its rating is arbitrarily taken as 100. Machinability ratings for other resulfurized and rephosphorized steels may exceed 100. For example, 1215 with these additives has a machinability rating of 136.

Once the design engineer has made the first cut in the materials selection process, more detailed machining information is necessary. A good source for such information is the Machinability Data Center (see references).

Table 6 Machinability Ratings of Selected Carbon[a] and Low-Alloy Steels[b]

AISI no.	Hardness Brinell	Machinability rating
1212	175	100
1010	105	55
1019	131	70
1040	170	60
1050	197	45
1075	192	48
1541	207	45
4130	187–229	70
4140	187–229	65
4150	187–240	55
4340	187–240	50
5130	174–212	70
5140	179–217	65
8645	184–217	65
8740	184–217	65

[a]Carbon steels are cold-drawn from hot-rolled condition.
[b]Alloy steels are from cold finished bars of a pearlite-ferrite microstructure.
Source: Adapted from F. W. Boulger, *ASM Handbook*, Vol. 1, 1990, pp. 597, 600.

6 MECHANICAL AND PHYSICAL PROPERTIES

Typical room-temperature mechanical properties as determined in tensile tests, along with hardness numbers, of selected carbon and low-alloy steels in the normalized and annealed conditions are listed in Table 7. Fracture toughness has not been included here since in these conditions toughness is good and usually not of concern. Table 8 lists typical mechanical properties in the quenched and tempered conditions. Two tempering temperatures have been included. The reader can mentally extrapolate how the properties might alter as the tempering temperature is changed. This data should allow the design engineer to determine the group of alloys on which attention should be focused. Specific values of strength, ductility, toughness, etc., can be obtained from more detailed handbooks and vendor literature.

Thermal properties for representative alloys are presented in Table 9. The room-temperature electrical resistivities for the plain carbon steels vary from 0.14 $\mu\Omega$-m for a 1010 steel to 0.18 for a 1095 steel. The common alloy steels have room-temperature resistivities of about of 0.2 $\mu\Omega$-m.

When the component of interest will be subjected to alternating stresses, the fatigue resistance of the steel must be considered. The fatigue properties are measured and expressed in many different ways, including crack growth rates, cyclic strain-controlled tests, and cyclic stress-controlled tests. The oldest, and still probably the most widely used, method for presenting engineering fatigue data is in

Table 7 Mechanical Properties of Selected Carbon and Alloy steels in Normalized and Annealed Conditions

AISI no.	Treatment[a] (°C (°F))	Y. S. MPa (ksi)	Elong. (%)	R. A. (%)	Hardness (RB)	Izod impact (J)	Izod impact (ft-lb)
1020	N 870 (1600)	346 (50)	35.8	67.9	131	118	87
	A 870 (1600)	295 (43)	36.5	60.0	111	123	91
1040	N 900 (1650)	374 (54)	28.0	54.9	170	65	48
	A 790 (1450)	353 (51)	30.2	57.2	149	44	33
1060	N 900 (1650)	421 (61)	18.0	37.2	229	13	10
	A 790 (1450)	372 (54)	22.5	38.2	179	11	8
1080	N 900 (1650)	524 (76)	11.0	20.6	293	7	5
	A 790 (1450)	376 (56)	24.7	45.0	174	6	4.5
1095	N 900 (1650)	500 (73)	9.5	13.5	293	5.4	4
	A 790 (1450)	379 (55)	13.0	20.6	192	2.7	2
1137	N 900 (1650)	396 (58)	22.5	48.5	197	64	47
	A 790 (1450)	345 (50)	26.8	53.9	174	50	37
1340	N 870 (1600)	559 (81)	22.0	62.9	248	93	68
	A 800 (1475)	436 (63)	22.5	57.3	207	71	52
3140	N 870 (1600)	600 (87)	19.7	57.3	262	54	40
	A 815 (1500)	423 (61)	24.5	50.8	197	46	34

4130	N	870 (1600)	436 9630	25.5	59.5	197	86	64
	A	865 (1585)	360 (52)	28.2	55.6	156	62	46
4140	N	870 (1600)	655 (95)	17.7	46.8	302	23	17
	A	815 (1500)	417 (61)	25.7	56.9	197	55	40
4340	N	870 (1600)	862 (125)	12.2	36.3	363	16	12
	A	810 (1490)	472 (69)	22.0	49.9	217	51	38
5140	N	870 (1600)	472 (69)	22.7	59.2	229	38	28
	A	830 (1525)	293 (43)	28.6	57.3	167	41	30
5150	N	870 (1600)	530 (77)	20.7	58.7	255	32	23
	A	825 (1520)	357 (52)	22.0	43.7	197	25	19
6150	N	870 (1600)	616 (89)	21.8	61.0	269	36	26
	A	815 (1500)	412 (60)	23.0	48.4	197	27	20
8620	N	915 (1675)	357 (52)	26.3	59.7	183	100	74
	A	870 (1600)	385 (56)	31.3	62.1	149	112	83
8650	N	870 (1600)	688 (100)	14.0	40.4	302	14	10
	A	795 (1465)	386 (56)	22.5	46.4	212	29	22
9255	N	900 (1650)	579 (84)	19.7	43.4	269	14	10
	A	845 (1550)	486 (71)	21.7	41.1	229	8.8	6.5
9310	N	890 (1630)	571 (83)	18.8	58.1	269	119	88
	A	845 (1550)	440 (64)	63.8	17.3	241	79	58

[a]N = normalized; A = annealed; temperature (austenitizing) is that to which piece was heated.

Table 8 Mechanical Properties of Selected Carbon and Alloy Steels in the Quenched-and-Tempered Condition

AISI no.	Tempering temperature °C	°F	Y.S (MPa (ksi))	Elong. (%)	R. A. (%)	Hardness (RB)
1040	315	600	593 (86)	20	53	255
	540	1000	490 (71)	26	57	212
1060	315	600	779 (113)	13	40	321
	540	1000	669 (97)	17	45	277
1080	315	600	979 (142)	12	35	388
	540	1000	807 (117)	16	40	321
1095	315	600	813 (118)	10	30	375
	540	1000	676 (98)	15	37	321
1137	315	600	841 (122)	10	33	285
	540	1000	607 (88)	24	62	229
1340	315	600	1420 (206)	12	43	453
	540	1000	827 (120)	17	58	295
4130	315	600	1379 (200)	11	43	435
	540	1000	910 (132)	17	57	315
4140	315	600	1434 (208)	9	43	445
	540	1000	834 (121)	18	58	285
4340	315	600	1586 (230)	10	40	486
	540	1000	1076 (156)	13	51	360
5130	315	600	1407 (204)	10	46	440
	540	1000	938 (136)	15	56	305
5150	315	600	1586 (230)	6	40	475
	540	1000	1034 (150)	15	54	340
6150	315	600	1572 (228)	8	39	483
	540	1000	1069 (155)	13	50	345
8630	315	600	1392 (202)	10	42	430
	540	1000	896 (130)	17	54	310
8740	315	600	1551 (225)	11	46	495
	540	1000	1138 (165)	15	55	363
9255	315	600	1793 (260)	4	10	578
	540	1000	1103 (160)	15	32	352

the form of *S-N* curves obtained in stress-controlled tests. Here a plot of the stress *S* is made against the number of cycles to failure, *N*, the latter being plotted on a log scale. The value of stress that is plotted is the alternating stress σ_a, which may or may not be superimposed on a mean stress σ_m. Most determinations have been made in reversed bending, often of a rotary type, where the mean stress is zero. The stress at which the steel will survive an infinite number of cycles is termed the ***fatigue limit*** or ***endurance limit***. Most steels show a well-defined limit after about 10^7 cycles. Many nonferrous metals do not show such a limit, and here the fatigue properties are described in terms of stress to cause failure after a given number of cycles—i.e., the ***fatigue strength*** at this number of cycles. Fatigue data show a wide scatter in results and must be treated in a statistical manner. Often

Table 9 Thermal Properties of Selected Carbon and Low-Alloy Steels

AISI no.	Condition	Thermal at 100°C	Conductivity (W/m · K) at 200°C	CTE (10^{-6}/°C) 20–100°C	20–600°C
1020		51.0	48.9	11.7	14.4
1040		50.7	—	11.3	14.4
1060		—	46.8	11.1	14.6
1080		—	46.8	11.0	14.7
1095		46.7	—	11.4	—
1141		50.5	47.6	—	—
4130	Q and T	42.7	—	12.2	14.6
4140	Q and T	42.7	42.3	12.3	14.5
4340		—	—	12.3	14.5
5140	Q and T	44.8	43.5	—	14.6
8622		37.5	—	11.1	—

the *S-N* curves will be plotted in terms of the probability showing the curves for 95%, 50%, and 5% failure or survival. Thus, when utilizing fatigue data one must know the mean stress, since with increasing mean stress the fatigue limit or fatigue strength decreases, and the probability used for this fatigue limit. The fatigue limit is very much dependent on the surface condition of the material, a factor that accounts for the wide scatter in the data. The most reliable fatigue data have been obtained on polished specimens. However, the material in use may not be as well polished as those used in the tests, requiring that a significant safety factor be introduced. Some tests are performed on both unnotched and notched specimens, the latter presumably representing the worst conditions to be encountered in use. Also, when possible, one should have fatigue data that have been determined on the actual lot of the steel to be used in the design.

The fatigue limit of steels is strongly dependent on the type of microstructure, with the microstructures that give the higher strengths also showing the higher fatigue limits. In smooth specimens it has been shown that the fatigue limit is roughly 50% of the ultimate tensile strength up to strength levels of about 600 MPa (87 ksi), decreasing somewhat at higher stress levels. These fatigue limits can be increased by surface treatments such as carburization, nitriding, and shot peening. In fact, decarburization during heat treatment or use at elevated temperatures is often a cause for fatigue failure when alternating stresses are present. Figure 6 shows the endurance limit of some common alloy steels as a function of hardness, the latter being related to strength, which can be approximated from hardness-strength tables. Some fatigue data that have been averaged from numerous values are presented in Table 10. In most cases a range of values is given. These values were results obtained on unnotched smooth specimens at zero mean stress. Unless specified, 50% survival values are normally stated. The data are strictly for guidance in making a rough approximation of fatigue strength of these alloys.

Carbon and low-alloy steels are sometimes used above room temperature. Chapter 9 discusses steels commonly used at elevated temperatures. These steels are called *heat-resistant*. They have one thing in common in that they contain

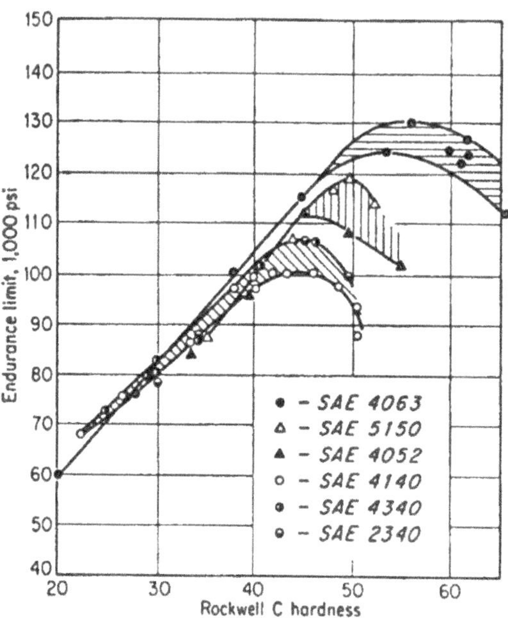

Figure 6 Fatigue limit of alloy steels as a function of Rockwell hardness. (From M. F. Garwood, H. H. Zurburg, and M. A. Erickson, *Interpretation of Tests and Correlation with Service*, ASM Int., Materials Park, Ohio, p. 12, 1951.)

Table 10 Typical Fatigue Limits of Selected Carbon and Low-Alloy Steels

		Ultimate T.S.		Fatigue limit	
Steel	Condition	MPa	(ksi)	MPa	(ksi)
1020	Normalized	390–435	(57–63)	188–195	(27–28)
1030	Normalized	450–510	(65–74)	230–240	(33–35)
1035	Q & T	—	—	214–228	(31–33)
1045	Pearlitic	725–752	(105–109)	324–331	(47–48)
1055	Q & T	710–1050	(103–152)	290–300	(42–44)
1141	Q & T	—	—	276 (40)	
1144	Pearlitic	690–718	(100–104)	331–338	(48–49)
1340	Quenched	1090–1125	(158–163)	490–511	(71–74)
4140	Q & T	—	—	379–435	(55–63)
4340	Q & T	1100 (159)		500–520	(72–5)
4340	Q & T	1400 (203)		580–610	(84–88)
4340	Q & T	1700 (246)		630–650	(91–94)
4340	Normalized	—	—	504 (73)	
8640	Q & T	—	—	538 (78)	

significant amounts of chromium for oxidation resistance. The Cr-Mo ferritic steels were included in Chapter 9, although they also fall into the carbon and low-alloy steel classification. Usually the Cr-Mo ferritic steels will be preferred to the low-carbon steels for most elevated temperature applications. The reader should refer to Chapter 9 for more creep and stress-rupture data on these Cr-Mo steels. The plain carbon–manganese steels, which were not included in Chapter 9, are frequently used in pressure vessels and in boiler tubes and pipes at elevated temperatures, primarily because of their low cost. Although for short time usage they find some application at temperatures to 540°C (1000°F), they most often are used at temperatures below about 400°C (752°F). Above these temperatures more heat-resistant alloys will be utilized.

Carbon and Cr-Mo ferritic steels are often identified by specifications listed by the American Society of Mechanical Engineers (ASME) under the ASME Boiler and Pressure Vessel Code. These specification numbers are prefixed by the letters SA, which are usually identical to the ASTM specifications that begin with the letter A. Some typical specifications and alloy compositions are listed in Table 11. Typical creep and stress rupture strengths for the carbon-manganese steels are listed as follows:

Composition	Temperature (°C (°F))	Creep strength (MPa (ksi)) for 1% elongation in 10^5 h
0.2C–1.2Mn	450 (842)	67 (9.7)
0.2C–1.2Mn	500 (932)	30 (4.35)

Composition	Temperature (°C (°F))	Stress for rupture in 10^5 h MPa (ksi)
0.1–0.2C	450 (842)	110 (16)
0.4–0.8Mn		
0.4–0.8Mn	500 (932)	55 (8)

The low-alloy steels such as 4140 and 4340, which are used where hardenability is important, are not considered as heat-resistant steels. Their strengths

Table 11 Specification Numbers of Selected Carbon-Manganese Steels for Elevated Temperature Service

ASTM spec.	ASME spec.	Product form	Nominal composition C	Nominal composition Mn	
A106	SA-106B	Seamless pipe	0.25 max.	0.3–0.9	
A106	SA-106B	Seamless pipe	0.3 max.	0.3–1.06	
A285	SA-285A	PV plate	0.9 max.	—	
A299	SA-299	PV plate	0.28 max.	0.9–1.4	0.3Si[a]
A204	SA-204A	PV plate	0.18 max.	0.9 max.	0.6Mo[a]

[a]Maximum.

Table 12 ASTM Specifications for Selected Steel Castings

Class or grade	Y.S. (MPa (ksi))	Min. elong. (%)	R.A. (%)	Composition
	ASTM A 27; carbon steel castings for general applications			
U60-30	205 (30)	22	30	0.25C; 0.75Mn; 0.80Si
70-36	250 (36)	22	30	0.35C; 0.70Mn; 0.80Si
70-40	275 (40)	22	30	0.25C; 1.20Mn; 0.80Si
	ASTM A148 carbon & alloy castings for structural applications			
80-40	275 (40)	18	30	comp. & heat treatment necessary to achieve specified properties
80-50	345 (50)	22	35	"
90-60	415 (60)	20	40	"
105-85	585 (85)	17	35	"
160-145	1000 (145)	6	12	"
210-180	1240 (180)	4	15	"
260-210L	1450 (210)	3	6	"
	ASTM A 217 alloy steel castings for pressure parts & hi-temp.			
WC4	275 (40)	20	35	0.20C; 0.50–0.80Mn; 0.60Si; 0.50–0.80Cr; 0.70–1.10Ni; 0.55Mo
WC11	345 (50)	18	45	0.18C; 0.50–0.80Mn; 1.1–1.75Cr; 0.5Ni; 0.45–0.65Mo
C12	415 (60)	18	35	0.20C; 0.35–0.65Mn; 1.0Si; 8.0–10Cr; 0.5Ni; 0.9–1.2Mo

decrease rather drastically at about 425°C (797°F). Their rupture strengths in 1000 h at 480°C (896°F) are less than 415 MPa (60 ksi).

7 CAST STEELS

Steel castings can be made from any of the same wrought carbon and alloy steels previously discussed. Steel castings are generally grouped into four categories as follows:

a. Low-carbon castings with less than 0.20% carbon
b. Medium-carbon castings with 0.02–0.50% carbon
c. High-carbon castings with more than 0.50% carbon
d. Low-alloy castings with alloy content less than 8%

For ferritic steel castings the strengths of cast, rolled, forged, and welded metal are virtually identical regardless of alloy content. The ductility of cast steels is nearly the same as that of forged, rolled, or welded steels of the same hardness. The impact values (or toughness) of wrought steels are usually those listed in the longitudinal direction and are generally higher than those for cast steels. However,

Table 13 Tensile and Fatigue Strength of Selected Cast Steels

Steel	Tensile strength (MPa (ksi)	Endurance limit Unnotched (MPa (ksi))	Endurance limit Notched (MPa (ksi))
	Normalized and tempered		
1040	648 (94)	260 (38)	193 (28)
1330	685 (93)	334 (38)	219 (32)
4135	777 (113)	353 (51)	230 (33)
4335	872 (126)	434 (63)	241 (35)
8630	762 (111)	372 (54)	228 (33)
	Quenched and tempered		
1330	843 (122)	403 (59)	257 (37)
4135	1009 (146)	423 (61)	280 (41)
4335	1160 (168)	535 (77)	332 (48)
8630	948 (137)	447 (65)	266 (39)
	Annealed		
1040	576 (84)	229 (33)	179 (26)

Values selected from *ASM Handbook*, vol. 1, 1990, p. 369, ASM Int., Materials Park, Ohio.

Table 14 Fracture Toughness of Cast Low-Alloy Steels at Room Temperature

Alloy type	Heat treatment[a]	Y. S. (MPa (ksi))	Toughness (MPa $\sqrt{m}$)
0.5Cr; 0.5Mo; 0.25V	NT	367 (53)	55
0.35C; 0.6Ni; 0.7Cr; 0.4Mo	NT	683 (99)	664
4335	QT	1090 (131)	105
Ni-Cr-Mo	QT	1207 (175)	98
4340	QT	1207 (175)	115
4325	QT	1263 (183)	90
Cr-Mo	QT	1379 (200)	84

[a]NT = normalize and tempered; QT = quenched and tempered
Values selected from *ASM Handbook*, vol. 1, 10th ed., 1990, p. 369, ASM, Int., Materials Park, Ohio.

since the transverse impact properties of wrought steels are typically 50% to 75% of those in the longitudinal direction, the impact properties of the cast steels, which are nondirectional, normally fall somewhere between the longitudinal and transverse impact properties of the same wrought steels. Tables 12–14 list some of the mechanical properties of a few selected cast steels.

REFERENCES

1. *ASM Handbook*, Vol. 1, Properties and Selection: Irons, Steels and High Performance Alloys, Materials Park, Ohio, 1990.

2. *Machining Data Handbook*, 3rd ed., Metcut Research Associates, Inc., 1980. From Machinability Data Center.
3. R. C. Rice, ed., *Fatigue Design Handbook*, Society of Automotive Engineers, 1988.
4. *Steel—Plate, Sheet, Strip, Wire*, vol. 1.03, Annual Book of ASTM Standards, American Iron and Steel Institute.
5. *Plates—Rolled Floor Plates; Carbon, High Strength Low Alloy and Alloy Steel*, AISI Steel Products Manual, American Iron and Steel Institute, 1985.
6. *Evaluating the Forgeability of Steel*, 4th ed., The Timken Company, 1974.

5
Ultrahigh Strength Steels

Thoni V. Philip and Thomas J. McCaffrey
Consultants
Reading, Pennsylvania

I INTRODUCTION

Ultrahigh strength steels are designed to be used in structural applications where very high loads are applied and often where high strength-to-weight ratios are required. In addition, they must also possess good ductility and toughness. The latter properties have been historically measured by tensile ductility, namely, percent elongation and reduction of area, and by impact resistance or energy absorbed in a notched impact test. However, in recent years the property of fracture toughness has become a significant factor in the acceptance of steels for critical high-performance structural applications.

The choice of a steel for a particular use is dependent on many factors. They constitute not only strength and toughness but also those requirements dictated by its environment such as temperature and corrosion or stress corrosion resistance. Most often the minimum life-cycle requirement is also a subject of consideration. In addition to all these, the commercial factors of availability and cost effectiveness are significant in the final selection of a steel for a particular use.

There is no universally accepted designation of an ultrahigh strength steel. The steels as described here must be of high strength, namely, a minimum 0.2% yield strength of 1380 MPa (200 ksi) at room temperature.

They must meet the criteria for the intended use such as adequate toughness and other property requirements dictated by the environment of the particular application. Criteria vary with the intended use. The aim of this chapter is to present the required properties such as strength, ductility, toughness and ancillary data. Variations in properties as affected by melting, processing and heat-treatment procedures will be discussed. The use of surface treatments such as carburizing, nitriding to obtain higher surface hardness, cold-working to increase the strength of lower strength steels, thermomechanical treatments such as ausforming, marforming, or hot-cold working to effect changes in the properties of high strength steels are not included. These treatments provide interesting possibilities in affecting final properties; however, these techniques normally must be tailored to the part being produced. If further information is of interest on any of these various techniques more detailed sources are available. The purpose of mentioning them here is to acquaint the reader with the fact that such treatments do exist.

There are numerous alloys used for high-strength applications that have slightly lower strength levels than the arbitrary limits imposed in this chapter. They would include alloys such as AISI 4330, 4335V, HP-9-4-20, Marage 200, and Hy Tuf. For information on these types of steels the reader should refer to sources such as *Aerospace Metals Handbook* [12].

Ultrahigh strength steels include a number of families of steels. However, in order to simplify the approach to understanding these alloys this chapter divides them into two categories: medium-carbon low-alloy/medium-alloy–air-hardening steels and high-fracture-toughness steels.

Generalizations

Medium-carbon low-alloy/medium-alloy steels

Older more established alloys
Less ductile (K_{Ic} 50 MPa$\sqrt{m}$ or less)
Lower price
Microstructure: acicular martensite, tempered
More susceptible to hydrogen embrittlement
Poor resistance to stress corrosion cracking
Lower fatigue life

High-fracture-toughness steels

Newer alloys (on going alloy development)
More ductile (K_{Ic} 100 MPa$\sqrt{m}$ or higher)
Higher price
Microstructure-lath martensite, aged
Less susceptible to hydrogen embrittlement
Better resistance to stress corrosion cracking
Higher fatigue life

2 MEDIUM-CARBON LOW-ALLOY STEELS/MEDIUM-ALLOY STEELS

2.1 AISI/SAE4340

In addition to arc melted-product this alloy can be purchased to higher quality levels including electroslag remelting (ESR), vacuum arc remelting (VAR), vacuum induction melting (VIM), double vacuum melted, and vacuum induction plus VAR Alloy 4340 has good strength, ductility, and toughness. It has high fatigue and creep resistance as well as deep hardenability.

The steel 4340 is usually forged at 1065–1230°C (1950–2250°F). After forging, the parts may be air-cooled or, preferably, furnace-cooled. The steel has good welding characteristics. Weld rods of the same composition should be used. Due to air hardenability, after welding, parts should be annealed or normalized and tempered.

The steel has deep hardenability. Although 4340 can be air-hardened in thin sections, it is normally oil-quenched in section sizes up to 75 mm (3 in) in diameter, above which it has to be water-quenched. The heat treatment usually consists of oil quenching from 800 to 845°C (1475–1550°F) after holding at temperature 15 min for each 25 mm (1 in) of section size (150 min minimum), and tempering at 200–650°C (400–1200°F) for the required strength level. Tempering at 260°C (500°F) (Table 2), produces the following typical properties: ultimate tensile strength, 1917 MPa (278 ksi); 0.2% yield strength, 1724 MPa (250 Ksi); elongation, 10%; reduction in area, 40%; Charpy V notch impact energy, 20ft-lb-fracture toughness K_{Ic}, 53 MPa $\sqrt{m}$ (48 ksi $\sqrt{in.}$). These properties are presented in Figs. 1–5.

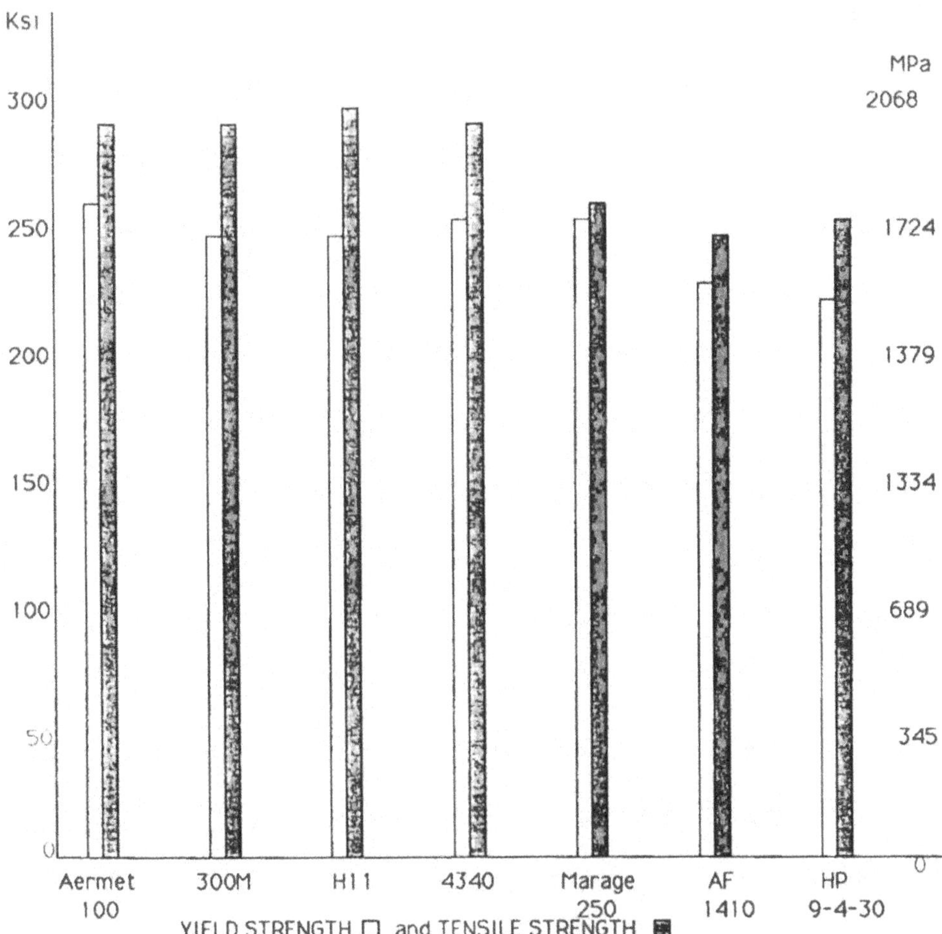

Figure 1 Yield and tensile strengths of various steels.

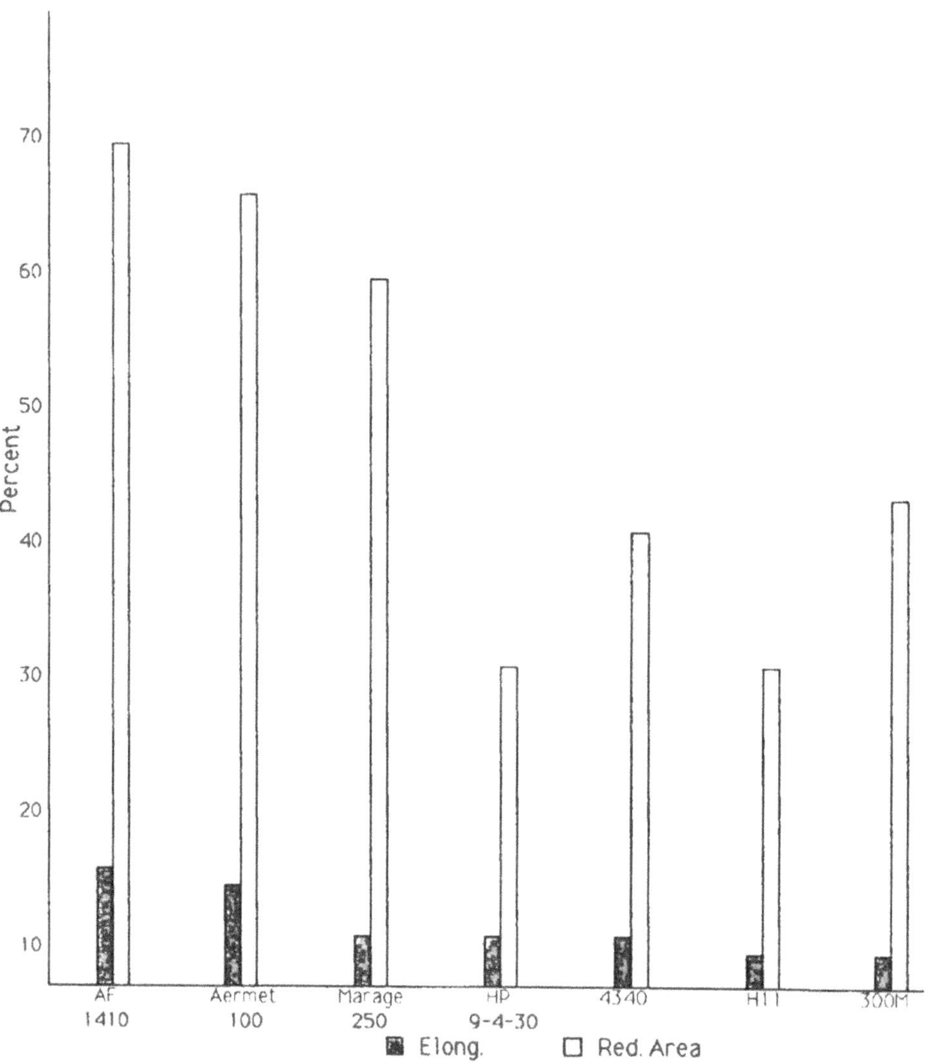

Figure 2 Ductility of various steels.

2.2 300M

300M has become a popular alloy for aircraft landing gears due to its combination of high-strength ductility and toughness and its availability in large sizes. The alloy may be considered as a modified 4340 steel containing about 1.6% Si and slightly higher carbon and molybdenum along with some vanadium. The steel has deeper hardenability and greater temper resistance than 4340. For similar strengths, 300M can be tempered at a higher temperature than 4340 steel, resulting in greater relief of quench stresses. Even at high tensile strengths of the order of 1724 to 2068 MPa (250 to 300 ksi), it has good ductility and toughness.

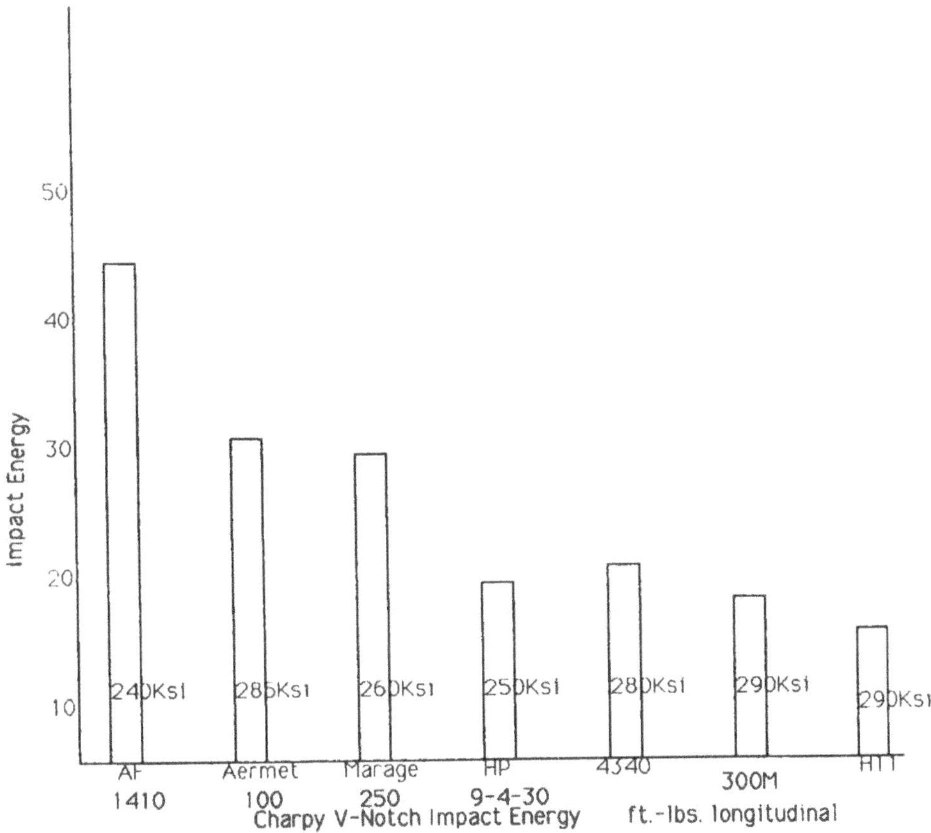

Figure 3 Impact energy of various steels.

Due to the combination of high silicon and molybdenum in its composition, this alloy is prone to decarburization and therefore adequate precautions should be taken during hot-working and heat treatment. The steel is oil-quenched or salt-quenched from 860 to 885°C (1575 to 1625°F) and when tempered at 316°C (600°F) (Table 2), typical properties obtained are: hardness, HRC 50; ultimate tensile strength, 1993 MPa (289 ksi); yield strength, 1689 MPa (245 ksi); elongation, 9.5%; reduction in area, 34%; Charpy V-notch impact, 18 ft-lb; and K_{Ic} fracture toughness, 59 MPa$\sqrt{m}$ (54 ksi$\sqrt{in}$). These values are presented in Figs. 1–5. In one RR Moore fatigue testing operation, fatigue strengths of about 800 MPa (116 ksi) and 585 MPa (85 ksi) respectively were observed on longitudinal and transverse specimens of air-melted 300M heat-treated to a tensile strength of about 2025 MPa (294 ksi). Toughness and ductility especially in the transverse direction are significantly increased by ESR or vacuum arc melting.

300M is forged at 1065 to 1095°C (1950 to 2000°F). Forging should not be continued below 925°C (1700°F). After forging, parts should be slowly cooled in a furnace or dry place. Although 300M can be readily gas or arc welded, welding

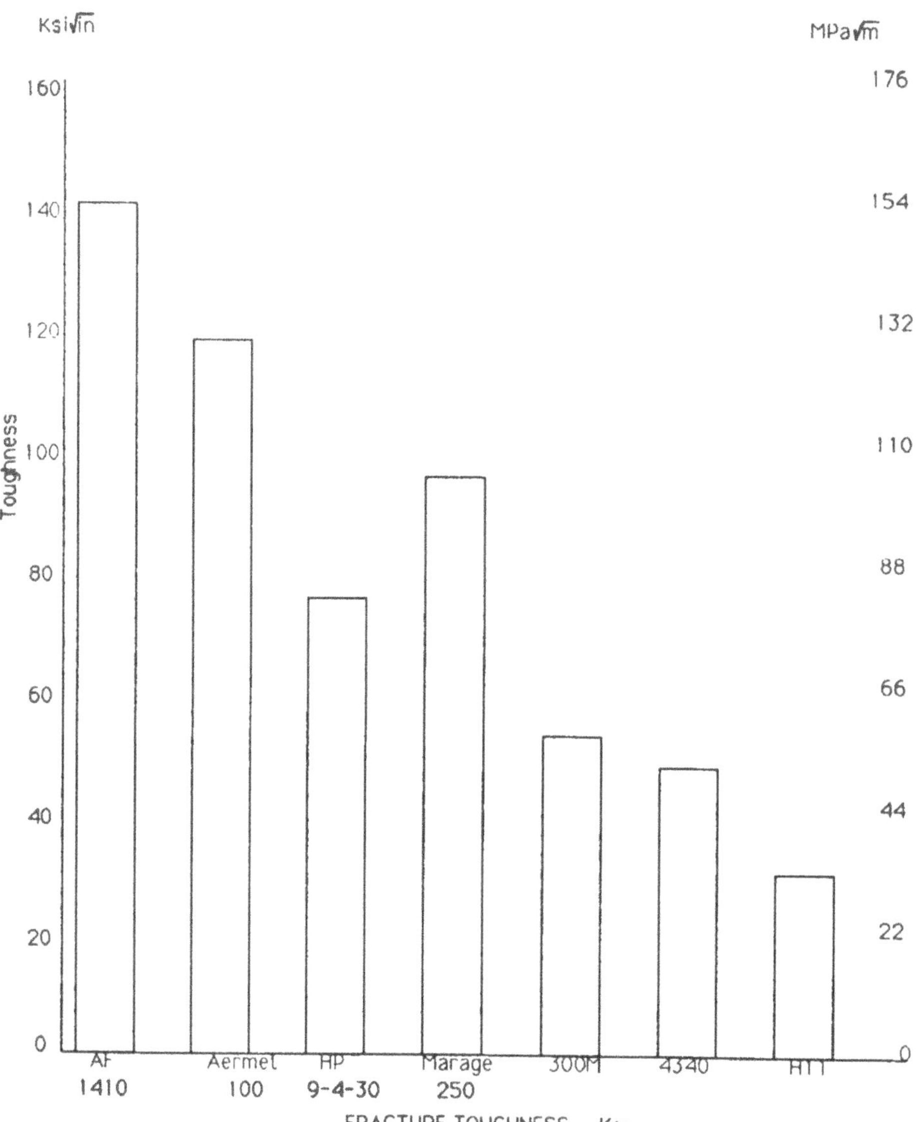

Figure 4 Fracture toughness of various steels.

is not generally recommended. If welded using welding rods of the same composition, parts should be annealed or normalized and tempered after welding.

The steel is available as bars, plates, sheet, wire, tubing, forgings and castings. Typical applications are aircraft landing gear, aircraft parts, fasteners and pressure vessels.

2.3 H11

The popular 5% chromium hot-work die steels H11 and H13 can be heat-treated to 0.2% yield strengths well above 1380 MPa (200 ksi). Therefore they both should

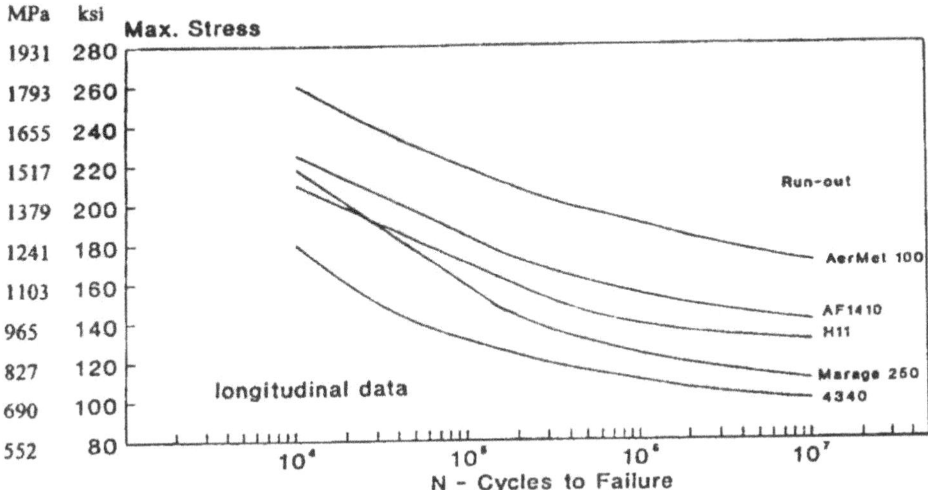

Figure 5 Stress versus cycles to failure of various steels.

be considered as ultrahigh strength steels. They are both secondary hardening steels and their heat treatments are similar. Also they have similar mechanical properties except that H13 has lower fracture toughness than H11. The fracture toughness of H11 steel in itself is low; it is lower than that of any other steel described in this chapter.

H11 is an air-hardening steel, and this results in minimal residual stress and dimensional change after hardening. Since it shows secondary hardening, the steel develops its optimum properties when tempered at temperatures above the secondary hardening peak of about 510°C (950°F). The required high tempering temperatures provide high stress relief and stabilization of properties so that parts made of this steel can be advantageously used at elevated temperatures up to about 55°C (100°F) below the original tempering temperatures. H11 steel can be used for parts requiring high levels of strength, ductility, toughness, fatigue resistance, and thermal stability in the temperature range −75 to +540°C (−100 to 1000°F). When used at elevated temperatures the surfaces of parts should be properly protected from oxidation.

For die applications the steel is usually air-melted; however, significant improvements in ductility and toughness are realized by ESR or VAR melting. H11 steel is readily forged from 1120 to 1150°C (2050 to 2100°F). Preferably it should be preheated at 790 to 815°C (1450 to 1500°F) and then heated uniformly to the forging temperature. Discontinue forging below 925°C (1700°F); reheat as often as necessary to continue forging. Being air-hardenable, after forging the part should be slowly cooled to room temperature and then fully annealed. H11 is readily weldable. Because the material is air-hardenable, all weldments should be slowly cooled to room temperature and fully annealed. To heat-treat H11, preheat to 760 to 815°C (1400 to 1400°F) and then raise the temperature to 995 to 1025°C (1825 to 1875°F), hold 20 min plus 5 min for each 25 mm (1 in) of section thickness and air-cool to room temperature. Although oil quenching from 995°C (1825°F) may be performed, air cooling is preferred since it produces less distortion. Tempering may be performed at the secondary hardening temperature of about 510°C (950°F) for maximum hardness. However, for improved ductility and toughness, parts should be tempered above this temperature. Preferably they should be double-tempered, usually 2 h plus 2 h with air cooling to room temperature after each temper. Typical properties obtained after air cooling from 1010°C (1850°F) and double-tempering 2 h each time at 540°C (1000°F), (Table 2), are: hardness, HRC 56; ultimate tensile strength, 2010 MPa (291 ksi); 0.2% yield strength, 1675 MPa (243 ksi); elongation, 9.6%; reduction in area, 30%; Charpy V-notch impact energy, 16 ft-lbs; K_{Ic} fracture toughness, 31 MPa$\sqrt{m}$ (28 ksi$\sqrt{in}$) and a fatigue strength of 896 MPa (130 ksi). These mechanical properties are shown in Figs. 1–5. Double-tempering at higher temperatures, for example, 595°C (1100°F), decrease hardness and strength but somewhat increases toughness, ductility, and fracture toughness.

H11 steel is available as bar, billet, rod, wire, plate, sheet, strip, forgings, and extrusions.

3 HIGH-FRACTURE-TOUGHNESS STEELS

The high-strength, high-fracture-toughness steels considered are commercial structural alloys capable of a minimum yield strength of 1380 MPa (200 ksi) and K_{Ic} of 100 MPa$\sqrt{m}$ (91 ksi$\sqrt{in}$). These steels also provide stress corrosion cracking resistance.

Four steels are considered here: HP9-4-30, Maraging 250, AF1410, and AerMet100. These alloys are of the Ni–Co–Fe type. They have a number of similar characteristics. They are all weldable. They all require a minimum of vacuum arc melting and control of residual elements to low levels to obtain maximum toughness.

3.1 HP-9-4-30

During the 1960s Republic Steel Corporation introduced a family of weldable steels. These steels had high fracture toughness when heat-treated to medium/high-strength levels. Except for the difference in carbon content, all four had the same composition. Only one containing 0.30% carbon HP9-4-30 became popular.

HP9-4-30 is capable of developing a tensile strength of 1520–1650 MPa (220–240 ksi) and a fracture toughness K_{Ic} of 100 MPa$\sqrt{m}$ (91 ksi$\sqrt{in}$). The steel has quite deep hardenability. Heat-treated parts can be readily welded, in the heat-treated condition. Neither postheating or postweld heat treating is necessary. After welding only a stress relief at about 540°C (1000°F) is required. There are no deleterious effects on strength or toughness. HP9-4-30 is usually electric arc plus vacuum arc remelted. Forging temperature should not exceed 1120°C (2050°F).

For heat treatment, HP9-4-30 is heated to 830–860°C (1525–1575°F) held 1 h per 25 mm (1 in) thickness or 1 h minimum, after which it is oil or water-quenched. It is then refrigerated for at least 1 h at −60 to −87°C (−75 to −125°F) and then air-warmed to room temperature and tempered at 200–600°C (400–1100°F) for the desired hardness or strength. The most popular tempering treatment is double tempering 2 h each at a temperature in the range of 540–580°C (1000–1075°F).

The minimum mechanical properties expected on double tempering at 540°C (1000°F) with a minimum hardness of HRC44 are: ultimate tensile strength, 1520 MPa (220 ksi); 0.2% yield strength, 1310 MPa (190 ksi); elongation, 10%; reduction in area, 35%; Charpy V-notch impact, 18 ft-lbs; and fracture toughness K_{Ic}, 99 MPa$\sqrt{m}$ (90 ksi$\sqrt{in}$). HP9-4-30 can be double-tempered to a higher hardness level of 49–53 HRC (Table 2). The average mechanical properties obtained are tensile strength, 1724 MPa (250 ksi); yield strength, 1413 MPa (205 ksi); elongation, 10%; reduction in area, 30%; Charpy impact energy, 18 ft-lb; and fracture toughness, 82 MPa$\sqrt{m}$ (75 ksi$\sqrt{in}$). These properties are shown in (Figs. 1–5).

HP9-4-30 is available as bar rod billet plate sheet, and strip. It has been used for aircraft structural parts, pressure vessels, rotor shafts for metal-forming equipment, drop hammer rods, and high-strength shock-absorbing automotive parts.

3.2 Maraging (250)

Maraging steels were developed by International Nickel Company in the late 1950s. The Maraging steels were the first to use a hardening reaction that does not involve carbon. In fact, carbon is an undesirable impurity. The steels are hardened by an age-hardening reaction involving precipitation of intermetallic compounds in iron-nickel lath martensite.

The heat treatment of solution treating and aging are simple treatments at temperatures from which they are aircooled. Also, these heat treatments are performed at rather low temperatures. Solution treated steels have a hardness of HRC 30–35, in which state they can be machined. The dimensional change on aging is small, an average of 0.012 to 0.025 mm (0.0005 to 0.0010) per 25 mm (1 in). The steels have excellent weldability; they produce soft ductile welds on cooling due to the very low carbon content.

Several Maraging steels are commercially available. The most popular family of maraging steels contains about 18% nickel. Their nomenclature is according to the nominal yield strength in ksi units. For example, the three most popular steels in this family, namely 18% nickel (200), 18% nickel (250), and 18% nickel (300) are normally hardened to a yield strength of 1380 MPa (200 ksi), 1724 MPa (250 ksi) and 2068 MPa (300 ksi) respectively. Only the composition of 18% nickel (250) is presented in Table 1.

Maraging steels are melted as clean and pure as possible. Most grades are air-melted and then vacuum arc remelted, or vacuum induction melted followed by vacuum arc remelted. Maraging steels used in critical aircraft and aerospace application should contain minimum residual elements namely carbon, manganese, sulfur, phosphorus, and the gases oxygen, nitrogen, and hydrogen. For this reason they are triple melted, namely arc melted, vacuum induction melted, and vacuum arc melted.

Maraging steels are prone to segregation. Therefore as part of the hot-working operation they are homogenized at temperatures of about 1260°C (2300°F) for several hours. Forging may be started at this temperature but should be stopped at 870°C (1600°F). After hot working, slow cooling from 750 to 1095°C (1380–2000°F) should be avoided because it will result in the precipitation of a thin titanium carbide film at the grain boundary which embrittles the steel.

Table 1 Compositions of the Ultrahigh Strength Steels Described in Text[a]

Designation trade name	C	Mn	Si	Cr	Ni	Mo	V	Co
4340	0.38–0.43	0.60–0.80	0.20–0.35	0.70–0.90	1.65–2.00	0.20–0.30	—	—
300M	0.40–0.46	0.65–0.90	1.45–1.80	0.70–0.95	1.65–2.00	0.30–0.45	0.05 min	—
H11	0.37–0.43	0.20–0.40	0.80–1.00	4.75–5.25	—	1.20–1.40	0.40–0.60	—
HP9-4-30[b]	0.29–0.34	0.10–0.35	0.20 max	0.90–1.10	7.0–8.0	0.90–1.10	0.06–0.12	4.25–4.75
18Ni(250)[c]	0.03 max	0.10 max	0.10 max	—	18.0–19.0	4.7–5.0	—	7.0–8.0
AF1410[d]	0.13–0.17	0.10 max	0.10 max	1.8–2.20	9.5–10.5	0.90–1.10	—	13.5–14.5
AerMet 100	0.21–0.27	0.10 max	0.10 max	2.5–3.3	11.0–12.0	1.00–1.30	—	13.3–13.5

[a]P and S contents may vary with steel-making practice. Usually these steels contain no more than 0.035 P and 0.040 S.
[b]HP9-4-30 is specified to have 0.10 max P and 0.10 max S. Some producers use narrower ranges.
[c]18Ni(250) is specified to contain 0.10 max P, 0.10 max S, 0.10 Al, 0.40 Ti, 0.003 B, 0.02 Zr, and 0.05 Ca.
[d]AF1410 is specified to have 0.008 P and 0.005 S.
[e]AerMet 100 is specified to have 0.003 max P and 0.002 max S as well as 0.01 max Al, 0.001 max N, 0.001 max O and 0.01 max Ti.

Maraging steels are usually solution-treated at 820°C (1500°F) for 1 h for each 25 mm (1 in) of section thickness and are then cooled to room temperature. Care should be taken to ensure that the part is cooled to room temperature before aging. This allows all the austenite to transform to martensite. Aging is normally performed at 480°C (900°F) for 3 to 6 h followed by air cooling to room temperature. This results in a hardness of HRC 50–54, (Table 2).

18% Ni (250) Maraging steel has the combination of high strength and toughness that fits the criterion of ultrahigh strength steels defined in this chapter. The mechanical properties of this alloy are ultimate tensile strength, 1758 MPa (255 ksi); 0.2% yield strength, 1703 Mpa (247 ksi); elongation, 11%; reduction in area, 57%; Charpy V-notch impact strength, 27 ft-lb, and fracture toughness K_{Ic}, 106 MPa$\sqrt{m}$ (96 ksi$\sqrt{in}$). These properties are shown in the figures.

Maraging steels are available as wire, bar, billets, forgings, strip, tubing and pipe. They have been used in a wide variety of applications such as missile cases, aircraft forgings, structural parts, cannon recoil springs, Belleville washers, transmission shaft, fans in commercial jet engines, couplings for hydraulic hoses, bolts and fasteners, punches and dies. There have been extensive uses in two general types of application, namely aircraft and aerospace, where the superior mechanical properties and weldability are important, and tooling applications where the mechanical properties and ease of fabrication, especially the lack of distortion during aging, are important.

3.3 AF1410

The U.S. Navy sponsored a program to produce advanced submarine hull steels. This program resulted in the HY series of low-carbon steels of the Fe–Ni–Co type alloys. These alloys were relatively low in strength but they had significantly high stress corrosion cracking resistance and very high toughness. The Air Force sponsored additional work, which resulted in AF1410. This alloy contains higher cobalt and carbon, resulting in an ultimate tensile strength of 1615 MPa (235 ksi). K_{Ic} was maintained at 154 MPa$\sqrt{m}$ (140 ksi$\sqrt{in}$) with high stress corrosion cracking resistance. The major application for this alloy is the arrester hook for the F/A A&B. The microstructure is Fe-Ni lath martensite. Melting practice requires that impurity elements remain low to insure good fracture toughness. The currently preferred melting practice is vacuum induction followed by vacuum arc remelt (VIM/VAR). Welding is done by continuous wave gas tungsten arc welding (GTAW) using high-purity wire. Oxygen contamination must be avoided.

AF1410 is prone to decarburization when heated for hot working and heat treatment. Appropriate protection should be used. In initial breakdown a maximum starting temperature of 1120°C (2050°F) may be used. Forging must employ 40% reduction below 899°C (1650°F) to maximize finish properties. Forgings should be air-cooled to room temperature, normalized and overaged/annealed. To normalize, heat between 880°C (1620°F) and 900°C (1650°F). Air-cool to room temperature. AF1410 is annealed by overaging at 675°C (1250°F) for a minimum of 5 h and air-cooled.

Originally a double austenitize 900°C (1650°F) plus 815°C (1500°F) treatment was used in the heat treatment of AF1410, but currently a single austenitize 815°C (1500°F) is usually applied followed by water, oil, or air cooling depending on section size. A deep freeze of −75°C (−100°F) is recommended followed by aging at 510°C (950°F) for 5 h min and air cooling, (Table 2). This treatment has produced properties of 1750 MPa (254 ksi) ultimate tensile strength, 0.2% yield strength of 1545 MPa (224 ksi), elongation of 16%, reduction in area of 69%, Charpy V-notch 48 ft-lb and K_{Ic} 154 MPa$\sqrt{m}$ (140 ksi$\sqrt{in}$). These values are presented in the figures.

Unfortunately AF1410 is not as readily available as the other alloys. However, it can be obtained in heat lot quantities in all the normal mill forms.

3.4 AerMet 100

Carpenter Technology Corporation developed AerMet 100 to meet the demands for a 280 ksi tensile strength alloy with a K_{Ic} of 100 ksi$\sqrt{in}$ fracture toughness for the landing gear for Navy aircraft landing on carrier decks. The alloy must also have superior stress corrosion resistance compared to the currently used 300M. The microstructure of AerMet 100 is Fe-Ni lath martensite, similar to AF1410. The first application for AerMet 100 is the landing gear and arrester hook for the F/A18C&D. AerMet 100 is only supplied in the VIM/VAR melted condition, as even tighter control of impurities is required to maintain high fracture toughness. AerMet 100 can be welded with electron beam techniques.

To forge AerMet 100, the initial breakdown may use a maximum starting temperature 1232°C (2250°F). Forging must finish below 899°C (1650°F) to optimize the final heat-treated properties. Forgings should be air-cooled to room temperature, annealed/overaged 677°C (1250°F) 16 h air-cooled, and normalized 899°C (1650°F) for 1 h and air-cooled. To obtain maximum softening for machining a second overage/anneal should be carried out. Proper precautions are necessary to prevent decarburization of AerMet 100.

Heat treatment of AerMet 100 is critical. The solution treatment temperature 885°C ± 14°C (1625°F ± 25°F) must be monitored by thermocouple in the load.

Table 2 Heat Treatment of Steels for Comparison of Mechanical Properties in Figs. 1–5

Steel	Heat treatment
AISI/SAE4340	Oil-quenched from 845°C (1550°F) and tempered at 260°C (500°F)
300M	Oil-quenched from 860°C (1575°F) and tempered at 316°C (600°F)
H11	Air-cooled from 1010°C (1850°F) and double-tempered, 2 + 2 h at 540°C (1000°F)
HP 9-4-30	Oil-quenched from 845°C (1550°F) refrigerated at −73°C (−100°F) and double-tempered at 205°C (400°F)
18 Ni (250)	Air-cooled from 820°C (1500°F) and aged for 3 h at 482°C (900°F), air-cooled
AF 1410	Oil-quenched from 899°C (1650°F), oil-quenched from 830°C (1525°F), refrigerated at −73°C (−100°F), air-warmed and aged at 510°C (950°F) for 5 h, air-cooled.
AerMet 100 480°C (900°F), air-cooled	Air-cooled from 885°C (1625°F), refrigerated at −73°C (100°F)

Proper quenching requires the alloy to be cooled from the solution treating temperature to 66°C (150°F) in 1 to 2 h. A thermocouple in the hottest part of the load must be monitored to insure a max 2 h cool to 66°C. To obtain full toughness after cooling to room temperature, cool to −73°C (100°F) and hold for 1 h. Then air-warm to room temperature. The standard aging to obtain a minimum 280 ksi UTS, 100 K_{Ic} is 482°C ± 6°C (900°F ± 10°F). The following properties, also shown in Figs. 1–5, resulted from this heat treatment, (Table 2); ultimate tensile strength, 1965 MPa (285 ksi); 0.2% yield strength, 1724 MPa (250 ksi); elongation, 14%; reduction in area, 65%; Charpy V-notch, 30 ft-lb; K_{Ic}, 126 MPa$\sqrt{m}$ (115 ksi$\sqrt{in}$). A 468°C (875°F) age will provide a UTS of approximately 300 ksi with lower ductility. Do not age below 468°C (875°F), as very low ductility will result. Applications include F22 landing gear, halfshafts for racecars, fasteners, railroad frog switches, and tooling. AerMet 100 is available as bar, billet, wire, strip, sheet, plate and hollow bar.

REFERENCES

The data presented in this chapter were extracted from numerous sources, including many commercial brochures and data sheets, various issues of Alloy Digest, Republic Alloy Steels, Aerospace Structural Metals Handbook (AFML-TR-68-115, Army Materials and Mechanical Research Center, Watertown MA, as well as that from the following references.

1. T. V. Philip and T. J. McCaffrey, Ultrahigh strength steel, *Metals Handbook*, vol. 1, 10th ed., ASM International, 1990.
2. A. M. Hall, Sr., *Introduction to Today's Ultrahigh Strength Structural Steels*, STP 498, American Society of Testing and Materials, 1971.
3. G. Sachs, *Survey of Low Alloy Aircraft Steels Heat Treated to High Strength Levels*, WADC-TR 53-254, Part 4, Wright Air Development Center, 1953.
4. J. J. Houser and M. G. H. Wells, *The Effect of Inclusions on Fatigue Properties in High Strength Steels, Mechanical Working and Steel Processing* vol. 14, Iron and Steel Society of the American Institute of Mining, Metallurgical, and Petroleum Engineers, 1976.
5. *Plane Strain Fracture Toughness (KIc) Data Handbook for Metals*, Army Materials and Mechanics Research Center, 1993.
6. K. Firth and R. D. Gargood, *Fractography and Fracture Toughness of 5% Cr-Mo-V Ultrahigh Strength Steels*, Publ. 120, Iron and Steel Institute, 1970.
7. T. V. Philip, ESR: a means of improving transverse mechanical properties in tool and die steels, *Met. Technol, vol. 2*, 554, (1975).
8. Kurt Rohrbach and Michael Smith, Maraging steels, *Metals Handbook*, vol. 1, 10th ed., ASM International, 1990.
9. S. Floreen, *Metal. Rev. 13*, no. 126 (1968).
10. M. L. Schmidt, *Maraging Steels: Recent Developments and Applications*, The Minerals, Metals and Materials Society, 1988.
11. R. M. Hemphill and D. E. Wert, *High Strength High Fracture Toughness Structural Alloy*, U.S. Patent No. 5,087,415 (1992).
12. *Aerospace Structural Metals Handbook*, 1991 edition, Metals and Ceramics Information Center, Battelle, Columbus, Ohio, 1990, section FeUH, several codes.
13. P. M. Novotny, An aging study of AerMet 100 alloy, Proc. Gilbert R. Speich Symp. on the Fundamentals of Aging and Tempering.

6
Tool and Bearing Steels

W. E. Burd
Consultant
Bernville, Pennsylvania

1 TOOL STEELS

The general title of tool steels covers steels used for cutting, forming, or shaping other materials. A listing of a variety of tool steels is shown in Table 1. The table includes the primary alloying elements of the steels as well as the most popular applications of each. In general, the properties of these steels include high hardness and abrasion resistance when they are heat-treated in the prescribed manner. They obtain these properties through the use of increased amounts of alloying elements, including carbon, when compared to the more common construction alloys. This increased alloy level significantly increases the difficulty and cost of producing these steels and therefore makes them more expensive to purchase. It should be pointed out, however, that the cost of the raw material is only a small portion of the cost of a finished complicated tool when it is put into service.

Tool steels are frequently classified by an AISI system as follows:

Group	Description	Symbol
1	Water-hardening	W
2	Shock-resistant	S
3	Cold-work steels	
	oil-hardening	O
	air-hardening	A
4	Hot-work steels	H
	chromium type	H1–H19
	tungsten type	H20–H39
	Molybdenum type	H40–H59
5	High-speed steels	
	tungsten type	T
	Molybdenum type	M
6	Special-purpose; low alloy	L
7	Mold steel	P

Table 1 Applications of Selected Tool Steels

Type	Composition	Applications
W 1	0.6–1.4C	Cold chisels, blanking tools, rivet sets, shear blades, punches, reamers, drills, mandrels, taps, dies
W 2	0.6–1.4, 0.25V	Similar to W 1
W 5	1.1C, 0.5Cr	Heavy stamping and drawing dies, mandrels
S 2	0.5C, 1.0Si, 0.5Mo	Forming tools, pipe cutters, rivet sets, shear blades, spindels, stamps
S 5	0.5C, 2.0Si, 0.8Mn, 0.4Mo	Chisels, shear blades, bending dies, screwdriver bits
S 7	0.5C, 0.7Mn, 0.30S, 3.25Cr 1.4Mo	High-impact chisels, rivet sets
O 1	0.9C, 1.0Mn, 0.5Cr, 0.5W	Blanking dies, trim dies, knives, taps, reamers, bushings, punches
O 2	0.9C, 1.6Mn	Blanking, stamping, trimming and threading dies, gauges, cutters, saws
O 6	1.5C, 1.0Si, 0.8Mn	Blanking dies, cold-forming rollers, forming punches, piercing dies, taps, wear plates, arbors, tool shanks
D 2	1.5C, 12Cr, 0.8Mo, 0.9V	A variety of cold tools, dies, punches and cutters
D 3	2.1C, 12Cr 0.5Ni	Cold-forming dies, drawing dies, forming tools, swaging dies
A 2	0.7C, 5.0Cr, 1.0Mo	Thread-rolling dies, extrusion dies, coining dies, embossing dies, stamping dies
A 6	0.7C, 2.0Mn, 1.3Mo, 1.0Cr	Punches, shear blades, coining dies, master hobs, retaining rings, spindles
H 11	0.4C, 0.9Si, 5.0Cr, 1.35Mo	Hot-extrusion dies, hot-heading dies, hot punches, forging dies
H 13	0.35C, 5.0Cr, 1.5Mo, 1.0V	Die casting dies, aluminum and brass extrusion dies, press liners forging dies, mandrels
H 42	0.6C, 6.0W, 5.0Mo, 4.0Cr, 2.0V	Hot upsetting dies, header dies, hot-extrusion dies, hot forming dies
T 1	0.7C, 18W, 4Cr, 1V	Drills, taps, reamers, broaches, chasers, cutters taps
M 1	0.8C, 8.5Mo, 4.0Cr, 1.5W, 1.1V	Drills, taps, end mills, milling cutters, saws, broaches, routers woodworking
M 4	1.3C, 5.5W, 4.5Cr, 4.5Mo, 4.0V	Broaches, reamers, lathe tools, checking tools, swaging dies
M 42	1.1C, 9.5Mo, 8Co, 3.7Cr, 1.5W, 1.1V	End mills, reamers, forming cutters, lathe and planer tools, hobs, twist drill

A brief description of the groups of alloys follows.

Water-Hardening Steels. These are inexpensive high-carbon steels that are hardened to shallow depths (low hardenability). They are somewhat prone to cracking and distortion during heat treatment. Small amounts of chromium and vanadium are added in some W-type steels to improve hardenability and wear resistance.

Shock-Resistance Steels. These steels are used for applications where repetitive impact stresses are encountered, and hence their carbon content is limited to about 0.5%. S1 and S2 steels are water-quenched, while S5, a popular low-price general-purpose tool steel, has a medium hardenability and is hardened by oil quenching. S7 has the ability to fully harden when quenched in still air. The shock-resisting properties (toughness) of most of these alloys are developed through the use of silicon as an alloying element.

Cold-Work Tool Steels. These steels are used for cold-work and for die applications where toughness and resistance to wear are both important. They are limited to applications that do not involve prolonged heating above room temperature. These include both air-hardening (A type) and oil-hardening (0 type) steels plus the high-carbon and high-chromium D type. The latter were originally developed for high-speed cutting operations but were later replaced by the better high-speed steels. The D type have now been found to be useful as cold-work die steels, hence the D symbol. D3 is an oil-hardening steel, while most other D types may be air-hardened.

Hot-Work Tool Steels. These steels contain 3% to 5% of chromium, tungsten or molybdenum in order to resist softening at elevated temperatures. Most of the chromium type are air-hardened. The tungsten types may also be air-hardened, but they are usually quenched in oil in order to minimize scaling. The only molybdenum hot-work steels of significance are the H42 and H43 steels, which are less expensive than the tungsten-type steels.

High-Speed Steels. These steels are named because of their ability to machine other hard materials at relatively high rates of speed. They are very complex iron-based alloys with high-carbon levels. Other alloying elements include chromium, vanadium, molybdenum, tungsten or cobalt in various combinations. The alloying elements form carbides that confer very excellent wear resistance and also resistance to softening at elevated temperatures. These carbides also impart a rather unique characteristic called secondary hardness. The term expresses the steel's ability to obtain a hardness as high or higher than the original as quenched hardness when they are tempered at a critical temperature. For example M42 high speed will show an as-quenched hardness of about 66HRC, but when tempered at 1000°F (538°C) will obtain a hardness of 69HRC.

Powder Metallurgy Tool Steels. As the alloying elements added to an iron-based material become very high and the structure becomes complex, it is extremely difficult, and sometimes impossible, to obtain a satisfactory product through the conventional processes. In view of this, the powder process was developed. The required alloy is prepared by blending the necessary components in powder form. It is then compacted through various methods until the needed size and form are obtained. A variety of high-speed steels and some highly wear-resistant tool steels are available through this method of manufacture.

All tool steels require a heat treatment in order to develop their optimum properties. A critical component of that heat treatment is the rate at which the steel is cooled from the hardening temperature. The fastest common medium is water into which salt has been added. A slower rate is obtained by quenching into an oil bath, while an even slower rate is through cooling in air. The slower rates of oil and air permit hardening with reduced tendency toward cracking and distortion. The increased alloy levels needed to allow the slower cooling of course helps increase the cost of these materials. Table 2 shows the quenchant required to fully harden the alloys listed in Table 1.

2 BEARING STEELS

A bearing is a component of a machine that transmits loads from one segment of the machine to another. A journal bearing operates such that the load is perpendicular to a shaft that is rotating relative to other segments. Two specific types of journal bearings are roller and ball bearings. In these cases the primary components (balls or rollers) are contained in collars and/or cages and produce a rolling contact as opposed to sliding. They are commonly referred to as antifriction bearings.

There are a wide variety of requirements for bearings throughout industry and therefore a wide variety of bearing materials to fulfill them. These include ferrous metals, nonferrous metals (Al, bronze, etc.). This discussion will concentrate on

Table 2 Quenching Medium

Alloy	Quenchant
W 1	water
W 2	water
W 5	water
S 2	water
S 5	oil
S 7	oil
O 1	oil
O 2	water
O 6	oil
D 2	air
D 3	oil
A 2	air
A 6	air
H 11	air
H 12	air
H 13	air
T 1	oil or salt
M 1	oil or salt
M 2	oil or salt
M 4	oil or salt
M 42	oil or salt

the ferrous metals. The highly loaded rolling element bearings are frequently carburized, which adds a hard case to a soft low-carbon steel.

Certain conditions must be considered when selecting a bearing steel. These include (a) type of load, (b) intensity of load, (c) speed of parts involved, (d) temperature of operation, (e) chemical environment, and (f) degree of lubrication. The properties exhibited by bearing steels must be closely coordinated with their application. In general, they must be able to be hardened to a high level, have good toughness, excellent wear resistance, high yield strengths, good temper resistance, i.e., the tempering temperature dictates the maximum working temperature, and a good rolling contact fatigue resistance.

Fatigue resistance is probably one of the most important properties of steels used in journal bearing applications. Fatigue cracks are caused by the rapid change in stress level in journal bearings from maximum load to almost zero load in a cyclical manner. The use of a very tough low carbon (0.10%) base steel with a case-hardened surface is one of the successful methods used to reduce the tendency for surface fatigue crack propagation. AISI 52100 (1.0% C) is a good journal bearing steel that does not require a surface treatment and performs well in many room-temperature applications. It can be hardened throughout a 1-in thickness to a hardness of about 60 HRC when tempered at 204°C (400°F). Larger sizes require some special hardening methods but can be hardened in the exterior regions without carburization treatments while maintaining a tough core.

For turbine engine applications AISI M50 (0.8C, 4.1Cr, 1.0V, 4.5Mo) has been adopted. This alloy is a modified high-speed steel which will harden to 60 HRC when tempered at 1100°F (593°C). It exhibits excellent resistance to softening at working temperatures of 700° for extended periods of time. This alloy has good oxidation resistance and a high compressive strength. It has good to excellent rolling contact fatigue resistance, the excellent state resulting when special melting techniques are used to minimize the presence of internal defects. These melting techniques may, and frequently are, used for many bearing steels. Bearing steel cleanliness is most often rated by microscopic techniques defined in ASTM A 295 for high-carbon steel and in A 534 for carburizing steels. These, as well as all other specifications require the testing for the presence of nonmetallic inclusions in bearing steels. These internal defects have been shown to have a negative impact on the fatigue resistance of the steel. Special processing techniques have been devised to significantly reduce the levels of these inclusions and, therefore, increase the life span of the bearings. These techniques include double melting under a vacuum. They, of course, increase the cost of the raw materials used in bearing manufacture.

For applications requiring more oxidation resistance, alloy AISI 440C (1.0C, 17.0Cr) may be specified. This steel can be hardened to 60 HRC when tempered at 300°F (150°C). An alloy that exhibits corrosion resistance similar to 440C but will operate at higher temperatures is CRB7 (1.10C, 14Cr, 2Mo, 1V, 0.25Co). This steel will maintain a hardness of about 60 HRC when tempered at 900°F (480°C).

The characteristics of selected bearing steels are summarized in Table 3. Typical comparative costs of various tool steels are listed in Table 4.

Table 3 Selected Bearing Steels

Steel	Nominal composition	Characteristics
8620	0.2C, 0.8Mo, 0.5Ni, 0.5Cr	Low-carbon core-carburizing steel
52100	1.0C, 1.5Cr, 0.3Mn	High-carbon heat-treated steel
ASTM A485	1.0C, 1.3Cr, 0.8Mn 0.2Si	High-carbon heat-treated steel
M50	0.9C, 4.1Cr, 4.2Mo 1.0V	Elevated temperature, 900°F (480°C)
440C	1.0C, 17.0Cr, 0.5Mo, 0.4Mn	Corrosion-resistant applications
440C mod.	1.0C, 14.0Cr, 4.0Mo	Corrosion resistance plus good temper resistance
CRB-7	1.1C, 14.0Cr, 2.0Mo, 1.0V	Corrosion resistance at elevated temperatures, plus high rolling contact fatigue resistance

Table 4 Typical Tool Steel Cost—American Metal Market August, 1995

Steel	Form	Cost ($/lb)
High speed	Round	3.50–8.83
"	Flat	8.99–12.32
Cold work	Round	2.11–2.70
"	Flat	2.41–3.56
Hot work	Round	1.63–5.56
"	Flat	2.49–3.05
S-7	—	1.78
A-2	—	1.77
H-13	—	2.05
D-2	—	2.66
O-2	—	1.52

REFERENCES

1. *ASM Metals Handbook, vol. I,* 8th Edition.
2. Technical Data Sheet, "Alloy Steels," Carpenter Technology Corp., p.5.

7
Stainless Steels

Paul T. Lovejoy
Allegheny Ludlum Corporation
Brackenridge, Pennsylvania

1 INTRODUCTION

Stainless steels combine a metal's fabrication and service advantages with good corrosion resistance to a variety of atmospheres and environments. The alloys resist structural deterioration in many locations and are used in many applications such as food and pharmaceutical equipment where even small amounts of product contamination would be a severe concern. These alloys are widely produced in all the usual shapes and forms.

On the other hand, stainless steels are not inert, or inherently noble materials, but are metals that "stain less" than many other options, due to the natural formation of a passive film separating the metal from the environment. The passive film, and hence the alloy, may react unfavorably to seemingly small details of fabrication, construction, or to seemingly minor excursions of service or cleaning conditions. The successful and economical use of stainless steel come from a good engineering background and knowledge of these details. Many successful examples exist in ordinary experience. A good source of useful information is the *Designers Handbook* published by the Speciality Steel Industry of North America (SSINA) [1].

The corrosion resistance of stainless steels results from the alloying of iron with chromium. After 10% chromium is added to iron one finds a remarkable improvement in atmospheric corrosion resistance compared to iron. However, the 10% chromium steel may discolor in service, and at this basic chromium level, the steel may not "hold water"; i.e., it is not suitable as a tank. By the 18% chromium level the metal will generally remain free of discoloration, and "many waters" are contained.

At the 18% chromium level there are niche austenitic alloys that will machine better (203, 303), are less (corrosion-wise) affected by the heat of welding (304L), are able to survive service at normally sensitizing temperatures (321, 347), or can be stretched or drawn more economically into complex parts (201, 301, 305).

Austenitic alloys beyond the 18% chromium level, branch to either better oxidation resistance (more chromium 309, 310), or have resistance against higher levels of chlorides and other corrosives (more molybdenum, sometimes in combination with nitrogen (316, 317, 317LN). Seawater requires special alloys. AL-6XN,® UNS NO8367 is one example frequently used for this purpose.

Other families of stainless steel start at the 10% chromium level and include the ferritic, martensitic, precipitation hardening (PH), and highly alloyed duplex grades. Within metallurgical limits, there are variations in strength, corrosion-oxidation resistance and fabricability.

2 CLASSIFICATIONS AND DESIGNATIONS

Stainless steels are classified based on their crystal structure, austenitic (face-centered cubic) or ferritic (body-centered cubic), along with the response of the austenite or ferrite or mixture of these two, to heat treatment. Corrosion resistance, although a logical thought, does not play a major role in classification of stainless steels. There is a continuous mixture of austenite and ferrite phases at high temperature, controlled by the elements present. Selected portions of this continuum are found useful and exploitable and are designated as specific alloys.

The austenite phase may be retained to room temperature or may transform to martensite while cooling. Ferrite is generally retained. In the highly alloyed duplex alloys ferrite may change into austenite during heat treatment. Advancements in steel mill processing have made it possible to exploit certain duplex mixtures of austenite and ferrite not previously utilized.

Historically, within some very broad limits seemingly arbitrary numbers were assigned by various authoritative bodies. The Society of Automotive Engineers (SAE) and the American Society for Testing and Materials (ASTM) agreed to a unified numbering system which assigns a five-digit number preceded by S to describe stainless steels. Many of the historically popular alloys have easily recognizable numbers carried over from the older systems. There are also ASTM and ASME specifications that are useful in specifying stainless steel to be used in certain product forms. Some alloys are patented or trademarked while others are generically available from many suppliers. Reference 2 has a large list.

The 200 and 300 series of stainless steels both start with the same high-temperature austenite phase as exists in carbon steel, but retain this structure down to room temperature and below. This retention results from the additions of nickel, manganese, and in some alloys nitrogen. The 200 series alloys rely mostly on manganese and nitrogen, while the 300 series utilize nickel. Austenite in both series has useful levels of ductility and strength. Fabrication and welding are readily done. In both series, alloys on the lean side of the retention elements, grades such as 201 and 301, which do cool as austenite, will transform to martensite when formed, which results in high-strength parts made by stretching a low-strength starting metal.

The 400 series of stainless steels includes both martensitic and ferritic alloys. In some cases it is possible to additionally strengthen the martensite phase by a later precipitation treatment in the range of 900–1150°F following fabrication. These are termed precipitation-hardening alloys (PH stainless steels). The ferritic alloys, high or low chromium, generally have titanium and/or niobium added to combine with carbon and nitrogen, thereby avoiding corrosion and ductility concerns after welding.

The duplex alloys have generally higher amounts of chromium and molybdenum for even higher levels of corrosion resistance. They are particularly noted for stress corrosion resistance. Fabricability and weldability come from controlled additions of nickel and nitrogen to maintain relatively equal amounts of austenite and ferrite. This equality offsets problems resulting from the variations in strength, ductility, and chemical partitioning between austenite and ferrite.

When one gets down to specific alloys, the composition may differ depending upon the form to be supplied. The best example of that is the comparison between casting alloys and the nearly, but not exactly the same, wrought versions. These differences are discussed later.

Table 1 lists chemical composition ranges for certain alloys that have been recognized by UNS designations. The corresponding AISI number is also listed, and where there is an older numerical type designation that number is also included. In the older number system three-digit numbers were used.

3 MECHANICAL AND PHYSICAL PROPERTIES

The physical properties of stainless steels of common interest describe the flow of heat and thermal expansion. Compared to carbon steel, heat does not flow readily in stainless steel. Numbers show conductivity rates only 28% of carbon steel at 212°F and 66% at 1200°F for 304. Thus, heat backs up and temperatures rise thus causing thermal expansion, and stress, and if the component is resisted, causing unanticipated distortion. On the other hand, low thermal conductivity is an asset in thermal insulation applications. There is also an ~3% density difference between the denser austenitic FCC alloys and the ferritic BCC alloys.

The thermal expansion is affected by both structure and chemistry. The ferritic and martensitic grades have the same BCC structure as carbon steel and expand at the same lower rates. The austenitic grades generally expand at about twice the rate of carbon steel. Nickel beyond the usual levels in stainless does reduce the thermal expansion. Higher nickel grades develop less thermal strain difference against oxides, and these alloys have higher cyclic temperature exposure limits, since the oxides spall less. The mixture of iron and 36% nickel without chromium is known commercially as Invar due to its low thermal expansion. Selected property values are shown in Table 2.

The mechanical properties of stainless steels are a function of the series and the product form (i.e., thin sheet, thick plate, large billets or bars). The austenitic steels start yielding at the 30,000 to 35,000 psi level and reach levels of 90,000 to 110,000 psi at their ultimate strength after total elongations of 40–60%. The ferritic alloys begin to yield at the same flow stress levels but stretch less (~20–40%) to lower ultimate strengths (~55,000–70,000) before fracture. The hardened martensitic grades reach the ~100,000 psi flow strength levels and high Rockwell hardness values typical of martensite. More heavily worked thin sections may be slightly

Table 1 Chemical Analyses of Wrought Stainless Steels

AISI	UNS #	C	Mn	Si	Cr	Ni	Mo	N_2	Other
201	S20100	0.15 max.	5.5/7.5	0.75	16/18	3.5/5.5	—	0.25	—
203	S20300	0.08 max.	5/6.5	1.00	16/18	5/6.5	—	—	Cu 2.0 S .18/.35
301	S30100	0.15 max.	2.00	0.75	16/18	6/8	—	0.10	—
303	S30300	0.15 max.	2.00	1.00	17/19	8.00/10.00	—	—	S .15 min.
304	S30400	0.08 max.	2.00	0.75	18/20	8.0/10.5	—	0.10	—
304L	S30403	0.03 max.	2.00	0.75	18/20	8.00/12.00	—	0.10	—
304LN	S30453	0.03 max.	2.00	0.75	18/20	8/12	—	0.10/0.16	—
305	S30500	0.12 max.	2.00	0.75	17/19	10.5/13.00	—	—	—
308	S30800	0.08	2.00	1.00	19/21	10/12	—	—	—
309	S30900	0.20	2.00	1.00	22/24	12.00/15.00	—	—	—
310	S30900	0.25 max.	2.00	1.00	24/26	19/22	—	—	—
316	S31600	0.08	2.00	0.75	16/18	10/14	2/3	0.10 max.	—
316F	S31620	0.08	2.00	1.00	16/18	10/14	1.75/2.50	0.10 max.	S 0.10 min.
321	S32100	0.08	2.00	0.75	17/19	9.0/12.0	—	—	Ti 5XC to 0.7
347	S34700	0.08	2.00	0.75	17/19	9/13	—	—	Nb 10X C to 1.0
AL-6XN	N08367	0.03	2.0	1.00	20/22	23.5/25.5	6/7		N0.22 Cu0.7
410	S41000	0.15 max.	1.00	1.00	11.5/13.5	0.75 max.	—	—	—
410s	S41008	0.08 max.	1.00	1.00	11.5/13.50	0.60 max.	—	—	—
409	S40900	0.08 max.	1.00	1.00	10.5/11.75	0.05 max.	—	—	Ti 6XC to 0.75
430	S43000	0.12 max.	1.00	1.00	16/18	0.75 max.	—	—	—
439	S43900	0.07 max.	1.00	1.00	17/19	0.50 max.	—	—	Ti 1.1 max.
444	S44400	0.025	1.00	1.00	17.5/19.5	—	1.75/2.50	—	Ti+N .8max.
2205	S31803	0.03 max.	2.00	1.00	21/23	4.5/6.5	2.5/3.5	0.08/0.20	—
Al294C	S44735	0.03 max.	1.00	1.00	28/30	—	3.6/4.2	—	Ti+Nb .30
450	S45000	0.05 max.	1.00	14/16	14/16	5/7	0.5/1.0	—	Nb8xC min.

Table 2 Physical and Mechanical Properties at Room Temperature

AISI	UNS #	Density	Thermal conductivity (W/m·K)	Min. Y.S. [ksi (Mpa)]	Min. UTS [ksi (Mpa)]	Min. % elong.	Tensile modulus [GPa (10^6 psi)]	CTE (10^{-6}/K @ 100°C)
201	S20100	7.75	16.3	38 (262)	95 (655)	40	197 (28.6)	15.7
203	S20300	7.860	16.3	35 (241)	85 (586)	45 Typical	—	—
301	S30100	8.027	16.3	30 (207)	75 (517)	40	193 (28)	17.0
303	S30300	8.027	16.3	35 (241)	85 (586)	50 Typical	193 (28)	17.2
304	S30400	8.027	16.3	30 (207)	75 (517)	40	193 (28)	17.2
304L	S30403	8.027	16.3	25 (172)	70 (483)	40	—	17.2
304LN	S30453	8.027	16.3	30 (207)	75 (517)	40	—	—
305	S30500	8.027	16.3	30 (207)	75 (517)	40	193 (28)	—
308	S30800			DNA used as weld wire			193 (28)	17.2
309	S30900	8.027	15.6	30 (207)	75 (517)	40	200 (29)	15.0
310	S31000	7.750	14.2	30 (207)	75 (517)	40	200 (29)	15.9
316	S31600	8.027	16.3	30 (207)3	75 (517)	40	193 (28)	15.9
409	S40900	7.750	24.9	30 (207)	55 (380)	22	200 (29)	11.1
430	S43000	7.750	23.9	30 (207)	65 (448)	22	200 (29)	10.4
439	S43900	7.750	—	30 (207)	65 (448)	22	200 (29)	10.4
444	S44400	7.750	26.8	40 (276)	60 (414)	20	200 (29)	10.0
2205	S31803	7.820	19.0	65 (450)	90 (620)	25	—	—
Al294C	S44735	7.660	15.2	60 (415)	80 (550)	18	—	—
450 Annealed	S45000	7.750	—	118 (814)	142 (979)	13	—	—
450 Aged 900°F	S45000	7.750	—	188 (1296)	196 (1351)	14	—	—

stronger. Table 2 lists many of the same repeated industry minimum values that apply for all product forms. Specific mechanical test data usually accompany each delivered order.

Variation in the ability to work-harden is the basic choice between types 201 and 301 which stretch farther, and reach higher flow stresses than 304, versus 305 which does not reach such high flow stress levels even though the elongation remains at reasonably high austenitic levels. The 201 and 301 alloys form martensite while deforming, whose high strength supports and maintains the continued strain to higher levels. The more highly alloyed stable austenitic alloys do not form martensite and their properties are similar to 304 as shown in Table 2. The effect of cold work on the mechanical properties of several steels is depicted in Fig. 1.

This work-hardening ability is behind the production of temper rolled austenitic alloys. These alloys can be produced as rolled strip items which have high initial strength but retain enough ductility to make many useful items. ASTM A167 and A666 minimum properties are listed below. The higher strength versions are generally made from 201 and 301. Techniques for the use of these strength levels are in Ref. 3.

Condition	Tensile psi	0.2% Yield psi	% Elong
Annealed	75,000	30,000	40
1/4 Hard	125,000	75,000	25
1/2 Hard	150,000	110,000	18
3/4 Hard	175,000	135,000	12
Full Hard	185,000	140,000	9

The PH stainless achieve high strength along with reasonable ductilities. Compositions and properties of selected PH steels are listed in Table 3. In the aged condition their strengths can exceed 200 ksi.

The fatigue characteristics of the stainless steels are generally good with the austenitic types having endurance limits in the range of 240–270 MPa (35–39 ksi).

Temperature Effects

The strengths of most metals decrease with increasing operating temperature and the stainless are no exception. Figure 2 shows the short-time strength versus temperature for a number of stainless steels and Fig. 3 depicts the general hot-strength characteristics of the stainless steels. Table 4 lists suggested maximum operating temperatures for several stainless steels. The creep and rupture characteristics of numerous alloys are listed in Table 5.

Stainless steels are also frequently used for low-temperature applications. In general, the austenitic steels are preferred for cryogenic applications. Their low temperature properties are listed in Table 6.

4 FORMABILITY AND MACHINABILITY

In order to make useful parts from stainless steel it is generally necessary to draw, stretch, bend, turn, forge or otherwise alter the shape of the mill product. In fact, many alloy designations are inherently intended to better accommodate such op-

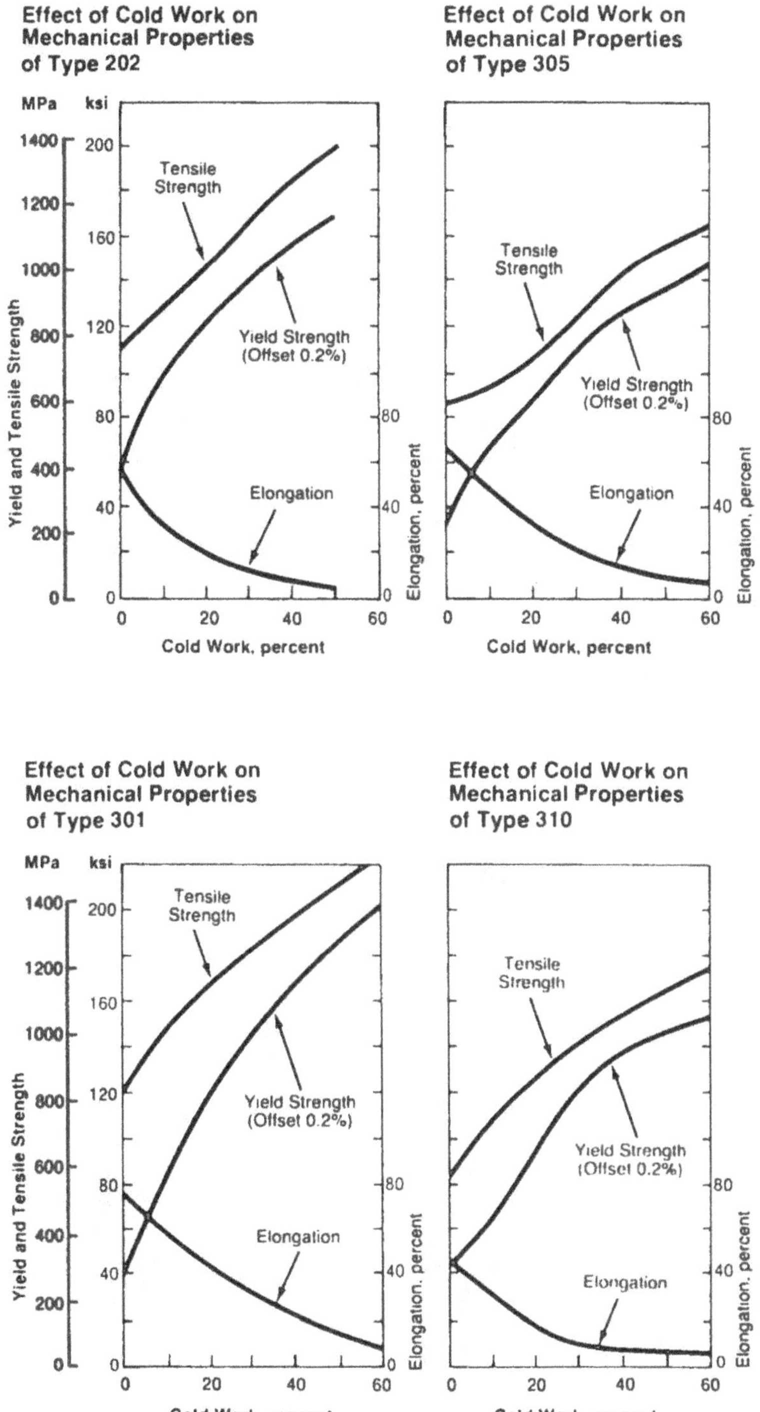

Figure 1 The effect of cold work on mechanical properties (From Ref. 1.)

Table 3 Precipitation Hardening Stainless Steels

Chemical analysis % (max. unless noted otherwise)									
Type	C	Mn	P	S	Si	Cr	Ni	Mo	Other
S13800	0.05	0.10	0.010	0.008	0.10	12.25/13.25	7.50/8.50	2.00/2.50	0.90/1.39 Al 0.010 N
S15500	0.07	1.00	0.040	0.030	1.00	14.00/15.50	3.50/5.50		2.50/4.50 Cu 0.15/0.45 Cb + Ta
S17400	0.07	1.00	0.040	0.030	1.00	15.50/17.50	3.00/5.00		3.00/5.00 Cu 0.15/0.45 Cb + Ta
S17700	0.09	1.00	0.040	0.040	0.040	16.00/18.00	6.50/7.75		0.75/1.50 Al

Nominal mechanical properties (solution-treated bar)[a]						
	Tensile strength		Yield strength (0.2% offset)		Elongation in 2" (50.80 mm)	Hardness
Type	ksi	MPa	ksi	MPa	%	(Rockwell)
S13800	160	1103	120	827	17	C33
S15500	160	1103	145	1000	15	C35
S17400	160	1103	145	1000	15	C35
S17700	130	896	40	276	10	B90

[a]Aged strengths exceed 200 ksi (1380 MPa).

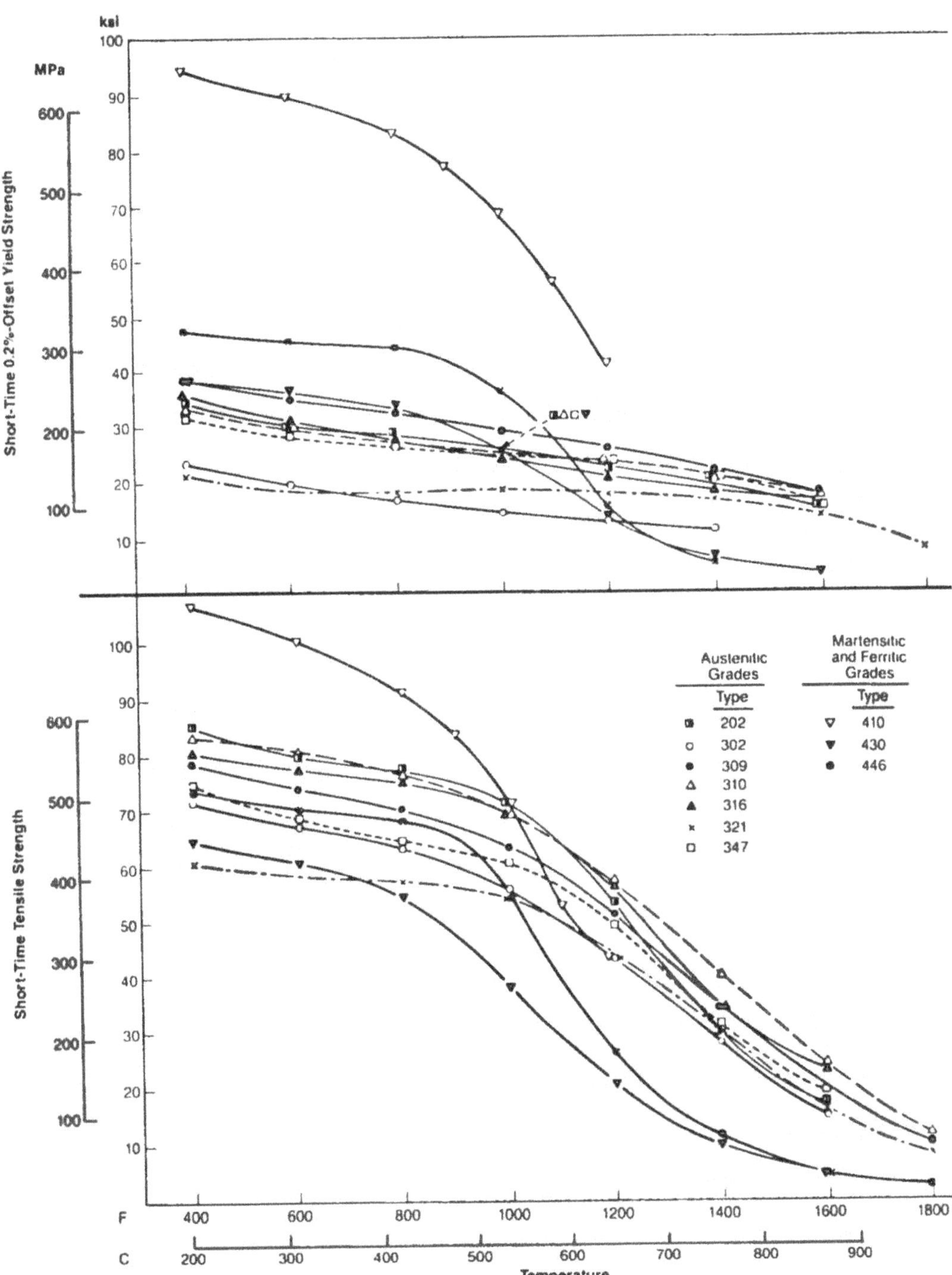

Figure 2 Typical short-time strengths of various stainless steels at elevated temperatures. All steels were tested in the annealed condition except for martensitic type 410 which was heat treated by oil quenching from 982°C (1200°F). (From Ref. 1.)

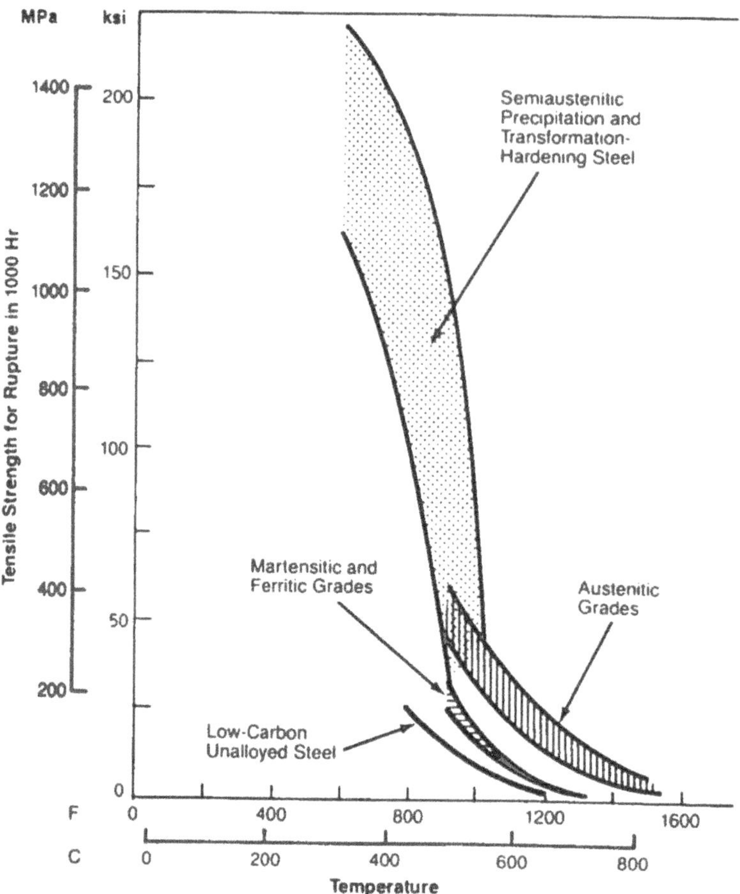

Figure 3 General comparison of the hot-strength characteristics of austentic, martensitic and ferritic stainless steels with those of low-carbon unalloyed steel and semiaustenitic precipitation and transformation-hardening steels. (From Ref. 1.)

erations. Once formed, the mechanical properties of the steel may be quite different. Annealing may or may not be needed to have a serviceable part.

The basic 304 alloy, the 18% chromium and 8% nickel alloy, is produced and used in the greatest tonnage, obviously meeting the greatest sum total of needs. This alloy can be machined, bent to angles, drawn into cups or cylindrical shapes, or stretched into parts where full use is made of the high total elongation (>50% in many cases). During bending the high levels of flow stress that are needed will result in a tendency for springback. Overbending is used to obtain specific angles.

The effect of high flow stresses or the basic strength levels is the root cause of other forming considerations. Lubrication can be quite helpful for ensuring that the high flow stresses do not get transmitted to tooling in a way that results in galling against a die and possible fracture. When machining, water soluble lubricants help remove the heat generated in these continuous operations. Machining also benefits from sharp rigid tooling and deep slow cuts. These techniques mini-

Table 4 Suggested Maximum Service Temperatures in Air

	Intermittent service		Continuous service	
AISI type*	°C	°F	°C	°F
201	815	1500	845	1550
202	815	1500	845	1550
301	840	1550	900	1650
302	870	1600	925	1700
304	870	1600	925	1700
308	925	1700	980	1800
309	980	1800	1095	2000
310	1035	1900	1150	2100
316	870	1600	925	1700
317	870	1600	925	1700
321	870	1600	925	1700
330	1035	1900	1150	2100
347	870	1600	925	1700
410	815	1500	705	1300
416	760	1400	675	1250
420	735	1350	620	1150
440	815	1500	760	1400
405	815	1500	705	1300
430	870	1600	815	1500
442	1035	1900	980	1800
446	1175	2150	1095	2000

mize the volume of metal reaching the maximum flow stresses, thus limiting energy-heat.

The principal alloying means to improve machinability is through the addition of sulfur. (Selenium is also used for this purpose.) The resultant manganese sulfide inclusions allow chips to break easily. The 18% chromium types 303 or 203 are the usual free-machining austenitic grades. At the 316 level the same sulfur addition is done within the basic 316 composition designated as 316F, 416 is the free-machining ferritic alloy, while 420F is the martensitic version. Even minor additions of sulfur to conventional grades gives noticeable improvements in machinability. Machinability ratings are compared in Fig. 4.

Difficulties in drawing 304 into any cylindrical shape can be addressed by selecting a 305 alloy which does not work-harden to the same high flow stress levels. If on the other hand, the problem is how to stretch the stainless to make a part, with a corresponding reduction in thickness, then a 201 or 301 alloy may be a better choice. When 201 or 301 are fully strained by stretching they can become ferromagnetic and may need a postforming, 5 min, 350°F level of heat treatment to drive off hydrogen.

There is little ability to control the flow stress of ferritic or martensitic stainless steels and there are no such alloys designed to have either a high or low tendency for work hardening.

Table 5a Rupture and Creep Characteristics of Chromium Stainless Steels in Annealed Condition

F	800		900		1000		1100		1200		1300		1400		1500	
C	427		482		538		593		649		704		760		861	
Type	ksi	MPa	ksi	MPa	ksi	MPa	ksi	MPa	ksi	MPa	ksi	MPa	ksi	MPa	ksi	MPa
						Stress for rupture in 1000 hours										
405	—	—	25.0	172	16.0	110	6.8	47	3.8	27	2.2	15	1.2	8	0.8	6
410	54.0	372	34.0	234	19.0	131	10.0	69	4.9	34	2.5	18	1.2	8	—	—
430	—	—	30.0	207	17.5	120	9.1	63	5.0	34	2.8	2.0	1.7	12	0.9	6
446	—	—	—	—	17.9	123	5.6	39	4.0	28	2.7	19	1.8	13	1.2	8
						Stress for rupture in 10,000 hours										
405	—	—	22.0	152	12.0	83	4.7	33	2.5	18	1.4	10	0.7	5	0.4	3
410	42.5	294	26.0	179	13.0	90	6.9	47	3.5	24	1.5	10	0.6	4	—	—
430	—	—	24.0	165	13.5	94	6.5	43	3.4	23	2.2	15	0.7	5	0.5	3
446	—	—	—	—	13.5	94	3.0	21	2.2	15	1.6	11	1.1	8	0.8	6
						Stress for creep rate of 0.0001% per hour										
405	—	—	43.0	296	8.0	55	2.0	14	—	—	—	—	—	—	—	—
410	43.0	296	29.0	200	9.2	63	4.2	29	2.0	14	1.0	7	0.8	6	—	—
430	23.0	159	15.4	106	8.6	59	4.3	30	1.2	8	1.4	10	0.9	6	0.6	4
446	31.0	214	16.4	113	6.1	42	2.8	20	1.4	10	0.7	5	0.3	2	0.1	1
						Stress for creep rate of 0.00001% per hour										
405	—	—	14.0	97	4.5	32	0.5	3	—	—	—	—	—	—	—	—
410	19.5	135	13.8	96	7.2	49	3.4	24	1.2	8	0.6	4	0.4	3	—	—
430	17.5	120	12.0	83	6.7	46	3.4	24	1.5	10	0.9	6	0.6	4	0.3	2
446	27.0	186	13.0	90	4.5	32	1.8	13	0.8	6	0.3	2	0.1	1	0.05	0.3

Table 5b Rupture and Creep Characteristics of Chromium-Nickel Stainless Steels

Type	Testing temperature		Stress: Rupture time, 100 h		Stress: Rupture time, 1000 h		Stress: Creep rate, 10,000 h		Stress: Creep rate, 0.01% h		Extroplated elongation at rupture in 10,000 h (%)
	°F	°C	ksi	MPa	ksi	MPa	ksi	MPa	ksi	MPa	
302	1600	871	4.70	33	2.80	19	1.75	12	2.50	17	150
	1800	982	2.45	17	1.55	11	0.96	7	1.30	9	30
	2000	1093	1.30	9	0.76	5	0.46	3	.62	4	18
309S	1600	871	5.80	40	3.20	22	—	—	3.50	24	—
	1800	982	2.60	18	1.65	11	1.00	7	1.00	7	105
	2000	1093	1.40	10	0.83	6	0.48	3	.76	5	42
310S	1600	871	6.60	45	4.00	28	2.50	17	4.00	28	30
	1800	982	3.20	22	2.10	15	1.35	9	1.75	12	60
	2000	1093	1.50	10	1.10	7	0.76	5	.80	6	60
314	1600	871	4.70	32	3.00	21	1.95	13	2.30	16	110
	1800	982	2.60	18	1.70	12	1.10	8	1.00	7	120
	2000	1093	1.50	10	1.12	7	0.85	6	.90	6	82
316	1600	871	5.00	34	2.70	19	1.40	10	2.60	18	30
	1800	982	2.65	18	1.25	9	0.60	4	1.20	8	35
	2000	1093	1.12	8	0.36	2	—	—	4.00	28	—

Table 6 Typical Mechanical Properties of Stainless Steels at Cryogenic Temperatures

Type	Test temperature		Yield strength 0.2% offset		Tensile strength		Elongation in 2"	Izod impact	
	°F	°C	ksi	MPa	ksi	MPa	%	ft-lb	J
304	−40	−40	34	234	155	1,069	47	110	149
	−80	−62	34	234	170	1,172	39	110	149
	−320	−196	39	269	221	1,524	40	110	149
	−423	−252	50	344	243	1,675	40	110	149
310	−40	−40	39	269	95	655	57	110	149
	−80	−62	40	276	100	689	55	110	149
	−320	−196	74	510	152	1,048	54	85	115
	−423	−252	108	745	176	1,213	56		
316	−40	−40	41	283	104	717	59	110	149
	−80	−62	44	303	118	814	57	110	149
	−320	−196	75	517	185	1,276	59		
	−423	−252	84	579	210	1,448	52		
347	−40	−40	44	303	117	807	63	110	149
	−80	−62	45	310	130	896	57	110	149
	−320	−196	47	324	200	1,379	43	95	129
	−423	−252	55	379	228	1,572	39	60	81

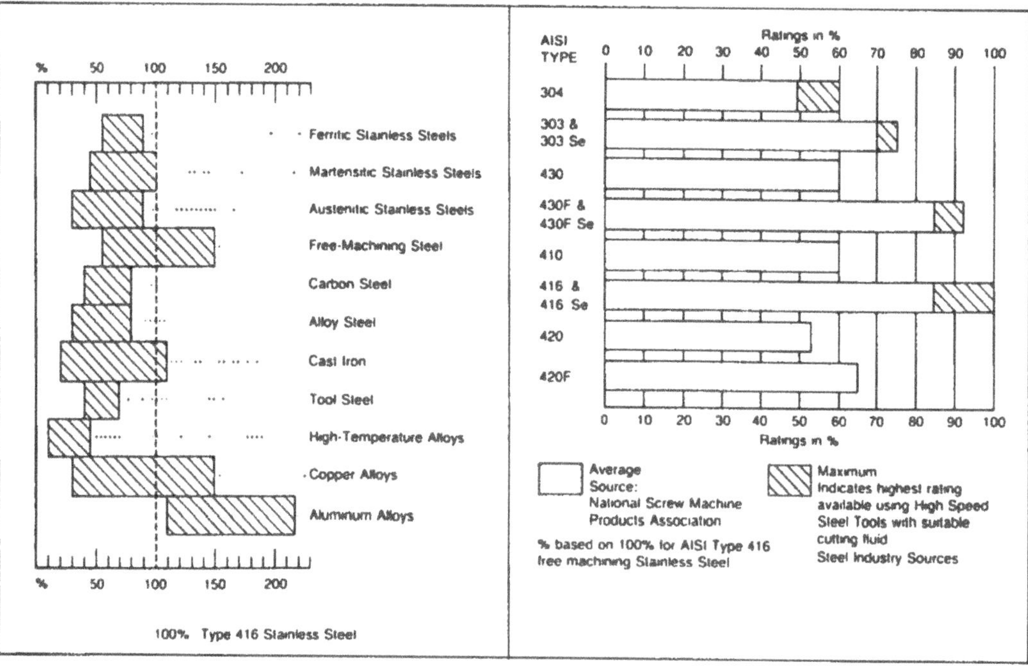

Figure 4 Comparative machinability of stainless steels and common metals. (From Ref. 1.)

5 JOINABILITY

Stainless steels are readily joined by the usual welding and brazing techniques. There are two basic focus areas in welding and one such area in brazing. When welding it is important to melt and solidify a small volume of metal without causing any hot cracks from thermal strains inherent with cooling. It is also important that the base metal adjacent to the weld is not adversely changed by heat from the passing adjacent molten weld metal. When brazing, the passive film, the basis of corrosion resistance, resists the capillary/surface tension driven flow of brazing alloys. Flux or low furnace atmosphere dewpoints are the usual cures. Brazing cycles can be made to act as heat treatment cycles. Pre-braze cleanliness and post braze cleaning are both very important.

The basic 304 composition is readily welded. If filler metal is needed, 308 is used. This composition is enriched in alloy content to avoid untempered martensite which could form in multipass 304 welds. Other, more highly alloyed grades, which do not form martensite, are generally joined with a matching filler. For all weld wires, the chemistry is controlled so the weld deposit contains a small percentage of delta ferrite, which limits the tendency for hot cracks to form.

The ferritic and martensitic grades are also welded, but there are more options. If strength and thermal expansion differences permit, many of these grades are joined with an austenitic alloy for ductility reasons, or to avoid post-weld heat treatment. Some ferritic and martensitic grades have matching compositions. For either weld route, the heat-affected zone should also be evaluated for ductility and corrosion resistance. Post-weld tempering may be needed.

For all grades of stainless steel, strong mechanical support during welding should be provided to control/guide thermal expansion. Also the detrimental effect of oxidation products on corrosion resistance needs to be considered. Weld oxides cause corrosion. Oxides are removed by grinding or acids. Wire brushing is not enough.

Stainless steel is often welded to carbon steel. This is generally done with 309 weld wire. The principal concern about such welds is martensite formed by melting (alloying) carbon steel and stainless. The extra 309 alloy content ensures that even diluted weld deposits solidify and remain austenitic rather than transform to brittle martensite. Corrosion is usually not a concern.

The weld-heat-affected zone obviously experiences temperatures between melting and ambient. The exposure time varies, based on thickness, energy input, and other mechanical factors. With austenitic grades, after-exposure times at 1200°F reach minutes, it is possible to form a continuous grain boundary network of chromium carbides through the metal thickness. This leaves a concurrent continuous volume of metal with reduced chromium content. That chromium is now in the carbides. The chromium depleted zone may easily corrode in an otherwise benign environment. When this zone spans the thickness cracks may occur. This condition is referred to as "sensitization."

There are two alloy solutions to this problem. The straightforward approach is to limit the carbon content as is done with the L grades. These grades still sensitize if put into service at high temperatures, but are not sensitized by the limited exposures encountered during welding. The second approach, utilizes a strong carbide former such as titanium, or niobium which forms a carbide at a temperature higher than does chromium. These grades (321 and 347) are also selected for service at sensitizing temperatures.

Another "solution" is to be sure that there is a problem. Welding does not automatically imply a need for a low-carbon grade. Metallurgical evaluation of the actual weld-heat-affected zones can reveal the specific level of carbide precipitation. Experience is also an excellent reference.

This same sensitization issue is addressed in ferritic and martensitic grades only by the additions of titanium or niobium. Carbon contents low enough to avoid sensitization cannot be melted.

6 APPLICATIONS

The applications of wrought stainless steel include the obvious ones which need metal and involve harsh corrosives. However, life-cycle considerations can broaden the range of use, versus coated metal products requiring maintenance or extra fabrication steps. The initial cost of stainless steel can also be offset by utilizing strength and design techniques to minimize weight, while retaining the corrosion attributes. Stainless steels do not need to be thicker to resist corrosion. A structural code for stainless steel buildings is listed in Ref. 3. This code takes full advantage of the strength attributes of stainless to reduce weight. Remember, strengths >100,000 psi are available.

A review of data from the Specialty Steel Industry of North America (SSINA) by country shows the United States of America uses the order of 17 lb of stainless steel per person. Sweden and Italy are just below 25 lb, and Japan, Germany, and South Korea are at the 30 lb per person level. Obviously other societies find added values in stainless steels and there is opportunity for more usage in the United States of America.

The corrosion applications for stainless should be approached by recognizing that the thin passive film that protects stainless steel is basically an oxide and hence oxidizing environments generally sustain this passive film. On the other hand, chemically reducing environments represent a direct attack on the passive film, and stainless steel may act no better than ordinary carbon steel. Sulfuric acid is the classic example. Chlorides, found in many environments, have a particular ability to locally attack the passive film causing pits. 316 and other alloys containing molybdenum resist greater levels of chloride, but seawater is generally too harsh for 316. Alloys with higher molybdenum contents such as the AL-6XN alloy (UNS NO8369) containing 6% Mo are required for good resistance to seawater.

The other broad corrosion area causing concern is water, from 140°F to boiling. In such an environment even modest amounts of tensile stress, and chloride, may result in branching transgranular stress corrosion cracks (SCC). While lowering the temperature or reducing the tensile stress may mitigate symptoms, the most effective cure is either a no-nickel ferritic alloy or an alloy having >50% nickel content. During service, alternate wetting and drying conditions can magnify otherwise benign environments to troublesome levels of chloride content. Crevices, either from design or from deposits, also aggravate seemingly benign environments. (See Chapter 15 for more complete data on corrosion resistance.)

Table 7 Chemical Compositions for Corrosion- and Heat-Resistant Castings

ACI	AISI equivalent	C	Mn	Si	Cr	Ni	Mo	Other
CA-6NM	—	.06	1.00	1.00	11.5/14.0	3.4/4.5	0.4/1.0	
CA-15	T410	.15	1.00	1.5	11.5/14.0	1.0	0.5	
CB-7 Cu	Armco 17-4PH	0.07	0.70	1.0	15.5/17.7	3.6/4.6	Cu 2.5/3.2	Cb 0.2/0.35
CE-30	T312	0.3	1.50	2.0	26/30	8/11	—	—
CF-3	T304L	.03	1.50	2.0	17/21	8/12	—	—
CF-8	T304	.08	1.50	2.0	18/21	8/11	—	—
CF-3M	T316L	.03	1.5	1.5	17/21	9/13	2/3	—
CF-8M	T316	.08	1.5	2.0	18/21	9/12	2/3	—
CF-8C	T347	.08	1.50	2.0	18/21	9/12	—	Cb 8XC 1.0 max
CF-16F	T303	0.16	1.50	2.0	18/21	9/12	1.5	.2 Se
CH-20	T309	0.20	1.50	2.0	22/26	12/15	—	—
CK-20	T310	0.2	1.5	2.0	23/27	19/22	—	—
HA	—	.20	0.35/0.65	1.00	8/10	—	0.9/1.2	
HC	T446	.50	1.00	2.00	26/30	4	0.5	
HH	T309	0.20/0.50	2.00	2.00	24/28	11/14	0.5	N 0.2
HK	T310	0.20/0.60	2.00	2.00	24/28	18/22	0.5	
HT	T330	0.35/0.75	2.00	2.50	15/19	33/37	0.5	
HW	—	0.35/0.75	2.00	2.50	10/14	58/62	0.50	
HX	—	0.35/0.75	2.00	2.50	15/19	64/68	0.50	

Table 8 Selected Mechanical Properties of Cast Stainless Steel

	Temperature (°F)	Stress ksi (Mpa)	Test type	
HA	RT	95 (656)	Tensile	
	1000	67 (462)	Tensile	
	1200	44 (304)	Tensile	
	1000	16 (110)	10^{-5}/h creep	
	1200	3.1 (110)	10^{-5}/h creep	
	1000	37 (255)	100 h rupture	
	1000	27 (186)	1000 h rupture	
HC	RT	70 (483)	Tensile	
	1400	10.5 (72)	Tensile	
	1400	1.3 (9)	10^{-5}/h creep	
	1400	3.3 (21)	100 h rupture	
	1400	2.3 (16)	1000 h rupture	
		Partially ferritic		Stress total austenite
HH	RT	80 (552)	Tensile	85 (586)
	1400	33 (228)	Tensile	35 (241)
	1600	18.5 (128)	Tensile	22 (152)
	1600	1.7 (12)	10^{-5}/h creep	4.0 (28)
	1600	6.4 (44)	100 h rupture	7.5 (52)
	1600	3.8 (26)	1000 h rupture	4.7 (32)
HK	RT	75 (518)	Tensile	
	1600	23 (159)	Tensile	
	1600	4.2 (29)	10^{-5}/h creep	
	2000	1.0 (7)	10^{-5}/h creep	
	1600	7.8 (54)	100 h rupture	
	1800	4.5 (31)	100 h rupture	
	1800	3.0 (21)	100 h rupture	
HW	RT	68 (469)	Tensile	
	1600	19 (131)	Tensile	
	1800	10 (69)	Tensile	
	1600	3.0 (21)	10^{-5}/h creep	
	1600	6.0 (41)	100 h rupture	
	1600	4.5 (31)	1000 h rupture	
HX	RT	66 (455)	Tensile	
	1600	29 (174)	Tensile	
	1800	10.7 (74)	Tensile	
	1800	1.6 (11)	10^{-5}/h creep	
	1800	3.4 (24)	100 h rupture	
	2000	1.7 (12)	100 h rupture	
	1800	2.2 (15)	1000 h rupture	
	2000	.9 (6)	1000 h rupture	

7 CAST ALLOYS

Cast alloy designations as developed by the Alloy Casting Institute start by dividing alloys into those intended for corrosive service, beginning with a C and those used at high temperature, beginning with an H. In the corrosive case, there may be wrought counterparts. However, in the high temperature arena, there are compositions with high hot strength and limited hot ductility that cannot be hot- and cold-worked to form wrought products by the usual techniques. Such alloys are excellent load-bearing parts at high temperatures.

The roles of ferrite and carbides are expanded in cast alloys as compared to wrought stainless steels. Castings, which may be considered as large welds, also benefit from the presence of delta ferrite to prevent cracks. A fine ferrite distribution also raises strength somewhat.

Carbides, which are more plentiful in the higher carbon castings grades, will precipitate first at the delta ferrite austenite boundaries. Since these boundaries are not generally continuous, there is less of a corrosion problem than might be ex-

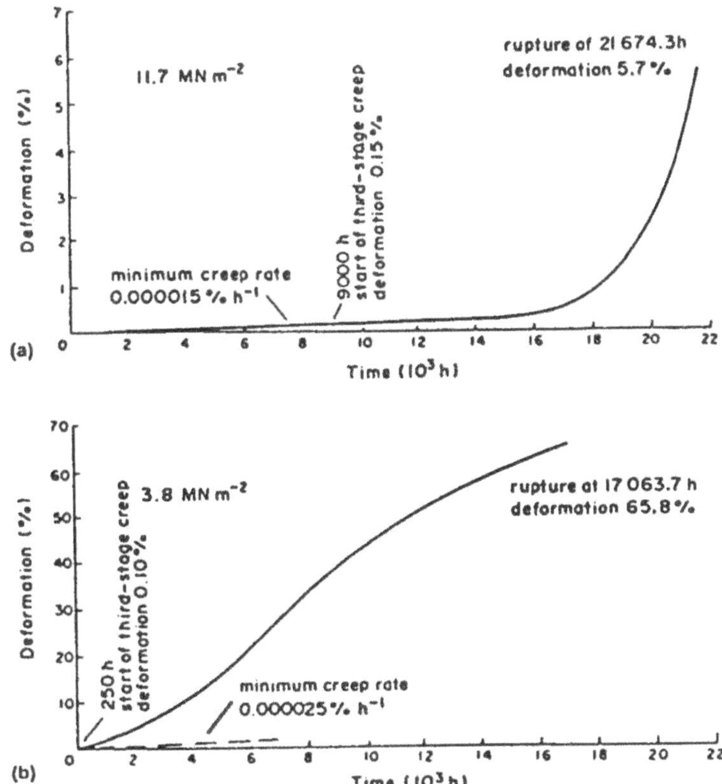

Figure 5 Time-deformation curves for creep-rupture test specimens of (a) cast HK-40 (0.44%C) and (b) wrought 310 (0.06%C) 25Cr-20Ni alloys at 982°C. Note the 310 vertical scale is 10 times the scale for the HK-40. (From Ref. 4.)

pected. Ferrite may also transform to a brittle sigma phase at temperatures above ~1400°F. Again lack of ferrite continuity is favorable.

7.1 Properties

Table 7 lists the chemical analysis ranges for many cast stainless steels. Table 8 lists short-term and longer term mechanical properties. Figure 5 shows the high-temperature mechanical property differences between cast and wrought alloys.

7.2 Joinability

Joining castings involve the same considerations as joining wrought compositions. The higher silicon content of most casting alloys does bring one extra consideration. Figure 6 shows an interaction between silicon and carbon content concerning weld ductility. Basically an excess of silicon can result in cracks while an excess of carbon results in less elongation.

7.3 Applications

The use of stainless steel castings for resisting corrosive environments follows the same basic trends as wrought alloys. The corrosion resistance still comes from a thin oxide film, strengthened by chemical oxidation and harmed by reduction. In those castings where there are already large amounts of carbon and silicon present

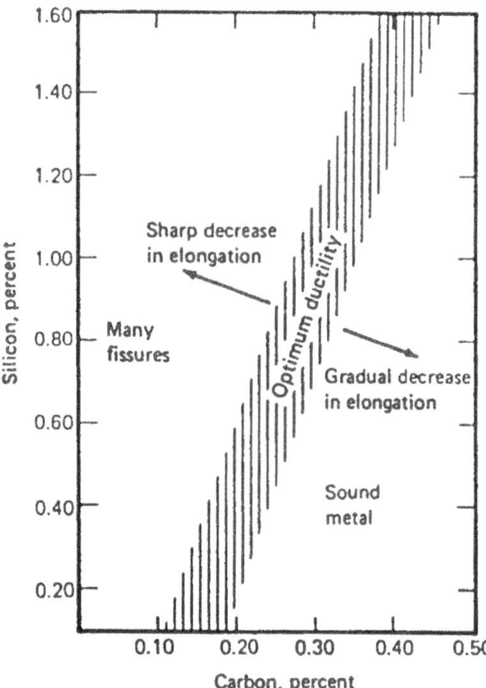

Figure 6 Effect of combined carbon and silicon contents on the ductility of welds in austenitic stainless steels.

there is good resistance to additional absorption of carbon and these alloys last in high-temperature carburizing furnace parts. It is in the highest temperature area where the use of castings offers strengths not available in wrought products.

7.4 Costs

The first costs of a stainless steel vary greatly by alloy and product form and are not easily listed in any exact form. Specific element costs are not the only factor. One data source listed prices for stainless sheet, strip, and plate from $0.70/lb to ~$4.00/lb, i.e., several times the cost of carbon steel. Inquiry should be made about the specific alloys, quantities, and sizes that are required for the application. Attention should also be paid to the total life-cycle costs. Up-front costs may be returned many fold by the longer life provided by the stainless steel.

Over the years stainless steel prices have grown far less than many other materials. One representation, starting at an index of 100 in 1967, found by 1994 stainless steel indices were 250, but finished carbon steel mill products, aluminum mill shapes, and industrial commodities in general, all had reached ~450.

REFERENCES

1. *Design Handbook*, Speciality Steel Industry of North America, Washington, DC.
2. *Steel Products Manual, Stainless and Heat Resisting Steels*, Iron and Steel Society, Warrendale, PA, 1990.
3. *Specification for the Design of Cold-Formed Stainless Steel Structural Members*, ANSI/ASCE-8-90, American Society of Civil Engineers, New York.
4. Peckner, Donald and I. M. Bernstein. *Handbook of Stainless Steels*, McGraw-Hill, 1977.
5. J. R. Davis, ed., ASM *Specialty Handbook, Stainless Steels*, ASM International, Materials Park, OH, 1994.

8
Cast Irons

Charles V. White
GMI Engineering and Management Institute
Flint, Michigan

1 INTRODUCTION

The term "cast iron" describes a number of alloys of iron, carbon and silicon with a wide range of properties and applications. The chemical composition and the subsequent microstructural make up of the alloys do not permit extensive mechanical working as with other ferrous products. Typically the carbon content of 1.8–4.0% and silicon of 0.5–3% produce a microstructure which is best described as a composite. This composition range describes all grades of cast irons with properties ranging from highly wear resistant hard materials to ductile energy-absorbing alloys suitable for high stress and shock-loading applications. The specifics of the individual alloy systems are detailed in this chapter, references are provided for additional study.

1.1 Casting and Solidification Characteristics

The properties of cast irons are developed in two stages during the solidification and subsequent cooling of the alloy. In general, the rate of solidification and chemical composition dictates the principal graphite form (a solidification event at the eutectic reaction) The cooling rate of the solid and the chemical composition determines the matrix characteristics (eutectoid reaction).

Eutectic Reaction

A eutectic reaction in an alloy is typically a reaction in which a molten liquid upon cooling below a specific temperature separates into two specific solid phases with each having unique properties. These two solids are intimately mixed and comprise the solid being cast. In the cast iron system, two variations of the solidification event (eutectic reaction) exist depending upon the chemistry of the liquid iron and the physical cooling rate of the part. This cooling rate is a function of the type of mold, green sand, dry sand etc. and the physical size of individual parts and the thermal mass of the casting configuration.

The results of higher silicon content or slower cooling rate will promote the formation of a graphitic form of carbon and a matrix of alloyed iron. Faster cooling rates and/or lower silicon contents promote the reaction to produce a carbide form of carbon and a matrix of alloyed iron. These factors strongly influence the design and manufacturing considerations of the cast iron product.

Eutectoid Reaction

In all the classifications of cast irons, the matrix structure is dictated by the rules applying to heat-treating an alloyed steel. Depending on the specific alloying elements, the hardenability of the matrix is generally high and responds well to thermal treatment. Irons are described according to the graphite type and the matrix form, such as ferritic ductile iron or pearlitic gray iron. Engineering properties are specified according to ASTM, SAE, Mil or other appropriate agencies.

1.2 Types of Cast Iron

Within the cast iron family are several varieties: gray, malleable, ductile, white and compacted. The name is descriptive of the microstructure features or the mechanical properties. Irons manufactured for the specific purpose of high temperature, corrosive or wear service are alloyed varieties of the basic format. Within each type of iron, property modifications are accomplished through alloying additions and heat treatment without changing the basic classification. (See Table 1.)

The sequence of the following discussion is based on the tonnage poured and does not follow an increasing carbon content. In general, the characteristics of cast irons are described by various means. One of which is the term "carbon equivalent" (CE). CE is calculated from the chemical analysis in one of the following ways.

$$\mathrm{CE} = \mathrm{C} + \frac{1}{3}\mathrm{Si}$$

$$\mathrm{CE} = \mathrm{C} + \frac{1}{4}\mathrm{Si} + \frac{1}{2}\mathrm{P}$$

$$\mathrm{CE} = \mathrm{C} + \frac{1}{3}(\mathrm{Si} + \mathrm{P})$$

Each method subscribed to by various authors is an attempt to describe the solid-

Table 1 Typical Cast Iron Compositions (% of Element)

Type	Carbon	Silicon	Mn	Sulfur	Phos
White	1.8–3.5	0.5–1.9	0.25–0.8	0.06–0.2	0.06–0.2
Malleable	2–2.9	0.9–1.9	0.15–1.2	0.02–0.2	0.02–0.2
Gray	2.5–4.0	1–3	0.2–1	0.02–0.25	0.02–1
Ductile	3–4	1.8–2.8	0.1–1	0.01–0.03	0.01–0.1
Compacted	2.5–4	1–3	0.2–1	0.01–0.03	0.01–0.1

ification characteristic of the iron. A 4.3 CE would be equivalent of a cast iron with a eutectic composition. CE values greater than 4.3 are considered *hypereutectic* and those below 4.3 are *hypoeutectic*. The significance of this division is discussed in more detail in later sections.

2 GRAY IRON

Gray iron is characterized by low casting temperatures, good castability and the least volume shrinkage of any cast iron during solidification. Gray iron gets its name from the characteristic gray surface of a fracture. The graphite is in flake form (Fig. 1). There are two principle ASTM designations for gray iron: A48 to describe the mechanical properties and A-247 to describe the graphite structure. The mechanical properties of gray iron vary with cooling rate and are measured from separately cast bars poured from the same metal as the casting. These bars are designated as A, B, C and S. The A, B, and C bars are listed in increasing diameter and are chosen for use by the relative size of the casting and its cooling rate. The S designation is a special bar agreed upon by the manufacturer and customer.

In addition, these bars are used to generate values of transverse breaking load and deflection when used in a three-point bending fixture as per ASTM A 438. This type of testing is used for more rapid results of the cast metal quality since it does not require the production of a machined test bar. The disadvantage of this test is that the surface finish of the as-cast bar can effect the test results.

Gray iron castings are generally not recommended for applications where impact resistance is required. In most applications, gray cast iron is used in a neutral or compressive application because the graphite in the flake form acts as an internal stress raiser. This graphite form offers significant advantages in machining, sound damping, and in heat transfer applications. Typical applications are soil piping, construction castings, engine blocks, manifolds housing and containers.

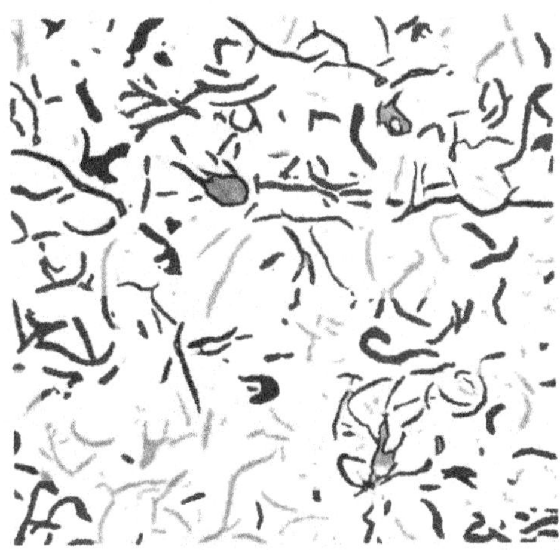

Figure 1 Typical ferritic gray iron microstructure. Nital etch approximately 100×.

Table 2 Grades of Gray Iron from SAE J431

Grade	Brinell hardness range	Test bar t/h[a]	Description
G1800	120 to 187	135	Ferritic-pearlitic
G2500	170 to 229	135	Pearlitic-ferritic
G3000	187 to 241	150	Pearlitic
G3500	207 to 255	165	Pearlitic
G4000	217 to 269	175	Pearlitic

[a]t/h = Tensile strength/Brinell hardness

2.1 Tensile Strength

Specifications are written as a Class of iron. *Class 30 A* gray iron would have a minimum of 30 ksi tensile in a separately cast *A* size bar. Typical designations run from Class 20 to Class 60 (138–414 MPa). In the SAE specification, *J 431* a tensile strength to Brinell hardness ratio is used to classify the irons for automotive applications. The term "class" is the common designation and the term "grade" as used in the SAE specification refers to a tensile hardness ratio. Table 2 lists grades of gray iron.

2.2 Compressive Strength

The strength characteristics are measured as a means of quality control under the proper sampling and statistical format. Gray cast iron, however, has considerably greater strength in compression than in tension (Table 3).

2.3 Damping Capacity

Cast irons are noted for their excellent damping capacity. A typical comparison is shown in Table 4.

3 DUCTILE IRON

Ductile iron, also referred to as nodular or spheroidal graphite iron (SG), has a chemical analysis similar to gray iron with small chemical modifications. These

Table 3 Compressive Strength/Tensile Strength

	Tensile strength		Compressive strength			E Tensile modulus	
Grade	MPA	(ksi)	MPA	(ksi)	CS/TS	GPA	$\times 10^6$ psi
20	152	22	573	83	3.7	69	10
30	214	31	752	109	3.5	96.6	14
40	393	57	966	140	2.45	124	18
60	431	62.5	1293	187.5	3.0	145	21

Table 4 Damping Capacity

Type of metal	Relative decrease in amplitude of vibration per cycle
Carbon steel	1.0–2.0
Malleable iron	3.3–6.3
Ductile iron	3.0–9.4
Hypoeutectic gray iron 3.2% C, 2.0% Si	40
Hypereutectic gray iron 3.7% C, 1.8% Si	126

chemistry changes alter the structure of iron to produce a microstructure in which the graphite form produced during the solidification process is spheroidal (Fig. 2). In this form the graphite does not act as a sharp notch as in gray iron, where it is in flake form and is therefore considerably more ductile in nature. The principle changes which must occur are the lowering of the sulfur (S) content and the addition of a small amount of magnesium (Mg) to the liquid metal. Nominally the S content is below 0.02% and a residual Mg content of 0.04% is sufficient to produce the nodular structure. Other elements can be used to produce the nodular graphite form such as yttrium (Y), calcium (Ca), and cerium (Ce). These are sometimes used in the production of ductile iron in conjunction with Mg. The production techniques and practices can be found in the references.

Property specifications for ductile irons fall under ASTM A 536 or SAE J 434 *60-42-18* is an example of a specification. This specification calls for a minimum

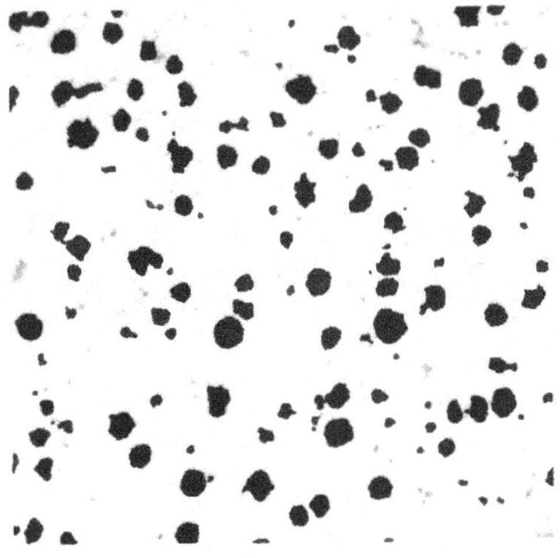

Figure 2 Typical ferritic nodular iron microstructure. Nital etch approximately 100×. Note the difference in the graphite shape between Figs. 1 and 2. In Figure 1 the gray iron flake graphite appear as long sharp ended particles. In Figure 2 the graphite has been rounded by the addition of a small amount of *Mg* and the reduction of the *S* content. Figure 1 is a Class 20 gray iron. Figure 2 is a 60-40-10 nodular iron.

tensile of 60 ksi, minimum yield of 40 ksi and a minimum elongation in 2 in of 18% (414 MPa/276 MPa/18%). These properties are usually obtained from separately cast test castings called "Y" or "keel" blocks. Property specification under J434 are listed as D 4018 specifying yield and elongation. Table 5 lists properties of ductile iron.

As in other cast irons the variation in grade is a result of the modification of the matrix either by cooling rate from the casting or by postcasting heat treatment. The graphite form is not changeable by heat treatment. *The only way to modify the graphite form is to melt and recast.*

Modification of the matrix is accomplished by adjusting the cooling rate from the region above the upper critical (austentite region). The exact temperature in this region (starting point) is dependent on the specific alloy content of the cast iron. The matrix behaves as a highly alloyed steel with the possibilities of producing ferritic, pearlitic, or martensitic structures. In general, the hardenability of cast irons are high; (i.e., they can be hardened to extensive depths. Because of its superior strength and ductility, ductile cast iron is used in competition with steel castings in some applications. Ductile iron is commonly used in automotive safety applications such as break and steering parts. It is also used in applications as valves, piping, pressure vessels and in applications requiring strength and impact tolerance.

Other modifications of the alloying elements result in two other major classifications of cast ductile iron, austenitic and austempered.

3.1 Austenitic Ductile Irons

The austenitic ductile irons (Table 6) are covered under ASTM A439. These castings contain a Ni content of 18–37% and Cr contents of 0.5–5.5%. They find applications in high or very low temperature applications, requiring a stable austenitic matrix. Obviously the cost of these alloys must be justified by the severeness of the application.

The examples of highly alloyed ductile iron cast materials are typically used for (1) high-temperature service like engine manifolds, turbocharger housings, and valves; (2) corrosion service as pumps and valve bodies in saltwater, chemical and

Table 5 Grades of Ductile Iron from ASTM A-536

Grade & heat treatment	Tensile strength minimum		Yield strength minimum		Percent elongation in 2"	Brinell hardness	Microstructure
	MPa	psi	MPa	psi			
60-40-18[a]	414	60,000	276	40,000	18	149–187	ferrite
80-55-06[b]	552	80,000	379	55,000	6	187–255	pearlite & ferrite
120-90-02[c]	828	120,000	621	90,000	2	240–30	tempered martensite

[a]May be annealed after casting.
[b]An as-cast grade with a higher manganese content.
[c]Oil-quenched and tempered to desired hardness.

Table 6 Austenitic Ductile Iron from ASTM A 439

	Tensile		Yield		
Grade	MPa	ksi	MPa	ksi	% elongation
D-2	400	58	207	30	8
D2-C	400	58	207	30	20
D-5	379	55	207	30	10
D-5S	449	65	207	30	10

paper processing industries; (3) dimensional stability and nonmagnetic applications; and (4) low-temperature service.

3.2 Austempered Ductile Iron (ADI)

Austempered properties are specified under ASTM A 897 (Table 7). This alloy-modified form of iron undergoes a specific programmed cooling to below the nose of the transformation curve commonly known as the bainitic region. The transformation products are not considered bainite, but are termed "ausferrite."

This type of iron is used for high-strength and toughness applications (lower strength and hardness grades) as well as high-wear application (high hardness grades) ADI property control is a function of the precise heat treatment below the nose of the transformation curve. Proper and consistent production requires a skilled heat treater with appropriately sized equipment.

4 COMPACTED IRON

Compacted graphite irons are a form of cast iron with a graphite shape between the true nodular (SG) and gray iron. The graphite form is termed "vernacular" and is a short blunt flake. This type of iron is noted for its high fluidity and a solidification shrinkage rate between gray iron and the higher shrinkage rate of nodular

Table 7 Austempered Ductile from ASTM A-897

	Minimum tensile strength		Minimum yield strength				
Grade	Mpa	ksi	Mpa	ksi	Elongation %	Impact Energy (ft-lb)	Hardness
125/80/10	863	125	552	80	10	75	269–321
150/100/7	1035	150	690	100	7	60	302–363
175/125/4	1207	175	863	125	4	45	341–444
200/155/1	1380	200	1070	155	1	25	388–477
230/150/0	1587	230	1276	185	—	—	444–555

iron. The true advantage is the combination of the fluidity, which allows pouring a more intricate casting with thinner sections and its mechanical properties. Typical values for a compacted iron with ferritic matrix are tensile strength in the 35–55 ksi range with a yield strength of 25–40 ksi and 6% elongation. This type of iron requires very careful control of the molten metal chemistry and solidification parameters because of this commercial applications are limited.

5 MALLEABLE IRON

Malleable iron has long been a staple in the use of a cast product with ductility. Prior to the invention of ductile iron in 1948, malleable iron was the only cast iron available with sufficient strength and ductility to be used in an engineering application requiring impact resistance.

Malleable iron is characterized by a carbon form referred to as "temper carbon," graphite with an alloyed steel matrix. This carbon form is not generated during solidification but rather during a heat treatment of the as cast product. Malleable iron is solidified as white iron, a carbidic form of iron. The chemical analysis of the iron is adjusted so that the an unstable form of iron carbide (Fe_3C) is produced during solidification. In this form the casting is very hard and brittle. To create the malleable form of iron used in engineering applications the iron carbide is annealed by holding the castings at 1500 to 1780°F (800–970°C) for up to 20 h. At this temperature the iron carbide breaks down and forms a graphitic nodule, called a temper carbon and austenite (Fig. 3). Upon subsequent cooling the austenite transforms to the various products of ferrite, pearlite, or martensite as dictated by the cooling rate. The number of graphite particle and the compactness dictate the strength characteristics of the final product.

Because of the need to solidify as a carbidic form of iron, the casting designs are limited in size and the chemistry is adjusted to ensure rapid enough cooling

Figure 3 Typical ferritic malleable microstructure. Nital etch approximately 100×. The graphite shape in this structure is more an irregular popcorn like shape rather than the spheroidal configuration of the nodular iron (Fig. 2). This is a 40018 grade malleable iron.

Table 8 Typical Malleable Iron Properties from ASTM A-220

Grade	Tensile MPa/ksi	Yield MPa/ksi	% Elongation in 2"/50 mm
40010	414/60	276/40	10
45006	448/65	310/45	6
60004	552/80	414/60	4
90001	724/105	621/90	1

rate so as to not produce any form of graphite on solidification. In addition the alloying of malleable iron is restricted to those elements which do not stabilize the carbides. Such alloying additions would increase the time or entirely prevent the graphitization heat treatment.

Malleable irons are specified under ASTM A-220 and SAE J158. The grade designation is stated as M 3210, which is interpreted as malleable iron with a 32,000 psi yield and 10% elongation in 2 in. This is typical of a fully ferritic iron. Properties are listed in Table 8.

The properties of the final cast product are developed to the above specification by matrix control after the treatment to create the temper carbon graphite form. The matrix heat treatment is usually done as a continuation of the graphitization process rather then a separate posttreatment. Typical uses of malleable iron include gears, connecting rods, housings, mounting brackets, bearing caps and cases.

6 GENERAL PROPERTIES

Cast irons have a unique set of physical properties (Tables 9 and 10). These properties allow the material to be used in many applications which result in unique solutions to engineering problems. These values are typically dependent on the amount and shape of the graphite.

7 MODULUS OF ELASTICITY

The modulus of elasticity for cast irons is in general lower then that of steel. This is the result of the low-strength graphite which is part of the matrix. In addition,

Table 9 Poisson's Ratio

Type	Ratio
Steel	0.3
Gray iron	.024 low-strength iron .027 high-strength iron
Ductile iron	0.26
ADI	0.26
Malleable iron	0.25–0.28

Table 10 Physical Properties of Cast Irons

Property	Unit	Ductile iron			Cast steel	Malleable gray[a]	
		120-90-02	80-55-06	60-4-18	0.3% C	Iron	Iron
Solidus temperature	2100°F		2,100	2,100	2,650	2,050	2,140
	1150°C		1150	1150	1454	1121	1115
Density gravity	lb/in^3	0.25	0.25	0.25	0.28	0.26	0.26
	g/cm^3	7.01	7.01	7.01	7.86	7.27	7.15
Mean linear thermal expansion	$10^{-6}/°F$	6.0	6.0	6.0	7.0	6.7	5.8
	$10^{-6}/°C$	—	—	—	11.7	12	10.5
Vibration damping	—	Good	Good	Good	Poor	Good	Excellent
Heat of fusion	BTU/lb	55	55	55	108	55	55
	kcal/kg	30.5	30.5	30.5	59.9	30.5	30.5
Specific heat	BTU/lb/°F	0.15	0.14	0.13	0.11	0.12	0.13
	J/kg/K	628	586	544	460	502	544
Thermal conductivity (near room temp.)	[b]	0.06	0.08	0.10	0.11	0.14	0.11
Maximum permeability at 5000 Gauss	Oersted	290	570	2,100	—	2,350	800

[a]A medium to high-strength gray iron was selected for comparison purposes.
[b]$Cal/cm^2/cm/°C/s$

Table 11 Modulus of Elasticity (E) for Cast Irons

Type	GPA	psi × 10^6
Gray	124–151	18–22
Malleable	169–176	24.5–25.5
Ductile	159–165	23–24 ferritic matrix
	172–179	25–26 pearlitic matrix
Compacted	117–124	17–18 ferritic matrix
	124–165	18–24 pearlitic matrix

the modulus will vary for a given type of iron depending on the volume fraction of graphite present. See Table 11.

RELEVANT SPECIFICATIONS FOR CAST IRONS

Gray Iron

ASTM A 48 Specifications for Gray Iron Castings
ASTM A 126 Specification for Gray Iron Castings for Valves, Flanges, and Pipe Fittings
ASTM A 159 Specification for Automotive Gray Iron Castings
ASTM A 319 Specification for Gray Iron Castings for Elevated Temperature for Non-Pressure-Containing Parts
ASTM A 377 Specification for Gray Iron and Ductile Iron Pressure Pipe
ASTM A 436 Specification for Austemitic Gray Iron Castings
ASTM A 438 Transverse Testing of Gray Cast Iron
SAE J431 Automotive Gray Iron Castings

Ductile Iron

ASTM A 395 Specification for Ferritic Ductile Iron Pressure-Retaining Castings for use at Elevated Temperatures
ASTM A439 Standard specification for Austenitic Ductile Iron Castings.
ASTM A 476 Specification for Ductile Iron Castings for Paper Mill Dryer Rolls
ASTM a 536 Standard Specification for Ductile iron castings.
ASTM A 571 Standard Specification for Austenitic Ductile Iron Castings for Pressure-Containing Parts Suitable for Low-Temperature Service
ASTM A 716 Specification for Ductile Iron Culvert Pipe
ASTM A 746 Specification for Ductile Iron Gravity Sewer Pipe
ASTM A 897 Standard specification for Austempered Ductile Iron

Malleable Iron

ASTM A 47 Specification for Malleable Iron Castings
ASTM A 197 Standard Specification for Cupola Malleable Iron
ASTM A 220 Standard Specification for Pearlitic Malleable Iron Castings
ASTM A 338 Specification for Malleable Iron Flanges, Pipe Fittings, and Valve Parts for Railroad, Marine and Other Heavy-Duty Service at Temperatures up to 650°F (345°C)

Cast Irons General

ASTM E 10 Test for Brinell Hardness of Metallic Materials
ASTM A 247 Recommended Practice for Evaluating the Microstructure of Graphite in Iron Castings
ASTM A 256 Standard Method of Compression Testing of Cast Iron
ASTM A 327 Impact Testing of Cast Iron
ASTM E 562 Determining Volume Fraction by Systematic Manual Point Count

REFERENCES

Tabular information adapted from the following references and ASTM and SAE standards.

1. K. Rohrig and W. Fairhurst, *Heat Treatment of Nodular Cast Iron, Transformation Diagrams*, GieBerei-Verlag Gmbh, Dusseldorf, 1979.
2. C. Walton and T. Opar, *Iron Castings Handbook*, Iron Casting Society, Inc., 1981.
3. ASM *Metals Handbook*, vol. 1, 10th ed.
4. B. Kovacs, *Ductile Iron Society News*, DIS project #26, Poisson's ratio in ADI, 1995
5. ASM, *Metals Handbook*, vol. 15, *Castings*, 9th ed.
6. H.T. Angus, *Cast Iron Physical and Engineering Properties*, 2nd ed., Butterworths, London, 1976.

OTHER SOURCES OF INFORMATION

1. American Foundrymen's Society Inc., 505 State Street, Des Plains, IL 60016-8399.
2. Iron Castings Research Institute, 2838 Fisher Rd. #B, Columbus, OH 43204-3538.
3. The Ductile Iron Society, 28938 Lorain Road, Suite 202, North Olmsted, OH 44040.

9
Heat-Resistant Alloys

G. T. Murray
California Polytechnic State University
San Luis Obispo, California

Heat-resistant alloys are those developed for high-temperature strength and oxidation resistance and may be classified as follows:

1. Cr-Mo-type ferritic steels: Typical composition of these steels include 1Cr–0.5Mo, 2.25Cr–1Mo, 0.5Cr–0.5Mo–0.25V, 1Cr–1Mo–0.75V + B and Zr, and 9Cr–1Mo. Mo and V improve the creep resistance, Cr the oxidation resistance, and B and Zr additions increase strength and notch toughness. The higher alloy content generally means a higher cost. These steels are widely used in the petroleum and chemical industries for piping, heat exchangers and pressure vessels from 500°C (932°F) to 650°C (1202°F). ASME, ASTM and AMS (aerospace materials specifications) are used to identify nominal compositions for use as boilers, pressure vessels and aerospace applications.
2. Precipitation hardening stainless steels:* These steels have exceptional high strengths at temperatures below 400°C (752°F) but lose this strength rapidly as the temperature increases beyond 425°C (800°F), their practical high-temperature limit. They are used for landing gear hooks, hydraulic lines and fittings, bearings, valves, miscellaneous hardware and aircraft applications. They are also used extensively at room temperature for hardware, golf club heads and assorted applications.
3. Martensitic stainless steels: The martensitic stainless steels have good short time strengths up to about 590°C (1094°F) but are not as corrosion resistant as the ferritic and austenitic stainless steels. They are used in steam and gas turbines and as other miscellaneous parts at temperatures up to 540°C (1000°F). The 12%Cr martensitic stainless steels are often referred to as "super 12 chrome" steels.

*The stainless steels are discussed more extensively in Chapter 7.

4. 12% Cr ferritic steels: These steels are often subdivided into three classes as follows: (a) 12%Cr, (b) 12%Cr + Mo, Nb, W, (c) 12%Cr + Mo, Nb, W, + Co. These steels are listed in order of increasing performance and find extended use for steam and gas turbine blades, turbine rotors, and boiler tubing.

5. Austenitic stainless steels: these steels may be subdivided as follows: (a) Modified standard austenitic stainless steels with additions of boron and which include 304H, 310H, 316H, 321H, and 347H. (b) Nonstandard austenitic stainless steels such as Armco 17Cr-14Ni + Cu, Mo and Ti. Other steels in this category include the high Cr (18–21%) plus Mn, Mo and nitrogen, e.g., the nitronic steels.

Both of these subgroups of austenitic stainless steels are used to 600°C (1112°F) for superheater tubes, steam pipes and pressure vessels, but only the nonstandard type have adequate stress-rupture strengths for use to 630°C (1166°F). The austenitic stainless steels as a class have exceptional toughness, ductility and formability and are more corrosion resistant than the ferritic or martensitic steels as well as having better high-temperature strength and oxidation resistance. The austenitic stainless steels also provide the highest limiting creep strength of all the stainless steels. However, because of their high nickel content, they are generally somewhat more expensive.

6. Superalloys: These high-chromium alloys may be subdivided as follows: (a) Iron-base alloys, which contain primarily iron, nickel and chromium. Trade names include Haynes 556, Incoloys, and A-286. (b) Nickel-base alloys, which include the trade names Inconels, Hastelloys, Nimonics, Rene 41 and 95, Udimets, Astroloy and Waspaloy. (c) Cobalt-base alloys, which include Haynes 188, L-605, Mar-M 918 and MP35N.

The superalloys are the best of the high-temperature strength and oxidation resistance alloy group. The nickel base are the most widely used of the superalloys. They are generally better performers than the iron-nickel-chromium type and less expensive than the cobalt alloys. The iron-nickel-chromium are the least expensive of the three and find considerable use for that reason.

For comparative screening purposes one can use the high-temperature yield strength of the heat-resistant alloys. This data for a number of wrought alloys of each group is presented in Table 1. However, for design purposes either creep or stress-rupture data must be employed. The stress-rupture data for typical heat-resistant alloys is presented in Fig. 1 and 2, the former showing the stress-to-produce rupture in 1000 h and the latter to produce rupture in 10,000 h. The design engineer must usually determine whether the component is limited by fracture or by the degree of deformation. The creep strength for some selected alloys is presented in Table 2. In general the alloys with the better stress-rupture properties will also show the best creep strengths.

The alloys discussed above are of the wrought and the mechanical alloyed (MA) types. The MA alloys contain a fine dispersion of oxide particles and are fabricated by powder metalurgy (PM) techniques. Several other PM alloys have replaced some of the forged alloys used as turbine disks. Although many powder metallurgy techniques have been applied to these alloys, production has been primarily confined to hot isotatic pressing processes and hot compaction followed by extrusion processes.

Many of the wrought alloys are also suitable for investment casting processes. In addition both nickel- and cobalt-base alloys have been developed to be used

Table 1 Yield Strength at Elevated Temperatures for Selected Heat-Resistant Alloys

Alloy Temperature °C (°F)	Y.S. MPa (Ksi)		
	400 (752)	600 (1112)	800 (1472)
1Cr-0.5Mo Ferritic steel, hardened	530 (77)	405 (55)	—
2.25Cr-1Mo Ferritic steel, hardened	580 (84)	460 (67)	—
1.25Cr-0.5Mo + V Ferritic, hardened	610 (89)	430 (62)	—
12Cr-0.6Mo Ferritic stainless	624 (91)	332 (48)	—
12Cr + Ni, Mo, V, Nb Ferritic stainless	700 (102)	410 (60)	270 (39)
PH 13-8Mo Prep. hardening stainless	1052 (152)	—	—
17-4PH Prep. hardening stainless	930 (135)	—	—
18Cr-8Ni Austenitic	110 (16)	100 (15)	—
18Cr-10Ni-Ti Austenitic	150 (22)	150 (22)	—
18Cr-12Ni + Mo Austenitic	480 (70)	405 (59)	—
Incoloy 909 Fe base superalloy	944 (137)	868 (120)	—
Incoloy 907 Fe base superalloy	1000 (145)	980 (142)	420 (61)
A286 Fe base superalloy	630 (91)	605 (88)	360 (52)
Inconel 718 Ni base superalloy	1060 (154)	1060 (154)	688 (100)
Rene 41 Ni base superalloy	1035 (150)	1010 (147)	830 (120)
Inconel X-750 Ni base superalloy	768 (111)	755 (110)	400 (59)
Nimonic 90 Ni base superalloy	688 (100)	676 (98)	—
Nimonic PK 33 Ni base superalloy	670 (97)	660 (96)	620 (90)
Astroloy Ni base superalloy	1000 (145)	960 (139)	840 (122)
MA 754 Ni base superalloy	580 (84)	570 (275)	275 (40)
Hastealloy X Ni base superalloy	280 (41)	230 (33)	—
L 605 Co base superalloy	240 (35)	260 (38)	—
Haynes 188 Co base superalloy	—	305 (44)	—

specifically as cast alloys. The properties of some typical polycrystalline cast nickel and cobalt alloys are listed in Table 3. Thermal properties of selected cast and wrought superalloys are listed in Table 4. Applications of the superalloys are presented in Table 5.

In the 1970s and 1980s nickel alloys designed specifically for directionaly solidified columnar-grain and single-crystal cast components emerged. These processes are highly specialized and the design engineers work closely with metallurgical engineers in alloy selection and processing. Design engineers for which this book is intended are seldom if ever involved in materials selection for these highly specialized applications and rather complex processes. The interested reader is referred to the ASM, Int. *Metals Handbook*, vol. 1, 10th ed., pages 993–1005.

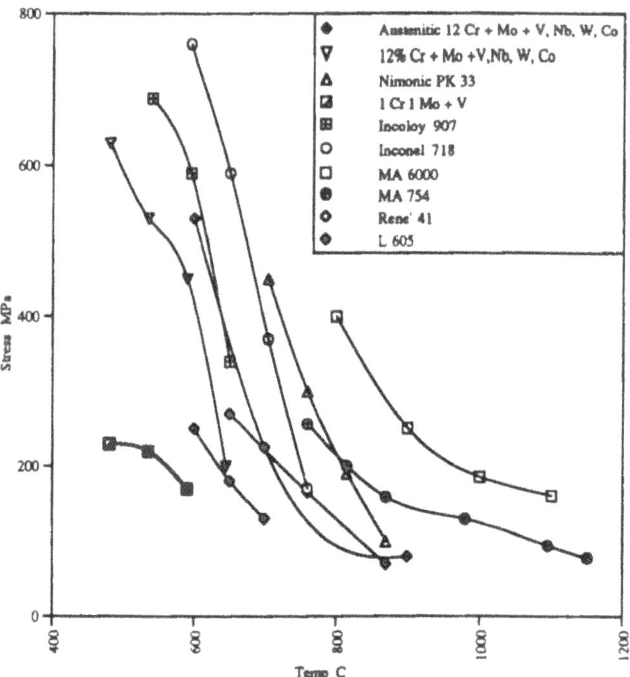

Figure 1 Stress-to-produce rupture in 1000 h.

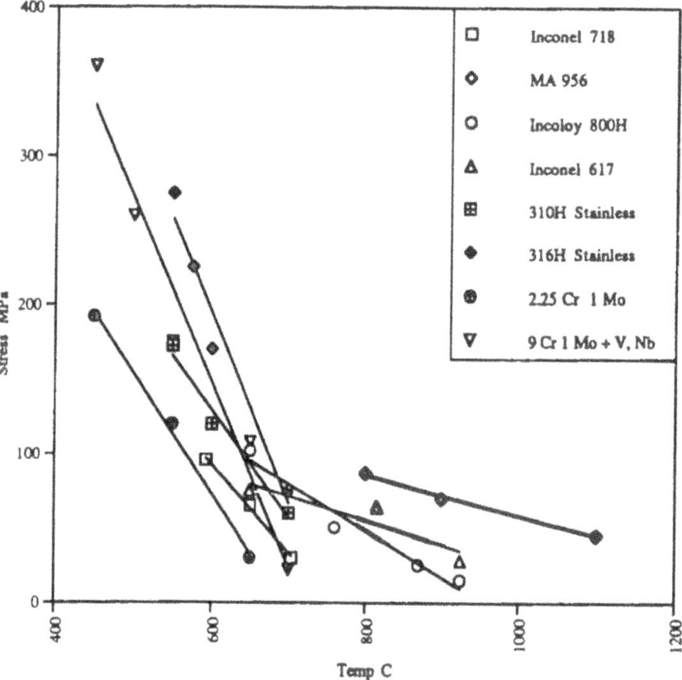

Figure 2 Stress-to-produce rupture in 10,000 h.

Table 2 Creep Strength (MPa) for 1% Elongation in 10^5 H

Alloy	450°C	500°C	550°C	600°C	650°C
2.25Cr–1Mo	170	100	62	27	—
9Cr–1Mo	—	—	—	39	20
12Cr + Mo + V	—	—	—	157	83
347H SS	—	220	170	120	70
12Cr Ferritic	170	85	50	23	—
Inconel 706	—	—	—	690	400
Inconel 718	—	—	—	—	450
Incoloy 800	—	275	300	120	60

Table 3 Elevated Temperature Strength and Stress Rupture of Selected Polycrystalline Cast Nickel-Base Superalloys

		1000 h Stress rupture	
Alloy	Y.S. [MPa (ksi)] at 538°C	870°C	980°C
Mar-M247	825 (120)	290 (42)	125 (18)
Mar-M246	860 (125)	290 (42)	125 (18)
Mar-M432	910 (132)	215 (31)	97 (14)
IN 100	885 (128)	260 (38)	90 (13)
B-1900	870 (126)	250 (36)	110 (16)
Udimet 500	815 (118)	165 (24)	—
CMSX-2	1245 (180)	345 (50)	170 (25)

Table 4 Thermal Properties of Selected Superalloys

	CTE (10^{-6}/K)		Thermal cond. (W/m·K)	
Alloy	538°C	871°C	538°C	871°C
In 718	14.4	—	19.6	24.9
In X750	14.6	16.8	18.9	23.6
Nimonic PK 33	13.1	16.2	19.2	24.7
Rene 41	13.5	15.6	18.0	23.1
Udimet 500	14.0	16.1	18.3	24.5
Waspaloy	14.0	16.0	18.1	24.1
Incoloy	16.4	18.4	20.1	—
A286	17.6	—	22.5	—
L-605	14.4	16.3	19.5	26.1
Mar-M246	14.8	—	18.9	—
In-100	13.9	—	17.3	—
B-1900	13.3	—	16.3	—

Table 5 Applications of Superalloys

Alloy type	Applications
Incoloy 800 series Fe-Ni-Cr	Catalytic cracking tubes, reformer tubes, aqueous corrosion applications, sulphuric and phosphoric acid environments, heat exchangers, industrial furnaces, steam generators.
Incoloy 902, 903, 907 Fe-Ni-Cr	Special physical properties such as controllable thermoelectric coefficient, low- and constant-coefficient thermal expansion.
Incoloy 909, 925 Fe-Ni-Cr	Low thermal expansion coefficient but with good notch-rupture and strength properties. Used for gas turbine casings, shrouds, vanes and shafts.
MA series Incoloy 956-Fe-Ni-Cr, Ni base Inconel MA 754, 758, 760, 6000	Oxide dispersion strengthened and produced by mechanical alloying. Good high-temperature strength and oxidation resistance to 1100°C (2000°F). Used in hot sections of gas turbines, vacuum furnaces, handling molten glass.
Inconel 617	Gas turbines, petrochemical processing, heat treating equipment, nitric acid production.
Inconel 718, X-750	Gas turbines, rocket motors, spacecraft, pumps.
Inconel 751	Exhaust valves in internal combustion engines.
Nimonic Ni-Cr and Ni-Cr-Co	Gas turbines, aerospace, internal combustion engines, metal working equipment, thermal processing.
L605 Co base wrought	Gas turbines, fabricated assemblies.
Mar-M Co base cast	Complex shapes, nozzle guide vanes.

REFERENCES

1. Inco Alloys International.
2. *ASM Metals Handbook, vol. 1*, (1990), ASM International, Materials Park, Ohio.

10
Properties and Applications of Polymers

G. T. Murray
California Polytechnic State University
San Luis Obispo, California

1 INTRODUCTION

The applications of polymers are enormous, ranging from an almost endless list of consumer uses, to high-strength and temperature uses, to high-tech electronics and space exploration uses.

Polymers are generally classified as those being thermoplastic, thermosetting or elastomers. Thermoplastic polymers consist of long-chain linear molecules that permit the plastic to be easily formed at temperatures above a critical temperature called the *glass transition temperature* (T_g). For some thermoplastic polymers (thermoplasts) this temperature is below room temperature; hence they are brittle at ambient temperatures. Thermosets have a three-dimensional network of atoms. They decompose on heating and cannot be reformed or recycled. Certain polymer materials fit into both the thermoplastic and the thermosetting categories. For example the unsaturated polyesters are thermosetting polymers while other polyesters are thermoplastic ones. Polyimides and polyurethanes can also fit into either category depending on their structure. Elastomers, better known as rubbers, are polymeric materials whose dimensions can be changed drastically by applying a relative modest force, but return to their original value when the force is released. The molecules are extensively kinked such that when a force is applied, they unkink or uncoil and can be extended in length up to about 1000%. In general, they must be cooled to below room temperature to be made brittle. Thermoplastic elastomers (TPE), which are block copolymers in which one component of the block is flexible and the other rigid, is a special class of polymers now considered by many to be a fourth class of polymers. One must now also consider the polymer alloys or blends. The term "alloy" is used here to describe miscible or immiscible blends which are usually blended as melts. Miscible blends are characterized by properties, such as glass temperature values, which lie somewhere between the two components. Immiscible blends are characterized by two or more phases. The most widely used blend is the blend of PPO and PS which is one of the "big five" engineering thermoplastics. Many of these blended polymers fall into the thermoplastic category.

Some authors have used the terms "engineering plastics," engineering polymers" and "high-performance plastics," which all mean the same thing, to describe the higher strength and better high-temperature resistance polymers and to distinguish them from the general purpose polymers. These engineering plastics generally compete in terms of load-bearing characteristics with metals and ceramics. They lend themselves for use in engineering design, such as for gears and structural members. In the thermoplastic polymer category the "general-purpose polymers" consist of those such as polyethylene, polypropylene, polyurethanes, acrylics and polyvinyl chlorides (some of these may be considered as engineering polymers in the reinforced or copolymer forms). The engineering thermoplastic polymers include the nylons, polycarbonates, acetals, polyphenylene ether and thermoplastic polyesters. The engineering or high-performance polymers also include the more recently developed high-temperature thermoplastics, which include polyimides (PA), polyamide-imides (PAI), polysulfones (PSU), polyetherketones (PEEK), and polyphenylene sulfides (PPS), which have service temperatures in excess of 200°C. These plastics are finding increasing applications, particularly for the under-the-hood automotive components. The thermoplastic engineering polymers and alloys can be either "neat" (unfilled") or reinforced with minerals, glasses or other fibers. The thermosetting molding compounds are almost always reinforced and considered to be engineering plastics. Elastomers are generally not considered as engineering polymers although many are used in engineering applications. No attempt will be made herein to subdivide the property data into engineering and other polymers.

The materials selection process where polymers are a viable material is a complex one because of the wide variety of plastics available. For many applications polymers are not used in their "neat" or pure form, i.e., without additives. Some common additives are fillers, reinforcements such as fibers and particulates of glass and minerals, plasticizers, and colorants. The properties depend on the types and quantities of additives used. Glass fibers are the most common reinforcement material used. Glass fibers increase tensile and impact strengths and the moduli. They generally improve the thermal conductivity and reduce the coefficient of thermal expansion (CTE). For the case of glass fibers there will be some overlap with Chapter 13 on polymer matrix composites, since glass-filled polymers are really composites.

2 THERMOPLASTIC POLYMERS

Table 1 lists a range of room temperature mechanical properties for the common groups of thermoplastic polymers. Such a listing is convenient for use in narrowing the field of potential polymers for use in a particular design. The design engineer can then examine the more detailed data supplied by vendors for particular polymers.

The heat deflection temperature (HDT) in Table 1 is defined in ASTM 648 test methods as the temperature at which a 125-mm (5-in.) bar deflects 0.25 mm (0.10

in.) under a specified stress, 1.82 MPa (0.264 ksi) being the most commonly used stress. In some tables this temperature may be listed as the "deflection temperature under load" (DTUL), both values being identical. Typical physical properties for selected thermoplastic polymers are given in Table 2.

The thermoplastic polymers have a wide range of applications. A compilation of these applications along with relative strengths and other characteristics are listed in Table 3.

2.1 Temperature Effects

In addition to the HDT or DTUL deflection under load temperatures, the maximum continuous use temperatures (MCUT), which is defined as the temperature at which the strength decreases to 50% of its original value in 11.4 years, may be used. What is more common to find in the literature is the continuous service or use temperature, which has not been so well defined. One should check with the manufacturer or vendor for the meaning of this temperature. Some quoted temperatures are listed in Table 4. In many cases these may be only the recommended temperature based on experience of the manufacturer.

Tensile creep and stress-rupture curves are also valuable for design and materials selection purposes. Examples for some common polymers are illustrated in Figs. 1 and 2.

2.2 Chemical Effects

Polymers are inert in many aqueous solutions and as such are frequently used to replace metal parts. Typical examples include ABS, polypropylene, and PVC as piping, tanks, scrubbers, and columns in processing equipment. Teflon has been used in heat exchangers, and nylons and acetals for small parts and gears. In the chemical industry polymers have been used alone, and as metallic coatings or liners where more strength is required. The chemical resistance and applications of the more common and commercially used polymers are listed in Table 5. This short compilation was made by J. H. Mallison in a book edited by P. A. Schweitzer. A far more complete listing of environmental effects is presented in Appendix A of this book, also compiled by P. A. Schweitzer.

2.3 Effects of Additives

The common additives used to enhance the performance of thermoplastic polymers are listed here. For a detailed description of additives see Ref. 2.

Antioxidants	protect polymers against atmospheric oxidation
Colorants	dyes and pigments
Coupling agents	used to improve adhesive bonds
Filler or extenders	minerals, metallic powders, and organic compounds used to improve specific properties or to reduce costs.
Flame retardents	change the chemistry/physics of combustion
Foaming agents	generate cells or gas pockets
Impact modifiers	materials usually containing an elastomeric component to reduce brittleness.
Lubricants	substances that reduce friction, heat, and wear between surfaces

Table 1 Typical Mechanical Properties of Thermoplastic Polymers at Room Temperature

Material	Tensile strength [MPa (ksi)]	Elong %	Tensile modulus [MPa (ksi)]	Hardness	Izod impact strength [J/cm (ft-lb/in)]	HDT (°C)[a]
Polyethylene						
High density	21–35 (3–5)	15–100	700–1400 (100–200)	R_r 65	3.0–10 (6.5–20)	40–44
Low density	7–21 (1–3)	50–800	100–250 (14.5–36.2)	R_r 10	no break	—
Polypropylene	30–40 (4.5–5.8)	150–600	1150–1550 (166–225)	R_r 90	0.2–0.6 (0.4–1.2)	49–60
Polystyrene	33–55 (5.1–8.0)	1–4	2400–3350 (348–486)	R_r 75	0.1–0.2 (0.2–0.4)	76–94
Polyvinylchloride	35–63 (5–9)	2–30	2000–4200 (286–600)	R_r 115	5–10 (10–20)	60–77
ABS	35–48 (5–7)	15–80	1750–2500 (250–357)	R_r 90–110	1.6–1.7 (3–5)	88–107
Polyamides						
Nylon 6/6	84 (12)	60–100	2070–3245 (300–470)	R_r 118	0.53–1.1 (1–2)	79–93
Nylon 6/12	62 (8.8)	150–340	2100 (300)	R_r 114	0.53–0.8 (1.0–1.6)	58
Polycarbonates	63 (9)	110	2400 (348)	R_r 118	6.4–8.5 (12–16)	132
PMMA Acrylic	55–75 (8–10.9)	5	2400–3100 (348–450)	R_r 130	0.15–0.3 (0.3–0.6)	—
Polyesters	56 (8)	300	2400 (348)	R_m 117	2.1 (4.2)	50–85

Acetals						
Homopolymer	69 (10)	50	3588 (520)	R_m 92	0.53–0.8 (1.0–1.5)	154
Polyimides	97 (14)	8	2070 (300)	R_m 97	2.0 (1.5)	154
Polysulfones	70.3 (10.2)	5–6	2482 (360)	R_m 69	0.7 (1.3)	192
Polyetherketone (PEEK)	100 (14.5)	>40	3900 (565)	R_r 123	0.6–0.8 (1.1–1.5)	167
Polyethylene terephthalate (PET)	45–145 (6.5–21.0)		2300–10300 (330–1500)	R_r 120	0.2–0.8 (0.4–1.5)	50–210
Polyvinylidene chloride (PVDC)	19 (2.8)	350	345–552 (50–80)	R_m 60–65	—	54–65
Acrylics cast	66 (9.6)	4–9	2622–3105 (380–450)	R_m 85–105	0.2–0.3 (0.35–0.55)	95
Cellulosics	14–15 (2.0–8.0	6–60	690–2100 (100–300)	R_r 50–115	1.1–4.5 (2.0–8.5)	65
Polyamideimides	125–185 (18–27)	5–12	4900 710	R_r 104	0.53 (1.0)	275
Polyarylates	69 (1.0)	50	16600 2400	R_r 125	22.4 (50)	175
Polyphenylene oxide (PPO)	55 (8)	50	2484–2622 (360–380)	R_r 114	2.7–3.7 (5.0–7.0)	100

[a]264 psi

Table 2 Typical Physical Properties of Selected Thermoplastic Polymers

Material	Specific gravity	Thermal conduct. (W.M·K)	Thermal expansion coefficient [10^{-5}/K (10^{-5}/°F)]	Volume[a] resistivity short time (ohm-m)	Dielectric strength [V/10^{-3} mm (V/mil)]
Polyethylene					
High density	0.96	0.44	11–13 (6.1–7.2)	10^{15}–10^{16}	20–19 (510–480)
Low density	0.92	0.35	13–20 (7.2–11.1)	10^{15}–10^{16}	39–18 (990–460)
Polypropylene	0.9	0.12–0.14	6–10 (1.0–5.5)	10^{15}	26–20 (660–510)
Polyvinylchloride	1.4	0.16	9–18 (5–10)	10^{13}	24 (610)
Nylon 6/6	1.14	0.24	8–9 (4.4–5)	10^{13}	16 (410)
Polycarbonates	1.2	0.20	6.8 (3.8)	2×10^{14}	16 (410)
PMMA (acrylic)	1.18	0.19	4.5 (2.5)	10^{13}–10^{14}	20–16 (510–410)
Acetal copolymer	1.4	0.23	8–10 (4.4–5.5)	—	—

Polyimides	1.43	—	5.4–8 (3–4.5)	10^{14}–10^{15}	22 (560)
Acetals	1.4	0.225	15.3–18.0 (8.5–10.0)	10^{15}	20 (500)
ABS	1.03	0.2–0.33	9.5 5.3	1–5 × 10^{16}	14–16 (350–400)
Acrylics	1.7–1.2	0.0.3	9–16 (50–90)	>10^{15}	18–21 (450–530)
PEEK	1.26–1.32	—	5.5 3.0	10^{16}	19 480
Polystyrene	1.04–1.09	0.1–0.14	11–14 (6–8)	10^{16}	20–28 (500–700)
PPO	1.06–1.10	0.22	5.4–7.2 (3.0–4.0)	10^{16}–10^{17}	16–24 (400–600)
Polysulfone	1.24	0.1	9.4–10 (5.2–5.6)	5 × 10^{16}	17 (425)
Cellulosics	1.1–1.2	0.16–0.2	18–36 (10–20)	10^{10}–10^{15}	14–20 (350–500)
Polyesters	1.3–1.4	0.2–0.3	11–16 (6–9)	10^{15}	23 (580)

[a]Volume resistivity increases with time.

Table 3 Properties and Uses of Thermoplastics[a]

Plastic material	First introduced	Strength	Electrical properties	Acids	Bases	Oxidizing agents	Common solvents	Product manufacturing methods	Common applications
Acetal resins	1960	H		P	P	P	G	Injection, blow, or extrusion molded	Plumbing, appliance, automotive industries
Acrylic plastics	1931		G	P	P	F-P	F-P	Injection, compression, extrusion, or blow molded	Lenses, aircraft and building glazing, lighting fixtures, coatings, textile fibers
Arc extinguishing plastics	1964		E					Injection or compression molded and extruded	Fuse tubing, lightning arrestors, circuit breakers, panel boards
Cellulose plastics Cellulose acetate	1912	M					P	All conventional processes	Excellent vacuum-forming material for blister packages, etc.
Cellulose acetate butyrate		H				F	F	Molded with plasticizers	Excellent moisture resistance-metallized sheets and film, automobile industry
Cellulose nitrate	1889	M						Cannot be molded	Little use today because of fire hazard
Cellulose propionate		H						All conventional processes	Toys, pens, automotive parts, ratio cases, toothbrushes, handles
Ethyl cellulose		H+						All conventional processes	Military applications, refrigerator components, tool handles
Chlorinated polyether	1959	M+	A+	VG	VG	VG	VG	Injection, compression, transfer, or extrusion molding	Bearing retainers, tanks, tank linings, pipe, valves, process equipment
Fluorocarbon (TFE)	1930	M	A	VG	VG	VG	VG	Molding by a sintering process following preforming	High-temperature wire and cable insolution, motor-lead insolution; chemical process equipment

Fluorinated ethylene prophylene (FEP)		M+	A	G	G	G	G	Injection, blow molding, and extrusion and other conventional methods	Autoclavable laboratory ware and bottles
Glass-bonded mica	1919		G	VG	G	G	G	Moldable with inserts like the organic plastics	Arc chutes, radiation generation equipment, vacuum-type components, thermocouples
Hydrocarbon resins	1960	M	A	A	A	A	A	Molding with transfer and compression process, coating	Used as lamination resins for various industrial laminates
Methylpentene polymers (TPX)	1965	M+	E	F	F	F	F	Most conventional processes	Used for electrical and mechanical applications
Parylene (poly-para-xylene)	1960							A monomer of the organic compound is vaporized and condensed on a surface to polymerize	Coating material for sensing probes.
Phenoxy plastics	1962	H-M	F	F	F	F	F	Injection, blow, and extrusion molding, coatings and adhesives	Adhesives for pipe-bonding compounds, bottles
Polyamide plastics Nylon	1938	H-M	A	P	P	P	VG	Injection, blow, and extrusion molding	Mechanical components (gears, cams, bearings), wire insulation pipe fittings
Polycarbonate plastics	1959	H-M	VG	G-F	G-F	F	F	All molding methods, thermoforming, fluidized-bed coating	Street light globes, centrifuge bottles, high-temperature lenses, hot dish handles
Polychlorotrifluoroethylene (CTEE)	1938	H	E	VG	VG	VG	VG	Molded by all conventional techniques	Wire insulation, chemical ware, pipe lining, pipe, process equipment lining
Polyester-reinforced urethane	1937	H		G	G	G	G	Compression molded over a wide temperature range	For heavy duty leather applications–industrial applications
Polyimides	1964	H-M	E	G-VG	P	P	G-VG	Molded in a nitrogen atmosphere	Bearings, compressors, valves, piston rings

Table 3 Continued

Plastic material	First introduced	Strength	Electrical properties	Acids	Bases	Oxidizing agents	Common solvents	Product manufacturing methods	Common applications
Polyolefin plastics Ethylene vinyl acetate (EVA)	1940	H		G	G	G-F	G-F	Most conventional processes	Molded appliance and automotive parts, garden hose, vending machine tubing
Polyallomers	1962	H	G-VG	F	F	F	F	Molding processes, all thermoplastic processes	Chemical apparatus, typewriter cases, bags, luggage shells, auto trim
Polyethylene	1939	H	VG	G-VG	G-VG	P	P	Injection, blow, extrusion, and rotational	Piep, pipe fittings, surgical implants, coatings, wire and cable insulation
Polypropylene plastics	1954	H-M	VG	VG	VG	F	F	Same as PVC	Housewares, appliance parts, auto ducts and trim, pipe, rope, nets
Polyphenylene oxide	1964	M	F-G	E	E	VG	VG	Extruded, injection molded, thermoformed and machined	Autoclavable surgical tools, coil forms, pump housings, valves, pipe
Polysulfone	1965	M	VG	VG	VG	F	P-F	Extrusion and injection molded	Hot-water pipes, lenses, iron handles, switches, circuit breakers
Polyvinylidene fluoride (VF_2)	1961	H	VG+	G-VG	G	G-VG	G-VG	Molded by all processes, fluidized bed coatings	High-temperature valve seats, chemical resistant pipe, coated vessels, insulation
Styrene plastics ABS plastics	1933	M-H	VG+	G-VG	G-VG	F-G	F	Thermoforming, injection, blow, rotational and extrusion molds	Business machine and camera housings, blowers, bearings, gears, pump impellers

Polystyrene	1933	M-H	VG+	G	G	F	P-F	Most molding processes	Jewelry, light fixtures, toys, radio cabinets, housewares, lenses, insulators
Styrene acrylonitrile (SAN)		H	VG	VG	G	G	G	Most molding processes	Lenses, dishes, food packages, some chemical apparatus, batteries, film
Urethane	1955	M-H+		G-VG	G	G	F-G	Extruded and molded	Foams for cushions, toys, gears, bushings, pulleys, shock mounts
Vinyl plastics Copolymers of vinyl acetate and vinyl chloride	1912	M+		G	G-F	G	G	All molding processes	Floor products, noise insulators
Poly(vinyl acetate)	1928	M		P-G	G	G	P	Coatings and adhesives	Adhesives, insulators, paints, sealer for cinder blocks
Poly(vinyl aldehyde)	1940	H		VG	VG	VG	VG	Most molding processes	Used for coatings and magnet wire insulation, interlayer of safety glasses
Poly(vinyl chloride) (PVC)	1940	M-H		VG	G	G	G	Extrusion, injection, rotational, slush, transfer, compression, blow mold	Pipe conduit and fittings, cable insulation, down spouts, bottles, film
PVC plastisols	1940	M-H		VG	G	G	G	Slush and rotationally molded, foamed, extruded	Used in coating machines to cover paper, cloth, and metal
Poly(vinylidene chloride)	1940	H		VG	VG	VG	VG	Same as PVC	Auto seatcovers, film, bristles, pipe and pipe linings, paperboard coatings

[a]A, average; F, fair; H, high; P, poor; E, excellent; G, good; M, moderate; VG, very good.
Source: Ref. 5.

Table 4 Continuous Use Temperatures of Unfilled Thermoplastic Polymers

Material	CUT [°C (°F)]
Acetals	91 (195)
Acrylics	60–80 (140–190)
ABS	85 (185)
Nylon 6/6	79–150 (175–300)
Nylon 6/12	79–150 (175–300)
Polyarylsulfone	190 (375)
Polycarbonate	120–135 (250–275)
Polyesters	93–120 (200–250)
PEEK	250 (482)
Polyethylene HD	80 (175)
Polyethylene LD	40 (104)
Polyether sulfone	170 (340)
PET	100 (212)
PPO	80–105 (175–220)
Polypropylene	70 (158)
Polystyrene	60–95 (140–205)

Optical brightners	organic substances that absorb UV radiation below 3000 Å and emits visible radiation below 5500 Å
Plasticizers	increase workability
Reinforcing fibers	increase strength, modulus, and impact strength
Processing aids	improve hot-pressing characteristics
Stabilizers	control for adjustment of deteriorative physicochemical reactions during processing and subsequent life

A list of the various fillers and the properties which they enhance is presented in Table 6.

Some polymers have highly useful properties without external modification, such as the polymethylmethacrylates (PMMA; plexiglas, arcylite, etc.) and polyethylene. At the other end of the spectrum polyvinylchloride (PVC) profits from all additives and is virtually useless in the "neat" or pure form. Other thermoplastic polymers share to varying degrees the need for a positive response to external additives. In the following data presentations only the impact characteristics, strengths, moduli, and HDT will be illustrated.

The impact characteristics of PVC as a function of the quantity of the additive in parts per hundred (phr) is depicted in Fig. 3. The modifier is of the paraloid core/shell methacrylate-butadiene-styrene type from Rohm and Haas. The K69, K62, and K55 numbers refer to decreasing molecular weights. Note that the brittle to ductile fracture transition occurs at lower quantities of modifier as the molecular weight of the PVC increases. Table 7 shows the effect of rubber additions on the impact strength off nylon 6/6 (Zytel, a Dupont nylon). Figure 4 shows the effect of the weight percent of the paraloid KM-653 (Rohm and Haas modifier) on the impact strength of polybutylene terephtalates (PBT) and likewise for Paraloid KM-

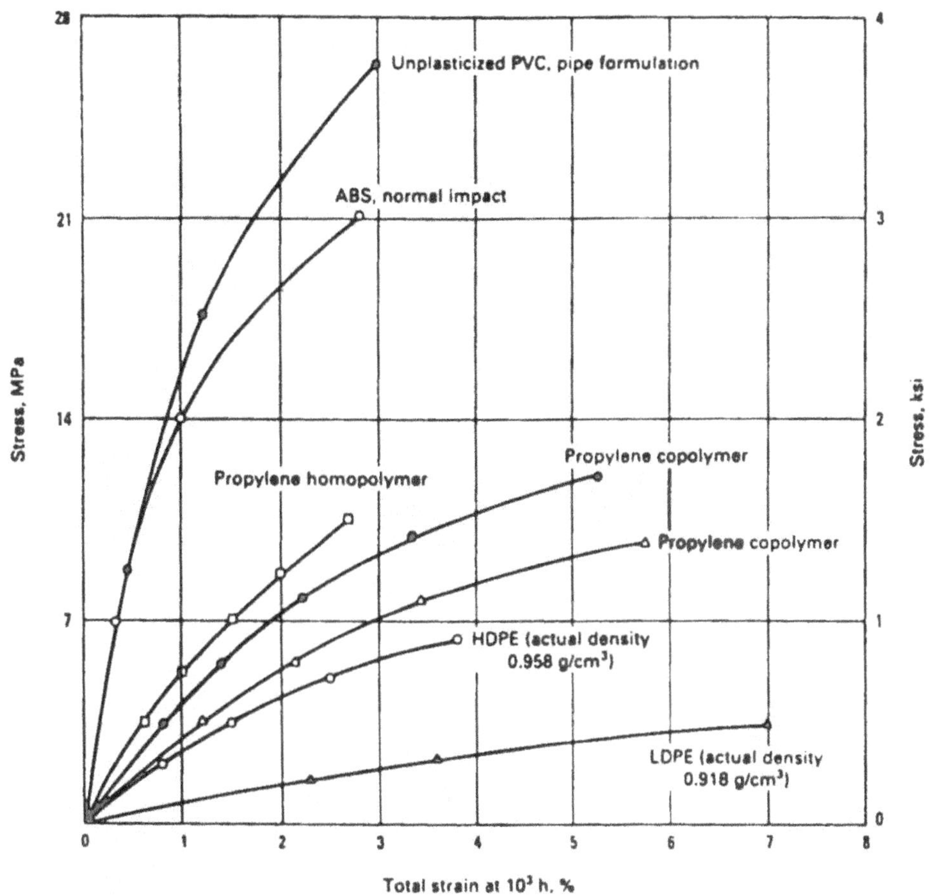

Figure 1 Tensile creep strain versus stress of various thermoplasts after 1000 h at room temperature. (From Ref. 10.)

330 on PET and copolyesters derived from PET (Kodar 6763 and Kodar A-150) in Fig. 5. Other thermoplastic polymers that are frequently improved by modifier additions in terms of impact behavior include polypropylene (20% improvement by styrene-butadiene-styrene), acetals (POM), polycarbonate and certain polymer blends. Fibers, which are primarily added to improve strength and stiffness, also tend to improve the toughness as measured by impact strength. Figure 6 shows the increase in impact strength of nylon, polycarbonate, polypropylene and polystyrene by the addition of 30 wt% of glass fibers.

Glass fibers effectively enhance strength and moduli of thermoplastic polymers. The effect of 30 wt% glass fiber additions on the tensile and flexural moduli of nylon, polycarbonate, polypropylene and polystyrene are depicted in Figs. 7 and 8. The effect of 30 wt% additions of glass fibers on the HDT of a number of polymers is shown in Tables 8 and 9. A more inclusive diagram of the effect of 30 vol% of glass fibers on both the continuous use temperatures and heat distortion temperatures for a large number of polymers is depicted in Fig. 9.

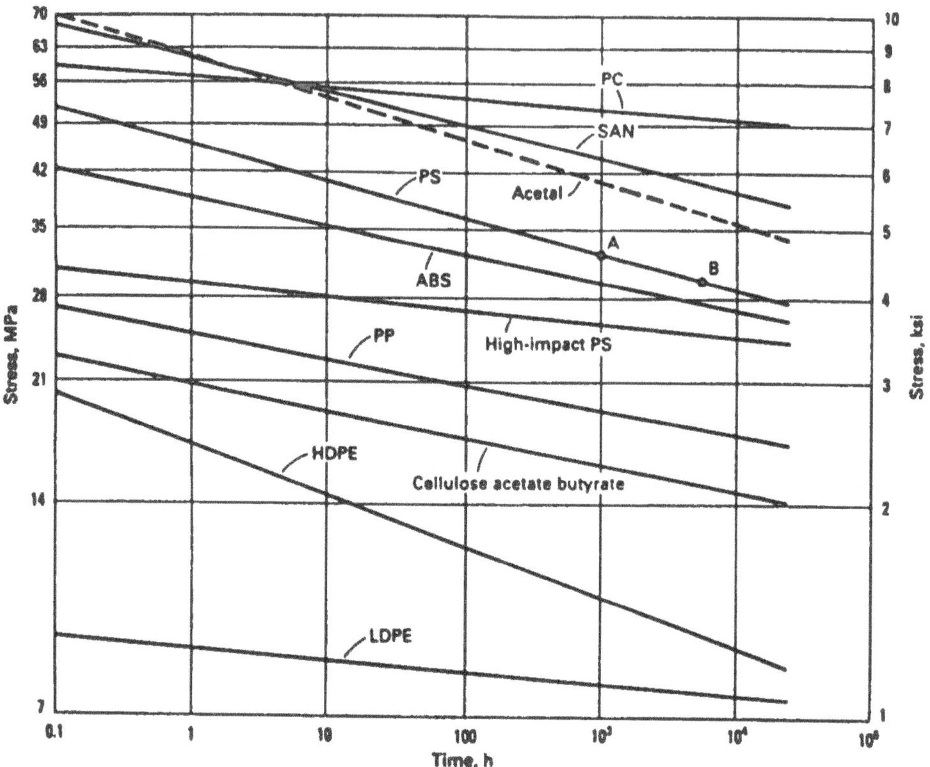

Figure 2 Tensile stress for rupture versus time for polymers. (From Ref. 11.)

3 THERMOSETTING POLYMERS

Thermosets are polymers that take on a permanent shape or *set* when heated, although some will set at room temperature. They are formed by a large amount of cross-linking of linear prepolymers or by the direct formation of networks by the reaction of two monomers. The classic reaction of this type is that of phenolformaldehyde. As the reactants combine, random three-dimensional networks are formed via branched structures until a giant molecule is formed. Often, these thermosets are partially polymerized and shipped in this state as a powder to the manufacturer, who completes the polymerization (curing) in the mold. In other cases the polymerization and shaping occur simultaneously. The principal thermosets are the phenolics, the amino acids, the unsaturated polyesters, the epoxies, and the cross-linked polyeurethanes. Some polymer technologists exclude thermosets from the engineering polymers, however, others do not and we will make no distinction herein.

The phenols are the most widely used thermosets and represent about 40% of thermoset use. They are relatively inexpensive and are readily molded with good stiffness. Most contain wood-flour or glass-flour fillers and often glass fibers for strength. Typical applications include pulleys, electrical insulators, connector plugs, housings, handles, and ignition parts.

Table 5 Summary of the Chemical Resistance of the Common Polymers and Their Applications

Type of plastic	Normal upper service limit °C	°F	Typical areas for use
Polyvinyl chloride	60	140	Piping: water, gas, drain, vent, conduit, oxidizing acids, bases, salts; ducts: (breaks down into HCl at high temperatures); windows plus accessory parts and machine equipment in chemical plants; liners with FRP overlay
Chlorinated PVC	82	180	Similar to PVC, but upper-temperaturre limit is increased
Polyethylene	60	140	Tubing, instrumentation (laboratory): air, gases, potable water, utilities, irrigation pipe, natural gas; tanks to 12-ft diameter
High-density polyethylene	82	180–220	Chemical plant sewers, sewer liners, resistant to wide variety of acids, bases, and salts; generally carbon filled; highly abrasion-resistant; can be overlaid with FRP for further strengthening
ABS	60	140	Pipe and fittings; transportation, appliances, recreation; pipe and fittings is mostly drain, waste, vent; also electrical conduit; resistant to aliphatic hydrocarbons but not resistant to aromatic and chlorinated hydrocarbons; formulations with higher heat resistance have been introduced
Polypropylene	82–104	180–220	Piping and as a composite material overlaid with FRP in duct systems; useful in most inorganic acids other than halogens; fuming nitric and other highly oxidizing environments; chlorinated hydrocarbons cause softening at high temperatures; resistant to stress cracking and excellent with detergents; flame-retardant formulations make it useful in duct systems; further reinforced with glass fibers for stiffening, increases the flex modulus to 10^6 and deformation up to 148°C (300°F)
Polybutylene	104	220	Possesses excellent abrasion and corrosion resistance; useful for fly ash and bottom ash lines or any lines containing abrasive slurries; can be overlaid with fiberglass for further strengthening when required
General-purpose polyesters	50	120	Made in a wide variety of formulations to suit the end-product requirements; used in the boat industry, tub showers, automobiles, aircraft, building panels

Table 5 Continued

Type of plastic	Normal upper service limit °C	°F	Typical areas for use
Isophthalic polyesters	70	150	Increased chemical resistance; used extensively in chemical plant waste and cooling tower systems plus gasoline tanks; also liners for sour crude tanks
Chemical-resistant polyesters	120–150	250–300	Includes the families of bisphenol, hydrogenated bisphenols, brominated, and chlorendic types; a wide range of chemical resistance, predominantly to oxidizing environments; not resistant to H_2SO_4 above78%; can operate continually in gas streams at 148°C (300°F); end uses include scrubbers, ducts, stacks, tanks, and hoods
Epoxies (wet)	150	300	More difficult to formulate than the polyesters; more alkaline-resistant than the polyesters, with less oxidizing resistance; highly resistant to solvents, especially when postcured; often used in filament-wound structures for piping; not commonly used in ducts; more expensive than polyesters
Vinyl esters (wet) (dry)	93–140	200–280	Especially resistant to bleaching compounds in chlorine-plus-alkaline environments; wide range of resistance to chemicals, similar to the polyesters; used extensively in piping, tanks, and scrubbers; modifications developed to operate continuously up to 140°C (280°F)
Furans (wet)	150–200	300–400	Excel in solvent resistance and combinations of solvents with oxidizing chemicals; one of the two resins to pass the 50 smoke rating and 25 fire-spread rating; carries about a 30–50% cost premium over the polyesters; excellent for piping, tanks, and special chemical equipment; does not possess the impact resistance of polyesters or epoxies

Source: Ref. 7.

Table 6 Various Fillers and Property Contributions to Plastics

Filler material	Chemical resistance	Heat resistance	Electrical insulation	Impact strength	Tensile strength	Dimensional stability	Stiffness	Hardness	Electrical conductivity	Thermal conductivity	Moisture resistance	Handleability
Alumina powder										×	×	
Alumina trihydrate			×				×				×	×
Asbestos	×	×	×	×		×	×	×				
Bronze							×	×	×	×		
Calcium carbonate		×				×	×	×				×
Calcium silicate		×					×	×	×			
Carbon black		×				×	×		×	×		×
Carbon fiber									×	×		
Cellulose				×	×	×	×	×				
Alpha cellulose			×		×	×						
Coal (powdered)	×										×	
Cotton (chopped fibers)			×	×	×	×	×	×				
Fibrous glass	×	×	×	×	×	×	×	×			×	
Graphite	×				×	×	×	×	×	×		
Jute				×			×					
Kaolin	×	×				×	×	×			×	×
Mica	×	×	×			×	×	×			×	
Molybdenum disulfide							×	×			×	×
Nylon (chopped fibers)	×	×	×	×	×	×	×	×				×
Orlon	×	×	×	×	×	×	×	×		×	×	
Rayon			×	×	×	×	×	×				
Silica, amorphour			×								×	×
TFE-fluorocarbon						×	×	×				
Talc	×	×	×			×	×	×			×	×
Wood flour			×		×	×						

Source: Ref. 5.

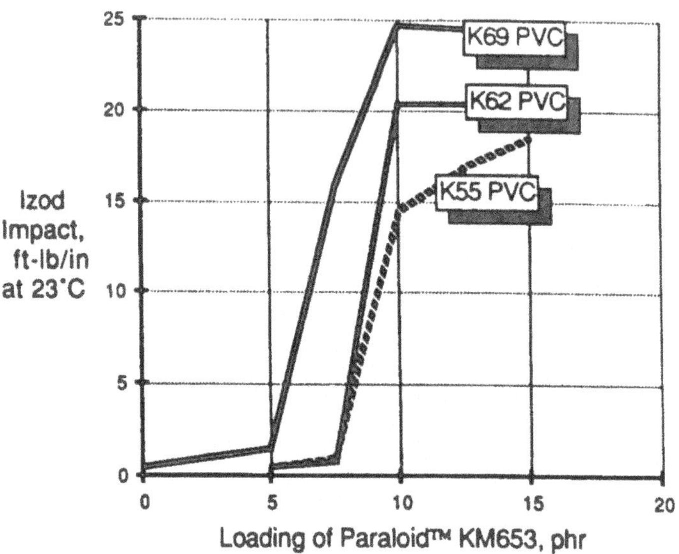

Figure 3 Relationship of notched Izod impact strength to impact modifier loading and matrix molecular weight for rigid PVC. (From Ref. 8.)

Amino (NH or NH_2 groups)-, urea- and melamine-containing thermosets are the second largest thermoset group. Their properties vary widely but they are generally hard and abrasion resistant. Applications of the ureas are in electrical and electronic components, while the melamines are used in dinnerwares and handles.

The epoxies are highly versatile but also somewhat more expensive than the other thermosets. They have excellent bonding properties and good chemical resistance and electronic properties. In addition to bonding applications, they find use in encapsulation of electronic components and for some tools and dies.

Polyester thermosetting resins are usually copolymers of a polyester and a styrene. Most are used in the reinforced condition as boat hulls, swimming pool fixtures, and lawn chairs. They have low moisture absorption and relatively high strength.

Table 7 Properties of Impact Modified Zytel Nylon 6/6

Property	Test temperature (°C)	Samples[a] Zytel 101 (unmodified)	Zytel 408	Zytel ST-801
Notched izod impact strength,	−40	0.6	1.3	3.0
(ft-lb/in.)	23	1.0	4.3	17
Tensile strength (psi)	23	12,000	8,800	7,500
Flexural modulus (psi)	23	410,000	285,000	245,000

[a]Sampling conditioning: dry as molded.
Source: Ref. 2.

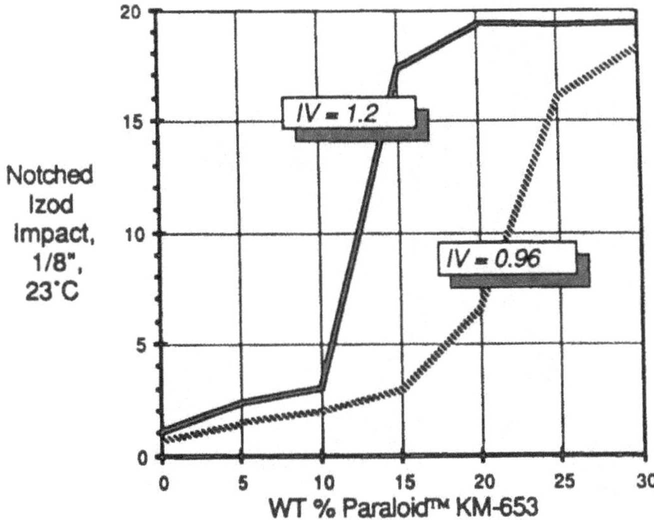

Figure 4 Impact modification of thermoplastic terepthalate with core/shell polymers. (From Ref. 8, p. 235.)

Polyurethane thermosets can be rigid or flexible, depending on the formulation. The flexible ones have better toughness. They are often used as foamed parts.

Thermosets in general are more brittle, stronger, harder and more temperature resistant than the thermoplastic polymers. Other advantages are better dimensional

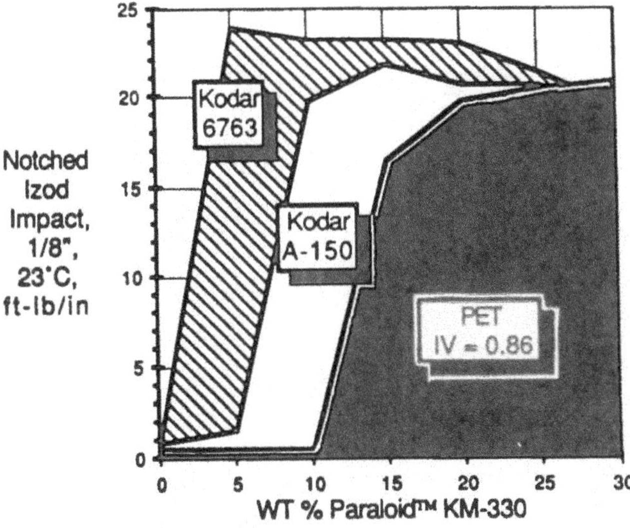

Figure 5 Modification of PET and related copolyesters with core/shell polymers. (From Ref. 8, p. 236.)

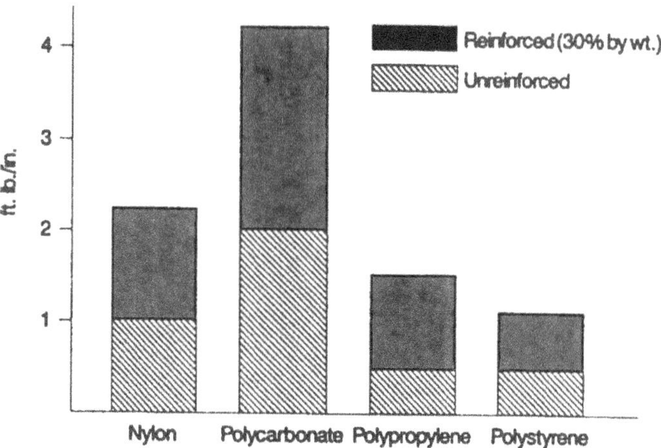

Figure 6 Izod impact strength change with glass reinforcement of thermoplastic polymers. (From Ref. 8, p. 420.)

stability, and good chemical resistance and good electrical properties. However, they are more difficult to process and are more expensive.

Typical mechanical and physical properties of selected thermoset polymers are presented in Tables 10 and 11 respectively. The properties in Table 10 show a considerable range of values because of different processing methods as well as different sources. In all cases the pure or "neat" forms, presumably absent of additives and reinforcements, are used in these tables. The thermosets generally have low impact energy values (0.5 to 2.0 ft-lb), although the bismaleimides are higher. Most of the engineering thermosets, are used in the reinforced form. Table

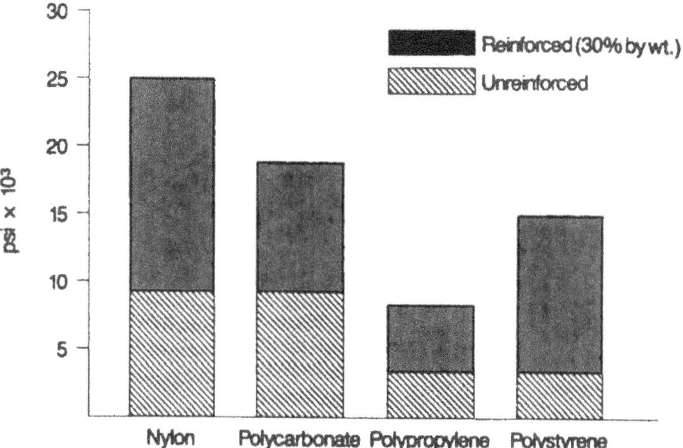

Figure 7 Increase in tensile strength with glass reinforcement of thermoplastic polymers. (From Ref. 8, p. 419.)

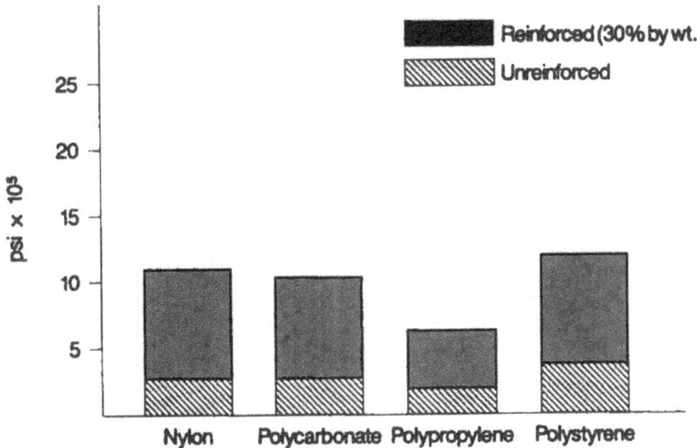

Figure 8 Increase in flexural modulus with reinforcement of thermoplastic polymers. (From Ref. 8, p. 419.)

12 lists properties of thermosets in this category. A list of typical applications of thermosets containing reinforcements is presented in Table 13.

3.1 Temperature Effects

As mentioned, thermosets are generally more heat resistance than the thermoplastic polymers both in the neat form and also with fillers. The continuous heat distortion temperatures for the latter were given in Table 12. Some additional useful temperature data for glass-filled thermosets is

DAP	HDT 232°C (450°F); CUT 175–205 (350–450)
Epoxies	HDT 150–340 (300–460)
Unsaturated polyesters	HDT 80–140 (175–265)

Table 8 DTUL Response for Crystalline Polymers

Polymer	DTUL at 264 psi, 20% glass (°F)	Increase over base polymer (°F)
Acetal copolymer	325	95
Polypropylene	250	110
Linear (HD) polyethylene	260	140
PCO-72 (modified polypropylene)	300	160
Thermoplastic polyester	405	250
Nylon 6	425[a]	305
Nylon 6/6	490[a]	330

[a]At 30% glass by weight.

Table 9 DTUL Response for Amorphous Polymers

Polymer	DTUL at 264 psi 20% glass (°F)	Increase over base polymer (°F)
Acrylonitrile-butadiene-styrene	215	25
Styrene-acrylonitrile	215	20
Polystyrene	220	20
Noryl	290	25
Polycarbonate	290	20
Polysulfone	365[a]	20

[a]At 30% glass by weight.
Source: Ref. 2.

BMIs	Service temperature 177–232 (350–450)
Phenolics	CUT 150 (300); HDT 175–260 (350–500)

3.2 Chemical Effects

The phenoloics, the most popular of the thermosets, have excellent resistance to humidity, acids, and oils but not to strong oxidizers. The epoxies in general show

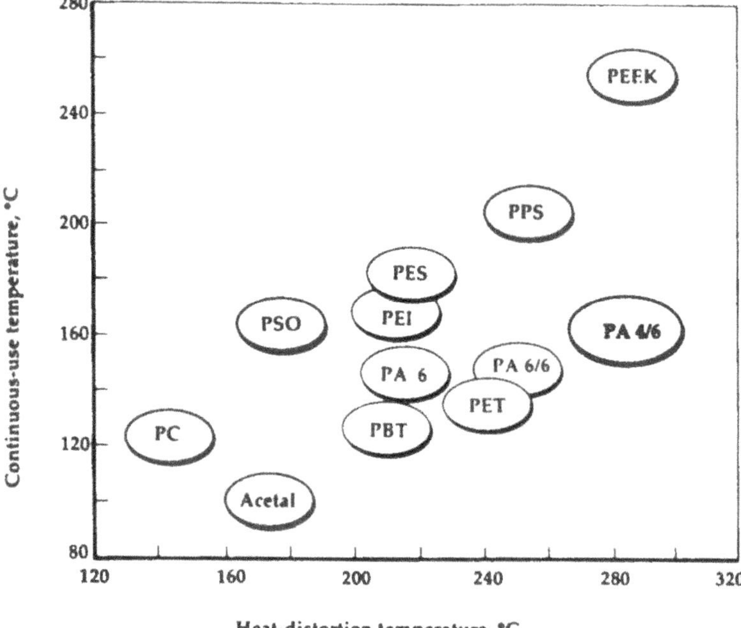

Figure 9 Continuous use temperatures (10,000 h) and heat-distortion temperatures of glass-filled (30 vol%) thermoplastic polymers. (From Ref. 9.)

Table 10 Typical Mechanical Properties of Selected Thermoset Polymers at Room Temperature

Material	Tensile strength [MPa (ksi)]	Elongation (%)	Tensile modulus [MPa (ksi)]	HDT (°C)	Hardness	Comments
Epoxies	50–140 (7.2–20.3)	1.5–1.8	3100–3800 (450–550)	150–240	R_m 106	Cast values are lower than molded
Unsaturated polyesters	40–70 (4–8)	1.3–3.3	2400–3450 (348–500)	80–130	Barcol 40, R_m 117	The saturated polyesters are thermoplasts
Phenol-formaldehyde	35–63 (5–9)	0	800 (114)	174	H_r 125, R_e 95	
Melamine-formaldehyde	52 (7.5)	—	9650 (1400)	—	R_m 120	Used as adhesives in pure form
Urea-formaldehyde	41–69 (6–10)	0.5	6900 (1000)		R_m 115	
Polyurethane high density	20–70 (2.9–10.1)	180–300	3000–6000 (438–870)	120	Shore D 39–64	Higher values are RIM material
Bismaleimides (polyimides)	97 (14)	1–8	4100 (594)	360	R_f 45–58	Izod impact 80 J/m

Table 11 Typical Physical Properties of Selected Thermosets

Material	Specific gravity	Thermal conductivity (W/m · K)	Thermal expansion coefficient (10^{-5} K^{-1})	Volume resistivity, short time (Ω · m)	Dielectirc strength (V/mil)
Epoxies	1.1–1.4	0.19	4.5–6.5	10^{14}–10^{16}	300–500
Unsaturated polyesters	1.04–1.20	0.17	11.3	—	380–500
Phenol-formaldehyde	1.24–1.32	0.09	6.8	10^{12}–10^{14}	250–400
Polyurethane	1.2	0.17	5.7 (3.2)	—	—
Polyimides	1.2	0.09	3.1	—	—

good chemical resistance to most all environments. The aminos have excellent resistance to fats, oils and waxes, but poor to strong alkalies. The unsaturated polyesters are noted for their low moisture uptake. For a more detailed list of environmental characteristics see Appendix A.

4 ELASTOMERS

Elastomers are defined in various ways in textbooks and related literature. From strictly a mechanical point of view they are polymers that exhibit large-scale extensions of 100% or more that is completely recovered on release of the load. By this definition the hard rubbers have much smaller elongations and are not true elastomers but are more akin to network polymers. Probably the reason for the confusing definitions is that some refer to elastomers as rubbers and vice versa. Elastomers should really include natural rubber (C_5H_{10}, isoprene), polybutadiene, styrene-butadiene, butyl rubber, polychloroprene, some polyurethanes, and synthetic polyisoprene, to name a few. Elastomers obtain their elasticity from the kinking or coiling of long-chain molecules. Application of a load causes unkinking to occur and the corresponding expansion. Likewise release of the load restores the molecules to the kinked form.

The butadiene-type molecules form many of the rubbers. The synthetic rubbers were modeled on natural rubber, being linear chain polydienes random copolymers with C═C double bonds which permit irreversible cross-linking with sulfur (Vulcanization). This cross-linking restricts the unkinking of the long-chain molecules under load. Cross-linking at about 10% of the possible sites gives the rubber mechanical stability. Large amounts of sulfur produce the hard rubbers. About 40% sulfur yields the *ebonites*, which are highly cross-linked, rigid nonelastomers. These are more like the three-dimensional network polymers. The technology that has been applied to natural rubbers has been extended to the polydiene and polyolefin synthetic rubbers. Vulcanization by sulfur has been applied to many of the synthetic rubbers.

The polyurethane elastomers form a group of heterochain polymer elastomer materials that have a variety of structures. They are strong and have exceptional

Table 12 Properties of Thermosetting Plastics

Resin material	Specific gravity	Tensile strength (psi × 10^3)	Compressive strength (psi × 10^3)	Continuous heat distortion (°F)	Thermal expansion (cm/°C × 10^{-5})	Thermal conductivity (cal/s/cm²/°C/cm × 10^{-4})	Volume resistivity (Ω-cm)	Dielectric constant (60 cycles)	Power factor (60 cycles)	Water absorption over 24 h (%)
Alkyd glass-filled general purpose	1.93–2.32	5–9	21–29	800	5.1–12.7	10–15	1×10^{15}	6.7	0.01–0.02	0.07–0.2
Alkyd mineral	2.17–2.24	3–8	16–20	275–400	5.1–12.7	15–25	10^{12}–10^{14}	5.7–6.3	0.030–0.045	0.05–0.12
DAP glass MIL-M-19833 GDI-30	1.57–1.86	7.0–9.5	24–45	365–465	6.9–8.9	6–8	10^{12}	4.5	0.010–0.071	0.05–0.25
DAP mineral (MDG) MIL-1-14F, MIL-P-4389	1.58–1.74	5.5–6.5	22–25	350–440	8.9–10.7	13.7	10^{13}+	5.2	0.06–0.40	0.2–0.5
DAP unfilled	1.27	4	22–24	350	—	—	2×10^{16}	3.6	0.008	0.09
Epoxy glass	1.8	11.5	30–38	300–500	2.7–3.0	8.5	3.8×10^{15}	—	0.015	0.03–0.05
Epoxy mineral	1.60–1.52	5–10	25–35	210	5.1–14.5	7.0–10.1	0.8–2×10^{12}	7.9–9.5	0.03–0.08	0.1–0.6
Melamine cellulose	1.47–1.52	5–10	25–35	210	5.1–14.5	7.0–10.1	0.8–2×10^{12}	7.9–9.5	0.03–0.08	0.1–0.6
Melamine cloth	1.40–1.50	7–10	30–35	250	6.4–7.6	10.6	1–3×10^{14}	8.1–12.6	0.1–0.34	0.3–0.6
Melamine glass MMI 30	1.9–2.0	6–10	20–32	300–400	3.0–5.1	11.5	2×10^{14}	9.7–11.1	0.14–0.23	0.09–0.60
Phenolic cloth										
CFI-10	1.36–1.40	6.5–7.0	20–25	275	5.1–7.6	9.3	5–8×10^{11}	6.1–21.2	0.16–0.64	0.8–1.0
CFI-20	1.37–1.40	6.5–7.5	22–25	340	3.8–7.6	7.0	6×10^{10}	5.2–2.1	0.64	0.9–1.0
Phenolic glass	1.70–1.90	7–11	14–35	335–450	2.0–4.1	9.7	1×10^{12}	5.6–7.2	0.02–0.05	0.5–1.0
Phenolic GP	1.34–1.45	6–9	25–35	300	7.6–11.4	4–8	2×10^{14}	5–10	0.05–0.5	0.2–1.0
Phenolic GPI-100	1.69	7.5–16.8	27–40	—	—	1–10	—	—	—	0.03–0.20
Phenolic MFE	1.70–1.85	9.2	32	—	2.8–4.6	10–14	—	—	—	0.03
Phenolic mineral										
MFG	1.72–1.86	4–7	17.5–20	325–425	3.8–6.4	8.16	10^{12}	40–60	0.01–0.40	0.12–0.40
MFH	1.6–2.0	5–8	25–35	400–450	3.8–5.1	8.16	10^{12}	8–15	0.08–0.20	0.07–0.20
MFI-20	1.76	8.7	23–34.5	475	3.8	14	1.6×10^{10}	45	0.28	0.04–0.40
Phenolic nylon	1.22	8	24	275	19.0	7.5–18	10^{12}	4	0.02	0.20
Phenolic rubber/flour	1.29–1.32	4	16–18	275	3.8–10.2	5.0	3.4×10^{9}	9	0.14	1.4–2.0
Urea cellulose	1.47–1.52	5–10	25–38	170	6.9	7.0–10.1	0.5–5×10^{11}	7.0–9.5	0.035–0.040	0.5–0.7
Silicone glass	1.88	4–8	10–13	450–700	2.0	7.5	3×10^{14}	4.35	0.003–0.02	0.10–0.30
Silicone mineral	1.85–2.82	2.5–4.4	11–18	400–700+	6.4–15.2	4.0–10.0	10^{14}	3.4–4.5	0.002–0.01	0.05–0.22

Source: Ref. 5.

Table 13 Application of Reinforced Thermosets

Epoxies	Electrical molding, adhesives, printed wiring boards, aerospace
Unsaturated polyesters	Appliance housings, automotive body panels, boats, shower stalls, fans, pipes, tanks, electrical
Phenolics	Motor housings, telephones, electrical fixtures, particle board, brake and clutch linings, electrical, decorative
Polyurethanes	Automotive parts such as steering wheels, head rests, and instrument and door panels
Bismaleimides	Elevated temperature applications, electronics, aerospace
Polyimides (general)	Excellent dielectric properties, wear- and heat-resistant applications
Vinylesters	Corrosion resistant uses, electrical equipment, chemical process equipment

abrasion and tear resistance along with good oxidation resistance. The silicone rubbers are also heterochain elastomers with service temperatures as high as 200°C.

During the 1970s the thermoplastic elastomers (TPE) began to be used in significant quantities. These elastomers contain sequences of hard and soft repeating units in the polymer chain. Elastic recovery occurs by the hard segments acting to pull back the more soft and rubbery segments. Cross-linking is not required. The copolymer styrene-butadiene-styrene has become the classic example where butadiene is the soft segment. There are about six generic classes of TPEs. In order of increasing cost and performance they are the styrene block copolymers, polyolefin blends, elastomeric alloys, thermoplastic polyurethanes, and thermoplastic polyamides.

The major use of elastomers is and will continue to be in tire construction. Some commercial elastomers and their applications are listed in Table 14 and some mechanical properties in Table 15. Properties of some thermoplastic elastomers (TPEs) are given in Table 16.

5 POLYMER BLENDS

Blends were defined earlier as miscible or immiscible mixtures of polymers, which are usually blended as melts. All sorts of mixtures are being attempted, some by more-or-less trial and error procedures. Some of the worthwhile blends contain mixtures of crystalline and amorphous polymers. Nylon-phenolic blends have been used as ablative heat shields on space vehicles. In the 1970s the term "polymer alloys" emerged to describe rubber-toughened materials which were a mixture of polymer types. The term "hybrids" has also been used to describe some polymer blends. Many of the new developments in elastomers are occurring by the alloying route.

PPO/PS is the most widely consumed blend. Polypropylene/synthetic rubbers (PP/EDPM) are a group of blends that are being used in the automotive industry for air ducts, side moldings, bumper covers, grills and wiring. Other blends finding considerable use include ABS/PVC, ABS/PC, and PBT/PET. Polymer blends have

Table 14 Commercial Elastomers and Their Applications

Material	ASTM Class	Applications
Natural rubber	R	Rubber
Styrene-butadiene rubber	R	Synthetic rubber: used in place of natural rubber tires, belts, and mechanical goods
Polybutadiene	R	Synthetic rubber
Polychloroprene	R	Synthetic rubber: gaskets, V-belts, cable coatings
Ethylene-propylene rubbers	M	Saturated carbon chain elastomer: copolymer-resistant to ozone and ultraviolet radiation, good chemical resistance to acids but not to hydrocarbons; cable coating, hoses, roofing, automotive parts
Poly(propylene oxide) rubbers	O	Polyester elastomer
Silicone elastomers	Q	Useful from −101 to 316°C: tubing, gaskets, molded products
Polyurethane polyester	Y	Thermoplastic elastomers: combination of a rubbery and a hard phase
Polyester polyether	Y	Thermoplastic elastomers: footwear, injected-molded parts
Elastomeric alloys	Y	Extruded, molded, blow-molded and calendered goods (thermoplastic)

also been widely used in business machines, appliances, and electronic applications. About 550 million pounds of blends were consumed in 1987.

Interpenetrating polymer networks (IPNs) consist of a combination of two polymers when at least one is synthesized and/or cross-linked in the immediate presence of the other. Some silicone thermosets are being added to a variety of thermoplastics polymers including nylon, PBT, and polypropylene to form IPNs, which result in improved elevated temperature properties.

Properties of some of the more popular polymer blends are presented in Table 17.

6 ADHESIVES

Adhesives are not polymers that are consumed in large quantities; nevertheless, they occupy an important role in our select group of engineering materials. ASM International has published a third volume of the *Engineered Materials Handbook* [3], and it is totally devoted to adhesives. Adhesives have been separated into five types.

1. Solutions of thermoplastics (including vulcanized rubber) which bond by loss of solvent
2. Dispersions of thermoplastics in water or organic liquids which bond by loss of solvent

Table 15 Room-Temperature Mechanical Properties of Some Typical Elastomers

	Pure gum vulcanizates		Carbon-black reinforced vulcanizates	
	Tensile strength (kg/cm^2)	Elongation (%)	Tensile strength (kg/cm^2)	Elongation (%)
Natural rubber (NR)	210	700	315	600
Styrene-butadiene rubber (SBR)	28	800	265	550
Acrylonitrile-butadiene rubber (NBR)	42	600	210	550
Polyacrylates (ABR)	—	—	175	400
Thiokol (ET)	21	300	85	400
Neoprene (CR)	245	800	245	700
Butyl rubber (IIR)	210	1000	210	400
Polyisoprene (IR)	210	700	315	600
Ethylene-propylene rubber (EPM)	—	300	—	—
Polyfluorinated hydrocarbons (FPM)	50	600	—	—
Silicone elastomers (SI)	70	600	—	—
Polyurethane elastomers (AU)	350	600	420	500

Source: Adapted from Ref. 6.

3. Thermoplastics without solvents (hot-melt adhesives)
4. Polymeric compositions which react chemically after joint assembly to form cross-linked thermoset polymers (e.g., epoxies)
5. Monomers which polymerize in situ

As one can see, adhesives do not fall into clearly defined categories and could be considered as subclasses of the others. The use of adhesives covers the range from

Table 16 Typical Mechanical Properties of Selected Thermoplastic Elastomers

Material	Tensile strength [MPa (ksi)]	Elongation %	Hardness Shore A
Acrylonnitrile butadiene (NBR)	18.6 (2.7)	300	72
Styrene-butadiene-styrene	31.7 (4.6)	880	71
EPM-PP blend	13.8 (2)	300	70
TPE-polyester	44.2 (6.4)	760	—
Polyether block amides	51.7 (7.4)	510	—

Table 17 Properties of Some Polymer Blends

Property	PVC-acrylic	Nylon 66-EPDM	PC-PBT	PPO+ 30% glass	PPO-nylon
T_g (°F)	220	—	—	120	—
HDT (264 psi °F)	160	160	220	290	370
MCUT (°F)	135	135	190	265	345
Tensile S. (psi)	10	8	8	17	13
% elongation	100	125	140	3	10
Tensile mod. (psi)	330	—	—	1200	200
Rockwell hard.	M100	R115	R116	R115	R119
Dielectric S. (V/mil)	400	450	475	2	500

aircraft and missile skins to food and beverage cans. Chemical bonding between the adhesive and adhered surface does not usually take place, although some believe there exists the possibility of some type of van der Waals bonding taking place in addition to some mechanical interlocking on rough surfaces. It is essential for the adhesive to wet the adhered surface during joint assembly. Surface cleanliness is important but not surface smoothness. Intentional roughening by abrasion, such as grit blasting, is recommended. There are many polymer and polymer systems that may be used for adhesives. Table 18 lists the properties and preparations of some common adhesives.

Table 18 Typical Adhesives and Their Properties

Material	Curing temp. [°C (°F)]	Service temp. [°C (°F)]	Lap-shear strength [MPa (ksi)]	Room-temp. peel strength [N/cm (lb/in.)]
Acrylics	RT (RT	To 120 (to 250)	17.2–37.9 (2.5–5.5)	17–105 (10–60)
Anaerobics	RT (RT)	To 166 (to 330)	15.2–27.6 (2.2–4.0)	17.5 (10)
Cyanoacrylates (thermosetting)	RT (RT)	To 166 (to 330)	15.2–27.6 (2.2–4.0)	17.5 (10)
Epoxy RT cure	RT (RT)	−51 to 82 (−60 to 180)	17.2 (2.5)	7 (4)
Epoxy HT cure	90–175 (195–350)	−51 to 75 (−60 to 350)	17.2 (2.5)	8.8 (5.5)
Epoxy-nylon alloy	120–175 (250–350)	−250 to 82 (−420 to 180)	41 (60)	123 (70)
Polyurethanes	149 (300)	To 66 (to 150)	24 (3.5)	123 (70)
Silicones	149 (300)	To 260 (to 500)	0.5 (0.04)	43.8 (25)

7 COMPARATIVE PROPERTIES

The strength and moduli of thermoplasts, thermosets, and elastomers are compared in Figs. 10 and 11. The fatigue characteristics for several polymers are depicted in Fig. 12. In general, the thermosets are stronger and harder than the thermoplasts. But if we take the top materials in each group we find that PEEK and the polyimide bismaleimides have about the same strength and moduli as do the thermosets. If the operating temperatures are of concern, certainly PEEK has the advantage, and in this case the polysulfones and PET must also be considered. The chemical resistivity may also be a factor for consideration, a topic that is addressed in detail for a number of polymers in a variety of environments in Appendix A.

DEFINITIONS

Ablative polymer: material that absorbs heat through a decomposition process (material is expendable).
ABS: acrylonitrile-butadiene-styrene copolymer. They are thermoplastic resins that are rigid and hard but not brittle.
Acetal copolymers: family of highly crystalline thermoplastic polymers.
Acetal homopolymer: highly crystalline chain polymer formed by the polymerization of formaldehyde with acetate end groups.
Acetal resins: copolymer thermoplastics produced by addition polymerization of aldehydes by means of the carbonyl function (polyformaldehyde and polyoxymethylene resins). These are strong thermoplastics.
Acrylic polymer: thermoplastic polymer made by polymerization of esters of acrylic acid or its derivatives.

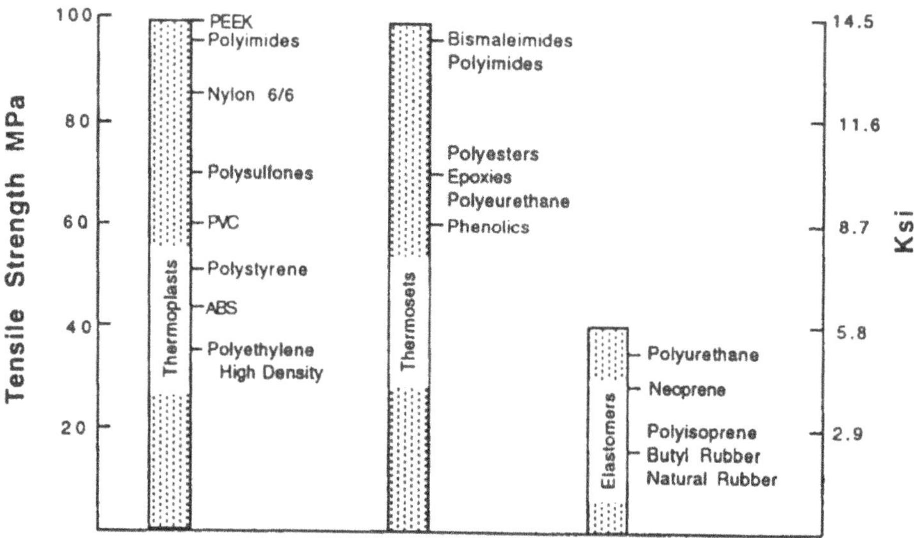

Figure 10 Comparison of room-temperature tensile strengths for thermoplasts, thermosets, and elastomers. (From Ref. 9.)

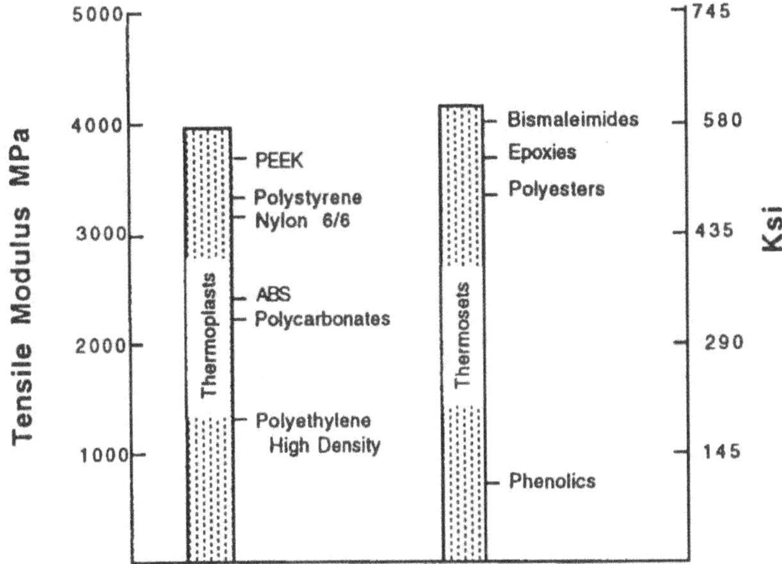

Figure 11 Comparison of room-temperature tensile moduli for thermoplasts and thermosets. (From Ref. 9.)

Acrylic resins: polymers of acrylic methacrylic esters, sometimes modified with nonacrylic monomers such as the ABS group.

Addition polymerization: chemical reaction, usually requiring an initiator, where unsaturated (bifunctional) double-carbon-bond monomer molecules are linked together.

Additive: substance such as plasticizers, fillers, colorants, antioxidants, and flame retardants added to polymers for specific objectives.

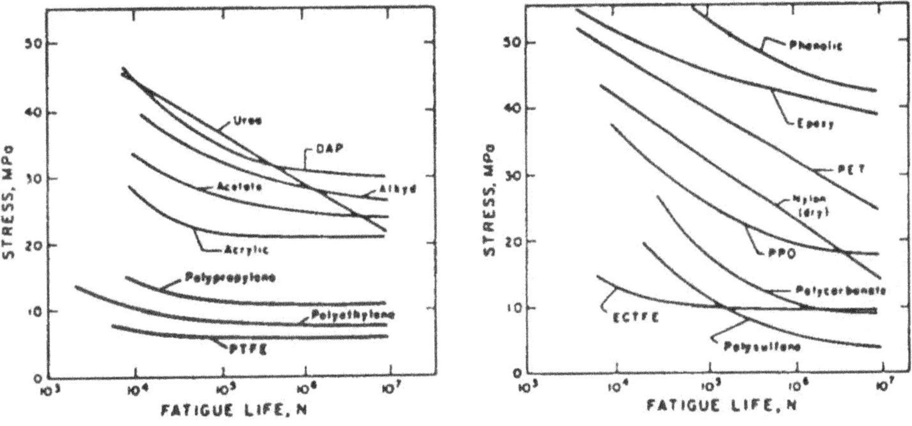

Figure 12 Representative S-N curves for several polymers. (From Ref. 9.)

Adhesive: bonding agent for bonding two surfaces of like or unlike materials together, or to seal a surface from its environment. Most are polymers. (Solders and brazing alloys are not normally defined as adhesives.)
Alkyd polymer: thermoset polymer based on resins composed principally of polymeric esters in which the ester groups are an integral part of the main chain and serve as the cross-links.
Alloy: in polymers, a blend of polymers or copolymers with other polymers.
Amino resins: resins made by polycondensation of amino groups such as urea or melamine.
Aramid: manufactured fiber of a long-chain synthetic aromatic polyamide.
Aromatic: unsaturated hydrocarbon with one or more benzene rings in the molecule.
Atactic stereoisomerism: chain molecule in which the side groups are positioned in a random fashion.
Bakelite: proprietary name for phenolics often phenol-formaldehyde. Named after L. H. Bakeland (1863–1944).
Bismaleimide (BMI): polyimide that is formed by addition rather than condensation.
Block copolymer: linear copolymer consisting of repeated sequences of polymeric segments of different chemical structures.
Branched polymer: main chain with attached side segments.
Butadiene: (CH_2:CH·CH:CH_2) widely used in forming copolymers.
Butene (1-butene): member of the olefin family used in the production of HDPE.
Butylene polymers: polymers based on resins made by polymerization or copolymerization of butene.
Butyl rubber: synthetic elastomer made by copolymerizing isobutylene with small amounts of isoprene.
Calendering: passing of sheet material between rollers.
Catalyst: substance that changes the rate of a reaction.
Cellulose (and its derivatives): natural polymer occurring in plant tissues. Modified cellulose includes *cellulose nitrate* (*celluloid*), *cellulose acetate*, *cellulose ester*, and *cellulose acetate butyrate*.
Cis stereoisomer: stereoisomer in which the side groups are on the same side of a double bond present in a chain of atoms.
Condensation polymerization: reaction of two monomers to produce a third, plus a by-product, usually water or alcohol.
Crazing: regions of ultrafine cracks that form in some polymers under sufficient stress.
Degree of polymerization (DP): number of mers per molecule.
Dielectric constant: ratio of the capacitance between two electrodes containing the subject material to that in air.
Dielectric strength: potentials per unit thickness (volts/mil) at which failure occurs by electric breakdown.
Engineering plastics: general term covering all plastics with or without fillers or reinforcements that have mechanical, chemical, and thermal properties suitable for use as construction materials.
EPM/EDPM rubbers: synthetic ethylene-propylene copolymer rubbers.

Epoxy resins: large chemically complicated class of thermosetting resins of great value as adhesives, coatings, and in composites.
Ethylene polymers: polymers or copolymers based on ethylene (C_2H_4).
Fluorocarbon plastics: polymers made with monomers composed of fluorine and carbon only (Teflon).
Fluorocarbons (general): include polytetrafluoroethylene, polychlorotrifluoroethylene, polyvinylidine, and fluorinated ethylene propylene.
Foamed plastics: resins in sponge form.
Glass transition: reversible change in an amorphous polymer at which the molecular mobility becomes sufficiently small to promote brittleness. Below T_g polymers are brittle.
Hardener: additive to enhance the curing action (mostly for adhesives).
Hardness scales: Barcol, Knoop, Mohs, Rockwell, and Shore.
HDPE: high-density polyethylene.
HDT: heat deflection temperature (see ASTM D648).
Hydrocarbon polymers: those based on polymers composed of hydrogen and carbon only.
Impact test: measure of toughness.
Interpenetrating polymer networks: polymer *X* (or its monomer) interpenetrating into an already cross-linked polymer network Y and then itself becoming cross-linked.
Injection molding and *injection blow molding*: methods of forming heat-softened polymers to desired shapes.
Inomer resins: group of linear chain thermoplasts that contains up to 20% of an acid monomer which is neutralized by a metal or quarternary ammonium ion. Contains both covalent and ionic bonds.
Kevlar: organic polymer composed of aromatic polyamides (aramids) often used as fibers in composites (DuPont trademark).
Liquid-crystal polymers: linear polymers with a stiff primary chain structure that align their molecular axes in solution in the melt to from liquid crystals. First commercial one was an aromatic polyamide.
Olefine: group of unsaturated hydrocarbons of the general formula C_nH_{2n} and named after the corresponding paraffins by the addition *-ene* or sometimes *-ylene* to the root.
Polyamides (nylons): tough crystalline polymers with repeated nitrogen and hydrogen groupings which have wide acceptance as fibers and engineering thermoplastics.
Polybutylene terephthalate (PBT): member of the polyalkylene terephthalate family, similar to polyethylene terephthalate.
Polycarbonate (PC): thermoplastic polymer derived from direct reaction between aromatic and aliphatic dihydroxy compounds with phosgene or by the ester reaction.
Polyether ether ketone (PEEK): linear aromatic crystalline with a continuous-use temperature as high as 250°C.
Polyether-imide (PEI): amorphous polymer with good thermal properties and continuous-use temperatures of about 170°C.
Polyether sulfone (PES): high-temperature engineering thermoplastic consisting of repeating phenyl groups linked by thermally stable ether and sulfone groups.

Polyethylene terephthalate (PET): saturated thermoplastic resin made by condensing ethylene glycol and terephthalic acid and used for fibers, films, and injection-molded parts.
Polyimide (PI): polymer produced by reacting an aromatic dianhydride with an aromatic diamine. It is a highly heat resistant resin (about 315°C).
Polymethyl methacrylate (PMMA): thermoplastic polymer synthesized from methyl methacrylate.
Polyolefines: polymers made from monomers of olefins only.
Polyoxymethylene (POM): acetal plastics based on polymers in which oxymethylene is the sole repeating unit.
Polypropylene (PP): tough, lightweight polymer made by polymerization of propylene gas.
Polystyrene (PS): homopolymer thermoplast produced by the polymerization of styrene (vinyl)-benzene.
Polysulfide: polymer containing sulfur and carbon linkages produced from organic dihalides and sodium polysulfides.
Polyurethane (PUR): large family of polymers based on the reaction product of an organic isocyanate with compounds containing a hydroxyl group.
Polyvinyl acetate (PVAC): thermoplastic polymer material composed of polymers of vinyl acetate.
Polyvinyl chloride (PVC): thermoplastic material composed of polymers of vinyl chloride.
Polyvinyl chloride acetate: thermoplastic material composed of copolymers of vinyl chloride and vinyl acetate.
Polyvinylidene chloride (PVDC): thermoplastic material composed of polymers of vinylidene chloride.
Reaction injection molding: molding process whereby liquids are mixed and reacted under pressure and then forced into a mold.
Resin: solid or pseudosolid organic material, usually of high molecular weight that tends to flow under stress. Most resins are polymers.
Sytrene-butadiene rubber (SBR): copolymer elastomer that is the most important of the synthetic rubbers (world production about 5 million tons in 1986). Competes with vulcanized natural rubbers.

REFERENCES

1. J. M. Margolis, ed., *Engineering Thermoplastics—Properties and Applications*, Marcel Dekker, 1985.
2. J. T. Lutz, Jr., ed., *Thermoplastic Polymer Additives—Theory and Practice*, Marcel Dekker, 1989.
3. ASM, International, *Engineered Materials Handbook*, vol. 2, *Engineering Plastics*, 1988.
4. R. B. Seymour, *Engineering Polymer Sourcebook*, McGraw-Hill, 1990.
5. N. P. Cheremisinoff, ed., *Handbook of Polymer Science and Technology*, Marcel Dekker, 1989.
6. R. B. Seymour and C. E. Carraher, Jr., *Polymer Chemistry*, Marcel Dekker, Inc., New York, 1988, p. 519.
7. J. H. Mallison, in *Corrosion and Corrosion Protection Handbook*, 2nd ed., Marcel Dekker, Inc., New York, 1989, pp. 348–350.

8. A. P. Berzinis, in Ref. 2.
9. G. T. Murray, *Introduction to Engineering Materials*, Marcel Dekker, Inc., New York, 1993.
10. R. Horsley, in *Mechanical Performance and Design in Polymers*, O. Delatychi, ed., Proc. of Applied Polymers Symposium, Vol. 17.
11. R. Kahl, *Principles of Plastic Materials Seminar*, Center for Professional Advancement, 1979.

11
Selection of Ceramic Materials and Their Applications

Isa Bar-On and R. Nathan Katz
Worcester Polytechnic Institute
Worcester, Massachusetts

1 INTRODUCTION

During the past several decades high-performance ceramics have emerged as enabling materials for many key technologies. Examples of where a property, or more usually the combination of several properties, of modern ceramics have been enabling include fiber optic data transmission based on low-optical-loss glass fibers; high-performance electronic packages based on the combination of dielectric constant, thermal conductivity and thermal expansion of ceramics such as aluminum oxide or aluminum nitride; advanced cutting tools based upon the hot hardness, chemical inertness, thermal shock resistance and toughness of silicon nitride or SiC whisker-reinforced aluminas. While, as these examples illustrate, ceramics may be chosen for their optical, electrical, magnetic, or mechanical properties, this chapter will focus on those properties of ceramics usually associated with their application as structural or wear components. There are several reasons for this emphasis. First, the readers of this book are most likely to have a mechanical rather than an electronic orientation. Second, the engineering difficulties in applying ceramics as optical and, increasingly, as electronic components are most often in areas such as attachment or thermal stresses, which are mechanical design issues. Lastly, the structural (mechanical load bearing) applications of ceramics are those projected to grow most rapidly over the next decade [1].

This chapter will briefly review ceramic processing. Following this, a brief introduction of the probabilistic design procedure utilized in the design of ceramic components (particularly those which will be subjected to tensile stresses) will be presented. Next, we will briefly review the room-temperature and high-temperature mechanical properties and the physical properties of selected ceramics. Finally we will review several examples of recent applications of high-performance structural ceramics, and discuss the combinations of properties which lead to the materials selection.

There are literally hundreds of high-performance ceramics presently on the market or in advanced development. We have chosen to provide data on representative materials from each of the important families of structural ceramics rather than to attempt to list every material. Data presented in this chapter are representative; they have frequently been averaged from several sources and rounded off. Data have primarily been compiled from the standard references in the bibliography at the end of this chapter. Such data was often compared and verified using data supplied by manufacturers. Where data appeared to be significantly different, such differences were resolved by invoking the authors' own experience and judgment. Presentation of such representative data is a reasonable approach since it is frequently the case that ceramic manufacturers modify the processing and composition of a given ceramic over the years, and thus, properties cited for a given material in 1996 may be somewhat different than those to be found in 2001, for example. This fact of life is often implicitly acknowledged in the ceramics literature where a material is often designated as Material Z obtained from Company Y in 19xx. *Data presented in this chapter are for comparison and preliminary design use only.* For detailed design purposes one should use precise data measured on the actual material and process used to manufacture the component. Such data are often available from the manufacturer and should be verified by the user.

In order to exercise engineering judgment on the utility of data for design purposes it is important to know how the properties were measured, especially in the case of the structural properties of ceramics. Therefore, we have included a listing of ASTM and JIS standards for ceramic testing at the end of the chapter. Prior to considering the properties of engineering ceramics, it is desirable to briefly consider how ceramics are processed (as the processing is what determines the properties) and how one designs with ceramics (which is somewhat different than how one designs with metallic materials). These topics are the subjects of the next two sections of this chapter.

2 PROCESSING OF ENGINEERING CERAMICS

The most common route for processing engineering ceramics is the "powder" processing route. Other methods for processing high-performance ceramics have been developed and commercially utilized, including: chemical vapor deposition (CVD), directed oxidation of the melt (the Lanxide process), controlled crystallization of glasses (glass ceramics), and sol-gel techniques [2]. Nevertheless, the powder route still accounts for over 95% of all engineering ceramics sold annually. Thus, this chapter will briefly describe the principal processes and variations of this processing route. It is important for the design engineer to have a basic understanding of the powder route for fabricating ceramic components, as the flaws which in practice define the strength distribution (as well as many other properties) arise at one or more stages of this processing route. Thus, variations in the processing route can lead to significantly differing levels of component reliability and cost. In the flow diagram in Fig. 1, there are three basic subprocesses common to all ceramic components fabricated via the powder route: (1) powder synthesis, pretreatment, and mixing; (2) forming a shaped, undensified powder assemblage

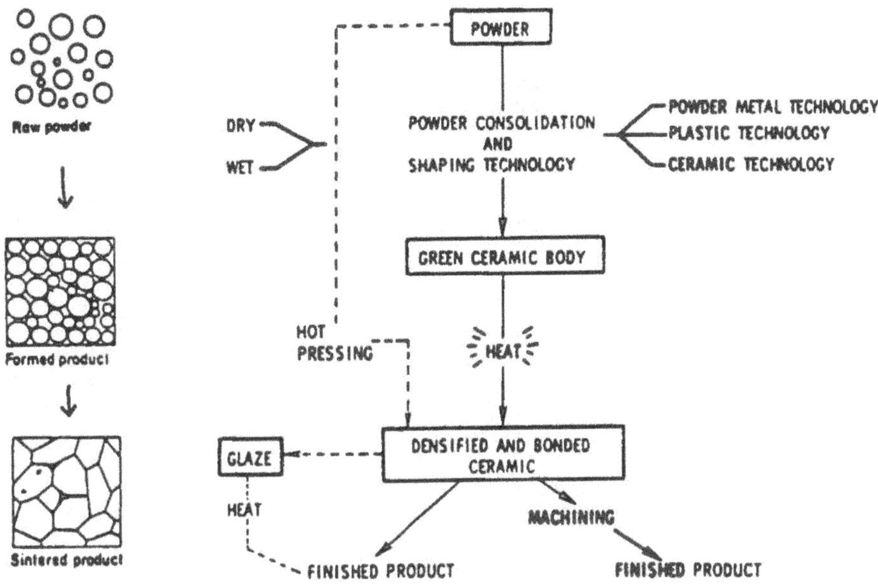

Figure 1 Powder (particulate) route for processing polycrystalline ceramics.

known in the ceramic industry as a "green" body; and (3) applying heat (or heat and pressure) to produce a dense, well-bonded component (a process called sintering). Many ceramic components also require some degree of machining, which is a fourth major subprocess. Machining can introduce surface, or near-surface, flaw populations which frequently dominate the strength statistics, and consequently the reliability of the ceramic component.

Figure 1 also illustrates the evolution of the microstructure of the ceramic as it progresses from a mass of powder, to a green body containing 50–60 vol% of unbonded particles, to an essentially 100% dense, well-bonded component as the result of the sintering process. We will now briefly describe some of the more common substeps that may be performed in each of the three principal subprocesses.

2.1 Powder Synthesis and Treatment

Many of the major phases used as the basis for high-performance engineering ceramics do not exist as naturally occurring minerals (i.e., SiC, Si_3N_4, TiB_2, B_4C, AlN, etc.) or if they do occur in nature are either rare (i.e., Al_2O_3, which occurs naturally as sapphire) or have varying purity (i.e., spinel). Additionally, to facilitate sintering and the achievement of a high level of microstructural control it is frequently desirable to use powders as fine as 0.5 to 5 μm. Consequently, most ceramic powders are synthesized from precursor chemicals and/or minerals by a variety of techniques, such as carbothermal reduction, vapor-phase reactions, or precipitation from aqueous solutions [3]. In the synthesis of oxide powders it is often the case that the product of the synthesis is a hydroxide or a nonstiochemetric oxide. In such cases a procedure called calcining, which is a high-temperature heat treatment in air or oxygen, is performed to yield the desired phase or oxygen content.

Once the desired powders are available, the powders for the dominant phase and any desired additives are milled in ball, rod or attrition mills. Milling reduces the powder size, changes the particle size distribution, and affects a uniform mixture of the major phase and additives. After milling, powders may be sized (screened, air classified, etc.), mixed with small amounts of organic binders (to provide sufficient strength to the future green body, so that it can be handled), and spray dried to form small spherical aggregations of powder which can readily flow and fill the cavities of dies used to form green bodies. Not all powders will undergo all of these pretreatments.

Flaws which can be introduced into the ceramic component at the powder preparation stage include

- Deleterious chemical impurities
- Wear debris from the milling, mixing, or sizing equipment
- Hard agglomerates, large well-bonded particles of the desired composition, but which do not breakdown during milling
- Nonuniform distribution of additives
- Airborne organic contaminants which may leave voids after sintering (i.e., lint, hair, insects, pollen, etc.)

Because of the near-zero tolerance for powder initiated flaws in many modern engineering uses of ceramics (i.e., space shuttle bearings, hip joint balls, etc.) high-performance powders (and green bodies) are now often prepared in clean room facilities, as opposed to traditional ceramic factory environments.

2.2 Green Body Formation

The second major process in fabricating a ceramic component is the consolidation of the powder mixture into a compacted shape which can be handled, and perhaps machined prior to densification by heating. Figure 1 indicates that this consolidation and shaping step can be performed "dry" or "wet." Dry processes include uniaxial pressing (the technology of making an aspirin tablet) or variants of it such as cold isostatic pressing (CIP). In these methods the powder mixture is poured into a die cavity or an elastic bag and sufficient pressure is applied to form a green part. The packing of powders in green bodies prepared by dry methods ranges from 45–55% of theoretical density. The schematic of the evolving microstructure on the left-hand side of Fig. 1 illustrates the ~50% void space typical of ceramic green bodies. Upon "firing" the ceramic to achieve densification this void space must diffuse out of the component. Achieving near full density, while simultaneously maintaining dimensional control of the final product as the void space is eliminated from the component is one of the principal challenges of ceramic engineering.

Wet processes for forming a green body include slip casting, injection molding, extrusion, and tape casting. What all of these methods have in common is that the shaping of the green body occurs via the flow of a particulate laden liquid into or through a forming die (or dies). Because the ceramic powder is "lubricated" by the liquid a slightly higher packing density of particles is achieved when wet processes are used as opposed to dry processes. The higher green density often leads to improved properties and dimensional control.

Flaws which may be introduced into the ceramic body during the formation of the green body include

- Density variations
- Cracking from die wall friction, pressure release phenomena, or unstable flow (mainly in extrusion)
- Knit and fold laminar flaws in injection molded pieces
- Wear debris from dies or mixing apparatus
- Formation of pores, blisters, or other flaws upon binder removal

An excellent review of the various powder consolidation and shape forming techniques, and the special issues related to each of them is available in Richerson [4].

2.3 Formation of the Dense, Well-Bonded Ceramic: Sintering

Sintering is the process in which the addition of heat, or heat and pressure, to the green ceramic body induces densification and the development of strongly bonded grains (strength) in the ceramic component. The most common form of sintering is "pressureless" sintering. In this case the thermodynamic "driving force" which causes sintering to proceed is the reduction of the surface energy that results when the individual particles surfaces become grain boundaries. The high temperature used in the sintering process enables the diffusion of vacancies out of the densifying ceramic. This is the fundamental mechanism by which densification occurs. Often sintering aids are utilized which either act in the solid state to promote diffusion, or to form a liquid phase which assists densification by facilitating solution/reprecipitation reactions or just to act as a high-temperature "adhesive".

Examples of temperatures typically utilized for sintering are 1050°C for Y-Ba-Cu oxide Hi-T_c superconducting ceramics, 1600°C for alumina, 1800–2000°C for silicon nitride, and 2200°C for SiC. Thus, the temperatures used in sintering most engineering ceramics are quite high. Notwithstanding these very high temperatures, most pressureless sintered ceramics do not attain end point densities greater than 98–99% of theoretical. To attain higher densities, or to achieve the same densities with lower amounts of sintering aids, several pressure assisted sintering methods have been developed.

Hot pressing is similar to cold pressing except that the operation takes place inside of a high-temperature furnace. Powder or a pressed preform are placed in the die cavity and temperature and pressure are applied. The details of the temperature-time and pressure-time schedules used in the hot-pressing process profoundly effect the resulting materials properties. Many small and intermediate size (up to ~12-in inner die diameter) hot presses are enclosed in chambers which can operate in a vacuum or with inert gases. Such hot presses frequently use graphite dies and plungers which limit the pressures that can be used, since the graphite dies are generally limited to about 35 MPa (5 ksi) and graphite composites to about 70 MPa (10 ksi) operating stress. Large industrial hot presses are often run with graphite dies in an ambient environment. In this case the dies are protected from oxidation by packing the die assembly in graphite powder. Hot pressing with graphite dies, even in a vacuum, creates a slightly reducing environment at the surface of the component being hot-pressed and this can cause property variations. For materials which must not be sintered in a reducing environment, air or oxygen hot presses have been developed. These often utilize alumina or silicon carbide dies and noble-metal heating elements. The rams in a hot press may be single acting (only one ram applies pressure) or double acting (both rams apply pressure). Double-acting hot presses generally provide more uniform density throughout the part. Hot pressing is generally restricted to symmetrical, flat components with a height-to-diameter ratio of less than 2 to 1. Parts with simple curvatures and shallow depressions (such as piston caps) can be hot-pressed, but parts with little symmetry and complex shapes are not easily hot-pressed. Hot-pressed material often has outstanding materials properties and 99% plus density. However, some materials (such as, beta silicon nitride) may exhibit anisotropic properties, due to preferential grain growth resulting from the uniaxial pressure applied by the hot-press rams.

To be able to retain the benefits of hot-pressed material, and also be able to produce isotropic materials and complex geometries, hot isostatic pressing (HIP) was developed. In HIP the pressure is applied uniformly in all directions by using a gas as the pressure transmitting medium. Since a porous green body will not support a pressure differential the parts to be HIPed are encapsulated in a glass or metal envelop which acts as a pressure-transmitting membrane. HIP is used for many commercial ceramic components where only the highest properties and reliability will suffice, such as silicon nitride turbocharger rotors, bearing balls, or zirconia hip joint components. HIPed components usually can be designed for very good dimensional accuracy with complex shapes. HIP requires inert gas pressures of about 2000 atm (30,000 psi), thus very elaborate containment chambers and furnaces are required.

Recently a technique which combines the relative simplicity of pressureless sintering and the benefits of HIP, has come into use. This technique is "reactive gas overpressure sintering" also referred to as gas pressure sintering (GPS) [5]. This technique is especially suited for densifying materials which may dissociate during sintering. By maintaining an overpressure of a gas such as nitrogen for nitrides, H_2S for sulfides such as $CaLa_2S_4$, or oxygen for oxides such as $BaTiO_3$, dissociation can be suppressed and stoicheometry maintained. After the ceramic has sintered to about 93–95% of theoretical density the porosity is closed and the ceramic part can now sustain a pressure differential. At this time the reactive gas overpressure can be increased and the ceramic component will act as its own encapsulant so that a type of low-pressure HIPing can occur. This process is referred to as sinter-HIP, or two-step GPS. The pressures used in two-step GPS are only 20 to 100 atm as compared to 2000 atm for conventional HIP.

Some powder preforms are converted into dense ceramics by a wide variety of reaction processes, such as reaction bonding of silicon nitride or reaction sintering of silicon carbide. These materials generally do not posses the strength of the pressure sintered materials within their families. However, they can be fabricated to near net shape and densified without shrinkage.

Flaws which may be introduced into a ceramic component during the densification/sintering stage of processing include

- Warping or cracking due to nonuniform shrinkage during sintering

- Nonuniform densification during hot pressing due to die friction
- Extraneous phases at or near the surface of the component due to reaction with the envelope material during HIP or the dies in hot pressing
- Large pores or pore coalescence during sintering

An excellent source of more detailed information on all of the processes described throughout Section 2, as well as other less common techniques which were not discussed here, can be found in Ref. 6.

As we have stated previously, the properties of a given ceramic are determined by the processing route. Thus, when materials will be referred to in the properties sections later in this chapter, we will, where possible, include a description to identify the processing route. These descriptions will usually be abbreviated as follows:

Process	Abbreviation	Example
Sintered	S	S-SiC
Hot-pressed	HP	HP-SiC
Hot isostatic pressed	HIP	HIP-SiC
Reaction sintered	RS	RS-SiC
Reaction bonded	RB	RB-Si_3N_4
Gas pressure sintered	GPS	GPS-Si_3N_4
Sinter-HIP	S/H	S/H-Si_3N_4

These abbreviations will be frequently found in papers and articles dealing with high-performance structural ceramics.

We turn next to a consideration of how one designs with structural ceramics and how this differs from traditional design practices used with metals.

3 DESIGN WITH CERAMIC MATERIALS

All ceramics are brittle. Brittle does not imply that these materials are weak (some structural ceramics exhibit strengths greater than 1 GPa). Brittle means that a ceramic will fail with a very small strain to failure, no plastic deformation or yielding, and usually with some fragmentation. Even ceramic matrix composites which may exhibit metal like stress-strain curves, still fail in a brittle manner. The successful design of high-reliability ceramic components must take into account the brittle nature of ceramics. This section will compare and contrast the differences in the ceramic and metallic design situations, provide a brief overview of the ceramic design process, and present guidelines for designing with ceramics.

Metals, even rather brittle ones, have sufficient capacity for plastic deformation to be able to redistribute loads in the presence of stress concentrations, at least at the microscopic level and usually, but not always on a macroscopic basis. Ceramics do not have this capability. Some modern "high toughness" ceramics use a variety of nonplastic deformation mechanisms to absorb energy at a crack tip, but ultimately they fail in a brittle mode. There is no yielding or gross deformation as in metals. As a consequence of ceramics lacking the capability to redistribute loads via plastic deformation, they are extremely sensitive to any condition which may give rise to a stress gradient where the *local* peak stress might exceed the strength of the material, even though the *average* stress would be well below the strength of the material. Therefore, ceramic materials are very sensitive to internal and surface flaws, point loads, impact loads, and thermal gradients (which in turn cause stress gradients). Ensuring the structural integrity of a ceramic component, therefore, requires that very close control be exercised on the development of the ceramic microstructure, machining or surface treatments, and both mechanical and thermal load transfer to the ceramic component.

In the case of metals, the primary fabrication process usually entails a cast or powder metal perform. Usually, this perform is converted to a shaped component though one or more subsequent thermal-mechanical treatments whereby many of the flaws initially present are removed or reduced. Such treatments also allow optimization of the microstructure (and consequently, properties) after primary fabrication. Since ceramics are not ductile, they cannot generally be subjected to such thermal-mechanical secondary processing. Thus, the microstructure and flaw distributions which result from the primary processing will be present for the lifetime of the component. Each primary ceramic fabrication technique (e.g., slip casting, injection molding, dry pressing or tape casting) will produce its own unique population of flaws. Therefore, the ceramic designer must be aware of and concerned with, the reproducibility of the processing technique by which the components are fabricated, and the quality assurance program by which the reliability is certified.

Generally, the most severe stress-concentrating flaws in ceramic components are cracks, particularly surface cracks. Machining is usually the dominant source of such flaws. Flaws, from whatever source, lead to a reduction in strength. Using the well-known relationship

$$K_{IC} \propto \sigma c^{1/2} \tag{1}$$

where K_{IC} is the fracture toughness (an intrinsic materials property defining the materials resistance to crack propagation), σ is the stress at fracture, and c is a measure of flaw (crack) size, it can be seen that a distribution of flaws sizes will produce a distribution of strengths throughout the component. Virtually all ceramic processes result in components containing distributions of flaws. And as we have discussed, these flaws cannot be eliminated or reduced in severity by secondary processing. These inherent flaw populations coupled with the ceramic material's lack of being able to undergo local plastic deformation to mitigate the severity of the flaws are responsible for ceramics, generally, having strength distributions much greater than commonly observed in metals. Thus, a corollary to brittleness is a comparatively large strength distribution. As it is important to define this strength distribution accurately, many more strength measurements are required in the case of ceramics (30 to 100) as compared with metals (10 to 20). The existence of these large strength distributions leads to an essential difference between ceramic and metallic design procedures: conventional metallic design is *deterministic*, whereas ceramic design is *probabilistic*. The tensile strength of most metals is treated as a fixed value (in those instances where it is recognized that some variability is likely, design allowables are set at an appropriate number of standard deviations below the average strength), whereas the tensile strength of ceramics are given as a probability function (and the design allowables then transforms into an acceptable probability of failure). In the limit both methods converge, since setting a design level based on some number of standard deviations of a fixed average value is also setting an implicit probability of failure. What is implicit in metallic design practice becomes explicate in ceramic design. Probabilistic design procedures correctly employed can produce ceramic components of extraordinarily high reliability. An example of this is the automotive turbocharger widely used in Japan. Of the several million of these highly stressed components in use, few if any failures have been recorded.

In the structural use of ceramics, regions or points of load transfer (whether mechanical or thermal) are most commonly observed to be the sites at which fracture originates. This is because the stress gradients in these regions are generally very severe and consequently difficult to control. Thus, it is at these areas that overstresses are most likely to occur.

A metallic structure has a strain tolerance (compliance) as a result of the ability of the metal to yield: a ceramic structure does not have this capability. The challenge to the ceramic designer is to shift the function of compliance from the material to the structure itself. A major difficulty with this approach is to design a compliant structure and at the same time contain all possible deflections within an allowable strain envelope (i.e., in a gas turbine the rotor and stator must not occupy the same space at the same time or the system will fail). Although this is a difficult task, there are sufficient examples to show that it can be accomplished for a variety of ceramic structures (i.e., the human skeleton, the turbocharger rotor, several ceramic heat exchanger designs, and the space shuttle thermal protection system).

The Ceramic Design Process

It is clear from the discussion that design with brittle materials requires a very precise knowledge of the state of stress at every point within the component under all applied loads. Therefore, successful brittle materials design in highly stressed applications begins with the careful application of two- or three-dimensional computerized finite-element thermal and stress analysis techniques. To a first approximation, the additional complexity that designing with a brittle material on a probabilistic basis brings to the design process is that for each element in the finite-element grid a probability of failure, as well as a stress prediction, must be made. In carrying out this element-by-element failure probability calculation the Weibull statistics of the strength distribution are almost universally utilized. (Weibull statistics will be discussed in Section 4.1) The strength of brittle materials are well known to be component-size dependent. Weibull statistics describe this volume dependence of strength analytically. Since the strength of ceramics is volume dependent so is the probability of failure. The probability of failure is a complex function of the stress level and the volume of material subjected to that stress [7]. Thus, an important and counterintuitive difference between ceramic and metallic design is that in a ceramic component the location of the maximum probability of failure is not necessarily the location at which the highest stress occurs. The design is iterated until an acceptably low probability of failure for the component is attained. Generally, a finer finite-element grid size is required when designing with ceramics as opposed to metals. This is due to the higher level of precision required in defining local stresses and stress gradients which is necessitated by the ceramic's lack of ductility. This requirement for greater precision also holds for specifying the thermal and mechanical boundary conditions. The brittle materials design process is schematically outlined in Fig. 2.

Thus far we have considered the design process with the assumption that the properties of the ceramic will be constant over time. If the ceramic component is to be used in a chemically active environment, at elevated temperatures, or in the presence of cyclic mechanical loads, various materials degradation processes will change the material's strength distribution, and thus lifetime prediction analyses must also be carried out. Lifetime prediction methodologies remain the most difficult and uncertain aspects of the ceramic design process. Nevertheless, considerable progress has been made in this area in the past decade, and concepts such as "fracture mechanism maps" (Fig. 3) have been used in the successful life prediction design of commercial silicon nitride automotive turbocharger rotors [9].

Guidelines for Ceramic Design

In designing with ceramics for high-performance mechanical applications, experience, codes of practice and procedures are not well developed. Even with the advanced computer techniques now available, designing still maintains a high content of art as well as science. However, there are rules of thumb which can be quite

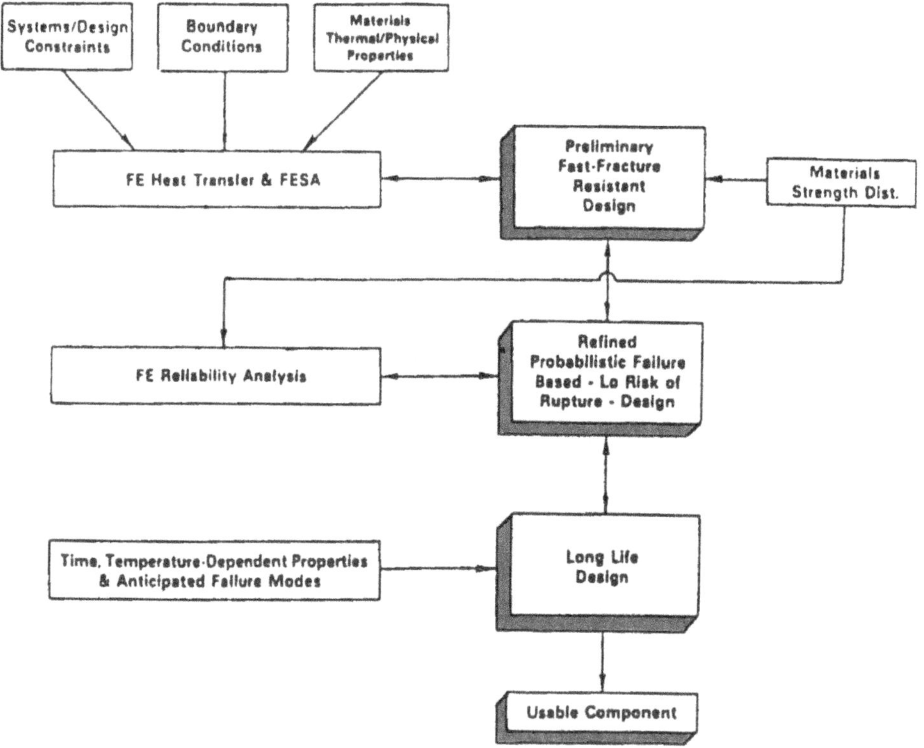

Figure 2 Outline of the brittle materials design process utilized for ceramics (Katz).

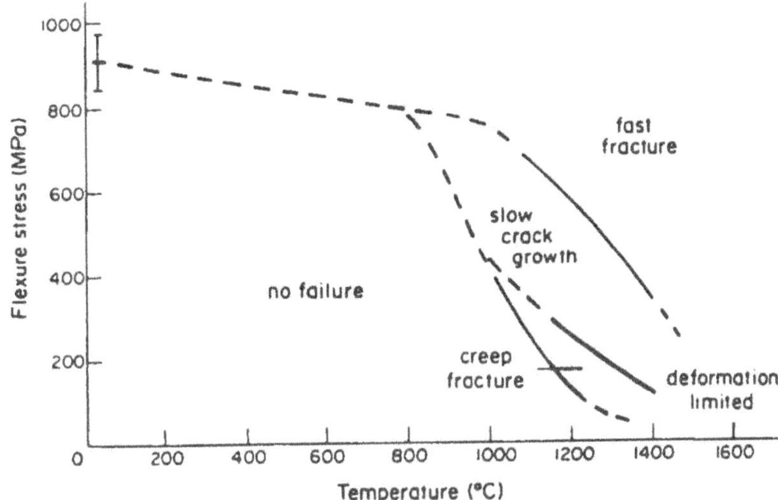

Figure 3 Schematic fracture mechanism map for hot-pressed silicon nitride. (From Ref. 8.)

useful. The following list provides a summary of many of these rules of thumb which are useful for checking preliminary designs.

(a) Point loads should be avoided to minimize stress where loads are transferred. It is best to use areal loading (spherical surfaces are particularly good); line loading is next best.
(b) Structural compliance should be maintained by using compliant layers, springs or radiusing of mating parts (to avoid lock-up).
(c) Stress concentrators—sharp corners, rapid changes in section size, undercuts and holes—should be avoided or minimized. Generous radiuses and chamfers should be used.
(d) The impact of thermal stresses should be minimized by using the smallest section size consistent with other design constraints. The higher the symmetry the better (a cylinder will resist thermal shock better than a prism), and breaking up complex components into subcomponents with higher symmetry may help.
(e) Components should be kept as small as possible—the strength and probability of failure at a given stress level are dependent on size; thus, minimizing component size increases reliability.
(f) The severity of impact should be minimized. Where impact (i.e., particulate erosion) cannot be avoided, low-angle impacts (20–30°) should be designed for. Note this is very different than the case of metals, where minimum erosion is at ~90°.
(g) If components are to be machined care must be taken to avoid surface and subsurface damage. Machining flaws are often found to be the principal strength-reducing defect.

Using the iterative brittle materials design methodology and guidelines discussed in this section, reliable design with structural ceramic components can be routinely achieved. The next section will focus on presenting the properties of the principal families of structural ceramics with a view to their relevance to the design process.

4 SELECTION BASED ON MECHANICAL PROPERTIES

4.1 Room Temperature Properties

The mechanical properties of ceramics are strongly dependent on the processing details. Thus a hot-pressed silicon nitride will have different properties than a reaction bonded or sintered silicon nitride even if the starting powders were the same. In addition, these properties can differ with the details of the starting powder(s) and the specific sintering additives. (See Section 2.)

The processing parameters can be selected to optimize a set of properties (room-temperature or elevated temperature properties) for a specific application. An example are the materials developed for the inlet and outlet engine valves as reported by Hamminger and Heinrich [10]. Here silicon nitride was selected for its good high-temperature properties, good wear resistance, and low weight. The materials were processed such that the silicon nitride used for the exhaust valve had superior room- and elevated-temperature strength to that used for the inlet valve, since the exhaust valve experiences higher thermal loads.

The abbreviations introduced in Section 2.3 will be used here to identify the processing route:

S = sintered
RS = reaction sintered
HP = hot pressed
HIP = hot isostaticallly pressed
RB = reaction bonded
GPS = gas pressure sintered
S/H = sinter -HIP

In many cases, though, the exact processing details are not published with the available data.

Tensile and Flexural Strengths

The tensile and/or flexural strength are measures of the load-bearing capability of the ceramic material. In most cases the strength denotes the strength at fracture, while for metallic materials it is more common to use the yield or tensile strength. The fracture strength is used since for most ceramics there is no significant deviation from linearity in the stress-strain relationship prior to fracture. The exception are a few new materials which show some deviation from linearity prior to fracture.

Ceramic materials are extremely strong when loaded in compression, but are perceived as "weak" when loaded in tension. This perceived weakness is due to the presence of strength-limiting flaws as discussed in Section 3. Thus the measured tensile or flexural strength is to a large extent a measure of the worst flaw present in a specific piece of material. This worst flaw is the location were fracture occurs in the same sense as the weakest link in a chain is the location of fracture. As the size, shape or processing details for a part change so will the probability of encountering a different and potentially worse flaw. Thus, the strength, tensile or flexural, has to be considered in a probabilistic sense. When determining the strength of a ceramic material a large number of specimens have to be tested in order to obtain a measure of the scatter of the results.

The strength values of ceramics are usually represented as two-parameter Weibull distributions, with the characteristic strength and the Weibull slope, m, as parameters. The Weibull slope, m, also frequently called the Weibull modulus, is a dimensionless number. The characteristic strength serves as a representative strength value, and the slope is an indication of the scatter. The *greater* the value of m the *less* the *scatter* of the distribution. (As a rule of thumb a Weibull slope of 20 is considered adequate for structural design purposes.) This is in contrast to the usage for the normal distribution, where a greater standard deviation indicates more scatter.

Tensile strength data for ceramics are more scarce than flexural strength data, because tensile tests on ceramics are more difficult to perform than flexural strength tests. The flexural strength data are usually obtained from four-point bend tests as described in ASTM C-1161, *Test Method for Flexural Strength of Advanced Ceramics at Ambient Temperature*. A variety of test techniques are in use for the determination of tensile strength. ASTM has published *Standard Practice for Tensile Strength of Monolithic Advanced Ceramics at Ambient Temperatures*, ASTM C-1273.

For a comparison of tensile and flexural strength data one has to remember that there are significant differences between the two types of tests. In both cases the failure occurs due to tensile loading. In the tensile test the specimen usually fails from anywhere in the uniformly stressed volume. In the flexural test the specimen usually fails from somewhere close to the tensile surface. Thus the volume exposed to the maximum tensile stress is vastly different for the two tests. This in turn results in different "strength values." In general, the strength is "highest" when measured by three-point flexure and "lowest" when measured in tension.

Characteristic strength values measured by one method can be converted to characteristic strength values for other geometries by use of Eq. 2, if the stressed volumes of both test specimens (or components) are known:

$$\sigma_{0\,A} = \sigma_{0\,B}\left(\frac{V_{eff\,B}}{V_{eff\,A}}\right)^{1/m} \tag{2}$$

where $\sigma_{0\,A}$ = characteristic strength of a component with a stressed volume $V_{eff\,A}$

$\sigma_{0\,B}$ = characteristic strength of a component with a stressed volume $V_{eff\,B}$

$V_{eff\,A}$, $V_{eff\,B}$ = stressed volumes of the two components; for MOR bars the stressed volumes can be calculated from Eq. 3

m = Weibull modulus

The effective volume for a four-point MOR bar is

$$V_{eff} = \frac{bh}{2}\left[\frac{mL_2 + L_1}{(m + 1)^2}\right] \tag{3}$$

where b, h are the thickness and height of the specimen, respectively, and L_1, L_2 are the outer and inner loading span, respectively.

It is clear that the exact strength data and specifically the strength distributions are very sensitive to the details of processing, finishing (manufacture), stressed volume and stress state. Thus the data given in Figs. 4 and 5 and Table 1 are representative for a given class of material. For design use one should obtain specimens cut from actual components made by the anticipated production process which will test stressed volumes anticipated in service. In Figs. 4 and 5, the "white" portion of the bar for each material shows the range of properties which may be encountered for the material. The horizontal line, if present, within the "white" portion of the bar indicates what the authors believe to be a "typical" value for commercial materials.

Elastic Properties

The elastic modulus, E, is a proportionality constant that relates normal stress to normal strain. Frequently it is called Young's modulus. Shear stress and strain are related through the shear modulus, G. Ceramic materials are stiff, which means that their elastic moduli have high values. The elastic modulus, E, of steel is about 200 GPa, of brass 100 GPa, and of aluminum 70 GPa. For ceramics E ranges typically from 250 to 400 GPa. Thus ceramics are a good choice when the elastic deflection is to be minimized. The elastic moduli are not very sensitive to processing variations, but they are a function of the density of the material: the lower the density of the material the lower its modulus.

Young's modulus is measured through the use of strain gauges or it is derived from ultrasonic measurements or through impulse excitation of vibration. The last two methods are standardized by ASTM as ASTM C-1198 and C-1259, respectively.

The data in Fig. 6 and Table 2 are most likely not based on the ASTM procedures in all details, since these standards have been published only in the past few years.

Hardness

Hardness is a measure of a materials resistance to permanent deformation under load. It is a measure of the scratch or abrasion resistance of materials. Accordingly, hardness is one of the most important materials properties in determining a material's wear behavior. Mineralogists developed the Mohs scale for measuring the relative hardness of minerals. This 10-point scale is used to define which mineral (ceramic) will scratch another. The higher Mohs-number materials will scratch those with a lower number and will not itself be scratched. This method is easy to use but is very nonlinear. Due to this nonlinearity many engineering ceramics have Mohs hardness values between 9 and 10 (Table 3). More quantitative, and more linear, values are obtained using various indentation methods such as Vickers indentation (the Vickers indentor is a diamond pyramid with equal diagonals) or the Knoop indentation (a diamond pyramid with one diagonal much longer than the other). Because Knoop indentations are shallower than Vickers indents, they are less likely to produce cracks in brittle materials during indentation; thus, Knoop hardness values, H_k are more often reported for ceramic materials than are Vickers hardness numbers, H_v. However, as can be seen from the existence of the JIS,

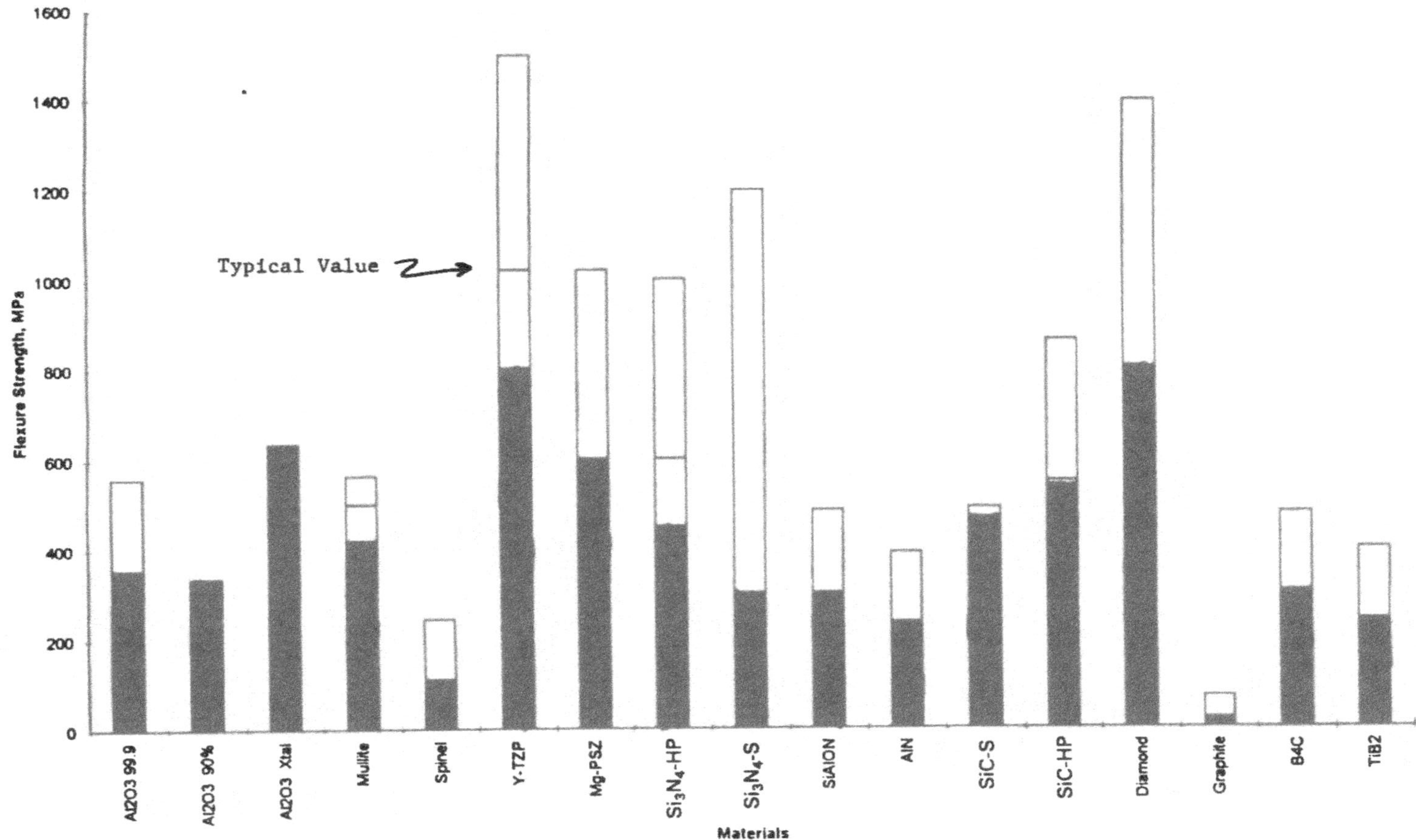

Figure 4 Range of room-temperature flexure strengths for selected engineering ceramics.

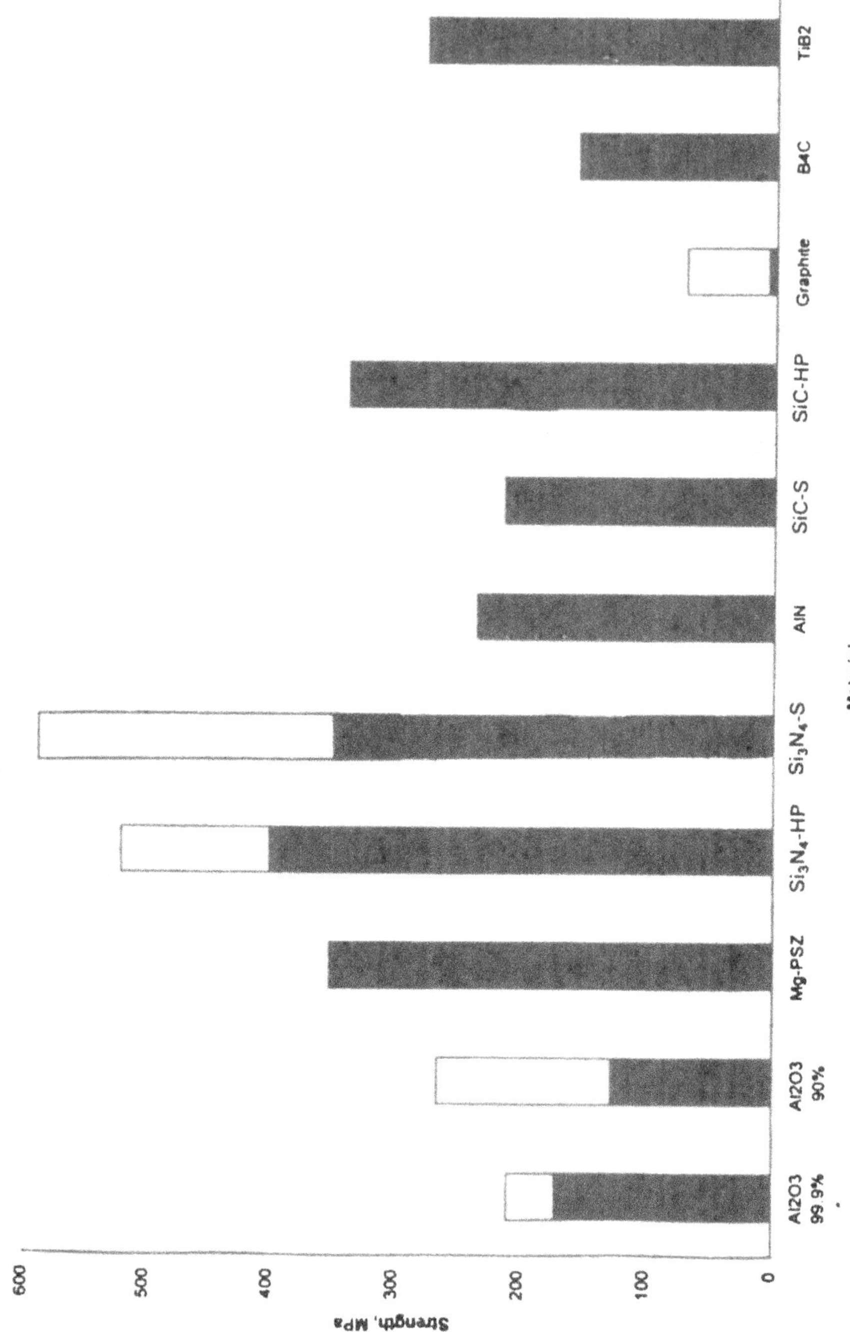

Figure 5 Range of room-temperature tensile strengths for selected engineering ceramics.

Table 1 RT Flexure, Tensile and Compressive Strength of Selected Engineering Ceramics

	MOR MPa	m	Tensile strength (MPa)	m	Compressive strength (MPa)	Density (g/cm^3)
Al_2O_3 99.9%	355–560		170–310	18.5	2550	
Al_2O_3 90%	335		125–265	13.3	1900–3790	
Al_2O_3 Xtal	635					
Mullite	420–565					
Spinel	110–245					
Y-TZP	800–1500	14			2900	
Mg-PSZ	600–1020	20	350		1760	
MgO	300–350					
Si_3N_4-RB	120–350	19–40				2.7
Si_3N_4-HP	450–1000	15–30	400–520	15		3.3
Si_3N_4-S	300–1200	10–25				3.3
Si_3N_4-HIP	600–1200					3.3
SiAlON	300–485	15				
AlN	235–390					
SiC-RB	390	10				
SiC-S	460–490	9	210	7.5	4600	
SiC-HP	540–860	10	335	11	4600	
Diamond	800–1400		21,600–32,400		4500–5800	
Graphite	20–70		6–70	14	16–185	
B_4C	303–480				2855	
TiB_2	240–400		275	15	5760	

Vickers hardness standard for high-performance ceramics (R 1610), Vickers testing is becoming preferred. Units for both H_k and H_v are the same, kg/mm^2 or GPa in SI units. Table 3 defines the Mohs scale and provides approximate H_k values for the Mohs numbers, as well as, Knoop and Vickers hardness values for selected advanced ceramics.

Fracture Toughness and Crack Growth Resistance

Fracture toughness is a material's resistance to the extension of a sharp crack and is measured in MPa$\sqrt{m}$. If the crack propagates in a stable manner the crack growth resistance is characterized by a crack growth resistance curve. (This curve is also called an R-curve or T-curve.)

Ceramic materials have low fracture toughness values, which means that they break under low loads when a crack(s) or defect is present. This and the lack of plasticity are the most problematic characteristics of ceramics. The fracture toughness of glass is about 0.7 MPa$\sqrt{m}$ and of silicon nitride 3.5 to 7 MPa $\sqrt{m}$. The fracture toughness of structural steels ranges from 35 to 100 MPa$\sqrt{m}$. Thus major efforts have been invested in making ceramics with higher fracture toughness. These efforts are currently continuing.

The fracture toughness of metallic materials, K_{IC}, is measured traditionally according to a standard test method, such as ASTM E-399 or equivalent. This method requires that a naturally sharp straight crack be introduced into a plate or

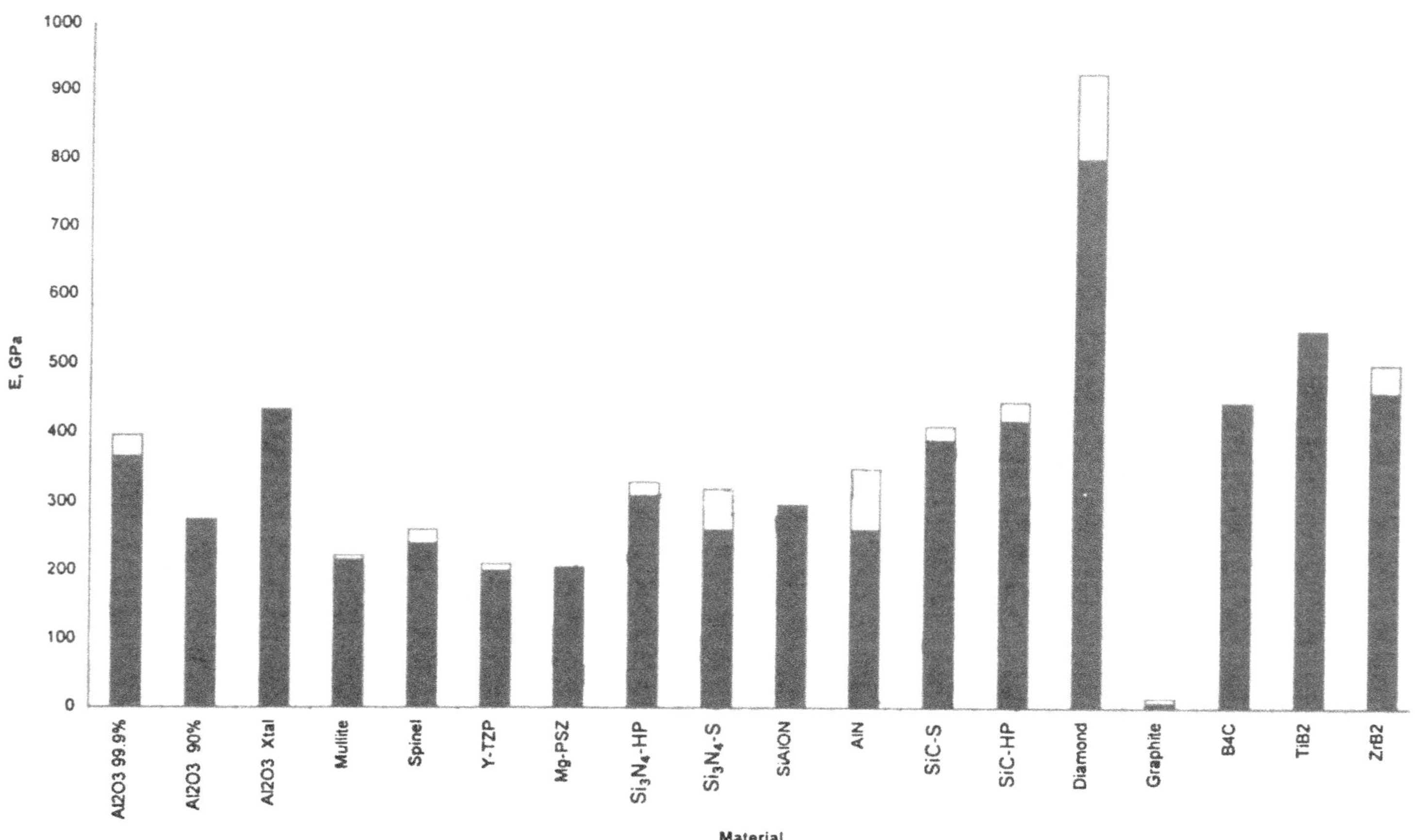

Figure 6 Elastic modulus of selected engineering ceramics.

Table 2 Elastic Moduli

	E(GPa)	Poisson's ratio	Density (g/cm^3)
Al_2O_3 99.9	366–396	0.22	
Al_2O_3 90%	275	0.22	
Al_2O_3 Xtal	434		
Mullite	216–223	0.27	
Spinel	240–260		
Y-TZP	200–210	0.31	
Mg-PSZ	205	0.22–0.3	
Si_3N_4-RB	120–250	0.2	2.7
Si_3N_4-HP	310–330	0.27–0.3	3.3
Si_3N_4-S	260–320	0.25–0.28	3.3
Si_3N_4-HIP		0.23	3.3
SiAlON	297	0.23	
AlN	260–350		
SiC-RB	382–413		
SiC-S	390–410		
SiC-HP	414–445	0.14–0.16	
Diamond	800–925		
Graphite	8–15		
B_4C	445	0.17–0.2	
TiB_2	550		
ZrB_2	460–500	0.11	

beam specimen. The cracked specimen is then loaded to fracture and the fracture toughness is calculated from the fracture load and the crack length at fracture.

This procedure is *not* readily applied to ceramic materials, since it is extremely difficult to introduce a sharp crack into a specimen in a controlled fashion. Consequently, over the past 25 years a variety of tests have been proposed as fracture toughness tests for ceramics. Careful comparison of the different methods has shown that not all of them give consistent and reliable results.

Currently ASTM is a developing a standard test method incorporating three test methods: the single-edge precracked beam (SEPB) for K_{Ipb}, the surface crack in flexure method (SCF) for K_{Isc}, and the chevron notched-beam method (CVN) for K_{Ivb}. The Japanese Industrial Standard uses a single-edge precracked beam, also.

To date, published fracture toughness data are based on a variety of methods. Some of these methods tend to give very high fracture toughness values. We have used our judgment in the exclusion of some of these data from Table 4.

Mechanical Fatigue

Until recently it was believed that ceramic materials do not lose their load-bearing capability due to mechanical fatigue. In recent years, however, it has been shown that ceramics are susceptible to fatigue failure. This is especially severe when sharp cracks or crack like defects are present, since these cracks propagate very rapidly under cyclic loading. Thus, over the past few years materials and components have been tested for fatigue resistance. These tests have been carried out on smooth

Table 3 Hardness Values of Advanced Ceramics and the Traditional Moh's Scale

Material	Chemical formula	Moh's hardness #	Knoop H (kg/mm^2)	Vicker's H (GPa)
Diamond	C	10	8000	
Boron carbide	B_4C		3000	
Silicon carbide HP β	SiC		2900	27.7
Silicon carbide sintα	SiC			27.5
Titanium diboride	TiB_2		2600	24.4
Tungsten carbide	WC		2100	
Sapphire	Al_2O_3-sing xstal	9	2000	
alumina HP polyxstal	Al_2O_3			18.4
Silicon nitride α	Si_3N_4		1900	20
Silicon nitride β	Si_3N_4		1600	16
Topaz	$SiAl_2F_2O_4$	8		
Spinel	Al_2MgO_4			12.7
Quartz	SiO_2	7	1000	
Zirconia PSZ	ZrO_2	6.5	1000	12.3
Mullite				11
Orthoclase	$KAlSi_3O_8$	6	600	
Apatite	$Ca_5P_2O_{12}F$	5	500	
Fluorite	CaF_2	4	200	
Calcite	$CaCO_3$	3	150	
Gypsum		2	50	
Talc		1	20	

Table 4 Fracture Toughness of Selected Engineering Ceramics

	Fracture toughness ($MPa\sqrt{m}$)
Al_2O_3 99.9%	3.85–4.5
Al_2O_3 90%	2.5–4.0
Mullite	1.9–2.6
Y-TZP	7.0–12.0
Mg-PSZ	7.0–10.0
MgO	1.1
Si_3N_4-RB	1.5–2.8
Si_3N_4-HP	4.2–7.0
Si_3N_4-S	5.0–7.0
Si_3N_4-HIP	4.2–7.0
SiAlON	6.0–8.0
AlN	3.0–4.0
SiC-S	2.6–4.0
SiC-HP	3.0–4.6
TiB_2	5.0–6.5
Diamond	6.0–10.0
Graphite	0.9
Glass	0.7

specimens for S-N curves or similar information, or on specimens containing sharp cracks to establish the crack propagation rates. This information can be found in the literature [11,12].

Impact Behavior

The impact behavior of high-performance structural ceramics are of particular importance in two major classes of applications. These are applications where wear resistance to particulate erosion or resistance to ballistic penetration by projectiles are the major issue. Resistance to particulate erosion is critical for applications such as mineral processing equipment or ceramic turbocharger rotors (where carbon particles or engine debris may be entrained in the exhaust and impact the rotor blades). Ballistically resistant ceramic armor is used to defeat high-velocity bullets or larger penetrators. In this case the ceramic will itself break up, but in the process will fracture or erode the impacting projectile, thereby stopping it. In both applications the performance of the ceramic component is a function of both the intrinsic properties of the ceramic material and also of the system design and the nature of the impacting particles/projectiles. We will briefly deal with each of these two generic applications.

Particulate Erosion Resistance. The particulate erosion resistance of ceramic materials is a complex function of the angle of the impacting particle; the mechanical and physical properties of the impacting particle; the size, shape, velocity and number of impacting particles; the gaseous or liquid medium in which the particle stream is entrained; and, of course, the mechanical and physical properties of the erosion-resistant ceramic itself. Thus, it is clear that erosion resistance is a system's specific behavior, not a fundamental materials property. Of all of the foregoing factors which influence the particulate erosion resistance of ceramics, only the angle of particulate impact and the erosion-resistant materials choice are amenable to modification once a basic systems concept has been selected, but before finalizing the design. Figure 7 illustrates the effect of the angle of incidence on the erosion

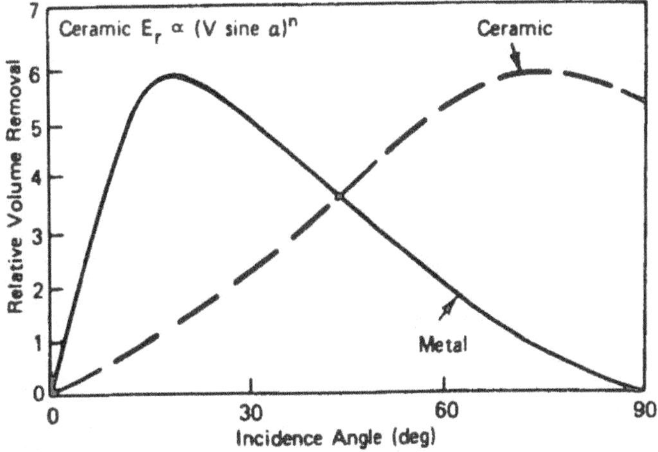

Figure 7 Comparison of ceramic versus metallic erosion resistance behavior. (From Ref. 13.)

rate of ceramics versus metals. Careful design of the particulate-laden fluid flow path can sometimes alter the angle of impact, thereby mitigating particulate erosion.

Materials selection offers the only route to mitigate erosion once the design has been set. An example of the effect of material properties on particulate impact on the selection of ceramic materials for a turbocharger application, where chipping due to one large impacting particle might cause failure (foreign object damage, FOD), has been provided by Matsuda et al. [14]. Since rust, wear debris, or carbon particles in an automotive exhaust gas stream may impact the tip of a turbocharger blade, Matsuda et al. used a gas gun to fire 1-mm-diameter steel balls at the edge of thin ceramic plates to determine the velocity at which chipping would be observed. They tested a variety of SiCs, TZP, Al_2O_3, and Si_3N_4s. The impact velocity at which chipping occurred ranged from a low of ~30 m/s for SiCs to ~60 m/s for the Al_2O_3, ~120 m/s for the TZP, and up to ~170 m/s (~MACH 0.45) for the best silicon nitride. For a given size, shape and material-impacting particle they found that the critical velocity for chipping, V_c, could be predicted by the equation

$$V_c = K_{IC}^{5/2} a^{-5/4} \tag{4}$$

where K_{IC} is the fracture toughness and a is the effective pore size of the material. Thus, to maximize resistance to FOD (wear resistance to chipping), one should select a material with high fracture toughness and minimum porosity.

Resistance to Ballistic Impact. Ceramic-faced armor systems have been used since the Vietnam conflict to defeat small arms ammunition, and have been extensively studied for application to larger threats [15]. As is the case with erosion or FOD due to small particle impact, the ballistic performance of a ceramic-faced armor system is a complex function of the materials, the specific armor design, and the specific projectile (threat) being utilized. Thus, the ballistic performance of a ceramic material is as much a systems property as a materials property. Although ceramic armor systems have been used for over 30 years and many models of ballistic penetration of ceramic armor systems have been developed, it is still not possible to select an optimum ceramic material for a given armor application solely based on the materials properties. Much empirical testing is generally required. For lightweight ceramic armor systems it is usually desired to have a high hardness (to shatter hard projectiles), a high elastic modulus to reflect the compressive shock wave back into the projectile as a tensile wave which will help fracture the projectile, and a low mass density to maintain light weight. Intuitively one might think that a high fracture toughness would be desirable; however, the limited data in Ref 15 for SiC ceramics indicate that ballistic performance seems to increase as fracture toughness decreases. This reference also describes several tests for quantitatively measuring ballistic performance. The so-called V_{50} test measures the velocity at which half of the targets will be penetrated. This probabilistic testing tool measures the performance of the armor system. The newer depth of penetration test, wherein projectiles are shot at varying thicknesses of ceramic mounted on an "infinite" metal backup, measures the effectiveness of the ceramic material in absorbing the penetrators kinetic energy. The shallower the depth of penetration into the backup, the more kinetic energy the ceramic has absorbed from the projectile.

4.2 Elevated Temperature Properties

Strength and Modulus

The tensile and/or flexure strength of ceramics varies with temperature. The strength of ceramics changes only slightly for several hundred degrees up to a temperature where the strength decreases significantly. This general trend is shown for oxide ceramics in Fig. 8. Figure 9 shows strength as a function of temperature for several nonoxide ceramics compared to the strength of superalloys. It can be seen that silicon nitride and silicon carbide retain strength at significantly higher temperatures than superalloys.

The strength decreases with temperature due to phase transformations, softening of grain boundary phases or oxidation. Thus the strength of a ceramic at a specific temperature depends strongly on changes in the microstructure. The data in this section are only representative values and trends. Specific values need to be established for each intended use and material.

Table 5 summarizes strength values or ranges for several materials over a range of temperatures. Some of the materials show a significant decrease in strength well below 1400°C. This is frequently due to melting of a constituent phase.

When executing a thermal-mechanical finite element analysis of a ceramic component, as described in Fig. 1, one of the key inputs are the elastic moduli as a function of temperature. The stress in a constrained volume element of a component in a temperature gradient is equal to $\Delta T \alpha E$, where ΔT is the temperature difference, α is the coefficient of thermal expansion, and E is the elastic modulus. As temperatures in a part used at elevated temperatures will vary with time, it is

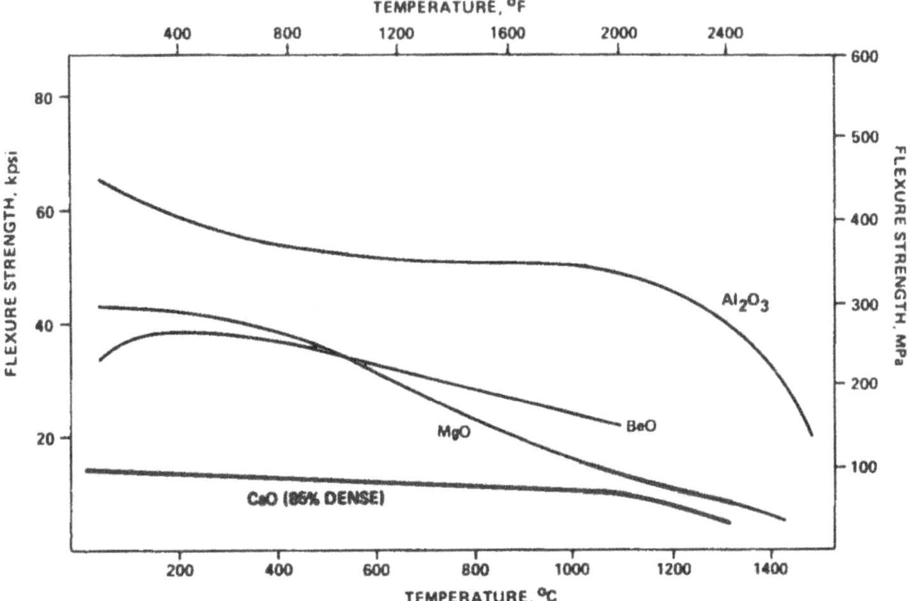

Figure 8 Strength vs. temperature for oxide and silicate ceramics. (From Ref. 2.)

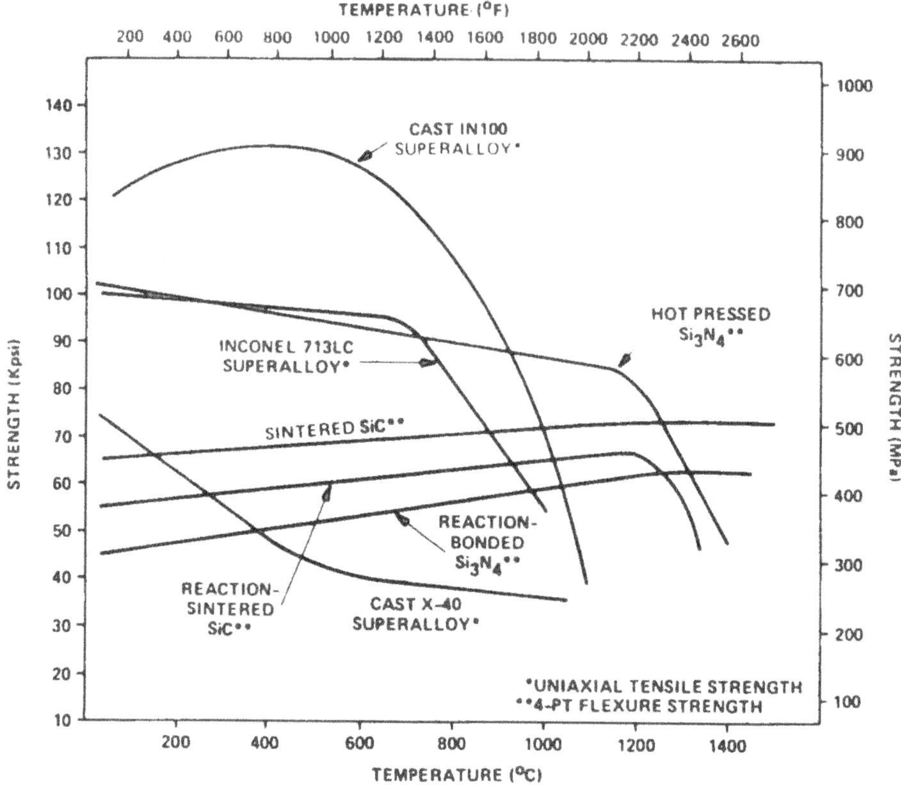

Figure 9 Strength vs. temperature for carbide and nitride ceramics and superalloy metals. (From Ref. 2.)

Table 5 Flexure Strength as a Function of Temperature

	RT	400°C	800°C	1200°C	1400°C
Al_2O_3	455	390	350	320	220
Al_2O_3 Xtal	635	500	380	380 (1000°C)	
MgO	350	275	170	75	40
BeO	230	260	190	150 (1000°C)	
CaO	100	90	75	50	
Y-TZP	1400	800	500		
Si_3N_4-RB	310	340	380	420	420
Si_3N_4-HP	700	670	610	570	320
Si_3N_4-S	600–1200	600–1200	600–1175	500–800	180–220
Si_3N_4-	840			650	560 (1375°C)
SiC-S	440	470	490	500	500
SiC-RS	380	400	425	460	310 (1350°C)
TiB_2	300	325	360	390	400

important to use the elastic moduli appropriate to the temperature regime being analyzed.

The measurement of the elastic modulus of ceramics above ~600°C is complicated by the lack of high-temperature adhesives for mounting of strain gauges and the complexity entailed in using high-temperature extensometry. New laser extensometric techniques are being used which resolve some of these difficulties. Figure 10 presents data on the temperature dependence of *E* for a variety of structural ceramics.

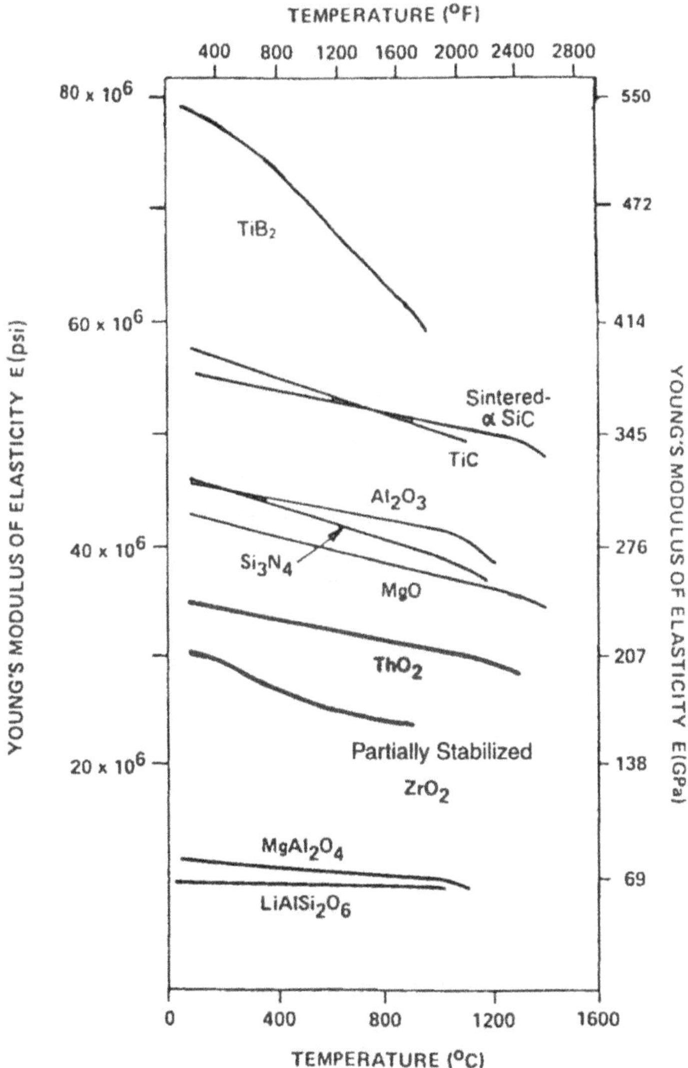

Figure 10 Young's modulus of selected ceramics as a function of temperature. (After Richerson)

Stress Rupture and Creep

If a ceramic component is to be used at high temperature for a significant period of time, it is essential to know if and how the strength and dimensional stability may change with time. Many structural ceramics will exhibit a tendency for the strength to diminish with time under load at high temperature. This behavior is measured by stress-rupture testing. Stress-rupture behavior is frequently referred to as static fatigue behavior in the ceramic literature. Although ceramics are brittle at room temperature, at elevated temperature they may elongate under load. This behavior is referred to as creep.

The stress-rupture behavior of ceramics is measured by subjecting a large number of specimens to a variety of loads and measuring the time to failure for each specimen. The results are generally plotted as the stress versus the time to failure on a log scale. When the data are plotted in this manner (Figs. 11 and 12), one can immediately compare the strength retention of one material versus another. Figure 11 presents data on the stress-rupture behavior of several types of silicon-nitride-based materials at 1200°C. Figure 12 presents data on several types of SiC material, also at 1200°C. It should be noted that at 1200°C the best superalloys only have lives measured in tens of hours at stresses of 70–100 MPa. Many silicon-nitride- and SiC-based ceramics and composites have stress-rupture lives of thousands of hours at appreciable stresses, at temperatures where superalloys melt. Stress-rupture life is clearly one behavioral regime where modern structural ceramics excel. Other than for silicon nitride and silicon carbide ceramics, very little

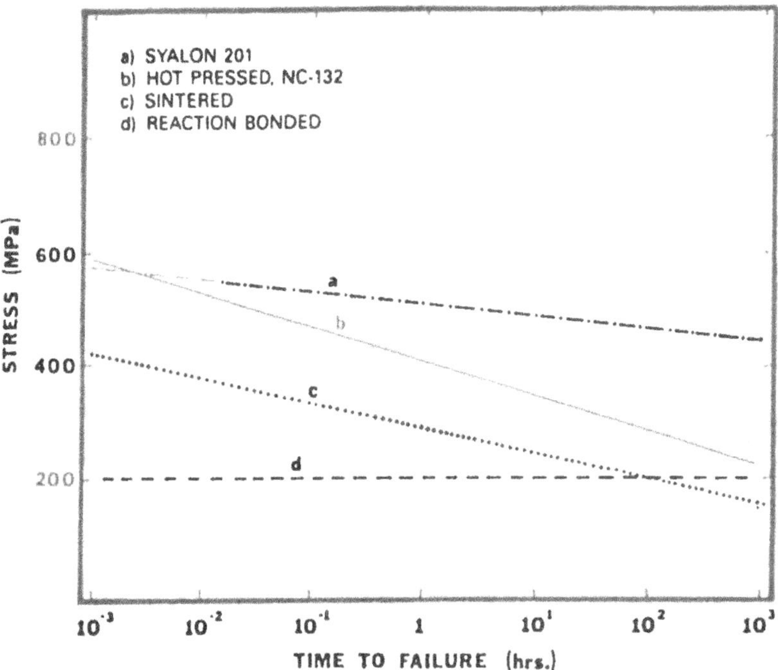

Figure 11 Stress-rupture behavior of various silicon-nitride-based ceramics, tested in air at 1200°C. (From Ref. 16.)

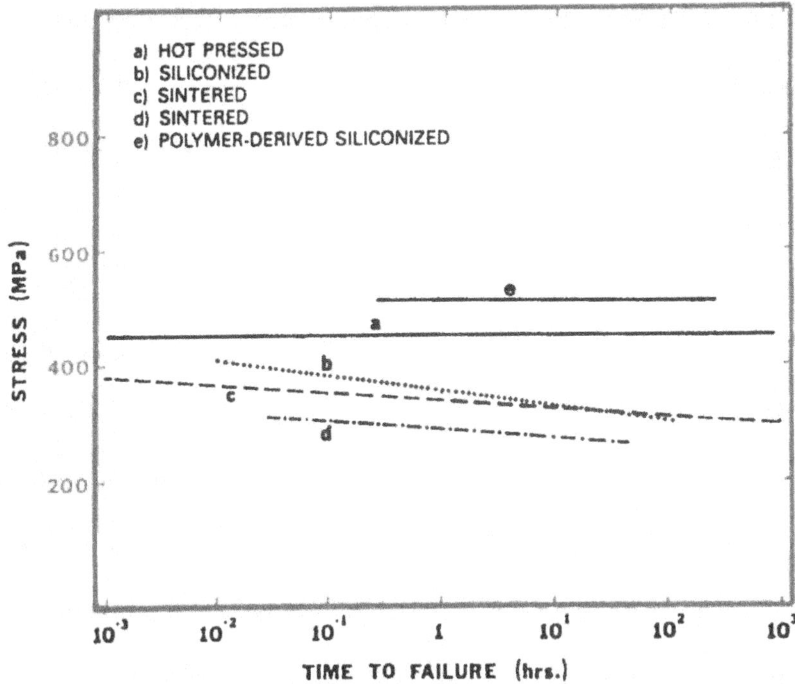

Figure 12 Stress-rupture behavior of various silicon carbide ceramics, tested in air at 1200°C. (From Ref. 16.)

stress-rupture data are readily available in the literature. One notable exception is the work of Swab and Leatherman, who have evaluated the stress rupture behavior of a variety of zirconia ceramics [17].

Stress-rupture testing of ceramics is most often carried out utilizing flexure specimens with four-point loading. Such specimens frequently creep during the stress rupture test, and consequently the stress is not constant over the duration of the test (the stress decreases). It is a commonly observed (though incorrect) convention to ignore this fall-off in stress (because it is difficult to precisely calculate) and to assume the initial stress level prevails throughout the test. To eliminate this complication and conduct stress-rupture testing in a rigorous manner, it is desirable to utilize uniaxial tensile tests. Such tests are very expensive and time consuming. Nevertheless, as ceramic materials are becoming more commonly utilized as high-temperature structural components, the use of tensile stress-rupture testing has started to grow and the growth will continue. Stress-rupture testing combined with fractographic analysis is the basis for deriving fracture mechanism maps (Fig. 3) which are increasingly utilized in the design of ceramic components for high-temperature structural components.

The creep of ceramics can be measured in tension or in compression. Compressive creep testing is rather straightforward. The uniaxial tensile creep testing of ceramics shares many of the experimental difficulties of elevated tensile or stress-rupture testing of ceramics. Thus, tensile creep tests are often carried out in flexure. In this case one measures the increase in length of the tensile surface of the bend bar as a function of time to establish the creep rate. Creep usually has three stages: initial or primary creep (stage I), steady-state creep (stage II, the most extensive stage), and tertiary (stage III), with an accelerating creep rate (this is not often observed in ceramics). When a creep rate is cited in the literature, it is usually the steady-state creep rate. Figure 13 presents steady-state creep rates versus stress for various ceramics at various temperatures.

5 SELECTION BASED ON THERMAL-PHYSICAL PROPERTIES

Often the primary selection criteria for the application of a ceramic material will be based upon one or the combination of several thermal-physical properties. One example would be a high-temperature heat exchanger tube. In this case the primary function of the component is rapid heat transfer, so thermal conductivity will play a dominant role. Issues of strength and erosion/corrosion resistance in hot combustion gases will play an important role in the eventual material's trade studies, but no material without a sufficiently high value of thermal conductivity will be selected for such studies. In ceramic packages for microelectronic circuits the coefficient of thermal expansion (CTE) is critical. If the difference in CTE between the package and the silicon chip is too great thermal fatigue leading to failure of the device will result. As the number of components on a chip has increased, so has the heat load that the package must dissipate. Thus, thermal conductivity is also a key property for microelectronic packaging. This has lead to AlN supplanting alumina for packaging in high-end microelectronics. In applications where inertia or centrifugal loads must be minimized, the mass density of a material becomes a prime selection criteria. Such applications include turbocharger rotors (cf. Section 7), rolling elements for bearings, valves and injector components, and gyroscope components. In an application such as a ceramic radome, the dielectric properties become the dominant selection criteria, with properties such as strength and rain erosion resistance being considered only after a material meets the dielectric properties requirements. The following sections briefly describe and list the thermal, physical and electrical properties of structural ceramics.

5.1 Thermal Properties

The thermal properties of ceramics of general importance in materials selection are heat capacity (c), thermal conductivity (k), CTE, and the thermal shock figure of merit for instantaneous thermal loads (R). Heat capacity, thermal conductivity and CTE are necessary to carry out finite-element or finite-difference computer codes which generate the transient and steady-state thermal analyses which are the starting point for the design process illustrated in Fig. 2. Such thermal analyses are not only carried out for high-temperature mechanical components but are now also routinely performed in the design of lower temperature components such as microelectronic packaging.

Heat Capacity

The heat capacity of a material is equal to the number of heat units (cals, BTUs) required to raise the temperature by 1 degree (K or F). Heat capacity is usually measured in cal/g K in SI units or BTU/lb · °F in English units. Heat capacity is

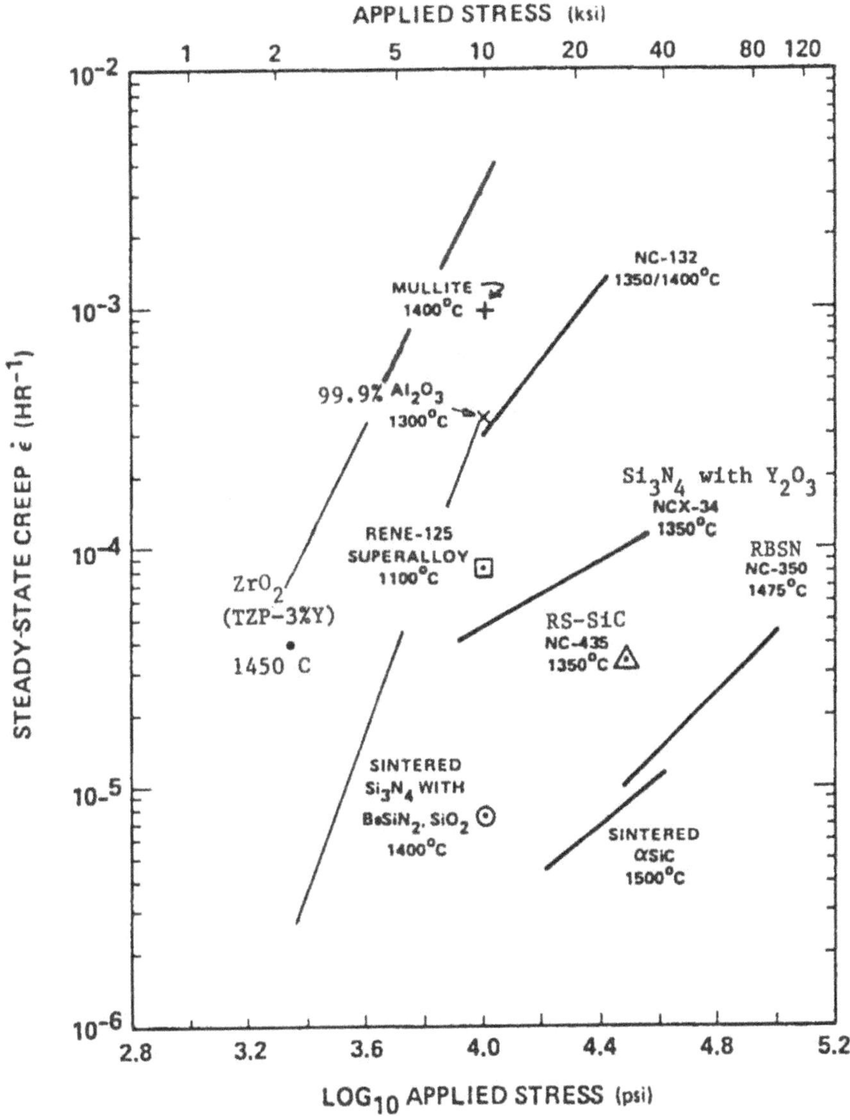

Figure 13 Steady-state creep rates of various ceramics. (From Ref. 17a.)

also frequently reported on a molar or gram atom basis. The units then become cal/mole · K or cal per gram atom K and can be converted to a weight basis by dividing by the atomic or molecular weights as appropriate. The term "specific heat" is defined as the ratio of the material's heat capacity to that of water at 15°C. Although the specific heat has no units since it is a ratio, the value of the specific heat is the same (or very nearly the same) as the heat capacity, as measured in units of cal/g · K. Thus, the terms are often used interchangeably.

Heat capacity is measured by a variety of calorimetric techniques. The most common of these is differential scanning calorimeter (DSC). ASTM standard

E-1264 describes the use of DSC to measure heat capacity. Table 6 lists the heat capacity of various representative structural ceramics. In preparing this table it became evident that values of heat capacity are not comprehensively reported in the standard data compilations. One surprising finding not shown is that the heat capacities of alumina, mullite, and silicon nitride matrix composites, reinforced with up to 30 vol% of SiC whiskers, are essentially unchanged from the matrix values.

Thermal Conductivity

The heat flux, or rate of heat flow through a material, is referred to as the thermal conductivity, k. The heat flux is driven by a temperature gradient dT/dx and k is defined as the constant of proportionality between the heat flux, Q, and the temperature gradient (the driving force). Thus,

$$Q = -k\frac{dT}{dx} = \left(\frac{1}{\text{area}}\right)\frac{dH}{dt}$$

where H is heat energy and t is time. This equation has the same form as Ohm's law for flow of electrical current, and Fick's first law for atomic diffusion. The SI units for thermal conductivity are $W/m \cdot K$. Units of $cal/s \cdot cm^2 \cdot C \cdot cm$ are also frequently utilized.

Thermal conductivity, $k = D\rho c$, where D is a term called thermal diffusivity, ρ is the density, and c is the heat capacity. Thus, if one measures the thermal diffusivity and knows c and ρ, one can calculate k. Thermal diffusivity can be readily measured by illuminating one side of a disk of the sample material with a laser pulse, and measuring the temperature rise on the opposite face of the disk as a function of time. The use of this "laser flash method" for determining thermal diffusivity and thermal conductivity is now the usual technique (see ASTM and JIS standards cited at the end of the chapter).

Table 6 Heat Capacity (cal/g K)

Material	RT	600 C (873 K)
Diamond	0.12	
ZrO_2	0.12	
Silicon nitride	0.16	0.27
SiAlON	0.17	
Mullite	0.18	0.28
Alumina	0.19	0.29
LAS	0.19	
AlN	0.20	0.24
Al-Titanate	0.21	
B4C	0.23	
TiB_2		0.25
Soda Glass	0.24	
SiC	0.25	0.26
SiO_2	0.27	

Table 7 lists the RT thermal conductivities for a variety of engineering ceramics. The data illustrate some general trends as well as values for specific materials. The data on sintered versus hot-pressed SiC illustrates the effect of grain boundary phases on k. Sintered SiC has less than 1% of sintering aid added to the material, and this sintering aid does not form a discrete second phase. By contrast, hot-pressed SiC has several percent of an oxide additive which forms a discrete grain boundary phase. Since this phase is alumina, which has a lower value of k than SiC, and surrounds each SiC grain, it acts as a thermal resistor or a thermal barrier and reduces heat flow, and consequently, the k of the SiC material. This is a general behavior; if a grain boundary phase of lower k than the matrix is used, it will decease the bulk thermal conductivity. SiAlON is a solid solution of silicon nitride and a few percent of alumina. The solid solution results in a lower value of k. AS 44 silicon nitride is a fully dense silicon nitride with a normal β silicon nitride microstructure. By contrast the AS 700 material is specially heat-treated to encourage some of the β silicon nitride grains to grow into elongated reinforcing "whiskers," so-called self-reinforced silicon nitride. For some as yet unexplained reason, the existence of the randomly oriented elongated grains significantly increases the value of k. This behavior is in marked contrast to the case of heat capacity where whisker reinforcement produced no effect. The data on diamond shows the effect of grain boundaries. Since atoms at grain boundaries are not in a "perfect" crystal lattice, their ability to conduct heat is reduced. Thus, a grain boundary may be viewed as a region of higher thermal resistance (i.e., a lower k).

Table 7 Thermal Conductivity of Engineering Ceramics at Room Temperature (RT)

Material	Thermal cond. (RT)
Fused silica (pure SiO_2 glass)	1
Aluminum titanate	2
Zirconia (TZP)	2
Pyroceram 9606	3
Mullite ($3Al_2O_3,2SiO_2$)	6
Spinel ($MgAl_2O_4$)	15
α-Al_2O_3	16
90% α-Al_2O_3	39
SiAlON	20
Si_3N_4 (AS 44)	25
Si_3N_4 (AS 700)	80
B_4C	28
TiB_2	70
SiC (sintered-α)	125
SiC (HP-2%Al_2O_3)	85
AlN	160
graphite (high density)	400
Diamond (polycrystalline)	540
Diamond (single cryst.)	1000

Thermal conductivity is strongly dependent upon temperature. Figure 14 illustrates this temperature dependence for several engineering ceramics. While most nonporous ceramics have a decreasing k versus T behavior, others show an increasing k versus T. For a discussion of the complex physical and microstructural rationale for this complex behavior see Richerson [17a].

Thermal Expansion

The thermal expansion, α, of a material is simply the fractional change in dimensions with increasing temperature. Typically the expansion of a rod is measured as temperature is increased in an instrument called a dilatometer, which is specifically designed to measure α. The linear CTE is

$$\alpha = \frac{\Delta l / l_0}{\Delta T}$$

where l_0 is the initial length of the rod (at 0°C), Δl is the change in length and ΔT is the change in temperature. It is rare, but possible, for the CTE to be negative. If an anisotropic ceramic should have a negative CTE along one direction, it will be positive along the other directions. LAS glass ceramics show an average slightly negative CTE over the 0 to 1000°C temperature range as shown in Fig. 15. In

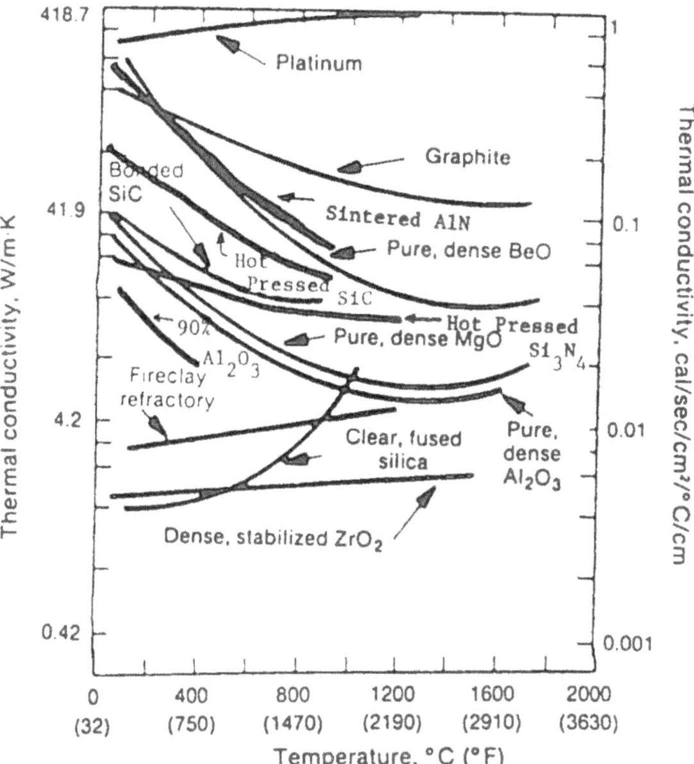

Figure 14 Thermal conductivity versus temperature for various engineering ceramics. (From Ref. 17a.)

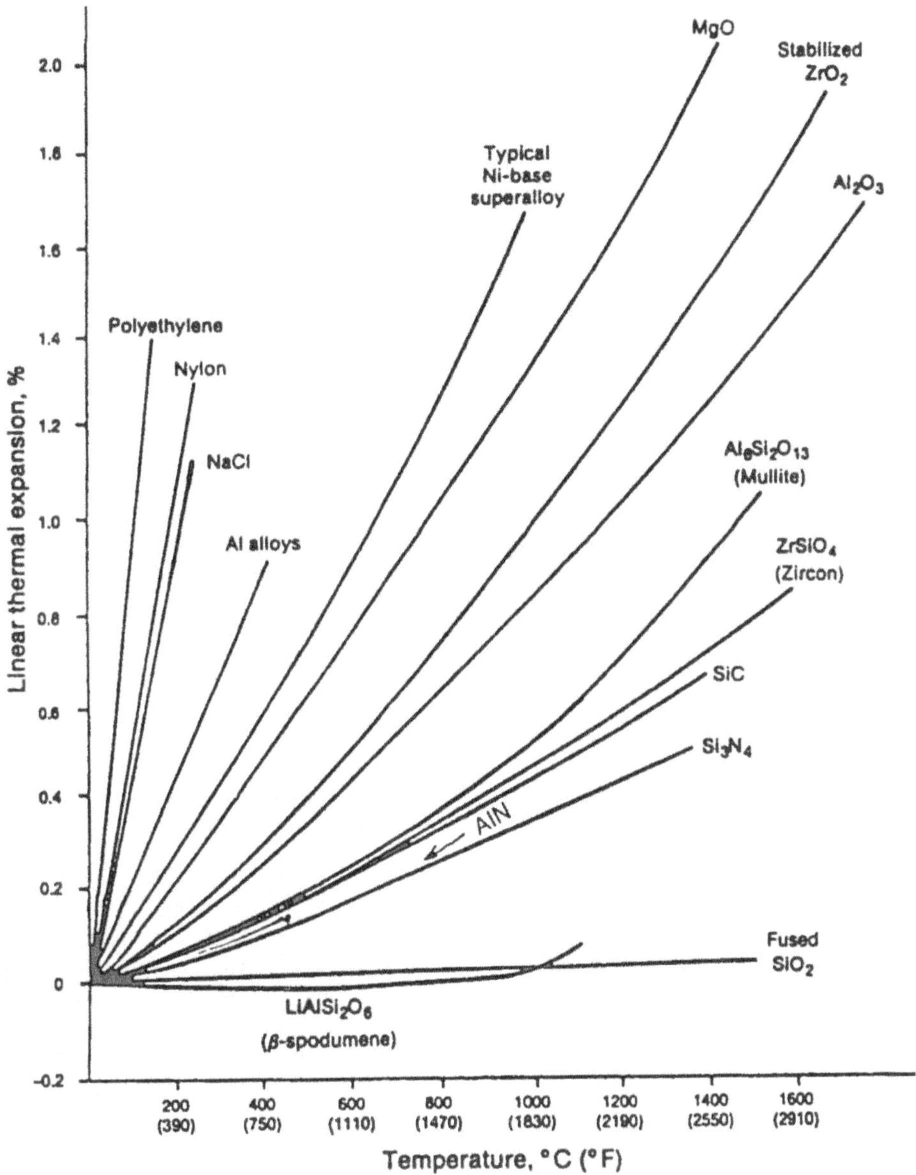

Figure 15 Thermal expansion behavior of engineering ceramics and typical metals and polymers. (From Ref. 17a.)

general, the lower the CTE the better the thermal shock resistance, as discussed in the next section. To avoid thermal fatigue failures where two materials are bonded or joined, it is critical to minimize the mismatch or difference between the CTEs of the mating materials. Figure 15, adapted from Richerson [17a], compares the thermal expansion behavior of several engineering ceramics and compares them with typical metals and polymers.

Thermal Shock Resistance of Engineering Ceramics

While heat capacity, thermal conductivity, and thermal expansion are fundamental properties of a ceramic with a given composition and microstructure, thermal shock resistance is not. It is a "derivative" behavior, derived from other material's properties, from the imposed heat-transfer conditions, and from the geometry of the component or specimen being subjected to the thermal shock. Thermal shock resistance is generally discussed and measured in terms of the maximum temperature differential, ΔT, a component can tolerate without failure. When discussing thermal shock the terms "up-shock" and "down-shock" are frequently utilized. An up-shock is a rapid raise in temperature (ΔT positive) and a down-shock is a rapid decrease in temperature (ΔT negative). A down-shock places the outer surface of a ceramic or a glass in tension (the cooler surface wants to contract but is restrained from doing so by the hotter interior; the constrained contraction results in a surface tensile stress). An up-shock places the surface in compression. Thus, since ceramics and glasses are have significantly lower tensile strengths than compressive strengths, ceramics are much more prone to failure on a down-shock than an up-shock. Accordingly, the tabulations of maximum ΔT for a material refer to a down-shock or a quench, unless otherwise noted.

The most common method for measuring ΔT, which is also known as the thermal shock resistance, is to quench bend bars of a material into water (or oil) from a series of elevated temperatures and to note the ΔT which causes the first noticeable strength degradation.

The most general equation for ΔT_{max} is

$$\Delta T_{max} = \frac{\sigma(1 - \mu)}{\alpha E} \frac{k}{r_m h} S$$

where σ, μ, α, E and k have their usual meanings, r_m is the half-thickness for heat flow, h is the heat transfer coefficient, and S is a shape factor totally dependent on specimen geometry. Thus it can be seen that thermal shock resistance, ΔT, is made up of terms wholly dependent on materials properties, dependent on heat-transfer conditions, and on geometry. It is the role of the ceramic engineer to maximize the former and of the design engineer to maximize the latter two terms. It has become the usual practice to report the material's related thermal shock resistance as the instantaneous thermal shock parameter, R, which is equal to $\sigma(1 - \mu)/\alpha E$. The value of R for selected ceramics is presented in Table 8.

Another frequently used parameter is R', the thermal shock resistance where some heat flow occurs: R' is simply R multiplied by the thermal conductivity k.

5.2 Electrical Properties

There are many cases when the selection of the appropriate ceramic material will require a knowledge of certain electrical, as well as mechanical properties. One example might be the selection of a silicon-nitride-based material for a wire-forming die. Usually silicon-nitride-based materials are insulators. However, a composite of silicon nitride with titanium nitride provides a sufficiently electrically conductive material for electrodischarge machining (EDM) to be performed, thus eliminating more costly diamond machining. The dielectric constant and loss fac-

Table 8 Calculated Thermal Shock Resistance of Various Engineering Ceramics

Material	Strength (MPa)	Poisson's ratio	CTE (cm/cm · K)	El. mod. (MPa)	*R* (K)
Alumina 99.9%	345	0.22	$7.4 \cdot 10^{-6}$	375000	97
AlN	350	0.24	$4.4 \cdot 10^{-6}$	350000	173
SiC α sintered	490	0.16	$4.2 \cdot 10^{-6}$	390000	251
PSZ	1000	0.3	$1.05 \cdot 10^{-5}$	205000	325
RBSN	295	0.22	$3.1 \cdot 10^{-6}$	200000	371
HPSN	830	0.3	$2.7 \cdot 10^{-6}$	290000	742
LAS	96	0.27	$5 \cdot 10^{-7}$	68000	2061
Al-titanate	41	0.24	$1.01 \cdot 10^{-6}$	11000	2819

tors (tan δ) are important in the design of electronic packaging and radomes. The speed at which a signal will propagate through a package is a function of the dielectric constant. The propagation speed, $T = 3.33\sqrt{\varepsilon}$, where ε is the relative dielectric constant [18]. Thus, to maximize the signal speed in a circuit contained in a package, the package material should have the lowest value of ε, provided it meets all other design criteria. A similar situation prevails in the design of radomes, where the lower the value of ε the easier it is to image the target. Tables 9–10 provide data on the electrical resistivity, dielectric constant, and loss tangent for several ceramic materials.

Table 9 lists several ceramics in order of increasing electrical resistivity. It is interesting to note the profound influence that a small amount of additive can exert on the resistivity of SiC. Changing the additive from 1% Al to 1.6% BeO changes the resistivity by ~13 orders of magnitude.

Table 9 Electrical Resistivity of Selecteed ENgineering Ceramics at Room Temperature (RT)

Material	Resistivity (ohm-cm)	Comment
Boron carbide	~0.5	semiconductor
Silicon carbide (1% Al)	~0.8	semiconductor
Zirconium oxide (@ 700°C)	$\sim 2 \times 10^{3}$	ionic conductor
Titanium diboride	$\sim 9 \times 10^{6}$	can be EDM'ed
Aluminum nitride	$\sim 10^{13}$	insulator
Silicon carbide (1.6% BeO)	$>10^{13}$	insulator
Silicon nitride	$>10^{13}$	insulator
Alumina	$>10^{14}$	insulator
Mullite	$>10^{14}$	insulator
Spinel	$>10^{14}$	insulator
Diamond (high purity)	$\sim 10^{16}$	insultor
Berylium oxide	$>10^{17}$	insulator
Silicon dioxide	$>10^{18}$	insulator

Table 10 Room-Temperature Dielectric Constants of Selected Engineering Materials

Material	Dielectric constant, ε	Tan δ
$BaTiO_3$ (10% $CaZrO_3$ + 10% $SrZrO_3$)	9500[b]	—
$BatiO_3$	1600[b]	—
Zirconia (TZP-3% Y_2O_3)	~24.0	~0.006
Alumina 99.9%	10.1	0.0004
Alumina 99.5%	9.7	0.003
Aluminum nitride	8.9	0.0001
Alumina 90.0%	8.8	0.004
Silicon nitride (HPSN-MgO)	~7.2[a]	~0.001
Mullite	6.6	0.006
Silica (fused)	3.8[a]	0.0001

[a]Measured at 10 GigaHz [19].
[b]Unknown frequency.
All other measurements at 1 MHz.

5.3 Density and Melting Point

In many cases the mass density of a component will be one of the major selection criteria (see Section 7). In other cases a melting point may impose a limitation on functionality. Tables 11 and 12 provide values of these physical properties for selected engineering ceramics.

6 TRIBOLOGICAL PROPERTIES

Tribology, derived from the Greek word for rubbing, is the study of the friction and wear of materials. However, the friction and wear behavior is not a pure materials property; it is a complex function of the two materials which are sliding or rolling over one another, the lubricant or chemical environment which surrounds them, the operating parameters such as pressure or speed, and the surface finish of the components. Thus, the tribological behavior of a material in a given situation is both a materials and a systems determined property (this is analogous to the previously described case of impact behavior). Since many of the structural applications of ceramics depend on their superior tribological behavior, it is important to briefly introduce this subject. A good reference for an in-depth study of this subject is Jahanmir [19a].

The effect of lubrication is critical. If a ball in a ball bearing operates only under conditions of elastohydrodynamic lubrication (always riding on a film of lubricant) there should be no wear. Once the lubricant film breaks down, even temporarily, wear will occur. The effect of solid lubrication by graphite on the sliding friction and wear of alumina and silicon nitride against steel (the steel wears, not the ceramics) is shown in Fig. 16. It is interesting to note that even unlubricated silicon nitride running against steel produces minimal wear. As we will see in

Table 11 Density of Engineering Ceramics and Selected Other Materials

Material	Density (g/cm3)
Graphite	1.4–1.9
SiO_2	2.2
Glass	
Sodium silicate glass	2.3
Lithium aluminosilicate (LAS)	2.3
B4C	2.5
Aluminum	2.7
Aluminum titanate	3
Mullite	3.1
AlN	3.2
S-a-SiC	3.2
HP SiC	3.3
S-Si3N4	3.3
HP Si3N4	3.3
Diamond	3.5
Spinel	3.6
Al_2O_3-90%	3.6
Al_2O_3-99.5%	3.9
Al_2O_3-99.9%	3.9
Ti B_2	4.6
TiN	5.4
Zirconia (PSZ)	5.8
Zirconia (TZP)	6
ZrB_2	6.1
Cast iron	7.1
Steel	7.8
UO_2	11
TaC	14.5
WC	15.8

Section 7, this behavior is exploited in hybrid ball bearings utilizing silicon nitride balls and steel races. Some lubricants will dissolve the surface of a ceramic to a slight extent, producing a smoother surface and reducing wear. This has been observed in silicon nitride sliding in water.

Frequently the wear of a ceramic is so-called abrasive wear. In this case small "hard" particles sliding across the surface of the ceramic "machine away" material. A variety of abrasive wear models exit. Most are of the form

$$V \propto K_{ic}^{-3/4} H^{-1/2} P^{4/5} N \tag{5}$$

where V = volume loss per unit distance of contact, H = hardness, N = number of abrading particles and P = applied load or pressure. Data have been collected by Wayne and Buljan which generally confirm this relationship [21]; see Fig. 17.

Table 12 Melting Point of Engineering Ceramics

Material	Temp (°C)
TaC	3540
ZrB_2	3040
TiB_2	2980
TiN	2927
UO_2	2878
ZrO_2	2850
SiC[a]	2300
WC	2627
BeO	2452
B_4C	2447
AlN	2200
Al_2O_3	2049
Mullite	1850
Si_3N_4[a]	1800
SiO_2-crystalline	1705
SiO_2-fused (glass)	1650
CaF_2	1423

[a]Material does not melt but dissociates.

7 APPLICATIONS OF MODERN STRUCTURAL CERAMICS

The growth in the number of successful structural ceramic applications over the past decade has been truly remarkable. Our definition of a successful application is one where a ceramic component has been introduced into a system which someone purchases and a manufacturer warrants. By this definition a successful application may include components with production rates of dozens to hundreds of thousands or even millions per year.

7.1 Cutting Tools

Many advanced ceramics have excellent hot hardness, chemical inertness and high-temperature strength retention. This combination of properties make them attractive candidate materials for single-point-turning cutting tool inserts. Fine-grained aluminum oxide ceramics were introduced into automotive manufacturing in the 1950s, for turning operations where gray cast iron is the workpiece. These materials received only limited market acceptance due to their tendency to chip and fail in an unpredictable manner. With the advent of new ceramics with higher fracture toughness in the 1980s, ceramic cutting tools have achieved significant levels of acceptance in several niche areas. Silicon carbide whisker-reinforced aluminum oxide composite cutting tools are widely utilized to machine Ni-based superalloys in the gas turbine industry. This application is a good example of what is meant by referring to ceramics as enabling materials. Although these tools are expensive compared to carbide tools, the productivity increases that are generated make the added costs inconsequential. Two case histories involving the machining of Inconel

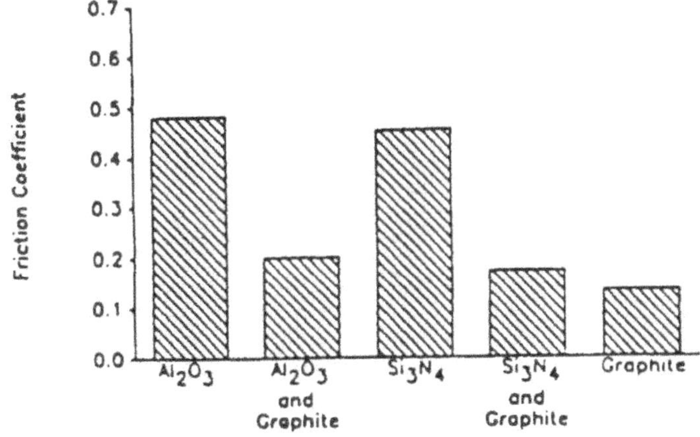

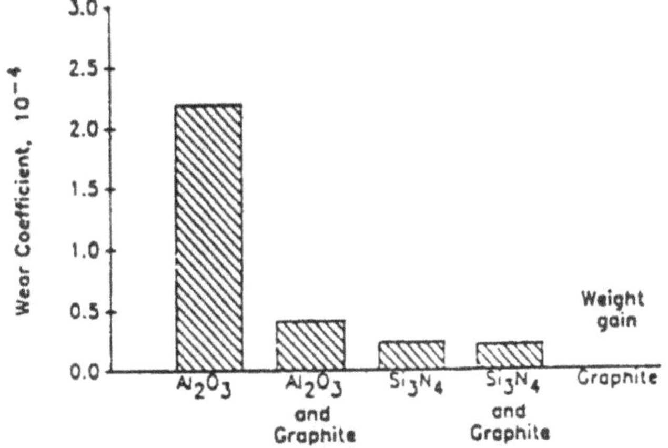

Figure 16 Wear of a steel ring against alumina and silicon nitride with and without graphite lubrication. (From Ref. 20.)

718, a nickel-based alloy, illustrating this point have been reported [22]. An operation which had used a SiAlON (a silicon-nitride-based alloy) roughing and a tungsten carbide finishing tool required a total of 5 h. A change to silicon carbide whisker-reinforced aluminum oxide composite inserts for both operations reduced the total machining time to only 20 min. This in turn resulted in a cost savings of $250,000 per year, freeing up 3000 machine hours, and avoiding the need to purchase a second machine tool. The second case study involved a four-axis machine tool with two cutting tools in engagement. The operation was the machining of a blade groove in a turbine disk. The silicon carbide whisker-reinforced aluminum oxide composite tool reduced machining time from 3 h to 18 s. Factors that accounted for this increase in productivity include increased cutting speed (up to 10×), lower wear rates which decreases the time spent in indexing, and highly predictable failure times which decrease unplanned down time.

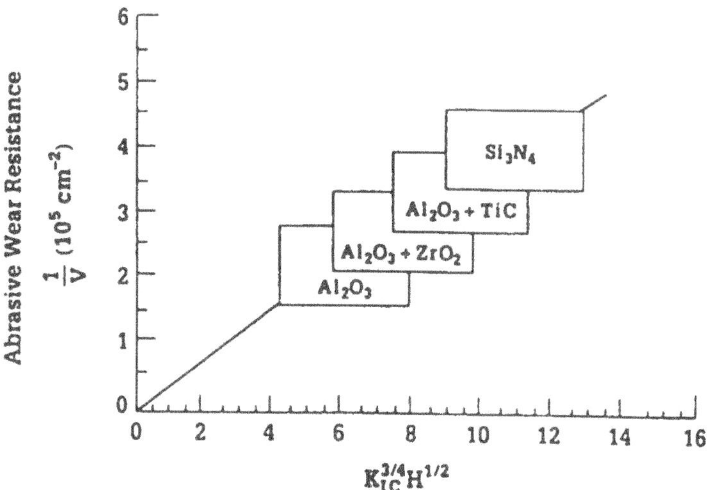

Figure 17 Abrasive wear resistance of structural ceramics vs. the wear parameter $K_{ic}^{3/4}H^{1/2}$. (From Ref. 21.)

Silicon-nitride-based ceramics, in particular hot-pressed Si_3N_4 (HPSN), continues to be the most cost-effective cutting tool for the single-point turning of gray cast iron. Most silicon nitride cutting tools possess whisker-like elongated β grains or are toughened with dispersed TiC particles. A report [23] comparing the economics of various cutting tools in outer diameter turning of an 11-in-diameter gray cast iron brake rotor showed the following results:

Polycrst. cubic boron nitride	SiAlON	HPSN
$0.024	$0.004	$0.002

Clearly HPSN has an overwhelming advantage in this area. Silicon nitride and SiAlON cutting tools are also used in interrupted cut operations such as the milling gray iron engine blocks and crankshafts. In one such case of interrupted cutting of a gray cast iron crankshaft with a SiAlON tool, the metal removal rate increased by 150% and the tool life increased 10-fold [24]. Interrupted cutting imposed severe impact on the cutting tool. That silicon-nitride-based cutting tools perform well in such an environment is a consequence of their combination of high toughness (for a ceramic material) and their thermal shock resistance.

Industrial acceptance of ceramic cutting tools continues to grow and the market continues to expand in spite of several negative factors not related to the cutting tools themselves. One issue is that in the United States, outside of the automotive and aerospace industries, few machine shops have machine tools with sufficient speed, stiffness, and horsepower to take full advantage of the potential of ceramic cutting tools. Another negative factor is the move away from gray cast iron engine blocks to aluminum blocks where silicon-nitride-based cutting tools lose their advantageous performance. However, new ceramic tools, such as diamond-coated silicon nitride end mills have been reported to give 300 times the life of micrograin tungsten carbide in machining SiC-reinforced aluminum metal matrix composites, an area of potential growth. In a similar vein, research on increasing the amount of alumina in SiAlON solid solution shows promise of increasing the wear resistance to enable the material to cut ductile iron effectively. Thus, it is quite likely that the application of ceramic cutting tools will grow well beyond their present market in the near future.

7.2 Bearings

Several properties of high-performance ceramics are very attractive for rolling-element bearing applications. These properties include high hardness and wear resistance, lower density than traditional steel bearing materials, high hertz strength, high upper-use temperature, and good corrosion resistance. Silicon nitride ceramics in particular have demonstrated rolling contact fatigue lives up to 10 times that of bearing steels, as well as a unique capacity to survive for long times in lubrication starvation situations [25]. Due to the difficulty of mating ceramic races on metal shafts, ceramics are usually utilized in "hybrid" bearings with silicon nitride balls and steel races. These hybrid bearings retain the outstanding lubrication starvation performance of all ceramic bearings. Surprisingly, it has been found that despite the fact that these ceramics possess lower fracture toughness values than the steels which they replace they consistently deliver higher performance (in terms of bearing speed) and longer lifetimes than the conventional materials.

At the present time silicon-nitride-based balls in hybrid bearings are widely used in machine tool spindle bearings, where their chief benefits are higher speed and greater stiffness; turbomolecular pump mainshaft bearings, where their ability to perform under marginal lubrication situations has greatly increased pump reliability and provided new flexibility in pump mounting orientations; and shaft bearings for dental drills, where the ability to operate unlubricated facilitates autoclaving for sterilization. Perhaps the most exotic new application for these bearing materials are their recent successful use in the space shuttle main engine fuel pump where they are lubricated by liquid oxygen. It is anticipated that silicon nitride balls will outlast steel by a factor of 10 in this aggressive environment. One major materials property of silicon nitride that has contributed to its acceptance by the bearing design community is its failure mode. Silicon nitride rolling elements do not fail catastrophically, they fail by spallation which although it proceeds by a different microstructural mechanism, manifests itself almost exactly as a steel bearing spall. Other ceramic materials when used as balls tend to shatter catastrophically when they fail. However, in some cases even these ceramics may perform successfully as bearings. One example is that of SiC used as seawater-lubricated pump bearings in North Sea oil rigs. In this case the loads on the bearings are relatively low and the resistance of the SiC to erosion and wear by the silica entrained in seawater is the key property in this materials selection.

7.3 Wear Applications

The hardness, compressive strength, erosion resistance, and relatively high fracture toughness (5 to 12 $MPa\sqrt{m}$) of high-performance ceramics such as transformation-toughened zirconia, in situ toughened silicon nitride or SiC-whisker-toughened alumina and some SiC's have enabled them to function in many wear applications.

Partially stabilized zirconia (PSZ) dies have been used in the drawing of copper and brass for over a decade. These dies last 10 to 15 times longer than steel dies and provide a superior surface finish due to the corrosion-erosion resistance and low coefficient of friction against the metals. These characteristics of PSZ have also led to its use in Asia as can rolling tools for steel food cans [26]. Aluminum beer cans in the U.S. are increasingly formed using ceramic drawing and forming dies made of zirconia- and alumina-based ceramics. Silicon-nitride-based ceramics are used in Japan in the hot rolling of a variety of steel alloys [27].

Alumina and silicon carbide ceramics have been widely used as face seals for rotary pumps. For over 20 years alumina seal rings have been standard in automotive water pumps, and sintered silicon carbide has entered commercial service several years ago, mainly in Europe. Alumina and silicon carbide seals are widely used in pumps for the chemical, petrochemical, food, pharmaceutical, mining and power generation industries. Hot-pressed silicon nitride seal rings are used as seals for the first-stage water circulation pumps of nuclear reactors in the United States and Europe. These are very large components (up to 66 cm diameter) require both excellent wear and corrosion resistance and must demonstrate the utmost in mechanical reliability [28].

There are many other wear applications of ceramics mainly in the textile, paper and mining industries where highly abrasive fibers or particulates are handled in high volumes and at high speeds. Examples include suction box covers in paper plants, typically alumina or transformation-toughened zirconia (TTZ), TTZ paper slitters, alumina pipe liners for mineral slurry transport, and alumina or silicon carbide cyclone liners.

7.4 Automotive Ceramics

During the past 15 years a substantial and increasing amount of ceramics has been utilized in automotive applications. A large "traditional" market has existed prior to this in areas such as windows and sparkplug insulators. The largest dollar value "advanced ceramic" market for automotive use has been in the area of sensors, such as oxygen sensors for control of engine fuel-to-air ratios, knock sensors, weather sensors and similar applications. However, in the past decade an appreciable market for structural ceramics for mechanical and thermal management components has emerged. Ceramic components such as turbocharger rotors, cam roller followers, fuel injector parts, diesel precombustion chambers, and exhaust port liners are now utilized in production engines. In general, these parts are selected based on a combinations of properties which provide the best available solutions to unique design requirements. Several examples will be briefly described.

Ceramic Turbocharger Rotors

Turbocharger technology is an effective means of enabling a small displacement engine perform like a larger engine. Small turbocharged engines are a particularly attractive alternative to larger engines in countries which tax automobiles according to the displacement of the engine. Turbocharger rotors typically operate in an exhaust gas stream of 800 to 900°C. However, if a sparkplug fails and uncombusted gas from that cylinder combusts in the exhaust, temperatures of about 1350°C could be encountered for short times. The design life of a passenger car is about 5000 h (150,000 miles/ave 30 mph). Thus, the first design goal is a material that can function for 5000 h at about 900°C with the possibility of short overtemperatures of up to about 1350°C. From a drivability standpoint "slow" spin up to operating rpm, known as turbocharger lag, has been a negative factor. Ceramics such as silicon nitride or silicon carbide, which have a specific gravity of only 3.2 compared to ~8 for high-temperature steel alloys, can reduce turbocharger lag by about a factor of 2. Turbocharger rotors often encounter impact from small sand or carbon particles emerging from the engine. Thus, a minimum level of impact resistance is required. Based on the combined design criteria of strength at temperature, specific gravity, and particulate impact resistance silicon nitride ceramic turbocharger rotors for light cars and trucks entered commercial use in 1985 in Japan. Since that time over 2 million ceramic turbocharger rotors have been placed in service. These devices have demonstrated exceptional reliabiity based on a combination of appropriate materials properties coupled with appropriate turbocharger design.

Ceramic Exhaust Port Liners

The function of an exhaust port liner is to keep heat entrained in the exhaust, thereby minimizing the need for water cooling and use the heat in useful ways. Exhaust port liners are used in the Porsche 944 engine in order to accelerate heating the catalytic converter and thus reduce emissions of hydrocarbons, CO and NO_x. In addition, the reduced heat rejection to the engine head allows a smaller radiator to be used, thus reducing costs and permitting a more aerodynamic front end design. Exhaust port liners must be cast into the aluminum manifold. While the port liner is subjected to relatively small mechanical loads in use, during the casting operation it must survive a significant thermal shock. Thus, the required properties for selecting an exhaust port liner material are low thermal conductivity and the ability to survive a very significant thermal shock. Aluminum titinate has a very low thermal conductivity and an essentially zero coefficient of thermal expansion, which enables it to sustain large thermal shocks.

Other areas where ceramics are utilized for a combination of properties where the structural or mechanical properties play a determining role include infrared and radar missile guidance domes, lightweight armor, and bone substitute materials. Descriptions of these applications and some of their property requirements may be found in Ref. 29.

8 COST CONSIDERATIONS

In general, high-performance structural ceramic components are more expensive than similarly shaped, traditional engineering metal components. However, as discussed in detail in the preceding section the *enabling materials* characteristics of structural ceramics in many applications will warrant paying the cost premium. When a designer procures prototype ceramic components in quantities of one to a dozen parts, the price is often many times the cost of comparable metallic prototypes. However, in large-scale production the cost of the ceramic component will usually decease dramatically, as has been the case with turbocharger rotors or automotive water pump seals. In some cases very complex shaped parts, which have severe electrical performance, severe thermal shock and vibratory mechanical loads (i.e., sparkplug insulators), and stringent reliability requirements, are manufactured in the hundreds of millions at a cost of a few pennies per part. It is also generally true that quality and reliability increase as components move from development to volume production

As a rule of thumb, for sintered or hot-pressed components the ceramic powder represents about 15% to 25% of the cost of the part. The other steps involved in making the fired part may account for 25–35%. Machining, quality assurance and allowance for scrap will often account for 50% of the cost in small-scale prototype or limited production runs. The costs for high-quality sinterable ceramic powders range, for example, from ~\$2–5/kg for alumina, \$10–40/kg for SiC, \$25–100/kg for silicon nitride, and \$20–180 for AlN. When looking at the per kilogram cost of the ceramics it should be recalled that because of the significantly lower mass density of the ceramics as compared to most metals a kilogram of ceramic can represent twice as many components as a kilogram of steel.

9 INDUSTRIAL STANDARDS FOR MECHANICAL AND PHYSICAL PROPERTY MEASUREMENTS OF ADVANCED CERAMICS

The following test standards are available from the American Society for Testing Materials (ASTM), 100 Barr Harbor Drive, Conshohocken, PA 19428-2959

- C-177-85(1993), Test Method for Steady State Heat Flux and Thermal Transmission by Means of the Gradient-Hot-Plate Apparatus
- C-1161-90, Test Method for Flexural Strength of Advanced Ceramics at Ambient Temperature
- C-1211-92, Test Method for Flexural Strength of Advanced Ceramics at Elevated Temperature
- C-1259-94, Test Method for Dynamic Young's Modulus, Shear Modulus and Poisons Ratio for Advanced Ceramics by Impulse Excitation of Vibration
- C-1273-94, Practice for Tensile Strength of Monolithic Ceramics at Ambient Temperatures
- C-1286-94, Classification for Advanced Ceramics
- E-228-85(1989), Test Method for Linear Thermal Expansion of Solid Materials with a Vitreous Silica Dilatometer
- E-289-94b, Test Method for Linear Thermal Expansion of Rigid Solids with Interferometry
- E-831-93, Test Method for Linear Thermal Expansion by Thermomechanical Analysis
- E-1269-94, Test Method for Determining Specific Heat Capacity by Differential Scanning Calorimetry
- E-1461-92, Test Method for Thermal Diffusion of Solids by the Flash Method

The following test standards are available from The Japanese Standards Association, 1–24, Akasaka 4, Minato-ku, Tokyo 107 Japan:

- Testing Method for Flexural Strength (Modulus of Rupture) of High Performance Ceramics; JIS (Japanese Industrial Standard) R 1601 (1981)
- Testing Methods for Elastic Modulus of High Performance Ceramics; JIS R 1602 (1986)

- Testing Methods for Flexural Strength (Modulus of Rupture) of High Performance Ceramics at Elevated Temperatures; JIS R 1604 (1987)
- Testing Methods for Elastic Modulus of High Performance Ceramics at Elevated Temperatures; JIS R 1605 (1989)
- Testing Methods for Tensile Strength of High Performance Ceramics at Room and Elevated Temperatures; JIS R 1606 (1990)
- Testing Methods for Fracture Toughness of High Performance Ceramics; JIS R 1607 (1990)
- Testing Methods for Compressive Strength of High Performance Ceramics; JIS R 1608 (1990)
- Testing Methods for Oxidation Resistance of Non-Oxide of High Performance Ceramics; JIS R 1609 (1990)
- Testing Methods for Vickers Hardness of High Performance Ceramics; JIS R 1610 (1991)
- Testing Methods of Thermal Diffusivity, Specific Heat Capacity, and Thermal Conductivity for High Performance Ceramics by Laser Flash Method; JIS R 1611 (1991)

BIBLIOGRAPHY

The following books and handbooks represent the primary sources of data presented in the chapter.

Ceramics and Glasses: *vol. 4, Engineered Materials Handbook*, ASM International, Materials Park, OH, 1991.

J. F. Shackelsford, W. Alexander, and J. Park, eds. *Materials Science and Engineering Handbook*, 2nd ed., CRC Press, Boca Raton, FL, 1994.

C. X. Campbell and S. K. El-Rahaiby, *Databook on Mechanical and Thermophysical Properties of Whisker-Reinforced Ceramic Matrix Composites*, Ceramics Information Analysis Center, Purdue University, West Lafayette, IN and The American Ceramic Society, Westerville, OH, 1995.

D. W. Richerson, *Modern Ceramic Engineering*, 2nd ed., Marcel Dekker, New York, 1992.

M. F. Ashby and D. R. H. Jones, *Engineering Materials 1 and 2*, Pergamon Press, Oxford, 1980.

D. R. Askland, *The Science and Engineering of Materials*, 3rd ed., PWS, Boston, 1994.

S. Jahanmir, ed., *Friction and Wear of Ceramics*, Marcel Dekker, New York, 1994.

References

1. T. Abraham, Current U. S. markets for advanced ceramics and projections for future growth, *Ceram. Eng. Sci. Proc 14*, 25–35 (1993).
2. D. W. Richerson, *Modern Ceramic Engineering*, 2nd ed., Marcel Dekker, New York, 1992, Chap. 11, pp. 577 ff.
3. D. L. Segal, Powders: chemical prepration, in *Concise Encyclopedia of Advanced Ceramic Materials* (R. J. Brook, ed.), Pergamon Press, Oxford, 1991, pp. 369ff.
4. D. W. Richerson, *Modern Ceramic Engineering*, 2nd ed., Marcel Dekker, New York, 1992, Chap. 10, pp. 418ff.
5. G. E. Gazza and R. N. Katz, Densification of ceramics by gas overpressure sintering, in MRS Symp. Proc. *251*, 1992, pp. 199–210.
6. *Ceramics and Glasses*: *vol. 4, Engineered Materials Handbook*, Sections 3 and 4, ASM International, 1991, pp. 123–304.

7. A. F. McLean and D. Hartsock, Design with structural ceramics, ibid., pp. 27–95.
8. G. D Quinn, in *Fracture Mechanics of Ceramics*, vol. 8, Plenum, New York, pp. 319–332.
9. M. Matsui, Status of research and development on materials for ceramic gas turbine components, *Mat. Res. Soc. Symp. Proc. 287*, 173–188 (1993).
10. R. Hamminger and J. Heinrich, Development of advanced silicon nitride valves for combustion engines and some practical experience on the road, in MRS Symp. Proc. *287*, 513–520 (1993).
11. T. Kawakubo and K. Komeya, Static and cyclic fatigue behavior of silicon nitride at RT, *J. Am. Ceram. Soc. 70*, 400–405 (1987).
12. I. Bar-On and J. T. Beals, Fatigue of silicon nitride bend bars, in *Fatigue 90: Proc. 4th Int. Conf. on Fatigue and Fatigue Thresholds*, vol. II (Kitagawa and Tanaka, eds.), 1990, pp. 793–798.
13. R. N. Katz, *Corrosion/Erosion Behavior of Silicon Nitride and Silicon Carbide Ceramics—Gas Turbine Experience*, Army Materials and Mechanics Research Center Report, AMMRC-MS 79-2, April 1979.
14. M. Matsuda, H. Tsuruta, T. Soma, and M. Matsui, Foreign object damage resistance of structural ceramics, in *Ceramic Materials and Components for Engines* (V. Tennery, ed.), American Ceramic Society, 1988, pp. 1031–1038.
15. D. J. Viechnicki, M. J. Slavin, and M. J. Kliman, Development and current status of armor ceramics, *Ceram. Bull. 70*, 1035–1039 (1991).
16. R. N. Katz, G. D. Quinn, M. J. Slavin, and J. J. Swab, Time and temperature dependence of strength in high performance ceramics, in *New Materials and Their Application* (S. G. Burnay, ed.), Inst. Phys. Conf. Series. No. 89, 1988, pp. 53–62.
17. J. J. Swab and G. L. Leatherman, *State Fatigue Behavior of Structural Ceramics in a Corrosive Environment*, U. S. Army Materials Technology Laboratory Report, MTL TR 90-32, June 1990.
17a. D. W. Richerson, *Modern Ceramic Engineering*, 2nd ed., Marcel Dekker, New York, 1992.
18. *Electronic Materials Handbook, vol. 1, Packaging*, ASM International, Materials Park, Ohio, 1989), p. 20.
19. D. R. Messier and P. Wong, *Silicon Nitride: A Promising Material for Radome Applications*, Army Materials and Mechanics Research Center report, AMMRC TR 74-21, September 1974.
19a. S. Jahanmir, ed., *Friction and Wear of Ceramics*, Marcel Dekker, 1994.
20. A. Gangopadhyay, S. Jahanmir and M. B. Peterson, Self-lubricating ceramic matrix composites, in *Friction and Wear of Ceramics* (S. Jahanmir, ed.), Marcel Dekker, New York, 1994, pp. 163–197.
21. S. F. Wayne and S. T. Buljan, Microstructure and wear resistance of silicon nitride composites, ibid., pp 261–285.
22. *Cutting Tool Engineer*, February (1987), pp. 56–57.
23. L. R. Anderson, *Cutting Tool Engineer*, April (1990), pp. 65–70.
24. C. W. Beeghley and A. F. Shuster, presented at the Conf. on Advanced Tool Materials for High Speed Machining, Scottsdale, AZ 25-27 Feb. 1987, ASM preprint 8701-002.
25. R. N. Katz, Ceramic materials for rolling element bearing applications, in *Friction and Wear of Ceramics* (S. Jahanmir, ed.), Marcel Dekker, New York, 1994, pp. 313–328.
26. *Newsletter of the Australian Ceramic Society*, No. 4 (1989) p. 17.
27. K. H. Jack, SiAlON ceramics: retrospect and prospect, *Mat. Res. Soc. Symp., Proc. 287*, 15–27 (1993).

28. R. N. Katz, Applications of silicon nitride ceramics in the U.S., *Mat. Res. Soc. Symp., Proc. 287*, 197–208 (1993).
29. R. N. Katz, Opportunities and prospects for the application of structural ceramics, in *Treatise on Materials Science and Technology*, vol. 29 (J. B. Wachtman, ed.), Academic Press, 1989, pp. 1–26.

12
Metal Matrix Composites

W. H. Hunt, Jr.
Aluminum Consultants Group
Murrysville, Pennsylvania

1 INTRODUCTION

The metal matrix composites (MMCs) considered herein consist of particulate ceramic compounds or ceramic fibers in metal matrices. These materials have been developed to obtain property combinations not possible with either unreinforced metals or ceramics alone. Specifically, the properties of metals which are improved by reinforcement are elastic modulus (especially on a density-compensated basis), strength at room and elevated temperatures, wear resistance, and control of physical properties, particularly the coefficient of thermal expansion.

Initially these materials were developed primarily for defense applications since relatively expensive, low volume processes could be justified to obtain the desired combinations of unique properties. In the recent decade, greater emphasis has been placed on cost-effective MMC processes and materials, which has expanded the range of possible commercial applications for these materials.

2 MMC CLASSIFICATION

Typically MMCs are classified according to the morphology of the reinforcement, specifically being divided into continuous fiber-reinforced MMCs, discontinuous fiber-reinforced MMCs, whisker-reinforced MMCs, and particulate-reinforced MMCs. This classification is due to the fact that the most substantial changes in properties achieved in MMCs are derived from the specifics of the reinforcement morphology. Continuous fibers include alumina, silicon carbide, boron and carbon, which are produced either as single large-diameter fibers or in fiber tows, and typically are relatively expensive as reinforcement materials. Discontinuous fiber-reinforced MMCs utilize chopped fibers of predominantly alumina or an aluminum-silica mixture with a focus on achieving reinforcement at a lower cost. Whiskers for reinforcement of metal matrices, primarily silicon carbide, were of interest in early products due to the high properties by the ceramic whiskers, but have fallen from favor recently due to their high cost and health and safety concerns. Finally particulate ceramics, including silicon carbide, alumina, silicon nitride, and boron carbide, have seen increasing use as reinforcement materials with the present emphasis on low-cost MMCs.

A second dimension of MMC classification is by the metal matrix involved. The largest efforts and range of commercial products are in MMCs with the light metals as the matrix, i.e. aluminum, titanium, and to a lesser degree magnesium. This is because reinforcement of these matrices enables these lightweight materials to have stiffness and elevated temperature properties comparable to heavier alloy systems, which is a strong driving force for application in weight-critical applications.

A final method of classification is by the method of fabrication of the primary composite. There is a substantial connection between the method of fabrication, morphology of the reinforcement, and matrix alloy. However, the types of formation processes can be roughly divided into solid-state and liquid-state categories. Solid-state processes in the form of molding are used in the manufacture of continuous fiber composites, where the fibers are placed between foils of the matrix alloy, which are diffusion bonded together. Another solid-state process used for the formation of particulate-reinforced MMCs is the powder metallurgy process. Powders of the metal matrix are blended with the ceramic reinforcement powders and bonded together under heat and pressure to form an essentially porosity-free product. This product can be used in the as-pressed condition or further fabricated to a variety of wrought product forms. Liquid-state processes again are used to produce composites for the full range of reinforcements. Infiltration of liquid metal into a packed bed of continuous fibers or particulate reinforcement is used to produce net shape parts using traditional casting processes such as permanent mold, die casting, etc., as well as specialized MMC casting processes. In addition, stirring of particulate ceramics is used to form a composite which can then be cast into an intermediate form such as billet or ingot and shaped in subsequent wrought product fabrication processes.

A listing of the current commercial producers of MMCs along with a synopsis of the products they offer and their method of manufacture is given in Table 1.

3 PROPERTIES

The extent of available MMC properties varies significantly from one company and product to another. Table 2 presents a summary of typical mechanical properties while Table 3 lists key physical properties. These properties illustrate the strong effect that reinforcement morophology has on properties. Continuous fiber-reinforced materials are characterized by high stiffness and strength when tested in the same direction as the fiber alignment (noted as the 0° direction in Table 2), with substantially lower properties when tested transverse to the fibers. Particulate-reinforced composites, on the other hand, show lower but isotropic increases in stiffness and strength relative to continuous fiber-reinforced composites. Another trend evident from the data is the increase in stiffness and strength with increasing volume fraction as expected from the rule of mixtures. Note that the specifics of

Table 1 Manufacturers of MMCs

Manufacturer	Type & Method and Potential Applications
Duralcan USA (Division of Alcan)	Al alloy-particulate SiC for gravity casting and aluminum-alumina for wrought materials. Driveshaft tubing, automotive front disc-brake rotor, golf club heads, bicycle frame tubes, and miscellaneous other parts. Both wrought and cast products are produced by stirring particulate reinforcement into liquid metal.
Lanxide Corporation	Pressureless molten metal infiltration of a ceramic particle bed or preform. Includes a variety of aluminum alloys with SiC and alumina particles. Net and near-net shapes by investment, die, and sand casting methods. Squeeze casting may also be applied to some products. Automotive brake calipers, disc brake rotors, connecting rods, and electronic applications.
Dynamet Technology, Inc.	Cold and hot isostatic processing of titanium alloys with TiC, TiAl and TiB_2 particulate. Medical implants, automotive, and sporting goods.
Alcoa	Pressure infiltration of ceramic preforms by aluminum to produce composites of about 70% volume reinforcement for net shape components for electronics and thermal management systems. A second family of products is produced by the blending of particulate ceramic particles (15–40%) with aluminum powders followed by powder metallurgy consolidation; for aerospace and automotive uses.
Thermal Ceramics (Babcock & Wilcox)	Discontinuous fiber MMC. Infiltration of alumina-silica (Kaowool) or alumina fiber preforms with liquid Al by squeeze casting.
Ametek, Inc.	Copper with 100 and 20 micron Mo particles. Used for heat sinks, substrates, and thermal spreaders.
SCM Metal Products, Inc.	Dispersion strengthened copper with alumina and also with 10% niobium additions. Internal oxidation process in which a dilute Cu-Al alloy is preferentially oxidized within a copper matrix. Mill products are produced using HIP and extrusion. Used for resistance welded electrodes and components requiring high conductivity and strength.
DWA Aluminum Composites, Inc.	Powder metallurgy processed MMCs with a range of ceramic particulates for automotive, aerospace, and specialized high-performance applications.
PCC Composites, Inc.	Pressure infiltration by a proprietary process of ceramic particle preforms to produce MMCs with high reinforcement volumes for thermal management and electronic applications.
Textron Specialty Materials	SiC continuous fibers are woven to produce fabric or drum wound for plasma spraying. Used to reinforce Al alloys by hot molding and titanium alloys by hot pressing with titanium foils. Aerospace and defense applications.
3M Company	Continuous Al matrix composite reinforced with alumina fibers and titanium matrix composite with SiC fibers. Applications same as for Textron material.

Table 2 Mechanical Properties of Selected MMCs

Composite	R.T. Y.S or UTS[a] MPa	(ksi)	Elong. % in 50 mm	Young's modulus GPa	(msi)	Hardness	Fatigue strength 10^7 cycles MPa	(ksi)	K_{Ic} (MPa$\sqrt{m}$)	Producer
A. Al-Si + 20 vol% SiC particulate, P.M. T6	324	(47)	0.4	98.7	(14.3)	77 HRB	151.8 $R = 0.1$	(22)	16	Duralcan
B. Al-Si + 20 vol% alumina T6, wrought	421	(61)	2.0	104.2	(15.1)	—	—		19.4	Duralcan
C. Al-Si + 10 vol% SiC T5 die cast	331	(48)	0.7	93.8	(13.6)	84 HRB	138 $R = 0.1$	(20)	—	Duralcan
D. Al-Si + 20 vol% SiC particulate, die cast	124	(17.9)	0.8	94	(14.2)	—	150 $R = 0.1$	(2.2)	—	Lanxide
E. Al-Si + 30 vol% SiC particulate T6, sand cast	274	(39.7)	0.4	125	(18.1)	—	83.5 $R = -1$	(12.1)	—	Lanxide
F. Al-Cu + 50% alumina particulate T6, infiltrated	239	(34.7)	0.3	164	(23.8)	—	—		14.6	Lanxide
G. Al-Si + 70% SiC particulate, infiltrated	225	(32.6)	0.1	265	(34.8)	—	150 $R = -1$	(21.8)	10	Lanxide
H. Ti-6Al-4V + 20 wt% TiC particulate	1035	(150)[a]	—	144.9	(21)	—	—		31.9	Dynamet

I. Al-Si + 70 vol%	192	(27.8)[a]	—	221	(32)	—	—		—	Alcoa, microelec.
7093 SiC/15p-T6	607	(14.5)	3.1	100	(14.5)	—	—		14.5	aero, auto
6113 SiC/25p-T6	438	(63.5)	3.0	121	(17.5)	—	—		16.2	aero, auto
2080 SiC/20p-T6	455	(66)	4.0	110	(16.0)	—	—		13.0	aero, auto
J. Al alloy 336 + 23 vol% Kaowool	110	(16)	—		—	—	—		—	Thermal cer.
K. Cu + 2.2 vol% alumina + 10 vol% Nb	550	(80)	9		—	95 HRB	—		—	SCM metal products
L. 85 Mo-15 Cu wrought powder	476	(69)	2.0	276	(40)	—	—		—	Ametek, Inc.
M. Al-65 vol% SiC particulate, die cast	249	(36)	<0.2	210	(306)	88 HRB	—		—	PCC Composites
N. Al-Si + 48 vol% SiC continuous	1550	(225) 0°	—	193	(28) 0°	—	—		—	Textron
	83	(12) 90°	—		—	—	—		—	
O. Ti-6Al-4V + 35 vol% continuous SiC, hot pressed	1725	(250) 0°	—	193	(28) 0°	—	—		—	Textron
	415	(60) 90°	—		—	—	—		—	
P. Al-2 wt% Cu + 60 vol% continuous alumina fibers (Nextel)	1500	(220) 0°	—	240	(35) 0°	—	—		—	3M Co.
	415	(220) 90°	—	160	(23) 90°	—	—		—	
Q. 6092/SiC/17.5p-T6, extrusion	448	(65)	6.0	105	(15.3)	—	241	(35)	55 (Kc)	DWA Aluminum Composites, Inc.

[a]UTS for materials which do not exhibit yielding.

Table 3 Physical Properties of Selected MMCs

Composite	Density [g/cc (lb/in³)]	Ther. cond. W/mK ( )[a]	CTE @ R.T. ppm °C (°F)	Elect. cond. % IACS
A	2.76 (0.1)	185 (107)	20.7 (11.5)	26.4
B	2.98 (0.108)	119 (68.8)	17.5 (9.7)	—
C	2.76 (0.1)	123.8 (71.6)	19.3 (10.7)	22.0
D	2.76 (0.1)	—	15.8 (8.8)	—
E	2.78 (0.1)	156 (90.2)	14 (7.8)	—
F	3.35 (0.12)	—	13.0 (7.2)	—
G	3.0 (0.11)	170 (98.3)	6.2 (3.4)	—
H	4.52 (0.16)	—	—	—
I	3.02 (0.11)	210 (121.4)	12.6 (7.0)	—
J	—	—	15.6 (—)	—
K	—	—	—	67
L	10.0 (0.36)	165 (95.4)	7.0 (3.9)	33.6
M	3.02 (0.11)	220 (127.2)	7.7 (3.9)	—
N	2.84 (0.103)	—	—	—
O	3.86 (0.14)	—	—	—
P	3.4 (0.123)	—	7 (3.9) 0°	—
P	—	—	15 (8.3) 90°	—
Q	2.8 (0.102)	—	—	—

[a]Btu/ft-hr-°F

the metal matrix alloy and heat treatment condition have the predominant effect on the strength of the composite with less influence on the stiffness. Certainly before any of these materials are selected for use the design engineer should verify the properties of the materials of interest.

4 APPLICATION OF MMCs

In selecting an MMC for a particular application, both cost and performance are important factors. While nonquantitative, Fig.1 provides a comparison for the main types of commercial aluminum matrix MMC products, which is representative of the relationships for other metal matrix systems. The continuous fiber-reinforced composites provide the highest property performance but also are the most expensive. As a result, applicaton of these materials has thus far been limited to high-performance applications in defense and sporting goods markets. Particulate-reinforced composites produced by the powder metallurgy method [labeled powder metallurgy DRA (for discontinuous reinforced aluminum) in Fig. 1] are lower in cost than the continuous fiber-reinforced composites but also lower in performance. This level of cost and performance enables application in commercial aerospace as well as high-performance automotive markets and also in selected defense and sporting goods applications. Lower cost powder metallurgy processes currently under development promise to open up the commercial automotive market for this product. The particulate-reinforced composites produced by molten metal processes

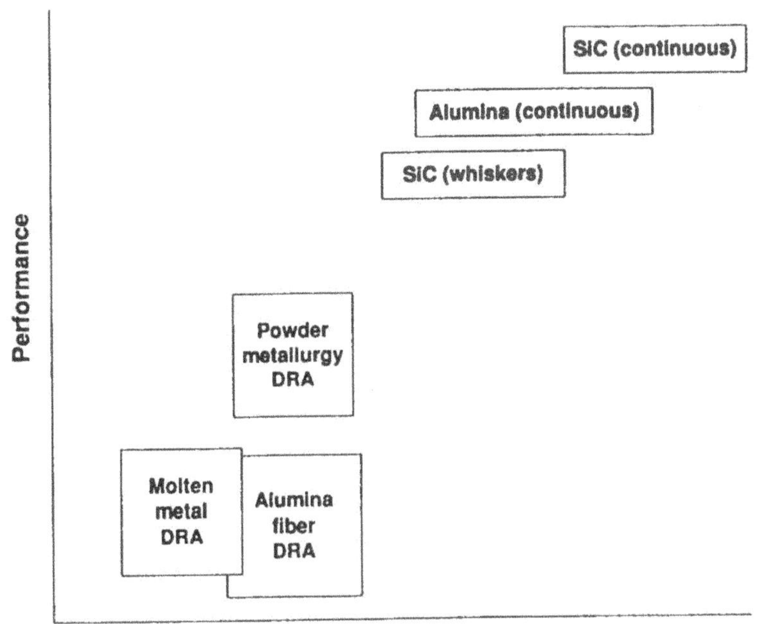

Figure 1 Performance versus cost.

(labeled molten metal DRA in Fig. 1) are the least expensive product and as a result have seen strong interest from the commercial automotive market. Finally composites reinforced with discontinuous fibers (labeled alumina fiber DRA in Fig. 1, which also includes other chopped fiber materials as reinforcements) can offer a low-cost product especially useful in cases where an approach known as selective reinforcement is used in producing properties in a local region of the component requiring improvement rather than reinforcement throughout the material.

13 Polymeric Matrix Composites

P. K. Mallick
University of Michigan–Dearborn
Dearborn, Michigan

Polymeric matrix composites (PMC) are fiber-reinforced polymers in which either a thermoset or a thermoplastic polymer is used as the matrix. The development of PMC for structural applications started in the 1950s, and they are by far the most common fiber-reinforced composite materials used today. Table 1 shows the shipment of PMC in the United States for a period of 10 years and their markets. While there have been some shifts in the utilization of PMC in various industries, the overall growth of PMC is 37.2% from 1985 to 1994.

One reason for the growing use of PMC is that the processing of PMC is relatively simple and does not require very high temperatures and pressures. The

Table 1 Polymeric Matrix Composites Shipment in the United States (in millions of pounds)

Applications/Market	1985	1989	1994
Aircraft/space/military	32	41	24.2
Appliance/business	133	151	160.7
Construction	445	470	596.9
Consumer products	142	158	174.8
Corrosion resistant	295	335	376.3
Electrical/electronic	191	229	299.3
Marine	335	405	363.5
Transportation	563	677	945.6
Other	82	76	101.8
Total	2218	2542	3043.1

Includes both filled and reinforced thermosets as well as thermoplastic composites.
Source: Annual Statistical Report, SPI Composites Institute, January 31, 1995.

equipment required for processing of PMC are also relatively simple and less expensive than those required for other types of composites. There is also a variety of processing techniques available for producing PMC parts. Depending on the type of PMC used, some of these techniques can be highly automated.

The most significant advantage of PMC derives from the fact that they are lightweight materials with high strength and modulus values. The lightweight of PMC is due to low specific gravities of its constituents. Polymers used in PMC have specific gravities between 0.9 and 1.5, and the reinforcing fibers have specific gravities between 1.4 and 2.6. Depending on the fiber and polymer type used and their relative amounts, the specific gravity of PMC is between 1.2 and 2 compared to 7.87 for steel and 2.7 for aluminum alloys. Because of their low specific gravities, the strength-to-weight and modulus-to-weight ratios of PMC are comparatively much higher than those of metals and their composites (Table 2). Although the cost of PMC can be high, especially when carbon fibers are used as reinforcements, their cost on a unit volume basis can be competitive with that of high-performance metallic alloys.

A second advantage for PMC is the design flexibility and the variety of design options that can be exercised with them. Fibers in PMC can be selectively placed or oriented to resist loads in any direction, thus producing directional strengths and/or moduli. The thermal properties of PMC, such as the coefficient of thermal expansion, can be controlled and/or varied relatively easily. PMC can be combined with aluminum honeycomb, structural plastic foam or wood to produce sandwich structures that are stiff and at the same time light weight. Two or more different types of fibers can be used to produce a hybrid construction with high flexural stiffness and impact resistance.

There are several other advantages of PMC which make them desirable in many applications. They have damping factors that are higher than those of metals, which means noise and vibrations are damped in PMC structures more effectively than in metal structures. They also do not corrode. However, there are disadvantages, namely, their properties may be affected by elevated temperatures, moisture, chemicals and ultraviolet light, mainly because of the adverse effects of these environmental factors on the polymer matrix.

In this chapter, we first consider the properties of the constituents used in PMC and briefly discuss the most important processing techniques used for structural PMC. We then examine the properties of structural PMC and the selection procedure of PMC in typical structural applications. For more information on polymer matrix composites, the reader may consult Refs. 1 and 2.

1 CONSTITUENTS IN PMC

1.1 Fibers

Fibers are the principal load-carrying constituent in continuous fiber PMC. The strength and modulus of PMC depend on the fiber type as well as the fiber volume fraction. In high-performance aerospace PMC, the fiber volume fraction is typically 50–60%. Lower volume fractions are used for random fiber PMC.

Table 2 Comparative Properties of Metals and Polymeric Matrix Composites

Material	Density (g/cm^3)	Modulus (GPa)	Tensile strength (MPa)	Yield strength (MPa)	Modulus-to-density ratio (10^6 m)	Tensile strength-to-density ratio (10^3 m)	% Elongation
SAE 1010 steel (cold drawn)	7.87	207	365	303	2.68	4.72	20
AISI 4340 steel (quenched and tempered)	7.87	207	1722	1515	2.68	22.3	—
6061-T6 aluminum alloy	2.70	68.9	310	275	2.60	11.7	15
7075-T6 aluminum alloy	2.70	68.9	572	503	2.60	21.6	11
AZ80A-T5 magnesium alloy	1.74	44.8	379	276	2.62	22.2	7
Ti-6Al-4V titanium alloy (aged)	4.43	110	1171	1068	2.53	26.9	8
High-strength carbon fiber/epoxy (unidirectional)	1.55	138	1550	—	9.07	101.9	1.1
High-modulus carbon fiber/epoxy (unidirectional)	1.63	215	1240	—	13.44	77.5	0.6
E-glass fiber/epoxy (unidirectional)	1.85	39.3	965	—	2.16	53.2	2.5
Kevlar-49/epoxy (unidirectional)	1.38	75.8	1378	—	5.60	101.8	1.8
Carbon fiber/epoxy (quasi-isotropic)	1.55	45.5	579	—	2.99	38.1	—
E-glass fiber/epoxy (random fiber SMC)	1.87	15.8	164	—	0.86	8.9	1.73

For unidirectional composites, the modulus and strength values are in the fiber direction.

A large variety of fibers are available commercially for PMC applications. A partial list of these fibers and their properties are given in Table 3. The following points should be noted about the characteristics of these fibers:

1. Fibers are very thin filaments with diameters ranging from 7 to 15 μm. In the usable forms of fibers, the filaments are bundled together to produce strands, rovings, tows or yarns. The number of filaments and how they are bundled can be varied. A larger number of filaments in the bundle gives higher yield in PMC processing, but may produce lower properties due to poor fiber wet-out by the polymer matrix.
2. Fibers in tension behave essentially in an elastic manner. They are brittle with low elongation to failure. They have high strengths, but the strength values show a wide variation. As seen in Table 3, some carbon fibers have modulus values that are much higher than that of steel. Glass and aramid fibers have lower modulus values than carbon fibers.
3. Carbon and aramid fibers are not isotropic, which is evidenced by the difference in their coefficients of thermal expansion in the longitudinal and radial directions.

Table 3 Properties of Some Reinforcing Fibers Used in Polymeric Matrix Composites

Fiber	Filament diameter (μm)	Density (g/cm^3)	Tensile modulus (GPa)	Tensile strength (GPa)	Percent strain to failure	CTE (10^{-6} per °C)
Glass						
E-glass	10	2.54	73	3.45	4.8	5
S-glass	10	2.49	87	4.3	5	2.9
PAN-carbon						
T-300[a]	7	1.76	231	3.65	1.4	−6(l)[f]
AS-4[b]	7	1.8	248	4.07	1.65	7–12(r)
IM-7[b]	5	1.78	301	5.31	1.81	—
HMS-4[b]	8	1.8	317	2.34	0.8	—
GY-70[c]	8.4	1.96	483	1.52	0.38	—
						—
Pitch-carbon[a]						
P-55	10	2	380	1.90	0.5	−1.3(l)
P-100	10	2.15	758	2.41	0.32	−1.45(l)
Aramid						
Kevlar-49[d]	11.9	1.45	131	3.62	2.8	−2(l) 59(r)
Polyethylene[e]						
Spectra-900	38	0.97	117	2.59	3.5	—

[a]Amoco
[b]Hercules
[c]Celanese
[d]DuPont
[e]Allied Signal
[f](l) and (r) represent the longitudinal and radial directions of the fiber, respectively.

E-glass fibers are the least expensive of all commercially available fibers, but their modulus value is not as high as that of carbon fibers and their strength is affected by moisture and abrasion. Carbon fibers are made in many varieties—high strength, high modulus, intermediate modulus and ultrahigh modulus. The density as well as cost of these fibers are also different, both generally increasing with increasing modulus. Carbon fibers made from pitch have modulus values greater than 350 GPa (50 Msi), but their strengths are lower than those made from PAN (polyacrylonitrile). Very high modulus carbon fibers are also extremely brittle and even though they can produce very high modulus in PMC, the impact energy will be low. Aramid fibers, such as Kevlar 49 and 149, are polymeric fibers with modulus values nearly twice that of E-glass fibers. Their density is the lowest, and consequently, their strength-to-weight ratio is the highest among the reinforcing fibers. Even though aramid fibers are brittle in tension, they show yielding in compression and bending. However, their compressive strength is low which is the reason for the low compressive strengths observed for aramid fiber composites. They are also not thermally stable above 160°C and they tend to absorb moisture when exposed to humid environments. Aramid fibers also exhibit creep at elevated temperatures, which is not observed with glass and carbon fibers.

1.2 Matrix

The role of the matrix in PMC is to keep the fibers in place, transfer load among the fibers and protect them from environments, such as moisture and chemicals. Compressive and shear properties of PMC are influenced greatly by the matrix properties, and any changes in the matrix properties due to increased temperature or moisture concentration are reflected in these matrix dominated properties.

The polymers commonly used in PMC and their key properties are listed in Table 4. Thermoset polymers, such as epoxies and polyesters, are the traditional matrix in continuous fiber PMC. The starting materials for these and many other thermoset polymers are low-viscosity liquid-like prepolymers. Long continuous fibers can be relatively easily coated and wetted by these prepolymers before starting the curing reaction which changes them into solid thermoset polymers. Thermoplastic polymers, on the other hand, have very high viscosities even at high processing temperatures. Wetting long continuous fibers with a thermoplastic polymer has been a problem for a long time and, even though several innovative processes, such as melt impregnation and powder coating, have been developed for coating long continuous fibers with thermoplastic polymers, high processing and material costs have prevented wider use of the continuous fiber thermoplastic composites. Short fibers are used more commonly with thermoplastic polymers, since they can be compounded with liquid thermoplastics and then injection-molded into desired shapes.

In addition to the processing advantage, thermoset polymers have several other advantages over thermoplastic polymers. Thermosets are more thermally stable and chemical resistant than thermoplastics. Their glass transition temperatures are much higher than most thermoplastics. A few thermoplastics, such as polyether ether ketone (PEEK) or polyether imide (PEI), have high glass transition temperatures and are considered high-temperature polymers, but they are very expensive compared to most thermosets, such as epoxies and polyesters. The advantages of ther-

Table 4 Properties of Some Polymers Used as Matrix in Polymeric Matrix Composites

Polymer	Density (g/cm^3)	T_g (°C)	Tensile modulus (GPa)	Yield or tensile strength (MPa)	Strain-to-failure (%)	HDT (°C) at 0.8 MPa	CTE (10^{-6} per °C)
Thermosets							
Epoxies	1.2–1.3	180–260	2.75–4.1	55–130 (T)	1.5–8	46–250	50–80
Bismaleimides	1.2–1.32	230–290	3.2–5	48–110 (T)	1.5–3.3	93–135	—
Vinyl esters	1.12–1.32	—	3–3.5	73–81 (T)	3–8	140	—
Polyesters	1.1–1.4	—	2.1–3.5	35–104 (T)	1–7	60–80	—
Phenolics	1–1.25	—	3–4	60–80 (T)	1.5–2	250	68
Polyurethanes	1.21	120–167	0.7	30–40 (T)	400	—	100–200
Thermoplastics							
PEEK	1.3–1.32	143	3.2	100(Y)	50	160	47
PPS	1.4	88	3.3	86	3–6	135	49
PEI	1.27	217	3	105(Y)	60	199	47–56
Polysulfone	1.24	190	2.6	70(Y)	50–100	174	56
Polyamide imide	1.42	275	4.5	150(T)	8	278	—
Polypropylene	0.9	−20	1.14–1.6	31–37(Y)	100–600	49–60	81–100
Nylon 6,6 (dry)	1.14	50	2	55(Y)	15–80	75–80	80

PEEK = polyether ether ketone
PPS = polyphenylene sulfide
PEI = polyether imide

moplastics are their high strains to failure, greater crack resistance and higher impact strength. They also have unlimited storage life and require lower processing time (although their processing temperatures are higher than common thermosets). Since thermoplastics can be heat-softened repeatedly, they can be postformed into various shapes and welded together instead of joining by adhesives.

1.3 Interface

An important parameter in developing composite action between the fibers and the matrix is a good bonding at the fiber-matrix interface. Increased bond strength is achieved by fiber surface treatment which helps in forming a chemical linkage between the fibers and the matrix across the interface. Ordinarily, a mechanical bonding is formed as the polymer matrix with higher coefficient of thermal expansion than the fibers cools down from its processing temperature. The mechanical bonding is disrupted relatively easily under stress and the fibers can debond from the matrix. Chemical bonding increases the stress at which debonding occurs and as a result the strength of PMC is improved.

The most common surface treatment for glass fibers involves the use of a silane coupling agent. Glass fibers are treated with an aqueous solution of silane before they are coated with the polymer. One end of the silane molecule bonds with the glass fiber surface in the form of a very thin layer, while the other end is reacted with the polymer at the time of processing of PMC. Carbon and aramid fibers are more difficult to surface treat; however, there are several techniques available for them as well [1].

2 TYPES OF PMC

One great advantage of polymeric matrix composites is the variety of ways in which fibers and matrix can be combined to achieve a wide range of properties. Fibers can be used either in continuous or discontinuous lengths. The discontinuous fibers can be either short (10 mm or less) or long (longer than 10 mm). The fiber orientation can be unidirectional, bidirectional or random (Fig. 1).

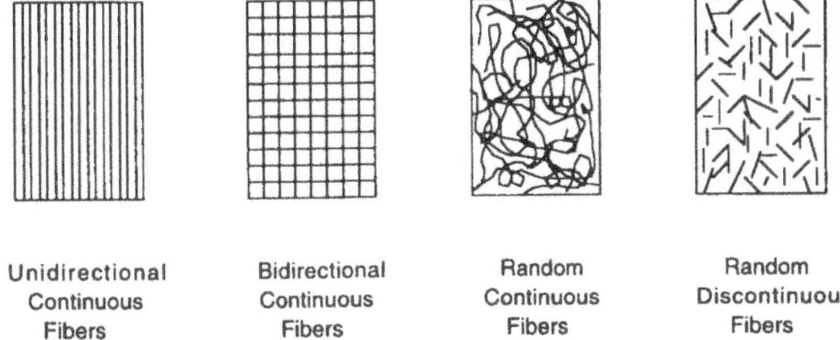

Figure 1 Types of PMC: (a) unidirectional continuous fibers, (b) bidirectional continuous fibers, (c) random continuous fibers, (d) random discontinuous fibers.

Continuous fibers are frequently used in laminated PMCs. Each layer in a laminate, called a lamina, contains either unidirectional continuous fibers or a bidirectional fabric. The layer thickness is 0.1 to 0.2 mm. The fiber orientation from layer to layer can be varied (Fig. 2). In addition to fiber orientation angles, the number of layers and the sequence in which the layers are stacked can also be varied. Thus, there is an unlimited number of options available to the designer to produce a wide range of properties by varying the laminate construction. For example, in a unidirectional laminate with the same fiber orientation angle in all the layers, the strength and modulus are very high in the fiber direction and lower in all other directions. Such a laminate is useful in uniaxial loading situations where the fibers may be oriented in the loading direction. If the loading is biaxial, the layers can be cross-plied with fibers oriented at 90° to each other in alternate layers. Woven fabrics are also used in biaxial loading conditions. Multidirectional laminates, such as 0/+45/−45/90/90/−45/+45/0 and 0/60/−60/−60/60/0, are used when there is a possibility that the loading direction may vary during the application of the laminate. The angle designation in a multidirectional laminate refers to the fiber orientation angle in each layer with respect to a preselected direction, which is either the major loading direction or the long dimension of the laminate.

Discontinuous fibers are mostly used in randomly oriented form as in a sheet molding compound (SMC) or in an injection-molded material. The most common form of sheet molding compounds contains 25-mm-long chopped fibers in a thin layer of a thermosetting resin, such as a polyester or vinyl ester resin. The fibers are oriented randomly in the plane of the sheet. It is also possible to combine

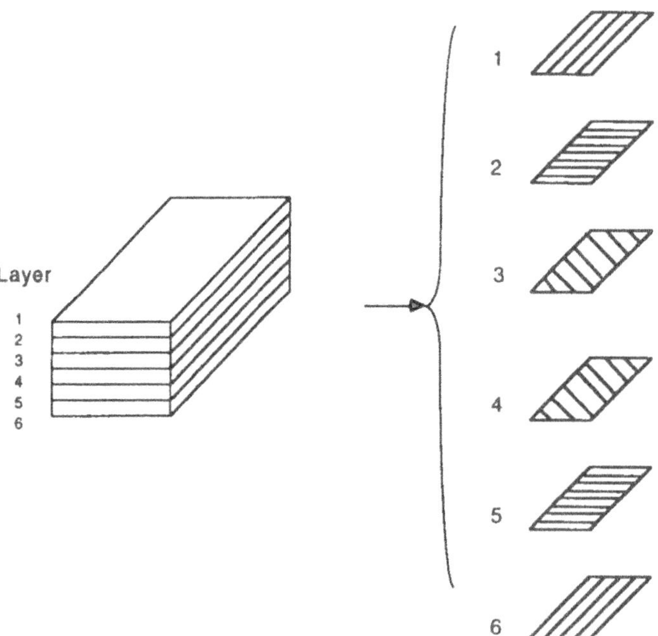

Figure 2 Laminate construction.

continuous fibers with random fibers in a sheet molding compound (Fig. 3); however, they are less common than completely random fiber SMCs. There are two ways in which the continuous fibers are combined with random fibers in an SMC. In SMC-CR, each layer of SMC contains unidirectional continuous fibers on one side and random discontinuous fibers on the other. In XMC, discontinuous fibers are interspersed with the continuous fibers that are oriented in an X-pattern with an included angle of 5–7° between them.

Injection-molded PMCs are the most widely used form of PMC. They can be made from either a thermoplastic or a thermosetting polymer. The fibers are 3 to 6 mm long and are oriented three dimensionally. Since the fibers are very short and randomly oriented, mechanical properties of injection-molded PMCs are not as high as continuous fiber-laminated PMCs. However, they can be produced relatively easily and the production time is not long. Thus, injection molding is amenable to mass production and injection molded PMCs have found wide ranging applications in consumer goods, automobile parts, appliance components, electrical and computer housings etc.

3 PROCESSING OF PMC

A variety of processing techniques exist for producing PMC parts. The most important of these techniques are compared in Table 5. Some of these techniques are highly labor intensive and can produce only one or two parts at a time, while others are either automated or semiautomated and can be adopted for continuous processing or mass production [3]. The selection of PMC processing depends on the shape, size and number of parts to be produced.

Some of the processes listed in Table 5 can be used for both thermoset and thermoplastic matrix composites. However, there are differences in their processing characteristics. For thermoset matrix composites, either uncured or partially cured resin is transformed into a cured solid by the application of heat. Pressure is used to flow the polymer before it is cured in order to consolidate the layers in the composite or to fill the mold and to expel the air or other volatile gases from the composite. Curing time is lower if higher temperatures are used. Generally the curing time for epoxies is higher than that for polyesters and vinyl esters. With some epoxies, the curing time may be several hours, which makes the epoxy matrix composites unsuitable for mass production. For thermoplastic matrix composites, high temperatures are used either to melt the polymer or to soften it. No polymerization reaction occurs during processing. High pressure is required to either consolidate the layers in the composite or to fill the mold cavity.

3.1 Bag Molding

Bag molding is a very common processing technique for producing epoxy or other thermoset matrix PMC laminates used in the aerospace industry. Bag molding has also been used with high-temperature thermoplastic matrix laminates. Although a slow and labor intensive process, it can produce parts with precise fiber orientation, low void content and controlled matrix volume fraction.

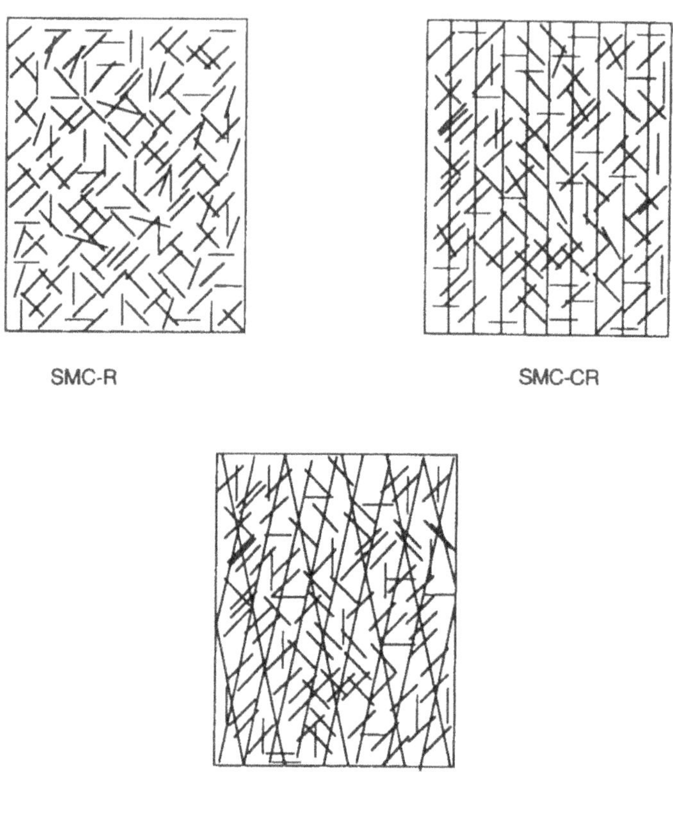

Figure 3 Sheet molding compounds (SMC).

The starting material in the bag molding process is a prepreg which is a thin layer of unidirectional continuous fibers in a partially cured (B-staged) thermosetting resin. Prepregs are made in a separate process and stored at 0°C prior to bag molding. In the bag molding process, several layers of prepreg are stacked on top of the mold surface, covered with a thin bag of high-temperature polymer film, sealed at the edges and cured in an autoclave at elevated temperatures. Vacuum is used to remove air and volatiles from the sealed bag as well as to discharge excess resin flowing out from the prepreg stack. Pressure in the range of 70–200 kPa is applied to consolidate the layers as the resin is cured. A typical cure cycle is demonstrated in Fig. 4.

3.2 Resin Transfer Molding

In the resin transfer molding (RTM) process, a stack of dry fiber layers or a dry fiber preform is placed in the mold cavity, the mold is closed and held in a clamping press. The liquid thermosetting resin mix, containing the thermosetting polymer (e.g., an epoxy or a polyester) and the curing agent or the catalyst, is injected into the mold at 70–700 kPa. As the resin flows through the dry fiber layers or the fiber preform, it wets the fibers and displaces air from the mold. Curing is completed

Table 5 Comparison of Processing Techniques Used for Polymeric Matrix Composites

Process	Common resins	Starting materials	Major equipment or tool	Typical process conditions	Cost[a] issues E/T/P	Application examples
Bag molding	Epoxy	Prepreg	Autoclave	125–175°C 0.7 MPa	H/L/H	Aerospace, sporting goods
Filament winding	Epoxy	Dry fiber tows, liquid resin	Filament winder	125–175°C 0.7 MPa	H/L/M	Pressure vessels, oxygen tanks, helicopter blades
Pultrusion	Polyester	Dry fiber rovings	Pultrusion die, puller		M/L/L	Structural beams, flat panels
Compression molding	Polyester, vinyl ester	SMC	Press, molds	150°C 2–14 MPa	H/M/L	Auto body panels, bumper beams, appliance doors
SRIM	Polyurethane	Dry fiber preform, liquid resin components	Mixing and dispensing heads, clamping press	50–120°C 0.5–1.5 MPa	H/L/L	Auto body panels
RTM	Polyester, vinylester	Dry fiber preform, liquid resin mix	Resin injection unit, clamping press	20–120°C 0.4–1 MPa	L/L/L	Auto front structure, aircraft components
Injection molding	Various thermoplastics, phenolics	Molding compounds	Injection molding machine	150–450°C 10–175 MPa	H/M/L	Numerous automotive and consumer goods

[a]E/T/P: Equipment/Tooling/Process Costs
H: High, M: Medium, L: Low

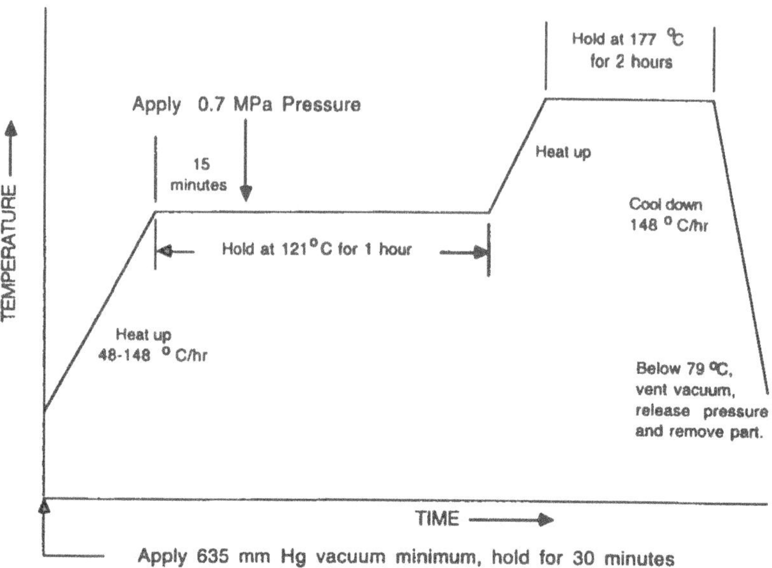

Figure 4 Typical cure cycle for an epoxy matrix composite.

either at room temperature or in air circulating oven. RTM has a higher production rate than the bag molding process. Since the resin is injected at a relatively low pressure, a high tonnage press is not required and the tooling cost is not very high.

3.3 Structural Resin Injection Molding

Structural resin injection molding (SRIM) is similar to the RTM process. It also uses a stack of dry fiber layers or a dry fiber preform, which is placed in the mold before injecting the liquid resin mix into the closed mold. The difference with RTM is in the resin reactivity, which is much higher for the SRIM resins, which are typically polyurethanes or polyureas. The chemical ingredients to make these resins are highly reactive and are mixed by a high-speed spray impinging method just before the mix is injected into the closed mold containing the dry fiber stack or preform. The curing reaction for the SRIM resins is very rapid and does not require temperatures greater than 120°C.

3.4 Compression Molding

Compression molding is a high-speed processing technique, used mostly for producing parts from SMCs. In this process, a stack of SMC layers is placed in the lower half of a preheated mold. The upper half of the mold is first quickly moved down to close the mold, contact the top of the stack, and then moved at a slower rate to apply pressure on the SMC. The heated SMC spreads under pressure and fills the mold cavity. The mold temperature is usually in the range of 130–160°C and, depending on the complexity of the part being molded, the pressure can be as high as 35,000 kPa. Usually high pressure is required for molding parts with deep ribs and bosses. The mold remains closed until the part is cured, which may vary from one to several minutes depending on the part thickness.

The tooling cost in compression molding is high, but it is capable of producing parts with good surface finish and close tolerances. It is possible to compression mold parts of complex shapes with ribs, bosses, flanges, holes and other geometric variations. The production rate is high and the process can be highly automated.

3.5 Filament Winding

Filament winding is a semicontinuous process of producing hollow structural parts, such as oxygen tanks, drive shafts, helicopter blades and spherical pressure vessels. In this process, a band of dry continuous fibers is pulled through a liquid resin tank containing a catalyzed resin and then wrapped around a rotating mandrel. The fiber band coated with the resin mix is traversed back and forth along the length of the rotating mandrel to create a helical winding pattern. The winding angle can be varied by controlling the mandrel speed as well as the traversing rate. The part is cured in an oven and the mandrel is removed to create a hollow shape. In some applications, such as oxygen tanks, the mandrel is used as part of the structure and, therefore, is not removed after curing.

3.6 Pultrusion

Pultrusion is a continuous process of producing long, straight structural members, such as I-beams, hollow rectangular beams and round tubes, containing long continuous fibers along their lengths. In this process, continuous fiber strands or rovings are pulled first through a liquid resin tank containing a catalyzed resin and then a long heated die where the resin-coated fiber strands are gathered and the cross-sectional shape of the structural member is produced. The resin is cured on the fiber strands as they move slowly along the length of the die. The cured shape is pulled out from the exit end of the die using a set of continuous belts or chains. A diamond-coated saw is used at the end of the pultrusion line to cut the pultruded member into desired lengths.

4 PROPERTIES

4.1 Mechanical Properties

The primary mechanical properties used in materials selection involving PMC are the modulus and strength of the material. For an isotropic material, such as steel, modulus as well as strength are the same in all directions and, if tensile properties are of interest, it is sufficient to quote only one modulus and one strength value, both determined in a tension test. Furthermore, if Poisson's ratio of an isotropic material is also determined in the tension test, its shear modulus can be calculated using the following equation:

$$G = \frac{E}{2(1 + \nu)}$$

where, G, E and ν represent shear modulus, tensile modulus and Poisson's ratio, respectively. Thus, there are only two independent elastic constants that must be determined experimentally to define the elastic properties of an isotropic material.

Unidirectional fiber-reinforced composites are not isotropic and their properties in the fiber direction are quite different from the properties in any other direction. As shown in Table 6, four independent elastic constants are required to define the elastic properties of a unidirectional composite, namely the longitudinal modulus (in the fiber direction), the transverse modulus (normal to the fiber direction), the major Poisson's ratio and the in-plane shear modulus. Elastic properties at any other fiber orientation angle can be calculated using these four elastic constants [1,4]. Elastic properties of a laminate can also be calculated using these elastic constants and the lamination theory [1,4].

Table 7 lists the static mechanical properties of several unidirectional fiber-reinforced polymeric matrix composites. As can be seen, longitudinal tensile modulus of unidirectional composites is much higher than their transverse tensile modulus. Likewise, the longitudinal tensile strength is also much higher than the transverse tensile strength. The longitudinal properties reflect the fiber properties and thus, for example, a higher modulus fiber, such as GY 70 carbon fiber, produces a higher longitudinal modulus for the composite than a lower modulus fiber, such as E-glass fiber. The transverse properties are influenced more by the matrix properties, and, therefore, any environmental condition, such as elevated temperatures or moisture, that changes the polymeric matrix properties will influence the transverse properties of PMC.

The longitudinal compressive properties of unidirectional epoxy matrix composites are also listed in Table 7. Unlike metals, the compressive strength and modulus of PMC can be different from the tensile strength and modulus, respectively. The longitudinal compressive properties are influenced by both fiber and matrix properties. For example, Kevlar-49 fibers have a low compressive strength, which is reflected in low longitudinal compressive strength of Kevlar-49/epoxy composite. When compressively loaded in the fiber direction, the matrix in a unidirectional composite provides lateral support to the fibers and resistance against fiber buckling, and therefore, the selection of a matrix with high modulus is important for obtaining high longitudinal compressive strength.

The in-plane shear properties of unidirectional fiber PMC are low compared to the shear properties of metals. For pure shear applications, a 45° fiber orientation with the shear stress direction gives the highest shear properties. The transverse

Table 6 Properties Needed for Characterization of Unidirectional Fiber-Reinforced Composites

Elastic properties		Strength properties	
E_L	= longitudinal modulus	S_{Lt}	= longitudinal tensile(t) strength
E_T	= transverse modulus	S_{Tt}	= transverse tensile(t) strength
ν_{LT}	= major Poisson's ratio	S_{Lc}	= longitudinal compressive(c) strength
G_{LT}	= in-plane shear modulus	S_{Tc}	= transverse compressive(c) strength
Coefficient of thermal expansion			
α_L	= longitudinal CTE	S_{LT}	= in-plane shear strength
α_T	= transverse CTE	ILSS	= interlaminar shear strength

Table 7 Mechanical Properties of Unidirectional Epoxy Matrix Composites

Property	E-glass/epoxy $v_f = 52\%$	T-300/epoxy $v_f = 60\%$	GY-70/epoxy $v_f = 62\%$	Kevlar 49/epoxy $v_f = 62\%$
Density (g/cm^3)	1.8	1.55	1.69	1.38
Tensile				
Strength (MPa)				
0°	1130	1447.5	586	1379
90°	96.5	44.8	41.3	28.3
Modulus (GPa)				
0°	39	138	276	76
90°	4.8	10	8.3	5.5
Compressive				
Strength (MPa)				
0°	620	1447.5	517	276
Modulus (GPa)				
0°	32	138	262	76
In-plane shear				
Strength (MPa)	83	62	96.5	60
Modulus (GPa)	4.8	6.5	4.1	2.1
Major Poisson's ratio	0.3	0.21	0.25	0.34

v_f = fiber volume fraction

shear properties of unidirectional composites are not commonly reported; however, they are also quite low. The interlaminar shear strength (ILSS), measured in a short beam shear test [1], is sometimes used as a measure of transverse shear strength of PMC; however, the use of ILSS is controversial and at best is good for quality control (QC) purposes.

It is important to note that the matrix in PMC plays important roles in its transverse tensile, compressive and shear properties. All of these properties are considered matrix-dominated properties. Since the properties of thermoplastic polymers are more sensitive to temperature variations than those of thermoset polymers, the matrix-dominated properties of thermoplastic matrix composites are affected more by temperature. Similarly, the time-dependent properties, such as creep and stress relaxation, of PMC will be affected more if the tests are conducted in the matrix-dominated directions and the temperature effect on creep and stress relaxation will be more for a thermoplastic matrix composite than for a thermoset matrix composite.

The properties of unidirectional composites depend on the orientation of fibers with the direction of loading (Fig. 5). As expected, the highest tensile strength and modulus are when the fibers are aligned in the loading direction. The highest shear strength and modulus are when the fibers are at a 45° angle with the loading direction. Poisson's ratio also depends on the fiber orientation angle.

The mechanical properties of several symmetric polymeric matrix laminates are given in Table 8. Symmetric laminates are the most common form in which the laminates are designed. The angles designated for each laminate refer to the fiber orientation angles in its various layers. Since in a symmetric laminate the

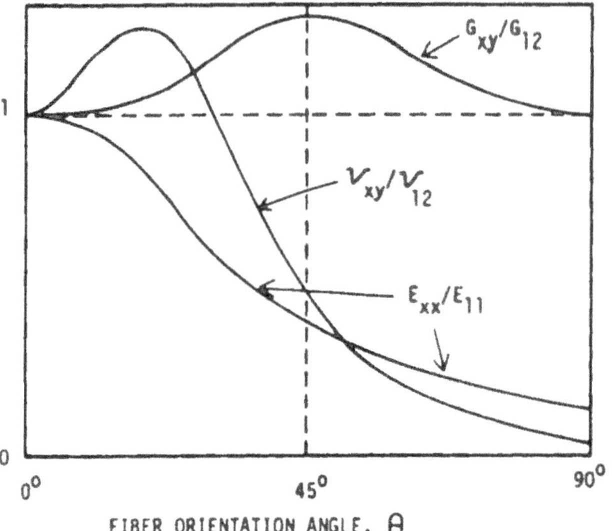

Figure 5 Effect of fiber orientation on the elastic properties of unidirectional PMC.

fiber orientation angles are symmetric about the mid-plane, fiber orientation angles in only one half of the laminate are specified. The subscript *s* at the end of the bracket indicates the symmetric nature of the laminate. As can be seen in Table 8, the static mechanical properties of PMC laminates depend on the laminate type which is controlled by the fiber orientations in various layers and stacking sequence. It should be also be noted that, in general, the elastic and strength properties of laminated composites are different in different directions. However, a few special laminate constructions, such as $[0/+45/-45/90]_s$ and $[0/+60/-60]_s$, produce equal elastic properties in all directions in the plane of the laminate. These laminates are called quasi-isotropic laminates.

Table 9 shows the strength and modulus of randomly oriented short fiber composites. Both strength and modulus of these composites are much lower than those of continuous fiber composites. However, for design purposes, random fiber com-

Table 8 Tensile Properties of High-Strength Carbon Fiber/Epoxy Symmetric Laminates

Laminate type	Tensile strength (MPa)	Tensile modulus (GPa)
0	1378	151.6
90	41.3	8.96
±45	137.8	17.2
0/90	447.8	82.7
$0_2/\pm45$	599.4	82.7
0/±60	461.6	62
0/90/±45	385.8	55.1

Table 9 Mechanical Properties of Random E-glass Fiber Composites

Material	Density (g/cm^3)	Tensile strength (MPa)	Tensile modulus (GPa)	Strain to failure (%)	Flexural strength (MPa)	Flexural modulus (GPa)
SMC-R25 (w_f = 25%)	1.83	82.4	13.2	1.34	220	14.8
SMC-R50 (w_f = 50%)	1.87	164	15.8	1.73	314	14
SMC-R65 (w_f = 65%)	1.82	227	14.8	1.67	403	15.7
Glass-reinforced injection-molded nylon 6,6 (w_f = 50%)						
(1) 4.8-mm-long fibers	1.57	220	—	2.5	320	15.2
(2) 10-mm-long fibers	1.57	240	15.8	4	400	15.8
Random fiber mat polyurethane SRIM (w_f = 46%)	1.6	228	12.6	—	285	11.1

w_f = fiber weight fraction

Table 10 Impact Properties of Polymeric Matrix Composites

Material	Notched Charpy impact energy (kJ/m^2)
Unidirectional (0°)	
Glass fiber/epoxy	420
Kevlar 49 fiber/epoxy	317
High-strength carbon fiber/epoxy	132
High-modulus carbon fiber/epoxy	23
Random discontinuous fiber	
SMC	20–40
4340 steel	214
6061-T6 aluminum alloy	153

posites are considered isotropic, even though there may be flow-induced preferred orientation of fibers in some of these composites [3].

4.2 Impact Properties

Impact properties of materials are commonly determined by either notched Izod or Charpy impact tests. Even though the Izod or Charpy impact data have limited use in design, they are useful in comparing materials that are tested in identical impact conditions. Table 10 gives Charpy impact energies of several PMC and metals. In general, impact energies of glass and Kevlar fiber PMC are equal to or better than those of steel or aluminum alloys. Carbon fiber PMC have lower impact energies; however, high impact energies can be obtained by combining carbon with glass or Kevlar fibers in an interply hybrid form (Table 11). In these laminates the carbon fiber layers are placed at the skins and the glass or Kevlar fiber layers are placed in the interior. High-modulus carbon fiber layers in the skins provide high stiffness for the laminate, while progressive delamination in the glass or Kevlar fiber layers provides energy absorption in impact.

Table 11 Impact Properties of Hybrid Polymeric Matrix Composites

Material[a]	Unnotched Charpy impact energy (kJ/m^2)
E-glass fiber	621.6
T-300 carbon fiber	186.9
GY-70 carbon fiber	12.3
Sandwich GY-70/E-glass/GY-70 hybrid	426.3
Sandwich T-300/E-glass/T-300 hybrid	434.7
Alternate GY-70/E-glass hybrid	217.9
Alternate T-300/E-glass hybrid	139.6

[a]The matrix is an epoxy in all of the laminates in this table.

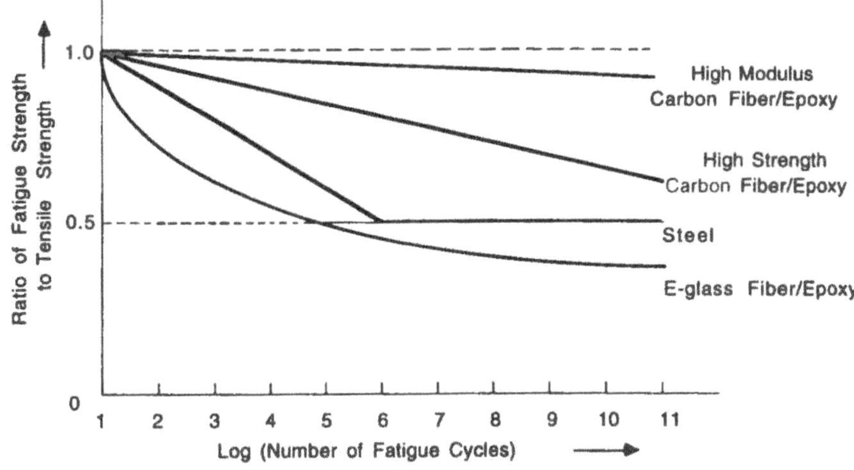

Figure 6 Schematic S-N diagrams of various PMC.

4.3 Fatigue Properties

Fatigue behavior of several unidirectional composites and common metals in tension-tension load cycling is compared in the S-N diagrams shown in Fig. 6. Unlike low-carbon steels, most PMC do not exhibit any endurance limit. Instead, the S-N diagram of a PMC may indicate increasing life with decreasing maximum cyclic stress level, which is similar to that observed with many aluminum alloys and polymers. In such cases, it is common to specify the fatigue strength of the material at very high cycles, say 10^6 cycles as indicated in Table 12.

A unique feature of many PMC is that they exhibit a gradual loss in stiffness (i.e., softening) as well as a gradual loss in residual strength (Fig. 7) due to the

Table 12 Fatigue Strength at 10^6 Cycles of Several Polymeric Matrix Composites

Material	Fatigue strength (MPa)	Ratio of fatigue strength to tensile strength
Unidirectional (0°)		
Fiber/epoxy ($v_f \approx 60\%$)		
(1) High-modulus carbon	650	0.9
(2) High-strength carbon	1010	0.7
(3) Kevlar 49	1000	0.725
(4) E-glass	690	0.6
Random E-glass fiber SMC		
(1) SMC-R25	40	0.49
(2) SMC-R50	63	0.38

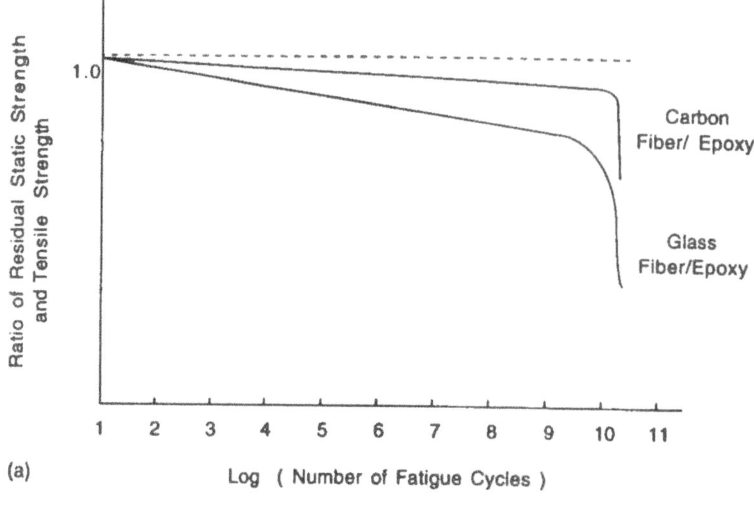

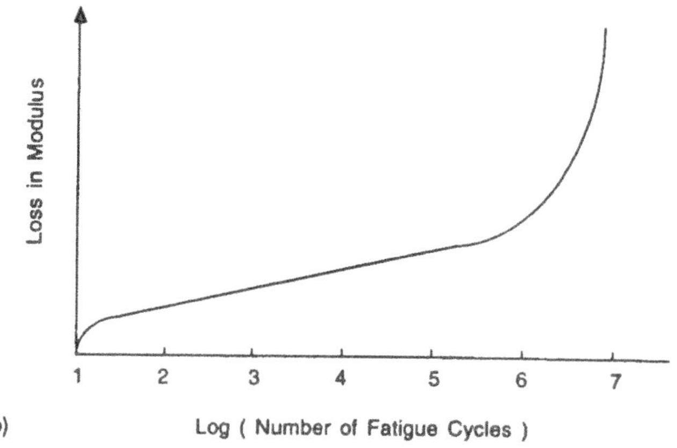

Figure 7 (a) Residual strength loss and (b) modulus loss in PMC due to fatigue loading.

appearance of damages in the form of matrix cracking, fiber-matrix debonding, fiber breakage and interply delamination. Some of these damages may appear at multiple locations in the first few hundred to thousand cycles and may be large enough to be visible. However, unlike metals, the propagation and accumulation of these damages occur in a progressive manner and complete failure does not occur immediately.

4.4 Fracture Toughness

Fracture toughness of a material indicates the material's resistance to uncontrolled crack growth or brittle failure. The Mode I fracture toughness of PMC is, in general, relatively high compared to other materials (Table 13). This is due to a number of

Table 13 Fracture Toughness of Polymeric Matrix Composites

Material	Fracture energy, G_{Ic} (J/m²)
Unidirectional (0°) ($v_f \approx 0.6$)	
E-glass/epoxy	525–1020
T-300 carbon/epoxy	60–254
T-300 carbon/toughened epoxy	530–690
AS-4 carbon/PEEK	1330–2890
Random E-glass Fiber	
SMC-R50	23,000

microscopic and macroscopic failure mechanisms that contribute to energy absorption in fiber-reinforced PMC. The major energy-absorbing failure mechanisms are fiber-matrix debonding, fiber pull-out, matrix cracking and interply delamination.

A unique failure mode observed in laminated PMC is the subsurface delamination and matrix cracking due to transverse impact loads. The laminate may appear undamaged from outside, but these internal damages may weaken the laminate and the postimpact residual strength of the laminate may be lower than its strength before the impact. In order to measure the resistance against delamination, new test methods have been developed which determine Mode I, Mode II or mixed delamination fracture toughness [4]. Figure 8 shows a comparison of Mode I delamination fracture toughness of various polymer matrix composites and its relationship to their post impact compressive strength. In general, the higher the fracture toughness of the polymeric matrix, the higher the delamination fracture toughness of the laminate.

4.5 Damping Properties

The measure of vibration damping in a material is the damping factor, which is listed for various polymer matrix composites in Table 14. PMC have a higher damping factor than metals. Higher damping is due to the viscoelastic nature of the polymeric matrix as well as the presence of fibers and the fiber-matrix interface. The damping factor depends on the polymer as well as the fiber type, fiber orientation angle and stacking sequence.

4.6 Coefficient of Thermal Expansion

The coefficient of thermal expansion (CTE) of PMC is controlled by the fiber type, fiber volume fraction, fiber orientation angle and the laminate construction (Table 15). In quasi-isotropic laminates, such as $[0/+45/-45/90]_s$ and $[0/+60/-60]_s$ type laminates, and random fiber laminates, the CTE values are the same in all directions in the plane of the laminate. However, for other laminates, the CTE values will be different in different directions. With proper fiber selection and laminate construction, the CTE can be made close to zero.

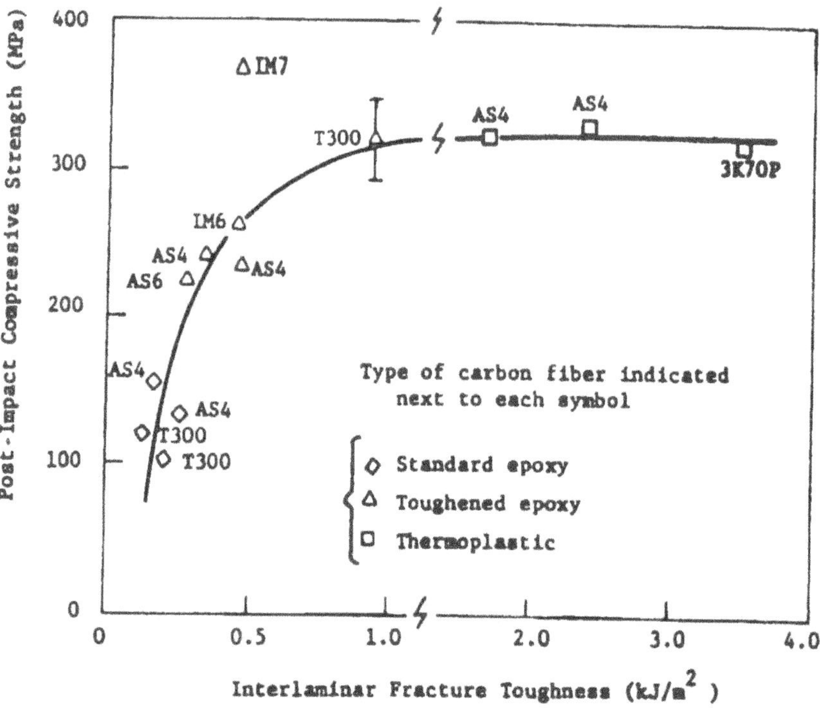

Figure 8 Effect of Mode I delamination fracture toughness on the postimpact compressive strength of carbon fiber-reinforced laminates (impact energy = 6.7 J/mm) with different matrix materials. (From: D. Leach, *Tough and Damage Tolerant Composites*, Symposium on Damage Development and Failure Mechanisms in Composite Materials, Belgium, 1987.)

4.7 Thermal Conductivity

The thermal conductivity of PMC depends on the fiber type, fiber volume fraction, fiber orientation angle and the laminate construction. A few representative values are given in Table 16. Carbon fiber PMC have the highest thermal conductivity

Table 14 Damping Factor of Various Polymeric Matrix Composites

Material	Fiber orientation	Modulus (GPa)	Damping factor
E-glass/epoxy	0°	35.2	0.007
Carbon/epoxy	0°	188.9	0.0157
	22.5°	32.4	0.0164
	90°	6.9	0.0319
	[0/22.5/45/90]s	69	0.0201
Low-carbon steel	—	207	0.0017
6061 aluminum alloy	—	70	0.0009

Table 15 Coefficient of Thermal Expansion of Various Polymeric Matrix Composites

	CTE (10^{-6}/°C)		
	Unidirectional (0°)		
Material	Longitudinal	Transverse	Quasi-isoptroic
E-glass/epoxy (v_f = 0.6)	7.13	32.63	12.6
Kevlar-49/epoxy (v_f = 0.6)	−3.6	54	−0.9 to 0.9
High-modulus carbon/epoxy (v_f = 0.6)	−0.9	27	0 to 0.9
Random E-glass fiber composites			
SMC-R25 (w_f = 0.25)	23.2		
SMC-R50 (w_f = 0.5)	14.8		
Injection-molded Nylon 6,6 (w_f = 0.5)	18		
Steel	11–18		
Aluminum alloys	22–25		

among the various PMC. As with the thermal expansion, thermal conductivity can also be direction dependent.

5 FIBER AND MATRIX SELECTION FOR A PMC

The selection of PMC for a structural application depends on many factors, such as the design requirements (load, stiffness, impact, etc.), operating environment, processability, cost and availability. Some of these factors are interrelated. For example, the high cost of carbon fibers or PEEK thermoplastics is related to the their low availability. The cost of epoxy matrix composites is high, since the epoxies used in structural laminates require long cure times. Long processing time is the primary reason for not selecting epoxies in automotive applications. Polyesters or vinyl esters are preferred as the matrix for automotive composites, such as sheet molding compounds, because they have a much shorter cure time.

Table 16 Thermal Conductivity of Various Polymeric Matrix Composites

	Thermal conductivity (W/m-°C)		
	Unidirectional (0°)		
Material	Longitudinal	Transverse	Quasi-isoptroic
S-glass/epoxy (v_f = 0.6)	3.46	0.35	0.346
Kevlar-49/epoxy (v_f = 0.6)	1.73	0.173	0.173
High-modulus carbon/epoxy (v_f = 0.6)	48.44–60.55	0.865	10.38–20.76
Steel	15.57–46.71		
Aluminum alloys	138.4–216.25		

5.1 Fiber Selection

In this section, we will consider the selection criteria for fibers and matrix in a polymeric matrix composite. For this, we will first consider a unidirectional continuous fiber composite (Fig. 9). The longitudinal modulus of this composite is given by the following "rule of mixture" equation:

$$E_L = E_f v_f + E_m v_m \tag{1}$$

where E_L = longitudinal modulus of the undirectional continuous fiber composite
E_f, E_m = fiber and matrix modulus, respectively
v_f, v_m = fiber and matrix volume fraction, respectively

Since $E_f >> E_m$, the matrix contribution to E_L is relatively small and, for all practical purposes,

$$E_L \approx E_f v_f \tag{2}$$

Equation (2) indicates that to obtain a high longitudinal modulus of the composite, a high modulus fiber should be selected. Carbon fibers have a much higher modulus than glass or aramid fibers and are, therefore, selected for high stiffness applications.

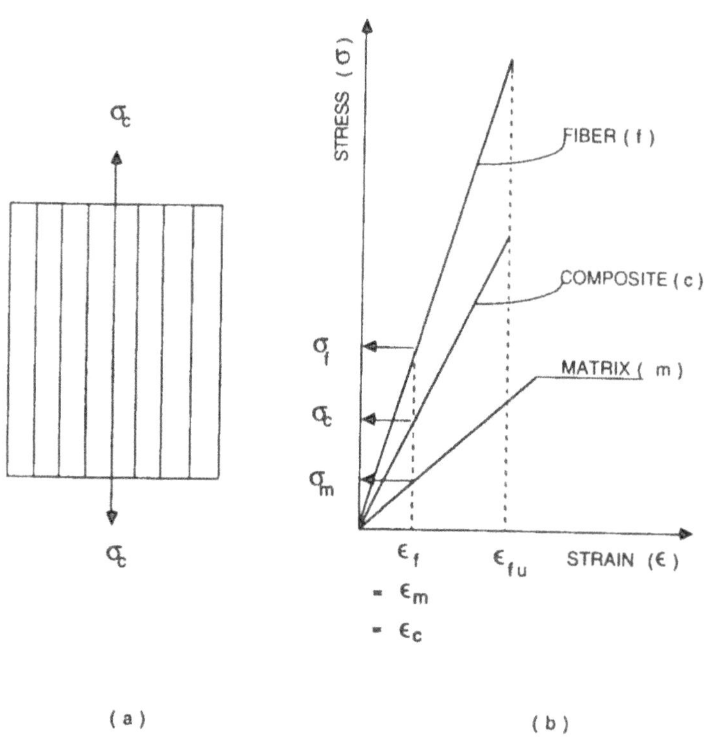

Figure 9 Stress application in the longitudinal direction of a unidirectional fiber composite.

The longitudinal tensile strength, S_{Lt}, of a unidirectional continuous fiber composite can be approximated by the following equation:

$$S_{Lt} \approx S_f v_f \tag{3}$$

where S_f is the tensile strength of the fiber. For high longitudinal tensile strength, a high-strength fiber is selected. Carbon fibers have higher strengths than glass or Kevlar fibers and are preferred for strength-critical applications. Although the strength of freshly produced glass fibers is higher than many carbon fibers, their strength is affected by abrasion and handling, and in practice, the strength of glass fibers is approximately one half the freshly produced strength. Aramid fibers are preferred over carbon fibers in tensile load applications if the strength-to-weight ratio is important; however, aramid fibers are not selected for compressive applications.

An important parameter for obtaining both high modulus as well as strength is the fiber volume fraction, which should be as high as possible (see Eqs (2) and (3)). Theoretically, the maximum fiber volume fraction is limited by how close they can be packed without touching each other, which can be in the range of 75–90%. In practice, the fiber volume fraction is limited to 60%, since above this value, it is difficult to wet the fibers with the matrix, which is required for good composite properties.

The transverse modulus and strength of a unidirectional continuous fiber composite are controlled more by the matrix modulus and strength, respectively. For example, the transverse modulus, E_T, can be approximated as

$$E_T \approx \frac{E_m}{1 - v_f} \tag{4}$$

As per Eq. (4), high matrix modulus as well as high fiber volume fraction is desirable for obtaining high transverse modulus of the composite. High matrix modulus also helps in preventing fiber buckling when the unidirectional composite is subjected to a longitudinal compressive load. Assuming fiber buckling as the failure mode in compression, the longitudinal compressive strength, S_{Lc} is predicted as

$$S_{Lc} = \frac{G_m}{1 - v_f} \tag{5}$$

where G_m is the shear modulus of the matrix.

The properties of unidirectional fiber composites are functions of the fiber orientation angle with the direction of loading (Fig. 5). For example, both tensile strength and modulus are the highest when the fibers are oriented in the loading direction. For slender rods or bars under uniaxial tension, fibers should be oriented along their lengths so that the fiber direction and the tensile stress direction coincide. To overcome the effects of fiber orientation angle in unidirectional composites, laminated composites can be formed by stacking unidirectional layers in a specified sequence of fiber orientations; however, the fiber orientation in each layer and the difference in fiber orientation in successive layers determine the failure mode, the order of failure in various layers and the properties derived from the laminate. For example, in a cross-plied laminate subjected to unidirectional tensile loading, matrix cracks appear in layers that have 90° fiber orientation to the applied load at relatively low loads and make these layers relatively ineffective. A common laminate family in aerospace applications is the symmetric quasi-isotropic [0/90/+45/−45] type. For maximum bending stiffness, the 0° layers are placed on the outside. The placement of 45 and −45 layers next to each other may cause delamination at low tensile loads, and therefore, other ways of arranging them should be considered. For applications involving high in-plane shear stresses, a +45/−45 combination is preferred.

Fiber length is an important parameter for short fiber composites since the longer the fiber length, the higher the tensile strength of the composite. On the other hand, processing becomes more difficult with longer fibers. However, for effective reinforcement, the fiber length should be greater than 10 times the critical fiber length, l_c, which is

$$l_c = \frac{S_f d_f}{2S_i} \tag{6}$$

where d_f is the fiber diameter and S_i is the fiber-matrix interfacial shear strength. Equation (6) also indicates the importance of improving the interfacial shear strength, since a higher S_i reduces the critical fiber length and increases the possibility of short fibers acting as reinforcements. If the fiber length is greater than the critical length, l_c, the longitudinal tensile strength of a unidirectional short fiber composite can be approximately calculated using the following equation:

$$S_{Lt} = S_f \left(1 - \frac{l_c}{2l_f}\right) v_f \tag{7}$$

where l_f is the fiber length. The importance of the fiber length in developing high strength for a short fiber composite is evident in Eq. (7) and is also demonstrated in Fig. 10.

5.2 Matrix Selection

The selection of the proper polymer for the matrix is extremely important in the determination and control of matrix dominated properties, such as the transverse tensile strength and the shear strength, of PMC. Since matrix cracking is one of the critical failure modes and can occur at relatively low stresses, it is important to select a matrix with high ductility as well as high fracture toughness. This is one area where thermoplastic polymers excel over thermoset polymers. But even with thermoset polymers, fracture toughness can be improved by adding rubber particles or other toughening agents; however, their modulus and temperature resistance may decrease.

Besides contributing to the matrix dominated properties of a PMC, the matrix has a special role in protecting the PMC against environmental degradation. The following environmental effects are controlled by the selection of the polymer:

1. Moisture absorption in humid environments, which increases the physical dimensions of the composite due to swelling, lowers the glass transition temperature of the polymer and changes the mechanical properties of PMC. The

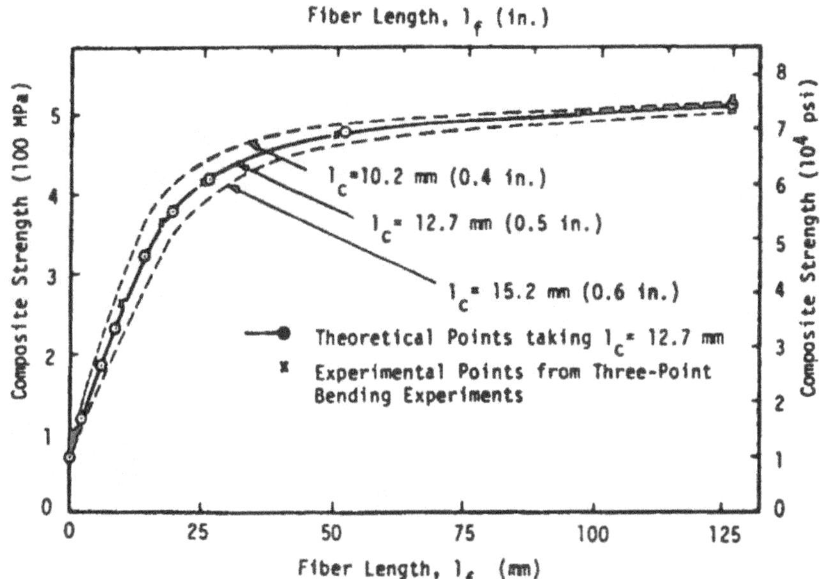

Figure 10 Fiber length vs. strength in short fiber composites. (From: Ref. 5.)

diffusion of absorbed moisture to the fiber-matrix interface may cause debonding at the interface, which reduces the composite's properties. The moisture absorption problem is particularly severe with many epoxy resins.

2. Absorption of organic solvents, which in addition to the effects similar to moisture absorption, can also attack the polymer and cause stress cracking in the matrix. The stress cracking problem is particularly severe in amorphous thermoplastics. The chemical resistance of polymers vary. It also depends on the nature of the chemical as well as the application temperature.
3. Absorption of ultraviolet (UV) light, which occurs when a PMC is directly exposed to long hours of sunlight. UV absorption may reduce the molecular weight of the polymer matrix, cause discoloration and a reduction in the properties of the matrix.
4. Elevated temperatures, which tend to reduce the modulus and yield strength of the polymer matrix. This is particularly true for thermoplastic resins; however, there are several high-temperature thermoplastics, such as PEEK and PPS, which retain their properties over a long temperature range. Creep of PMC at elevated temperatures is not low if the applied stress is in the fiber direction; however, high creep can be observed if the stress is in the matrix-dominated directions.
5. Low temperatures, which tend to reduce the strain-to-failure of the polymer and embrittle the polymer matrix. The low-temperature problem becomes severe if it is exposed to temperatures that are much below the glass transition temperature of the polymer.
6. Fire, which may create flame, toxic fume and health hazard. Among the thermosets, phenolic resins have inherently good fire performance than polyesters

or epoxies. Phenolics form a char on the surface when they come in contact with the flame, thereby cutting off the oxygen supply to the resin underneath. Fire retardants, such as aluminum trihydrate, are added to other polymers to reduce their combustibility.

A qualitative summary of the characteristics of different thermoset and thermoplastic polymers is given in Table 17. In general, a polymer with high strength, modulus and ductility should be selected for achieving a good transverse strength for the composite. In many applications, thermal or some other properties may also be deciding factors. In addition, the compatibility with the fiber and ease of processing are important considerations in polymer matrix selection for a PMC.

6 SELECTION OF PMC

In previous sections, we have reviewed the properties of PMC and the selection of its constituents. In this section, we will consider the issues in the selection of PMC over other materials.

The primary reason why a polymeric matrix composite is often considered for an application is the possibility of weight saving, since PMC have low specific gravities. However, in selecting PMC over other structural materials, especially metals, several important differences in the mechanical characteristics of PMC and metals should be kept in mind:

1. PMC are anisotropic and inhomogeneous materials, whereas metals, for all practical purposes, can be considered isotropic and homogeneous.
2. The strength and modulus of PMC are high only in the fiber direction. The transverse and shear properties of the PMC can be low.
3. PMC are often used in laminated forms. Properties can be tailored in a laminate as per the design requirements. However, other than the quasi-isotropic laminates, the properties of a laminated PMC are different in different directions. Additionally, the interlaminar strengths of laminates can be low.
4. Unlike ductile metals, PMC do not exhibit yielding. However, if properly designed, PMC can exhibit several unique failure modes arising from its heterogenous structure that can provide high fracture toughness and low notch sensitivity for the PMC. Selections of fiber, matrix, laminate type, fiber volume fraction, fiber orientation and fiber-matrix interface are all very important factors in controlling the failure modes in PMC.

The above discussion indicates that PMC are different from metals in many design aspects. The processing techniques of PMC are also different from those used for metals. Therefore, a one-to-one substitution of a metal with PMC is not recommended. To take full advantage of PMC, a good understanding of its characteristics relative to other materials must be developed. If a PMC part is used in an assembly where other materials are used, it becomes even more important to understand the differences as well as interactions between these materials before a successful design can be achieved.

Table 17 Comparison of Common Polymeric Matrix Materials

Polymer	Processing temperature (°C)	Mold shrinkage (%)	Continuous use temperature (°C)	Chemical resistance	Cost	Principal application market	Comments
Epoxies	120–177	1–5	80–180	Fair	Medium	Aerospace	High moisture absorption
Polyesters	60–150	5–12	70–130	Fair	Low	Automotive	
Vinyl esters	100–150	5–12	80–150	Good	Medium	Automotive	Higher fatigue resistance than polyesters
Phenolics	90–125	0.5–1	120–260	Fair	Low	Aerospace	Excellent fire resistance
Polyurethanes	60–120	2	80	Good	Low	Automotive	
PEEK	350–430	1.1	120–250	Good	Very high	Aerospace	
PPS	315–430	0.6–0.8	100–200	Good	High	Aerospace	
Nylons	260–330	0.7–1.8	140	Good	Medium	Automotive	High moisture absorption
Polypropylene	200–290	1–2.5	110	Good	Low	Automotive	

7 EXAMPLES OF PMC APPLICATIONS

In this section, we will review several examples where polymeric matrix composites have been successfully applied. These examples are drawn from several industries to demonstrate the variety of applications of PMC.

7.1 Leaf Springs

Leaf springs, made of unidirectional continuous E-glass fiber-reinforced epoxy, are used in many cars, trucks and trailer suspension systems. These monoleaf PMC springs are 25–75% lighter than the multileaf steel springs that they replace. On a unit weight basis, the PMC springs are able to store nearly 14 times more elastic energy than the steel springs, which is a distinct advantage in spring applications.

Glass fiber/epoxy springs have been used by General Motors in Corvettes since 1981. Glass fibers are selected for their low cost. Glass fibers have a lower modulus than carbon fibers, which helps in storing more energy in the spring. Although glass fiber/epoxy has a lower fatigue strength than carbon fiber/epoxy, it shows a progressive fatigue failure and a better impact damage tolerance. Epoxy is used as the matrix. Even though epoxies are more expensive than polyesters and vinyl esters, they provide better fatigue damage tolerance as well as higher interlaminar shear strength than the latter.

The leaf spring is a good application for unidirectional fiber PMC, since for all practical purposes, it is a beam application with normal stresses acting only in the length direction of the spring. Since the continuous fibers are oriented in the length direction of the spring, it provides the most efficient use of fibers in a composite.

The Corvette glass fiber/epoxy spring is designed for uniform maximum stress throughout the length. This is accomplished by varying the height and width of the spring along the length, while maintaining the same cross-sectional area. The Corvette spring is approximately 15 mm thick and 86 mm wide at each end, and 25 mm thick and 53 mm wide at the center. They are manufactured by a combination of filament winding and compression molding.

7.2 NGV Fuel Tanks

Natural gas vehicle (NGV) fuel tanks, manufactured by Lincoln Composites of Nebraska, are filament-wound using carbon fiber/epoxy over a high-density polyethylene (HDPE) liner. These thin-walled cylindrical tanks are designed to store natural gas at an internal gas pressure of 10 to 170 MPa, which creates high hoop and axial tensile stresses in the tank wall. High-strength carbon fibers are selected for their lower density as well as higher tensile strength compared to glass fibers. Carbon fiber tanks are more than 50% lighter in weight than glass fiber tanks. Epoxy resin is selected since it provides higher transverse tensile strength and environmental resistance than many other resins. HDPE liners are selected over metal liners, since they can be easily blow-molded and the cost of HDPE liners is 50% of the metal liner cost. HDPE provides more than adequate resistance against natural gas permeation through its wall.

One major concern in using all carbon NGV tanks in commercial automotive applications is their low impact damage tolerance. For improved damage tolerance, glass fiber/epoxy layers 1.3 mm thick are overwrapped on the outside of the carbon fiber/epoxy layers.

7.3 Golf Clubs

PMC are used in many sporting good applications, including golf club shafts and heads. In one design of PMC golf clubs, the shaft utilizes a 0° carbon fiber/epoxy layers along the length with glass fiber/epoxy overwraps in the 90° direction. Carbon fibers provide high flexural strength and stiffness in the length direction of the shaft, while the 90° glass fibers prevent longitudinal splitting between the carbon fibers. The golf club heads use randomly oriented chopped carbon fibers in an impact resistant thermoplastic resin, such as ABS and polycarbonate. The heads are injection-molded.

7.4 Oil Field Sucker Rods

Sucker rods are used in lifting underground oil to the surface. Each sucker rod is approximately 11 mm long and 19 to 30 mm in diameter. Several sucker rods are joined together in series to form the lifting mechanism in a beam-type oil lifting unit. Two major problems with steel sucker rods are their weights and poor corrosion resistance. Unidirectional E-glass fiber/polyester sucker rods, produced by the pultrusion process, are 60–70% lighter than steel sucker rods. The lighter weight of the glass fiber/polyester sucker rods not only reduces the stresses in the rods, but it also decreases the amount of energy needed for pumping each barrel of oil to the surface. Glass fiber/polyester is corrosion resistant, since both glass fibers and polyester resin are not affected by the oil. Steel sucker rods corrode easily and need to be replaced every three or four months, which decreases production rate and increases maintenance cost. No such problem exists with the glass fiber/polyester sucker rods.

7.5 Aircraft Applications

The development of structural PMC with advanced fibers, such as carbon, started in the aerospace industry. Structural PMC was introduced first in military aircrafts and then in secondary structures of commercial aircrafts. Many of the PMC aircraft structures are in service for 15–20 years and a significant amount of information has been collected on their performance, durability etc. Here we briefly examine one of these structures, namely the inboard ailerons of an L-1011 aircraft.

The inboard ailerons is a multicomponent box-type structure, containing upper and lower covers, front and rear spars, several reinforcing ribs and miscellaneous fittings and fairings. It is hinged to the wing trailing edges to control the roll of the aircraft. The upper and lower covers in the composite construction are thin sandwich panels of T-300 carbon fiber/epoxy skins and a core of hollow glass microsphere filled epoxy. The front spar is a channel section of T-300 carbon fiber/epoxy laminate, while the ribs are T-300 carbon bidirectional fabric/epoxy. The rear spar is an 7075-T6 clad aluminum. The use of PMC for the rear spar is not cost-effective considering the the small amount of weight saving to be achieved over the aluminum construction. The aileron components are assembled with titanium screws and stainless steel collars. To prevent galvanic corrosion, all aluminum parts are anodized, primed with epoxy and painted with a polyurethane top coat. All PMC components are also painted with a polyurethane top coat. The composite aileron is 23.2% lighter than the earlier aluminum construction and contains 50% fewer parts and fasteners.

REFERENCES

1. P. K. Mallick, *Fiber Reinforced Composites*, 2nd ed., Marcel Dekker, New York, 1993.
2. N. L. Hancox and R. M. Mayer, *Design Data for Reinforced Plastics*, Chapman and Hall, London, 1994.
3. P. K. Mallick and S. Newman, eds., *Composites Materials Technology: Processes and Properties*, Hanser, Munich, 1990.
4. I. M. Daniel and O. Ishai, *Engineering Mechanics of Composite Materials*, Oxford University Press, New York, 1994.
5. P. Hancock and R. C. Cuthbertson, Effect of fibre length and interfacial bond in glass fibre-epoxy resin composites, *J. Materials Sci.*, *5*, 762 (1970).

14
Ceramic Matrix and Carbon-Carbon Composites

Sarit B. Bhaduri
University of Idaho
Moscow, Idaho

1 INTRODUCTION

In recent years it has become clear that advancements in system and device performance are critically dependent upon development of advanced materials with enhanced physical and mechanical properties [1,2]. No longer can metallic materials meet all of these requirements, particularly when high hardness and exposure to high temperature are necessary. For such challenging applications, ceramic materials are especially suitable. Compared to metals, ceramics typically have higher melting points and lower density. However, conventional ceramics have very low fracture toughness (i.e., brittle) leading to their poor reliability. To utilize the unique capabilities of ceramics, a great deal of research is going on in order to improve upon their reliability by fabricating composites with the addition of second or more phases in the matrix. By "composites" we mean an intimate mixture of materials, formed into a desired shape. The composite contains a "primary" or "matrix" phase (in the present case a ceramic called a "ceramic matrix composite," CMC) in which other phases with appropriate properties are added; here "appropriate" means, for example, materials of enhanced levels of properties such as higher strength or stiffness. By doing this, an enhanced combination of properties are obtained in the composite material, compared to those of the original matrix material.

To put these materials into perspective, a comparison of the characteristics of Al, monolithic Al_2O_3, and composite Al_2O_3 (produced by the Lanxide process) is shown in Table 1. This table shows that the approximate hardness numbers for Al and Al_2O_3 are 2 GPa and 14 GPa respectively, while the equivalent toughness values are 40 $MPa \cdot m^{1/2}$ and 3.8 $MPa \cdot m^{1/2}$. A typical Lanxide material consisting of 80 vol% Al_2O_3 and 20 vol% Al has hardness and toughness values intermediate between those of the ceramic and the metal [3].

Table 1 Comparison of Properties of Al, Al_2O_3, and Al_2O_3 + Al (Lanxide) Composite

	Hardness (GPa/ksi)	Toughness (MPa$\sqrt{m}$/ksi$\sqrt{in}$)	Density (g/cc)	E (GPa/ksi × 10^3)
Al	2/580	40/36.7	1	70/10.150
Al_2O_3	14/2030	4/3.7	3.75	320/46.4
Al_2O_3/Al (20% vol)	5[a]/725 17.3[b]/2508	9.5/8.71	3.53	230/33.35

[a]Al, rich microstructure.
[b]Al_2O_3, rich microstructure.

It is the purpose of this chapter to discuss the following topics toward building an understanding of the CMCs including a description of potential monolithic matrices, categorizing the types of reinforcements, enhancement of physical and mechanical properties, processing techniques and examples of applications. A small portion of the chapter is devoted to carbon-carbon composites.

2 CONSTITUENTS OF CMCs

2.1 Potential Matrices

Table 2 shows the various advanced ceramic materials and their physical and mechanical properties, e.g., Young's modulus, toughness, coefficient of thermal expansion, density, and melting point. The list includes glass, glass ceramics, oxide materials and nonoxides. The data refer to glass ceramic materials subjected to standard heat treatments, glasses in an annealed condition, and the ceramics in a fully dense condition. In the case of ZrO_2 two of the most commercially important materials are Mg-PSZ (magnesia partially stabilized zirconia) and Ce-TZP (ceria tetragonal zirconia polycrystal) [4]. Among the monolithic materials ZrO_2 exhibits the highest levels of toughness [4].

2.2 Geometric Reinforcements

The reinforcements can be shaped into various well-defined architectures [5]. These are discrete (discontinuous), continuous, planar interlaced (two-dimensional) and fully integrated (three-dimensional) configurations (Table 3).

The discrete materials have no continuity, e.g., particulate, whiskers or short fibers. It is difficult to orient these discrete phases in specific directions, although some success has been achieved recently [5]. The second category is the continuous or unidirectional (0°) reinforcement. This has the highest level of continuity and linearity and is suitable for filament winding or tape lay-ups. The third category is the planar interlaced and interlooped system which consists of two to three yarn diameters in the thickness direction with the fibers oriented in the x-y phase [6]. The fully integrated configuration consists of continuous fibers oriented in various in-plane and out-of-plane directions. The most important feature in this case is the

Table 2 Properties of Various Ceramic Matrices

Materials	Strength (MPa/ksi)	Density (g/cc)	*E* (GPa/ksi)
Soda-lime glass	100/14.5	2.5	6.0/870
Borosilicate glass	100/14.5	2.3	6.0/870
Lithium aluminosilicate g.c.[a]	100–150/ 14.5–21.75	2.0	10.0/4500
Silica glass	48/6.96	2.2	7.2/1044
Magnesium aluminosilicate g.c.[a]	110–170/ 15.95–24.65	2.6–2.8	11.9/1725.5
Mullite	83/12.035	9.8–10.0	14.3/2073.5
ThO_2	97/14.065	5.6–5.75	14.5/2102.5
ZrO_2	113–128/ 16.385–18.56	3.6	17–25/ 2465–3625
MgO	97–130/ 14.065–18.85	3.2	21–30/ 3045–4350
Si_3N_4	410/59.45	3.0	30.7/4451.5
BeO	130–240/ 18.85–34.8	3.9–4.0	29.5–38/ 4277.5–5510
Al_2O_3	250–300/ 36.25–43.5	3.2	36–40/ 5220–5800
SiC	310/44.95	6.4–6.9	40–44/ 5800–6380
ZrC	110–210/ 15.95–30.45	9.7	19.5–48/ 2827.5–6960
HfO_2	69/10.005	15.6–15.8	56.5/8192.5
WC	345/50.025		54–70/ 7830–10150

[a]Glass ceramic.

additional reinforcements in the through thickness direction which make the composite virtually delamination free [6].

2.3 Reinforcing Materials

Table 4 shows the reinforcements including ceramic and metallic materials which are currently commercially available. The ceramic reinforcements can be categorized into oxide as well as nonoxide materials as described in the following.

2.4 Oxide Fibers

Essentially three types of oxide fibers are available for reinforcement. Alumina is one of the materials, which is available in the fiber form. This material is usually produced by heating aluminum hydroxide or other aluminum containing salts, which lose water, and undergo several crystal transformations before becoming the stable α-alumina or corundum structure. The original small-diameter alumina fiber was developed by DuPont in 1979 [7]. The fiber was made by blending the alumina

Table 3 Fiber Architecture for Composites

Reinforcement system	Textile construction	Fiber length	Fiber orientation	Fiber entanglement
Discrete	Chopped fiber	Discontinuous	Uncontrolled	None
Linear	Filament yarn	Continuous	Linear	None
Laminar	Simple fabric	Continuous	Planar	Planar
Integrated	Advanced fabric	Continuous	3-D	3-D

in powder form in water and other compounds, which could be spun. The as-spun fiber is coated with a thin layer of silica so as to increase its strength, which is retained up to 1000°C. Yet another development at DuPont is PRD-166 which is Al_2O_3-ZrO_2 composite fiber [8]. The addition of ZrO_2 stabilizes the microstructure of Al_2O_3. PRD-166 has a higher strength value than FP, but the strength falls off around 1000°C.

Other fibers containing mostly alumina are supplied by Sumitomo (Japan) and ICI (U.K.) with commercial names of Alf and Safimax respectively. The Sumitamo

Table 4 Properties of Reinforcements (brittle and ductile)

Material	Composition	Strength (MPa/ksi)	Density (g/cc)	E (GPa/ksi × 10^3)
Nicalon (Nippon carbon)	Si, C, O	3200/464	2.55	190/27.55
Tyranno (Ube)	Si, Ti, C, O	3000/435	2.4	7200/1044
SCS-6 (Textron)	SiC on C	3900/565	3.0	400/58
MPDZ (Dow Corning/ Celanese)	Si, C, N, O	1900/275	2.3	185/26.825
HPZ (Dow Corning/ Celanese)	Si, C, N, O	2250/326	2.35	165/23.925
MPS (Dow Corning/ Celanese)	Si, C, O	1350/195	2.6	185/26.825
Nextel 312 (3M)	Al_2O_3,B_2O_3,SiO_2	1750/253	2.7	150/21.75
Nextel 440 (3M)	-do-	2100/304	3.05	180/26.1
Nextel 480 (3M)	-do-	2275/330	3.05	225/32.625
FP (Dupont)	Al_2O_3	1400/203	3.9	385/55.825
PRD 166 (Dupont)	Al_2O_3,ZrO_2	2250/326	4.2	385/55.825
Niobium	Nb	750/108	8.6	80–120/ 11.6–17.4
Copper	Cu	400/58	8.9	120/17.4
Tantalum	Ta	1350/196	16.6	180/26.1
Molybdenum	Mo	2200/320	10.0	320/46.4
Tungsten	W	3400/490	19.3	380/55.1

fibers have Al_2O_3 in the α-phase and contains about 15% SiO_2, which reduces its Young's modulus as compared to α-Al_2O_3-based fibers. The Sumitamo fiber is fabricated from alkyl aluminum, which is then polymerized with the addition of water, making it more viscous for easier spinnability [9]. The role of 15% SiO_2 addition is to stabilize the α-Al_2O_3 structure. The fibers show no noticeable loss of their virgin strength up to 1000°C, beyond which they degrade.

The Safimax fiber is a long staple fiber where the SiO_2 content is much less (<4%). The fiber is produced from aluminum chlorohydroxide, which on hydration becomes thickened and can be spun. The addition of SiO_2 is believed to suppress the transformation from γ to α-alumina, while SiO_2 simultaneously crystallizes to mullite which stops further grain growth of alumina [10].

Aluminosilicates serve as the base material for Nextel family of fibers, as developed by 3M Corp. [11]. Several commercial grades are available, e.g., 312, 440, 480, 710, etc. The important aluminosilicate placed in these fibers is mullite. All of these fibers retain 75% of their original strength at a temperature of 1000°C.

2.5 Nonoxide Fibers

These fibers are essentially produced by the pyrolysis of organometallics such as polycarbosilene which gives rise to SiC as the predominant phase. The major producer of this fiber is Nippon Carbon Company in Japan, based on the pioneering research of Yajima and co-workers [12–14]. Basically, the structure of polycarbosilane consists of cycles of six atoms arranged in a similar way to the cubic structure of β-SiC. The fiber drawing of polycarbosilane is difficult and may contain full carbon and excess SiO_2. The as-supplied fibers can retain their original strength up to 1000°C, above which there is grain growth in SiC and the fibers degrade. Newer versions of this fiber are somewhat less susceptible to degradation.

A competitor to Nicalon fibers is the Tyranno fibers produced by Ube Chemicals in Japan [15]. As opposed to Nicalon fibers, Tyranno fibers have added Ti, which inhibits crystallization and further growth of grains.

Other approaches to production of such fibers have been reported by Dow Corning and Celanese. These fibers are prepared by melt-spinning amorphous thermoplastic polymers followed by a cross-linking step to gel the fibers. The Si-C-N-O fibers were derived from methylpolydisilylazane (MPDZ), The fibers of Si-N-C were derived from hydridopolysilazane [16,17].

Whiskers or platelets are single-crystal materials of very high strength in which grain growth is not a problem, so strength is retained to high temperatures. The most important whisker materials are also SiC based, produced by the VLS (vapor-liquid-solid) process [18]. This process utilizes a molten metal catalyst which helps in the heterogeneous nucleation of the solid phase from the vapor phase [19,20]. There have also been some recent developments in the production of oxide whiskers, both alumina [21] based and mullite based [22].

3 PHYSICAL PROPERTIES

Before getting into an elaborate discussion of the enhancement of the mechanical properties, it is worthwhile to briefly examine how the physical properties of the matrix phase change with the addition of a second phase. Frequently, the physical properties are additive and depend on the relative abundance of the constituents. An example of such a property is density. Equation (1) relates the density of the composite d_c, to the properties of the constituents:

$$d_c = d_m V_m + d_r V_r \tag{1}$$

where d_m and d_r are the densities of the matrix and the reinforcement and V_m and V_r their respective volume fractions. As mentioned later, this law of addition is also true for Young's modulus.

However, in the case of electrical conductivity, if the composite is fabricated from aligned fibers, the electrical resistivity along the direction of the fibers is

$$\frac{1}{\rho_c} = \frac{V_m}{\rho_m} + \frac{V_r}{\rho_r} \tag{2}$$

This means that the resistivity is dictated by the phase of lowest resistivity. If, however, the resistivity is measured perpendicular to the direction of the fibers, the composite resistivity is

$$\rho_c = \rho_m V_m + \rho_r V_r \tag{3}$$

where ρ is the resistivity and the subscripts c, m, r refer to the composite, matrix and reinforcements respectively. These rules are somewhat changed if particulates are used as the second phase and in high volume fraction whereby the particulate may link up in a continuous fashion, thus altering the properties of the composite altogether.

Thermal conductivity of the composite is yet another property that behaves in an opposite way to electrical resistivity [23]. The relations in this case are given by

$$K_{tc} = K_{tm} V_m + K_{tr} V_r \tag{4a}$$

$$\frac{1}{K_{tc}} = \frac{V_m}{K_{tm}} + \frac{V_r}{K_{tr}} \tag{4b}$$

where K_t is the thermal conductivity, and the subscripts c, m and r represent the composite, matrix and the reinforcement phases respectively. It may be noted here that some fibers, notably carbon fibers, have anisotropic thermal properties which will also be reflected in the property of the composite.

The thermal expansion coefficient along a direction parallel to continuous unixial fibers in a composite is

$$\alpha_c = \frac{\alpha_m V_m E_m + \alpha_r V_r E_r}{E_r V_r + E_m V_m} \tag{5}$$

where α is the thermal expansion coefficient of constituting phases and E is the respective Young's moduli. The thermal expansion coefficient perpendicular to the direction of the fibers is

$$\alpha_c = \alpha_m V_m + \alpha_r V_r \tag{6}$$

4 MECHANICAL PROPERTIES

As mentioned, the most important property that is substantially enhanced by the introduction of reinforcements is fracture toughness resulting in improved reliability. The process of enhancement in fracture toughness is referred to as toughening. This section is divided into two broader subsections. To begin with, micromechanical models of stress-strain behavior of composites will be discussed. This will be followed by various mechanisms of toughening, specifically, bridging and process mechanisms.

4.1 Micromechanical Stress-Strain Behavior

It is logical to begin this discussion with long unidirectional fibers incorporated into the matrix. At any composite strain ε_c prior to failure the stresses in the fiber, σ_f and σ_m, can be obtained from these curves or from

$$\sigma_r = E_r\varepsilon_c \tag{7a}$$

$$\sigma_m = E_m\varepsilon_m \tag{7b}$$

where E_r and E_m are Young's moduli for the reinforcements and the matrix. The composite stress is

$$\sigma_c = \sigma_r V_r + \sigma_m V_m \tag{8}$$

The axial Young's modulus of the composite is obtained by combing Eqs. (7) and (8).

When the composite is loaded to the fracture strains of the matrix, the stresses in the two phases are

$$\sigma_{mf} = E_m\varepsilon_{mf} \tag{9}$$

where the subscript f denotes fracture, and the composite stress at this point is

$$\sigma_c = \sigma_{mf}V_m + \sigma_r V_r \tag{10}$$

where σ_r is the stress in the fiber at $\varepsilon_r = \varepsilon_{mf}$, and noting that the axial Young's modulus is

$$E_c = E_m V_m + E_r V_r \tag{11}$$

At sufficiently low fiber fractions, the fibers cannot sustain the transfer of load and the composite failure occurs simultaneously. However, if a large volume fraction of fibers is incorporated the fibers can carry the load and the final composite failure is

$$\sigma_{cf} = \sigma_{rf}V_r \tag{12}$$

When the matrix fails, it shrinks from the fracture surface, thus relaxing the stresses near the fracture surface. However, the fracture stresses of σ_{mf} are attained at a distance away from the fracture surface, which is usually proportional to the strength of the matrix/fiber bond. This results in fracture at a regular spacing throughout the composite, a phenomenon known as multiple fracture. The approximate multiple crack spacing is

$$1 = \frac{\sigma_{mf} R V_m}{2\tau V_r} \tag{13}$$

where R is the radius of the fiber and τ is the interfacial stress, assumed to be constant.

Even though these equations predict initial multiple matrix cracking that is observed to occur prior to final failure in some composites, they do not consider the presence of fibers on the fracture strain of the matrix, the effect of matrix cracking on the fracture of fibers and the effect of scatter in fiber strength.

Aveston, Cooper, and Kelly realized that the presence of fibers inhibit matrix cracking [24]. The fibers limit the matrix strain which leads to reduction in fracture. The matrix failure strain ε_{mf} is related to

$$\varepsilon_{mf} = \left[\frac{6\Gamma E_r V_r^2}{E_m^2 E_c R V_m}\right]^{1/3} \tag{14}$$

where Γ is the specific fracture surface energy of the matrix. Budianisky et al. have extended this analysis further to incorporate shrinking of the matrix due to stress relaxation [25]. The foregoing models both predict that both the effective matrix strength and strain are enhanced in the composite beyond a certain critical reinforcement fraction. Finally, the reinforcements, especially the fibers, eventually fail in a brittle manner. Consequently, some sort of distribution of strength must be incorporated into the models. Further refinements are possible via the stress intensity fracture analyses.

4.2 Toughening and Related Mechanisms

The philosophy followed recently to toughen ceramics is to create microstructures such that the Griffith theory no longer dictates behavior, so flaws do not control the fracture process [26–28]. Instead, as the flaw propagates the microstructure resists propagation and toughness progressively increases to reach a plateau value. This situation is shown in Fig. 1 where the crack length is plotted against stress intensity factor (toughness); such diagrams are called R-curves (resistance curves). Also shown are the applied stress intensity curves. The condition for failure is reached when the two curves intersect tangentially. C^* and K_R^* are the critical flaw size and stress intensity factors respectively. This implies that a preexistent flaw can grow until it reaches C^*, upon which failure occurs.

Several important aspects of R-curve behavior should be considered. First, some processing related flaws can be present and yet the material does not fail. Second, the scatter in strength values is considerably reduced, making designing with these materials relatively simple since a single value of failure strength can be used by the designer. Further, R-curve behavior is not a material-related property but is dependent on geometry. Finally materials can be divided into two different classes—high toughness/low strength, and low toughness/high strength. The former type usually shows a gradually rising R-curve with failure being governed by the R-curve behavior. The latter type of material has a steeply rising R-curve with failure occurring in accordance with Griffith theory. The first class of materials is clearly more desirable than the second class, because in the former case R-curve behavior controls the fracture.

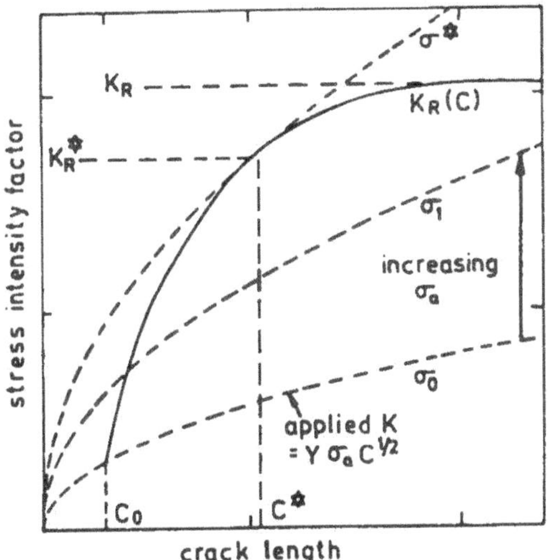

Figure 1 Typical resistance curve (R-curve) plot.

The crack tip behavior is substantially different in ceramics to that observed in metals. In the case of metals, the plastic zone ahead of the crack tip is important, whereas in ceramics it is the wake zone behind the crack tip that controls fracture behavior. The criticality of the wake zone was clearly demonstrated by Steinbrech et al. [29], who removed the wake zone by sawing off this part of the specimen, with a resultant reduction in toughness.

Key to increasing the toughness of ceramics materials is the incorporation of nonlinear phenomena into the matrix. These phenomena can be generically described in terms of the hysteresis behavior of the stress-strain relationship. As the area within the hysteresis loop increases, the toughness in enhanced. The specific mechanism occurring can be divided into two categories: process mechanism (Figs. 2a,b) and bridging mechanism (Fig. 2c).

4.3 Process Mechanism

The process mechanism occurs in ZrO_2-based materials and can be subdivided into phase transformation toughening, and microcracking toughening [30–33] (Figs. 2a,b respectively). Both mechanisms can occur both in monolithic zirconias and in ZrO_2-reinforced oxide and nonoxide composites. The process mechanism is essentially controlled by the onset of stress nonlinearity in the stress-strain curve. Specifically, this means the stress necessary to cause the phase transformation to occur or the stress needed to generate microcracks (discussed later).

Transformation Toughening

ZrO_2- and HfO_2-based materials are particularly amenable to a unique toughening mechanism known as phase transformation toughening, which may occur with or without concomitant microcracking [30–33] (Figs. 2a,b respectively). Therefore, these two processes will be discussed together. The transformation-toughening

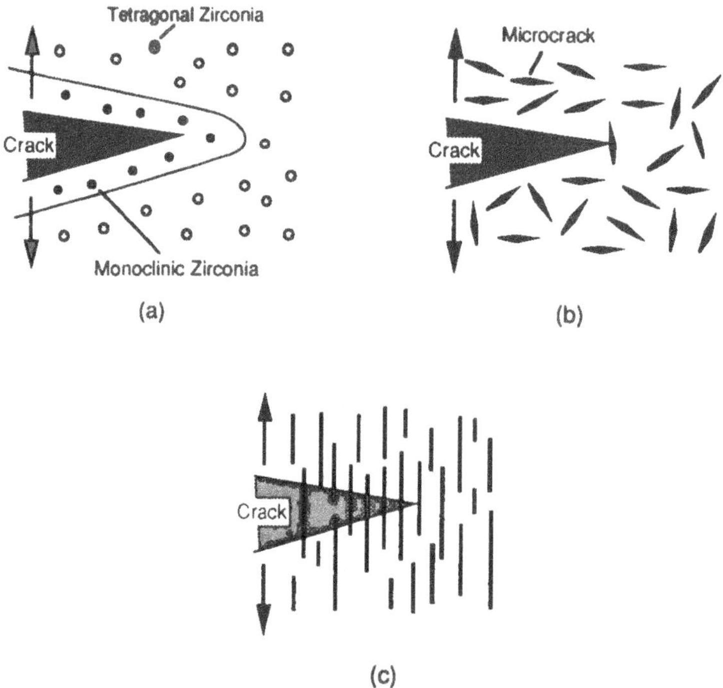

Figure 2 Microstructural features of ceramics toughened by process mechanism (a and b) and by bridging mechanism (c).

mechanism is related to the diffusionless transformation of metastable tetragonal ZrO_2 to a stabler monoclinic phase which has a higher specific volume. This increase in specific volume effectively reduces the extent of crack propagation. The mechanism requires a significant amount of the tetragonal phase to be present in the microstructure by suitable alloy additions, e.g., MgO, CaO, Y_2O_3, or CeO_2. When MgO or CaO are used as an alloying material, the microstructure contains large grains containing regions of the tetragonal phase and is called partially stabilized zirconia (PSZ). With Y_2O_3 and CeO_2 additions and using very fine powders tetragonal zirconia polycrystals (TZP) is produced. Examples of matrices that can be reinforced with ZrO_2 are Al_2O_3, Si_3N_4, TiB_2, mullite, and various glasses.

To describe the aforesaid behavior quantitatively, it is assumed that a steady transformation zone of half-width h develops around a growing crack. In the simplest case, it is assumed that V_p of the transformable particles is consistent over the whole zone and the boundary of the zone is dictated by the critical transformation strain ε_t. The increase in toughness (or reduction in stress intensity) is

$$\Delta K_c = \frac{AEV_p\varepsilon_t h^{1/2}}{1 - \nu} \tag{15}$$

where ΔK_c is the reduction in stress intensity factor, A is a constant, ε_t is the transformation strain. However, it is noted that the value of h is variable which starts with a small value but grows to a larger extent leading to R-curve behavior.

The stress needed for the phase change produces a nonlinearity in the stress-strain curve and results in a reduced stress intensity near the crack tip. If we distinguish between two different stress intensity factors K_I^a, the stress intensity far from the crack tip, and K_I^{local} in the vicinity of the crack tip, the difference between the two gives the increase in toughness:

$$\Delta K_I = K_I^a - K_I^{local} \tag{16}$$

The net result is that the crack tip is shielded from the applied stress intensity factor. A quantitative expression for the amount of shielding (i.e., increase in toughness) can be obtained by calculating the J integral in the vicinity of the crack tip. Figure 3a shows a frontal zone around the crack tip, and Fig. 3b a wake zone around the tip. It can be shown that as the crack begins to propagate, the increase in toughness is minimal but as the wake zone develops after some crack growth has taken place, there is an increase in toughness which gives rise to R-curve behavior [32].

Microcrack Toughening

This may occur either alone or in conjunction with a phase transformation. The stress-strain curve is similar to that for the transformation-toughened materials. The only difference is a slight reduction in Young's modulus due to microcracking. Consequently, compared to transformation toughening a somewhat greater degree of enhancement of toughness may be expected during crack initiation. During subsequent crack propagation the wake zone develops and R-curve behavior occurs.

This form of toughening has been analyzed in detail by Hutchinson [34]. To be effective the microcracking must occur only in response to the stress field around the crack tip in conjunction with residual stresses and be well dispersed so as to avoid their linkage. This form of toughening has two components, e.g., the formation of a zone of lower Young's modulus and dissipation of local energy at the crack tip. The second component is responsible for the departure from linearity in the stress-strain curve and a crack-tip-shielding concept, such as phase transformation toughening, can also be developed.

Hutchinson also considered the case where short fibers and whiskers are incorporated into the matrix. These cases also satisfy the criteria put forth earlier and enhance the overall toughness. The increase in toughness is

$$\Delta K_c = BE\delta H^{1/2} + Av \tag{17}$$

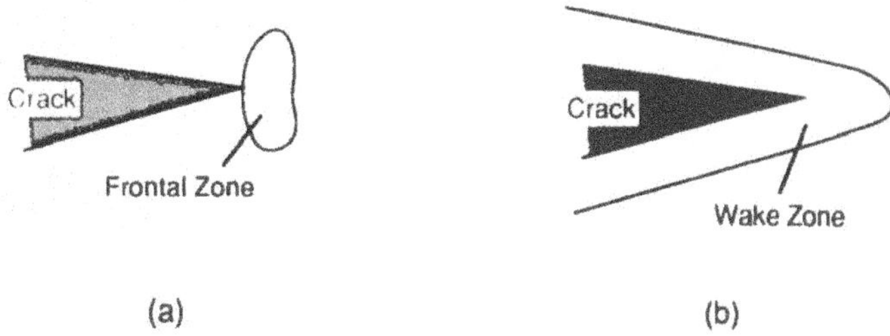

Figure 3 (a) Frontal zone and (b) wake zone about a sharp crack tip.

where B is a constant dependent on the zone shape, δ is the fractional dilation due to penny-shaped microcracks. A is a constant related to the modulus reduction, and v is an increase of microcrack density, which can be determined from microstructural characterization.

Besides phase transformation toughening and microcrack toughening, an important mechanism of fiber pull-out can also be classified as a processing mechanism since it is associated with the bridging mechanism.

4.4 Bridging Mechanism

Bridging mechanism involves reinforcements which form bridges behind the crack tip (Figs. 2c and 4). These ligaments can be grains in noncubic monolithic materials such as alumina [35], while in composites, they may be a ductile metal [36,37], strong whiskers [38,39], fibers, etc. [40,41]. The increase in toughness is governed by the amount of compressive stresses that the intact ligaments generate, thereby restraining crack opening.

The bridging reinforcements can be ductile or brittle. In the former case, the tougher and more ductile metal reinforcements deform plastically thus contributing toward the enhancement of toughness. The brittle reinforcements can be in the form of high-strength whiskers or long fibers, which remain intact restraining the crack opening.

Ductile reinforcement toughening can be further divided into three classes: isolated ductile reinforcements [42], interpenetrating networks [43], and a continuous ductile phase [44]. Use of metallic wire, such as Nb to toughen TiAl is an example of isolated ductile reinforcement [45]. Lanxide materials represent the second type of reinforcement [43], while cemented carbides are an example of the third class [44]. In the first two classes the plastic strain is limited by the elastic strain of the brittle matrix. The maximum utilization of the bridging mechanism is with a continuous ductile phase since in this case the ductile reinforcements can stretch (bridge) to a substantial extent behind the crack tip. If the bonding between the reinforcement and the matrix is strong, the peak stress in the reinforcement is dictated by necking. However, if there is any interfacial debonding, the constraint of the matrix may be reduced and the bridging zone can extend further. The tough-

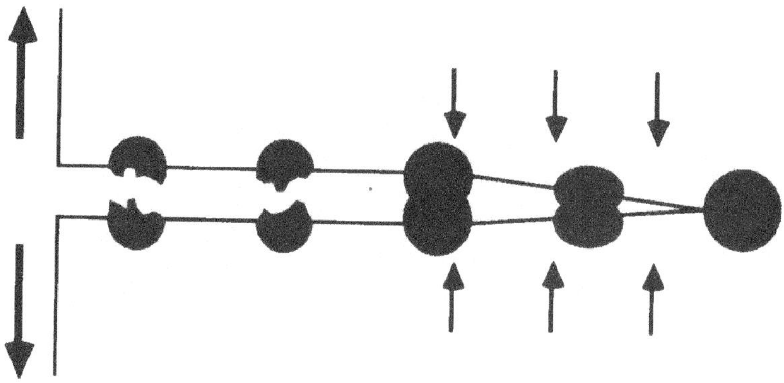

Figure 4 The bridging mechanism in a CMC reinforced with a ductile second phase.

ness can then be correlated with yield strength, the maximum stretch, and the radius of the second phase.

The mechanism of bridging with brittle second phases is somewhat different from the ductile reinforcement situation in that no necking of the bridging ligaments occurs. In the brittle case the interfacial bonding should be weak so that cracks are partially deflected around the reinforcement leaving behind intact ligaments (Fig. 4). As the crack extends, further debonding occurs and eventually the bridging ligaments fail. Alternatively, in a monolithic anisotropic material, thermal expansion mismatch in adjacent grains, (because different crystallographic directions are oriented in the same direction in the different grains), can suppress local crack propagation and intact ligaments may form.

Figure 5 shows a zone of bridging reinforcements adjacent to a crack tip. The length of the bridging zone is x_0. The crack opening displacement is u_0 at the end of the zone. Each reinforcement is characterized by a debond length. The stress on the reinforcement σ_f is maximum at the end of the zone and decreases linearly over the debond length to a value corresponding to the general. The toughening produced by such a bridging zone can be estimated either by a stress intensity route or by the strain energy release route.

The increase in toughness can be calculated either by the stress intensity factor approach or by the energy approach. It is expected that both the strategies should lead to equivalent results. However, values of stresses acting on the reinforcements as a function of the distance behind the crack tip must be known a priori. Results can differ depending on the relationship between this stress as a function of distance from the crack tip. For example, for short-fiber-reinforced ceramic matrix composites, Evans and McMeeking assumed that the debond length increased linearly with distance behind the tip [36]. On the other hand, Becher et al. [38] assumed that the debond length is constant. Both models satisfactorily explained the increase in toughness in the particular composites that the respective groups studied. Similar formalisms can easily be incorporated into ductile phase bridging as well.

The pull-out model is shown in Fig. 6. It is a simplified model containing a system of fully debonded fibers, aligned uniaxially and normal to the crack plane. They are assumed to have a uniform length, l_r, that is less than l_c, the critical length of fracture, so that all are pulled out of the matrix. At an intermediate point, a maximum stress σ_r' will be experienced by the fibers, which can be written as

$$\sigma_r' = \frac{2\tau}{R(l_r/2 - u)} \tag{18}$$

Incorporating this value into the energy criterion gives

$$\Delta G_c = \frac{\tau V_r l_r^2}{6R} \qquad (\text{where } l_r < l_c) \tag{19}$$

Since l_c is the longest length that can be pulled out there is a limit to the increase in toughness by this mechanism.

5 PROCESSING

The major fabrication technique used for ceramics is the powder route since ceramics are extremely difficult to deform. This requires powders as the starting

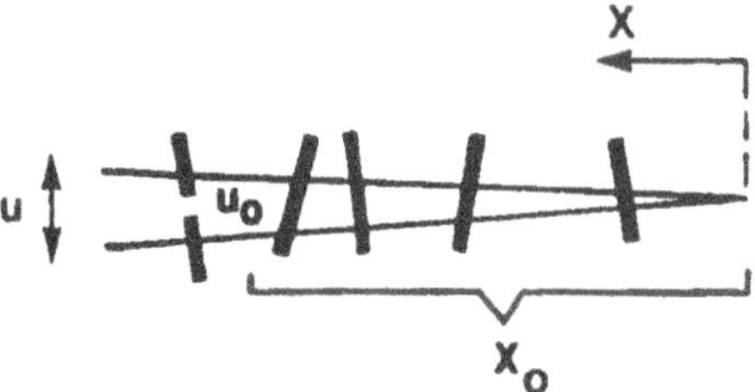

Figure 5 A schematic for modeling crack bridging mechanism.

material, followed by green forming of the powders by application of pressure, and finally exposure to temperatures at which diffusion of atoms can take place (sintering). Alternatively hot compaction in a shaped container effectively combines the cold press and sinter steps together.

5.1 Green Forming

The simplest green forming method is die pressing, where the advantages include simplicity, automation and low capital investment. The main disadvantages are an inhomogeneous green density due to the die to powder friction and limitations in the complexity of the shapes which can be produced. Both these drawbacks can be circumvented by cold isostatic pressing. A further alternative is to use traditional ceramic processes such as "slip casting." In this approach initially, a "slip" is formed by dispersing the ceramic in a liquid (e.g., water) and the mixture is then introduced into a gypsum mold, which absorbs the liquid, leaving the particles attached to the walls of the mold. This process is, therefore, particularly suitable for producing thin-walled hollow shapes [47]. Recently, it has been used to make monolithic shapes such as turbocharger rotors from Si_3N_4 [48].

Injection molding is an alternative technique being developed for production of complex shapes [49–51]. This method is used for making small parts at a high volume production rate and a high degree of tolerance. Like metal injection molding (MIM) the process is quite different from polymer injection molding in the sense that the input material contains mainly ceramic powder and a small but optimized amount of binder (in the polymer case the charge is almost entirely polymeric material). In this process the binder, either waxes or polymers, and fine

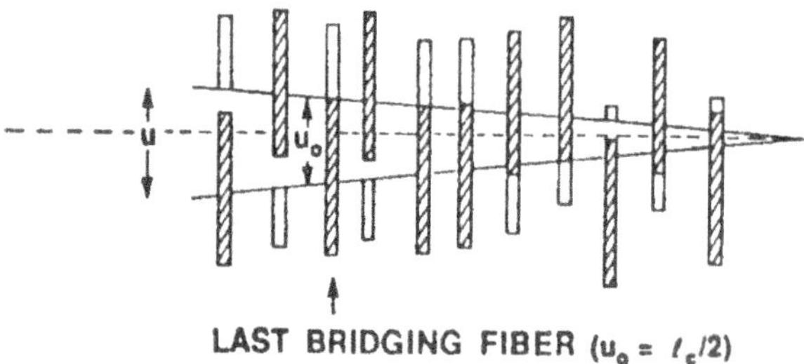

Figure 6 A schematic for modeling pull-out mechanism.

ceramic powders are mixed together, followed by injection (application of pressure) of this mixture into the mold. Subsequently the binder is removed (pyrolysis) and the product is subjected to a final sintering operation.

Green shapes can also be formed using the "sol-gel" technique. A "sol" is a dispersion of solid particles (the disperse phase,) in a liquid called (the dispersion medium) in which at least one dimension of the particles is between 1 nm and 1 μm. A sol converts to a gel when it loses the liquid and the viscosity rises. The process involves preparation of the sol and the sol-to-gel transformation. For production of monolithic ceramics, alcohol containing gels (Alcogel) are often used. These undergo hydrolysis by reaction with water, followed by condensation. The second reaction leads to rejection of alcohol (which must be carefully controlled) and formation of a rigid polymeric structure [52,53]. Optimally rigid monoliths are formed, however, if the process is carried out too fast, the structure can crumble to powder [54].

Green processing of composites is somewhat different from producing monolithic ceramics. If discontinuous reinforcements are to be used they must be carefully deagglomerated and redispersed into a slurry of matrix particles. This can be further facilitated by the application of ultrasonics or by controlling the pH of the slurry [55,56]. When continuous fibers are to be incorporated in the matrix, fabrication occurs by infiltrating the yarn or a woven cloth with a slurry of the powdered matrix. The resultant tape may then be stacked and sintered [57].

Following the green specimen preparation, samples can be subjected to sintering processes where densification occurs by interdiffusion of atoms. The process is similar to that which occurs in metals. Several processes of diffusion may occur during sintering, e.g., plastic flow, evaporation condensation, volume diffusion, surface diffusion, etc. [58].

5.2 Hot Forming

The process of sintering can be enhanced by the application of pressure either uniaxially or isostatically. If a shaped cavity is used, the green forming step can be essentially eliminated and a dense geometric shape formed in one step. The problem with uniaxial pressure is that the shape complexity of the compact is restricted. In a hot isostatic pressing (HIPing), pressure is applied isostatically using a hot gas allowing complex net shapes to be produced [59,60]. A new development in HIPing of ceramics is the commercial availability of HIP units with an oxygen environment [61]. This is particularly beneficial for materials like ZrO_2 and for high-temperature superconductors due to the improved control of metal-to-oxygen-atom ratios.

Superplastic forming is now firmly established as a viable method for net shape processing of ceramics [62–64]. Initial work involved yttria-doped tetragonal zirconia polycrystals (Y-TZP), and more recent studies have been extended to include MgO, Al_2O_3 + Y-TZP, Ce-TZP, Ce-TZP + Al_2O_3, $BaTiO_3$, hydroxyapatite, ZnS, Si_3N_4, etc. Up to 500% strain has been reported for Y-TZP [62] and dome shapes have been formed out of Si_3N_4 [64]. The primary requirements for such processing is fine grains, a high homologous temperature, and slow strain rates. The mechanisms of superplastic deformation in ceramics include phase transformations, solution reprecipitation, and other diffusion-controlled mechanisms [62].

5.3 Novel Techniques

A number of fabrication techniques do not fit the conventional categories of processing classifications described. While some of the techniques are specific to ceramic processing, others can also be used in producing net-shaped objects from metals. The techniques involved are glass ceramic processes, reaction processes, combustion synthesis processes, and slurry infiltration processes.

The glass ceramics process is unique to the processing of ceramics [65]. This method involves making a batch of glass and subsequently forming articles by the normal well-established techniques of glass forming. The formed articles are heat-treated to induce crystallization. The resulting material may therefore be defined as a glass-crystal composite. The success of the process depends on a good knowledge of crystallization T-T-T diagrams for the various systems. Systems that are of importance are SiO_2-LiO_2 [66], Li_2O-Al_2O_3-SiO_2 [67], MgO-Al_2O_3-SiO_2 [68].

Reaction-based processing is current being used extensively for producing ceramics and CMCs in situ. Normally the process involves multiphase reactions between gas-solid, gas-gas, and gas-liquid [69]. A prime example of reaction-based processing is the conventional reaction-bonded silicon nitride material (RBSN) [70]. A recent development in this area is the Dimox process producing Lanxide composites [71]. The process requires a molten alloy which is directionally oxidized outward from the metal surface. The product is an intimate mix of the metal and its oxide, e.g., Al-Al_2O_3. Other composites such as Ti-TiN, Zr-ZrN can also be formed. The reaction process is capable of producing net-shaped components.

In combustion synthesis (CS), or self-propagating high-temperature synthesis (SHS), an exothermic reaction is ignited in a mixture of powders. This exothermic process generates substantial heat, which can be utilized to produce refractory materials. If the heat produced is very high, the product is porous and can be crushed easily to produce powders [72,73]. If the desired product is a fully dense compact, the heat generated must be optimized at a lower level, and application of pressure is also required. Reinforcements can also be added in the initial mixture. Commercially one variation of this technique is known as the XD process [74,75]. The simplicity, small energy requirement, and ease in scaling up make CS very attractive.

In infiltration processes, a porous skeleton of the matrix material is infiltrated with the potential reinforcement producing an interpenetrating network of the matrix and the reinforcements. The porous skeleton is produced by partially sintering a compact to achieve sufficient strength for handling. The second phase can be a ceramic in which case a sol can be used as an infiltrant [76]. The second phase can also be a metal, in which case high temperatures are necessary so that the metal is molten [77]. A further requirement in this case is that the molten metal should wet the skeletal structure [78].

6 OVERALL CHARACTERISTICS

Table 5 summarizes the mechanical characteristics of various CMC materials (but excluding C-C composites). One example from each of the various types of CMCs is highlighted in Table 5. These vary from a material perspective such as oxide matrix/oxide particulate reinforcement [79]; oxide matrix/nonoxide whisker rein-

Table 5 Examples of Mechanical Properties of CMCs

Type	Example	Strength (MPa/ksi)	Toughness ($MPa\sqrt{m}/ksi\sqrt{in}$)
1. Matrix/reinforcement oxide/oxide (particulate) [79]	Al_2O_3/ZrO_2	600/87	10/9.17
2. Oxide/nonoxide (whiskers) [80]	Al_2O_3/SiC	600–800/ 87–116	9/8.25
3. Nonoxide/nonoxide (particulate) [81]	SiC/TiC	280–430/ 40.6–62.35	3.8–5.7/3.48–5.23
4. Glass ceramics [82,83]	Mg-Al-silicate/ Cordierite	250/36.25	2.25/2.06
5. Glass/fiber [84]	LAS/SiC	400/58	9.0/8.25
6. Oxide/metal [3]	Al_2O_3/Al	400/58	9.5/71

forcement [80]; nonoxide matrix/nonoxide particulate reinforcement [81], glass matrix/crystallite reinforcement [82,83], glass matrix/fiber reinforcement [84] and oxide/metal interpenetrating composites [3]. In all of these cases, the enhancement in mechanical properties is significant. In general, in most of the cases, the toughness values are enhanced approximately twice that of the original matrix material, thus making ceramic materials more reliable.

7 APPLICATIONS

Ceramics such as aluminum oxide, silicon nitride, and silicon carbide are in commercial use for wears parts, cutting tool inserts, bearings, and coatings [1]. However, ceramics only occupy about 5% of the market share in these applications; and substantial growth is projected, with the U.S. market by itself reaching as much as $2 billion per year by the turn of the century. Ceramics are also in semicommercial production in Japan in certain engine components such as turbochargers, glow plugs, rocker arms, precombustion chambers, and a limited number of consumer products.

Table 6 shows the current applications of advanced engineering ceramics in a variety of industries. Different characteristics of ceramic materials make them at-

Table 6 Typical Current Uses of Advanced Ceramics

Industry	Uses
Chemical	Seals, valves, pump impellers, heat exchangers
Manufacturing	Wire-drawing die, cutting tools, bearing
Aerospace	Rocket nozzle, aircraft journal, bearing
Automotive	Turbocharger rotor, pushrod tips, rocker arm, swirl chamber, piston liner
Defense	Gun liners, ceramic armor

tractive for different applications. For example, in chemical industries the chemical inertness of these materials is very important. For heat exchanges, the thermal conductivity should be high. In machining applications tribological properties are very significant, and generally as hardness increases tribological properties are enhanced. In aerospace applications, the materials must be light and exhibit inertness and good elevated temperature behavior. In the automotive industry, the ability to run an engine hotter is very critical and high-temperature properties are again important. A further interesting application is in gun liners and armor, where the most significant characteristic is very high hardness.

The use of ceramics in advanced engines could revolutionalize performance, with increased fuel efficiency due to increased operating temperatures, more compact design, and reduction or elimination of the cooling system [1]. However, a solution to the "forgiveness" and cost must be found for full-scale application. For the short term increased use in bearings, bioceramics (ceramics inside the body), construction applications, heat exchanges, electrochemical devices and military hardware such as radomes, laser mirrors, and rail gun components will be realized.

8 CARBON-CARBON (C-C) COMPOSITES

Just like the CMCs, the carbon-carbon (C-C) composites have potential uses at high temperatures (2000°C and beyond) [85,86]. The unique property of carbon-carbon composites include excellent strength retention at high temperatures, high stiffness, thermal expansion coefficient match, and very low density (1.5 to 2 g/cc). As the name implies, both the matrix as well as the reinforcement are carbon. For continuously reinforced C-C composites, the mechanical strength is provided by the fibers and the carbon matrix provides the high-temperature properties.

The carbon fiber reinforcements in the C-C composites play a significant role in the development of composites. There are three types of commercially available precursors, namely rayon, polyacrylonitrile (PAN) and pitch [87]. Rayon-based fibers were very popular in the 1970s. However, the carbon yield is low. Conversion from rayon to graphite takes place in several stages at up to 2000°C. PAN-based fibers are more important these days [87]. Several steps are needed to produce the fiber with the desired properties. These include drawing and spinning of polymeric fibers, stabilization in air between 200 to 300°C, followed by graphitization at a temperature up to 3000°C. Pitch-based fibers are produced by heat-treating a petroleum or coal tar pitch feed stock to produce a liquid crystal precursor, termed "mesophase" [88]. This precursor fiber is converted into a carbon fiber in a process analogous to the PAN-derived fibers. The advantages of pitch-derived fibers is the high yield, high modulus/density ratio, and high thermal conductivity, etc.

Just like the various geometrical forms of reinforcements available in the case of CMCs, the carbon fiber reinforcements could also be classified into two major classifications: discontinuous and continuous [85,86]. The processing of the composites depends on the geometrical classification [85,86]. In the case of discontinuously reinforced C-C composites, the most common technique uses a carbonized rayon felt substrate with a pyrolytic carbon matrix and short chopped fibers in a pitch-based matrix. Other processes include pulp molding, isotropic casting, spray lay-up, etc. Continuous fiber C-C composites are processed by forming a preform of a woven or braided fabric to build a structural shape. This preform is then impregnated by carbon by means of several different processes. These include conversion of a liquid resin or liquid pitch precursor, chemical vapor infiltration or combination of these to obtain the desired properties.

The major drawback of C-C composites is the lack of oxidation resistance [89]. Carbon's oxidation resistance decreases dramatically above 600°C. Therefore, to retain high-temperature properties of these composites, proper coatings should be developed. For successful coating development three criteria must be satisfied: the coating should provide an effective barrier to avoid diffusion of oxygen; it should have low volatility to prevent excessive erosion in high-velocity gas streams; and it should be adherent to the substrate. Also, the internal layers must prevent outward carbon diffusion to prevent carbothermic reduction of oxides in the external layers. SiC and Si_3N_4 provide effecting coatings at temperatures lesser than 1800°C. Beyond a temperature of 1800°C, the reaction products are gaseous and tend to disrupt the coatings. For exposures greater than 1800°C, borides (such as HfB_2, HfC) are useful since they resist oxidation at very high temperature. However, for prolonged exposures, oxides such as ZrO_2, HfO_2, Y_2O_3, etc., are useful. For the best possible protection at very high temperature, a graded coating may be desirable. The coating is constituted of such materials which satisfy the aforementioned strategies. A typical example of such a graded coating is a refractory oxide/SiO_2 glass/refectory oxide/refractory carbide coating.

C-C composites are utilized in challenging applications such as rocket nozzles, nose cones for reentry vehicles, heat shields, brakes for aircrafts, etc. [86]. While in most of the cases structural and thermal properties are crucial, in some cases, such as in aircraft brakes and in furnace insulations, the requirements are not stringent [86].

In future, C-C composites will find use in gas turbines, in hypersonic vehicles, in nuclear power plants as neutron-absorbing material, as crucibles for handing molten metals, in high-temperature bearings, etc. [86].

9 CURRENT STATUS AND FUTURE TRENDS

In the present paper the attributes of monolithic ceramics and CMCs have been presented. The strength, particularly at elevated temperature, and inertness make both classes of materials very attractive for an array of present and potential applications. However, it is also clear that even utilizing the toughening mechanism discussed that "forgiveness" is still much lower than desired and is preventing use in many demanding applications. Fatigue (cyclic) properties of these materials remain abysmal even for materials with R-curve behavior. Thus, this area requires substantial enhancement. Similarly improvements are also required in tribological properties; even though the high hardness is very desirable from a tribological point of view, further improvements in wear and erosion characteristics are required for more widespread use. A further factor preventing more widespread use of ceramic-based materials is the high cost compared to traditional ceramics and traditional metals. Here it appears that a reduction in cost will take place as "novel processing" routes are introduced. For example, it is well established that the polymer pyrolysis route yields extremely fine nonoxides. Therefore, if ways can be found to cheaply produce the starting raw materials (polymers in this case), cost will be reduced. For expanded commercialization there must be effective communication between research groups, manufacturing organizations and end users. Further, if full advantage is to be taken of these costly materials, both acquisition cost and cost of ownership must be considered in specifying a material for a particular use.

REFERENCES

1. *Sci. Am. 255*, no. 4 (Oct. 1986).
2. T-I Mah, M. G. Mendiratta, A. P. Katz and K. S. Mazdiyasni, *Bull. Am. Ceram. Soc. 66*, 305 (1987).
3. M. K. Aghajanian, N. H. McMillian, C. R. Kennedy, S. J. Luszcz and R. Ray, *J. Mater. Sci. 24*, 658 (1989).
4. R. Stevens, *Zirconia and Zirconia Ceramics*, Magnesium Elektron Publ. 113, Twickenham, UK, 1986.
5. F. K. Ko, *Bull. Am. Ceram. Soc. 68*, 401 (1989).
6. T. W. Chou and F. K. Ko, *Textile Structural Composites*, Elsevier, Essex, UK, 1988.
7. A. K. Dhingra, *Philos. Trans. R. Soc. London A294*, 411 (1980).
8. J. C. Ronnie, *Ceram. Eng. Sc. Proc. 8*, 755 (1987).
9. Y. Abe, S. Horikiri, R. Fujimura and E. Ichiki, in *Progress in Science and Engineering Composites* ICCM-V (T. Hayashi, ed.), Tokyo, 1982, p. 1427.
10. A. R. Bumsell, *Mat. Forum, 11*, 78 (1988).
11. NEXTEL (3M, Minneapolis-St. Paul, MN) product literature.
12. S. Yajima, T. Shishilo and K. Okamura, *J. Am. Ceram. Soc. 59*, 324 (1976).
13. S. Yajima, *Bull. Am. Ceram. Soc. 62*; 893 (1983).
14. S. Yajima, T. Iwai, T. Yamanura, K. Okamura and Y. Hasegawa, *J. Mater. Sci. 16*, 1349 (9181).
15. Ube Chemicals (Japan) product literature.
16. W. H. Atwell, W. E. Hauth, R. E. Jones and N. R. Langley, AFWAL TR-4061 (1984).
17. G. Legrow, T. F. Lin, J. Lipowitz and R. S. Roch, *Bull. Am. Ceram. Soc. 66*, 363 (1987).
18. J. J. Petrovic, J. V. Milewski, D. L. Rohr, and F. D. Gac, J. Mater. Sci., *20*, (1985) 1167.
19. T. Kida and M. Yamamoto, U.S. Patent 4,690,811 (1987).
20. P. D. Shalek, U.S. Patent 4,702,901 (1987).
21. S. Jagota and R. Raj, U.S. Patent 4,806,198 (1989).
22. M. G. M. U. Ismail, H. Arai, Z. Nakai and T. Akiba, *J. Am. Ceram. Soc. 73*, 2736 (1990).
23. D. P. H. Hasselman, in *Tailoring Multiphase and Composite Ceramics* (R. E. Tressler, ed.), Plenum Press, New York, 1986, p. 731.
24. J. Aveston, G. Cooper, and A. Kelly, in *Properties of Composites*, IPC Science and Technology Press, U.K., 1971, p. 15.
25. B. Budianisky, J. W. Hutchinson, and A. G. Evans, *J. Mech. Phys. Solids, 34*, 167 (1986).
26. D. B. Marshall and A. G. Evans, *Acta. Met. 37*, 2567 (1989).
27. A. G. Evans, *J. Am. Ceram. Soc. 73*, 187 (1990).
28. Y-W Mai and B. R. Lawn, *Ann. Rev. Mater. 16*, 415 (1986).
29. R. Steinbrech, R. Knehans and W. Schaarwachter, *J. Mater. Sci. 18*, 265 (1983).
30. R. C. Garvie, R. H. J. Hannink and R. T. Pascoe, *Nature*, 258, 703 (1975).
31. A. G. Evans and R. M. Cannon, *Acta Met. 34*, 761 (1986).
32. A. H. Heuer, *J. Am. Ceram. Soc. 70*, 689 (1987).
33. D. J. Green, R. H. J. Hannink and M. V. Swain, *Transformation Toughening of Ceramics*, CRC Press, FL, 1989.

34. J. W. Hutchinson, *Acta Met. 35*, 1605 (1987).
35. P. L. Swanson, C. J. Fairbanks, B. R. Lawn, Y-W Mai and B. J. Hockey, *J. Am. Ceram. Soc. 70*, 279 (1987).
36. A. G. Evans and R. M. McMeeking, *Acta Met. 34*, 2435 (1986).
37. B. D. Flinn, M. Ruhle and A. G. Evans, *Acta Met. 37*, 3001 (1989).
38. P. F. Becher, C-H Hsueh, P. Angelini and T. N. Tiegs, *J. Am. Ceram. Soc. 71*, 1050 (1988).
39. P. F. Becher, *J. Am. Ceram. Soc. 74*, 255 (1991).
40. K. M. Prewo and J. J. Brennan, *J. Mater. Sci. 15*, 463 (1980).
41. K. M. Prewo and J. J. Brennan, *J. Mater. Sci. 21*, 3590 (1986).
42. V. Krstic, P. S. Nicholson and R. G. Hoagland, *J. Am. Ceram. Soc. 64*, 499 (1981).
43. H. C. Cao, B. J. Dalgleish, H. E. Deve, C. Elliot, A. G. Evans, R. Mehrabian and G. R. Odette, *Acta Met. 37*, 2969 (1989).
44. L. S. Sigl and H. F. Fischmeister, *Acta Met. 36*, 887 (1988).
45. C. K. Elliot, G. R. Odette, G. E. Lucus and J. W. Sheckhard, *MRS 120*, (1988) 95.
46. R. J. Gray, *J. Mater. Sci. 19*, 861 (1984).
47. R. E. Cowan, Treatise Mater. Sci. Tech. 9, 154 (1976).
48. R. N. Katz, *Treatise Mate. Sci. Tech. 29*, 1 (1989).
49. M. J. Edirisinghe and J. R. G. Evans, *Int. J. High Tech. Ceram. 2*, 1 (1986).
50. B. C. Mutsuddy and D. K. Shetty, in *Engineering Applications of Ceramic Materials* (M. M. Schwartz, ed.), ASM, Ohio, 1985, p. 89.
51. R. M. German, *Powder Injection Molding*, Metal Powder Industry Federation, NJ, 1990.
52. B. Yoldas, *J. Am. Ceram. Soc, 65*, 387 (1982).
53. C. J. Brinker and G. W. Scherer, *Sol-gel Science*, Academic Press, 1990.
54. K. S. Mazdiyasni, C. T. Lynch, and J. S. Smith, *J. Am. Ceram. Soc. 49*, 286 (1966).
55. T. N. Tiegs and P. F. Becher, *Bull. Am. Ceram. Soc. 66*, 339 (1988).
56. M. D. Sacks, H-W. Lee, and O. Rojas, *J. Am. Ceram. Soc. 71*, 370 (1987).
57. K. M. Prewo, J. J. Brenan and G. K. Layden, *Bull. Am. Ceram. Soc. 65*, 305 (1987).
58. R. M. German, *Powder Metallurgy Science*, MPIF, NJ, 1984.
59. A. S. Helle, K. E. Easterling and M. F. Ashby, *Acta Met. 33*, 2163 (1985).
60. K. Uematsu, K. Itakura, N. Uchida, K. Saito, A. Miyamoto, and T. Miyashita, *J. Am. Ceram. Soc. 73*, 74 (1990).
61. K. Nakajima and T. Masaki, pp. 625–633 in *Advances in Ceramics*, vol. 24 (S. Somya, N. Yamamoto and N. Yanagida, eds.), The American Ceramic Society, 1986.
62. T. G. Langdon, *J. Metals, 42*, 8–13 (1990).
63. Y. Maehara and T. G. Langdon, *J. Mater. Sci. 25*, 2275 (1990).
64. I-Wei Chen and L. A. Xue, *J. Am. Ceram. Soc. 73*, 2585 (1990).
65. P. W. McMillian, *Glass Ceramics*, Academic Press, London, 1964.
66. R. H. Doremus, *Glass Science*, Wiley, New York, 1973.
67. J. F. Kay and R. H. Doremus, *J. Am. Ceram. Soc. 57*, 480 (1974).
68. T. I. Barry, J. M Cox and R. Morrell, *J. Mater. Sci. 13*, 594 (1978).
69. M. E. Washburn and W. S. Coblenz, *Bull. Am. Ceram. Soc. 67*, 356 (1988).
70. A. J. Moulson, *J. Mater. Sci. 14*, 1017 (1979).
71. M. S. Newkirk, A. W. Urquhart, H. R. Zwicker and E. Breval, *J. Mater. Res. 1*, 81 (1986).
72. Z. A. Munir, *Bull. Am. Ceram. Soc. 67*, 342 (1988).
73. H. C. Yi and J. J. Moore, *J. Mater. Sci. 25*, 1169 (1990).
74. J. M. Brupbacher, L. Christodoulu, and D. C. Nagle, U.S. Patent 4,710,348 (1987).
75. D. C. Nagle, J. M. Brupbacher, and L. Christodoulu, U.S. Patent 4,774,052 (1988).
76. B. R. Marple and D. J. Green, *J. Am. Ceram. Soc. 71*, C471 (1988).
77. C. Toy and W. D. Scott, *J. Am. Ceram. Soc. 73*, 97 (1990).
78. D. C. Halverson, A. J. Pysik and I. A. Aksay, *Ceram. Eng. Sci. Proc. 6*, 736 (1985).

79. N. Claussen, J. Steeb and R. F. Pabst, *Bull. Am. Ceram. Soc. 56*, 559 (1976).
80. P. F. Becher and G. C. Wei, *Bull. Am. Ceram. Soc. 64*, 298 (1985).
81. M. A. Janney, *Bull. Am. Ceram. Soc. 65*, 357 (1986).
82. R. Morena, K. Niihara and D. P. H. Hasselman, *J. Am. Ceram. Soc. 64*, 298 (1985).
83. S. Baskaran, S. B. Bhaduri and D. P. H. Hasselman, *J. Am. Ceram. Soc. 68*, 112 (1985).
84. K. M. Prewo and J. J. Brennan, *J. Mater. Sci. 17*, 1201 (1982).
85. J. D. Buckley, *Bull. Am. Ceram. Soc. 67*, 364 (1988).
86. *High Performance Synthetic Fibers for Composites*, NMAB-458, The National Research Council, 1992.
87. P. Bracke, H. Schurmans, and J. Verhoest, *Inorganic Fibers and Composite Materials*, Pergamon Press, 1984.
88. B. Rand, *Handbook of Composites*, (W. Watt and B. B. Peror, eds.), Elsevier, North-Holland, 1985, pp. 496–595.
89. J. R. Strife and J. E. Sheehan, *Bull. Am. Ceram. Soc. 67*, 369 (1988).

15
Environmental Degradation of Engineering Materials

Philip A. Schweitzer
Consultant
Fallston, Maryland

1 INTRODUCTION

During the preliminary design of any structure or facility the potential problem of corrosion should be considered for two important reasons. The first is economics and the second is conservation.

Economic losses, both direct and indirect, result from the weakening or sudden failure of piping, tanks, structures, metal components of machines, ship's hulls, marine structures, bridges, towers, etc. As can be seen all types of construction and/or manufacturing facilities have the potential of being affected. These losses sustained by industry, by governmental agencies, and by the military amount to billions of dollars annually.

Direct losses are those costs (both labor and material) associated with the replacement of corroded structures and machinery or their components, such as condenser tubes, pipelines, metal roofing, and similar items. In addition to these costs are those associated with routine preventive maintenance, such as repainting of structures to prevent corrosion, inspections, and the upkeep of cathodically protected structures and pipelines.

Indirect losses include items such as shutdown, loss of product, loss of efficiency, and contamination of product. While the actual replacement cost of an item may be minimal, the loss of production resulting from the need to shut down an operation to permit the replacement may amount to hundreds of dollars per hour. When a pipeline or tank corrodes and a leak develops, product is lost. Since the leak may go undetected for time; the value of the lost product could be substantial. In addition, contamination could result from the leaking material, which requires cleanup. These cleanup costs could be considerable. When corrosion takes place, corrosion products build up, which, in the case of heat exchangers, reduce the efficiency of heat transfer. In pipelines, reduced flows result. Both situations increase the operating costs. Corrosion products from piping or tankage may be detrimental to the quality of the product being handled. This can result in discarding valuable products.

Premature failures due to corrosion can also be responsible for human injury such as in the failure of bridges or structures. There have been instances where such failures have resulted in the loss of life.

When all of these factors are considered, it is obvious why the potential problem of corrosion should be considered during the design state. In order to evaluate the potential corrosion problem some understanding of corrosion mechanisms is required.

One other factor must also be taken into account. With the many technological advances in the development of new alloys and polymeric materials, there is a wide choice in the selection of a material of construction for specific applications. It is only natural that there are vast differences between the initial costs of the various materials. However, consideration must be given to the fact that the initial cost of the material itself is seldom the true cost of process equipment. The material with the least expensive initial cost may not be the material with the true least cost.

The true cost of process equipment consists not only of the initial material cost but items such as labor of fabrication and installation, annual maintenance costs, anticipated life, salvage value, and cost of corrosion control, if any. Usually the initial material cost is insignificant in terms of the overall total cost; therefore it is prudent to specify the material that will do the job. A complete discussion of methods of calculating the true cost is covered by NACE International of Houston, Texas, in their NACE Standard RP-0272, *Direct Calculations of Economic Appraisals of Corrosion Control Methods.*

2 METALLIC CORROSION

Most of the commonly used metals are unstable in the atmosphere. These unstable metals are produced by reducing ores artificially; therefore they tend to return to their original state or to similar metallic compounds when exposed to the atmosphere. Exceptions to this are gold and platinum, which are already in their metal state.

Metals and their alloys will tend to unite with chemical corrodents to form stable compounds similar to those found in nature. When this occurs the compound formed is referred to as the corrosion product and the metal surface is said to be corroded. In some instances the corrosion product will actually form a protective film on the surface of the metal. This film prevents further corrosion as long as it remains intact.

All structural metals corrode to some extent in natural environments. Bronzes, brasses, stainless steels, zinc, copper, and aluminum corrode so slowly under the service conditions in which they are normally placed that they are expected to survive for long periods of time without protection. However, when these same metals are placed into contact with more aggressive corrodents they may suffer chemical attack and are degraded. Corrosion of structural grades of iron and steel, however, proceeds very rapidly unless the metal is protected.

2.1 Forms of Corrosion

The several forms of corrosion to which a metal may be subjected are

1. Electrochemical corrosion
2. Uniform corrosion
3. Intergranular corrosion
4. Galvanic corrosion
5. Crevice corrosion
6. Pitting
7. Erosion corrosion
8. Stress corrosion cracking (SCC)
9. Biological corrosion
10. Dezincification (dealloying)
11. Concentration cell
12. Embrittlement
13. Filiform
14. Corrosion fatigue
15. Fretting
16. Graphitization

All of these forms of corrosion are not present in all applications, but it is possible to have more than one form present. Understanding when each of these forms of corrosion could be potentially present will permit the designer to take steps to eliminate the condition or to limit the corrosion to acceptable limits.

Electrochemical Corrosion

Corrosion of metals is caused by the flow of energy (electricity). This flow may be from one metal to another, or from one part of the surface of one metal to another part of the surface of the same metal, or from one metal to a recipient of some kind. This flow of electricity can take place in the atmosphere, underwater, or underground as long as a moist conductor or electrolyte, such as water, especially saltwater, is present.

The differences in potential that causes the electric currents is mainly due to contact between dissimilar metallic conductors, or differences in concentration of the solution, generally related to dissolved oxygen in natural waters. Any lack of homogeneity on the metal surface may initiate attack by causing a difference in potentials that results in localized corrosion.

The flow of electricity (energy) may also be from a metal to a metal recipient of some kind, such as soil. Soils frequently contain dispersed metallic particles or bacterial pockets that provide a natural electrical pathway with buried metal. The electrical path will be from the metal to the soil, with corrosion resulting.

The presence of water is a key factor for corrosion to take place. For example, in dry air, such as a desert location, the corrosion of steel does not take place, and when the relative humidity of air is below 30% at normal or lower temperatures, corrosion is negligible.

Uniform Corrosion

Metals resist corrosion by forming a passive film on the surface. This film is formed naturally when the metal is exposed to air for a period of time. It can also be formed more quickly by chemical treatment. For example, nitric acid, if applied to an austenitic stainless steel, will form this protective film. These films are actually a form of corrosion, but once formed they prevent further degradation of the metal as long as the film remains intact. They do not provide an overall resistance to corrosion since they may be subject to chemical attack. The immunity of the film to attack is a function of film composition, temperature, and aggressiveness of the chemical. Examples of such films are patina formed on copper, rusting of iron, tarnishing of silver, "fogging" of nickel, and high-temperature oxidation of metals.

When exposed to a corrosive medium, metals tend to enter a chemical union with the elements of the corrosive medium, forming stable compounds similar to those found in nature. When metal loss occurs in this manner, the compound formed is referred to as the corrosion product and the metal surface is referred to as being corroded. An example of such an attack is that of halogens, particularly chlorides. They will react with and penetrate the film on stainless steel, resulting in general corrosion. Corrosion tables are developed to indicate the interaction between a chemical and a metal. This type of chemical attack is termed uniform corrosion.

Uniform corrosion of a metal in contact with a specific corrodent is measured in metal loss in mils per year (mpy) or inches per year (ipy) (1 mil = 0.001 in.). Losses are generally classified into three groups as follows:

1. <5 mpy (<0.005 ipy). Metals in this category are considered as having excellent resistance against chemical attack.
2. 5–50 mpy. Metals in this group are considered satisfactory, providing a high corrosion rate can be tolerated.
3. >50 mpy. Usually not a satisfactory choice. In most corrosion tables, rates are given as <5 mpy, <20 mpy, <50 mpy, or >50 mpy. A metal with a rating of <5 mpy is an excellent choice for use in that specific media. Metals with ratings of <20 mpy, while not as resistant as those with a lower rating, may still be used. Usually any metal with a loss greater than 20 mpy is not recommended for the application.

Tabulated corrosion rates also assist in determining how much corrosion allowance should be included in the design, based on expected life of the equipment.

Intergranular Corrosion

Intergranular corrosion is a localized form of attack taking place at the grain boundaries of a metal with little or no attack on the grains themselves. This results in loss of strength and ductility. The attack is often rapid, penetrating deeply into the metal and causing failure.

In the case of austenitic stainless steels the attack is the result of carbide precipitation during welding operations. Carbide precipitation can be prevented by using alloys containing less than 0.03% carbon, or by using alloys that have been stabilized with columbium or titanium, or by specifying solution heat treatment followed by a rapid quench, a process which will keep the carbides in solution. The most practical approach is to use either a low carbon content or a stabilized austenitic stainless steel.

Nickel-base alloys can also be subjected to carbide precipitation and precipitation of intermetallic phases when exposed to temperatures lower than their annealing temperatures. As with the austenitic stainless steels low-carbon-content alloys are recommended to delay precipitation of carbides. In some alloys, such as alloy 625, niobium, tantalum, or titanium is added to stabilize the alloy against precipitation of chromium or molybdenum carbides. These elements combine with carbon instead of chromium or molybdenum.

Galvanic Corrosion

Galvanic corrosion is sometimes referred to as dissimilar metal corrosion and is found in the most unusual places, often causing the most painful professional headaches. The Galvanic Series of Metals provides details of how the galvanic current will flow between two metals and which metal will corrode when they are in contact or near each other and in the presence of an electrolyte. Table 1 lists the Galvanic Series.

Note that several materials are shown in two places in the Galvanic Series, indicated as either active or passive. This is a result of the tendency of some metals and alloys to form surface films, especially in oxidizing environments. These films shift the measured potential in the noble direction. In this state the material is said to be passive.

The way in which a material reacts can be predicted from the relative position of the material in the Galvanic Series. When it is necessary to use dissimilar metals, two materials should be selected which are relatively close in the Galvanic Series: the further apart they are, the greater the rate of corrosion.

The rate of corrosion is also affected by the relative areas between the anode and cathode. Since the flow of current is from the anode to the cathode, the com-

Table 1 Galvanic Series of Metals and Alloys

Corroded end (anodic)	Muntz metal
Magnesium	Naval bronze
Magnesium alloys	Nickel (active)
Zinc	Inconel (active)
Galvanized steel	Hastelloy C (active)
Aluminum 6053	Yellow brass
Aluminum 3003	Admiralty brass
Aluminum 2024	Admiralty bronze
Aluminum	Red brass
Alclad	Copper
Cadmium	Silicon bronze
Mild steel	70-30 cupro-nickel
Wrought iron	Nickel (passive)
Cast iron	Inconel (passive)
Ni-resist	Monel
13% chromium stainless steel (active)	18-8 Stainless steel type 304 (passive)
50-50 lead-tin solder	18-8-3 Stainless steel type 316 (passive)
Ferritic stainless steels 400 series	Silver
18-8 Stainless steel type 304 (active)	Graphite
18-8-3 Stainless steel type 316 (active)	Gold
Lead	Platinum
Tin	Protected end (cathode)

bination of a large cathode area and a small anode area is undesirable. Corrosion of the anode can be 100–1000 times greater than if the two areas were equal. Ideally the anode area should be larger than the cathode area.

The passivity of stainless steel is the result of the presence of a corrosion-resistant oxide film on the surface. In most natural environments, they will remain in the passive state and tend to be cathodic to ordinary iron or steel. When chloride concentrations are high, such as in seawater or in reducing solutions, a change to the active state usually takes place. Oxygen starvation also causes a change to the active state. This occurs where there is no free access to oxygen, such as in crevices, and beneath contamination on partially fouled surfaces.

Differences in soil conditions, such as moisture content and resistivity, can be responsible for creating anodic and cathodic areas. Where there is a difference in concentration of oxygen in the water or in moist soils in contact with metal at different areas, cathodes will develop at relatively high oxygen concentrations, and anodes at points of low concentrations. Strained portions of metals tend to be anodic, unstrained portions cathodic.

When joining two dissimilar metals, galvanic corrosion can be prevented by insulating the two metals from each other. For example, when bolting flanges of dissimilar metal together, plastic washers can be used to separate the two metals.

Crevice Corrosion

Crevice corrosion is a localized type of corrosion occurring within or adjacent to narrow gaps or openings formed by metal to metal or by metal to nonmetal contact. When this occurs, attack is usually more severe than on surrounding areas of the same surface. This form of corrosion can develop as the result of a deficiency of oxygen in the crevice, acidity changes in the crevice, buildup of ions in the crevice, or depletion of an inhibitor.

Prevention can be accomplished by proper design and operating procedures. Nonabsorbent gasketing material should be used at flanged joints, while fully penetrated butt-welded joints are preferable to threaded joints. In the design of tankage, butt-welded joints are preferable to lap joints. If lap joints are used, the laps should be filled with fillet welding or a suitable caulking compound designed to prevent crevice corrosion.

Pitting

Pitting corrosion is also a form of localized corrosion, with the rate of attack being greater in some areas than in others. The main factor that causes and accelerates pitting is electrical contact between dissimilar metals or between areas of the same metal where oxygen or conductive salt concentration in water differ.

Pitting develops when the anodic (corroding) area is small in relation to the cathodic (protected) area. For example, it can be expected where large areas of the surface are generally covered by mill scale, applied coatings, or deposits of various kinds, but breaks exist in the continuity of the protective material. Pitting may also develop on bare clean metal surfaces because of irregularities in the physical or chemical structure of the metal. Localized dissimilar soil conditions at the surfaces of steel can also create conditions that promote pitting.

If appreciable attack is confined to a small area of metal acting as an anode, the developed pits are described as deep. If the area of attack is relatively large the pits are called shallow. The ratio of deepest metal penetration to average metal penetration as determined by weight loss of the specimen is known as the pitting factor. A pitting factor of unity represents uniform attack.

Performance in the area of pitting and crevice corrosion is often measured using critical pitting temperatures (CPT), critical crevice temperatures (CCT), and pitting resistance equivalent number (PREN). As a general rule the higher the PREN, the better is the resistance. Alloys having similar values may differ in actual service. The pitting resistance number is determined by the chromium, molybdenum, and nitrogen contents.

$$\text{PREN} = \%\text{CR} + 3.3 \times \%\text{Mo} + 30 \times \%\text{N}$$

Prevention can be accomplished first by proper material selection, followed by a design that prevents stagnation of material, and then alternate wetting and drying of the surface. Also if coatings are to be applied, care should be taken that they are continuous and without "holidays."

Erosion Corrosion

Erosion corrosion is also referred to as impingement attack and is caused by contact with high-velocity liquids and results in a pitting type of corrosion. It is most prevalent in condenser tubes and piping fittings such as elbows and tees. Prevention can be accomplished by one or more means:

1. Reduce the velocity.
2. Select a harder material.
3. Properly design the piping system or the condenser.

Stress-Corrosion Cracking (SCC)

Stress-corrosion cracking occurs at points of stress. Usually the metal or alloy is virtually free of corrosion over most of its surface, yet fine cracks penetrate the surface at the points of stress. This type of attack takes place in certain mediums. All metals are potentially subject to SCC. The conditions necessary for stress-corrosion cracking to occur are

1. A suitable environment
2. A tensile stress
3. A sensitive metal
4. Appropriate temperature and pH values

An ammonia-containing environment can induce SCC in copper-containing alloys, while with low-alloy austenitic stainless steels a chloride-containing environment is necessary. It is not necessary to have a concentrated solution to cause SCC. A solution containing only a few parts per million of the critical ion is sufficient. Temperature and pH are also factors. There is usually a threshold temperature below which SCC will not occur and a plus or minus pH value before cracking will start.

Normally stress-corrosion cracking will not occur if the part is in compression. Failure is triggered by a tensile stress which must approach the yield strength of the metal. These stresses may be induced by faulty installation, or they may be residual stresses from welding, straightening, bending, or accidental denting of the component. Often pits which act as stress concentration sites will initiate SCC.

Alloy content of stainless steels, particularly nickel, determines the sensitivity of this metal to SCC. Ferritic stainless steels which are nickel free and the high nickel alloys are not subject to stress-corrosion cracking. An alloy with a nickel content greater than 30% is immune to SCC. The most common grades of stainless steel (304, 304L, 316, 316L, 321, 347, 303, 302, and 301) have nickel contents of 7–10% and are the most susceptible to stress-corrosion cracking.

Examples of stress-corrosion cracking include the cracking of austenitic stainless steels in the presence of chlorides, caustic embrittlement cracking of steel in caustic solutions, cracking of cold-formed brass in ammonia environments, and cracking of Monel in hydrofluosilicic acid. Table 2 contains a partial listing of alloy systems subject to stress-corrosion cracking.

It is important that any stresses which may have been induced during fabrication be removed by an appropriate stress-relief operation. The design should also avoid stagnant areas which could lead to pitting and the initiation of stress concentration sites.

Biological Corrosion

Corrosive conditions can be developed by living organisms as a result of their influence on anodic and cathodic reactions. This metabolic activity can directly or indirectly cause deterioration of a metal by the corrosion process. This activity can

1. Produce a corrosive environment
2. Create electrolytic concentration cells on the metal surface
3. Alter the resistance of surface films
4. Influence the rate of anodic or cathodic reaction
5. Alter the environment composition

Because this form of corrosion gives the appearance of pitting, it is necessary to first diagnose the presence of bacteria. Once established, prevention can be accom-

Table 2 Alloy Systems Subject to Stress-Corrosion Cracking

Alloy	Environment
Carbon steel	Anhydrous liquid ammonia, HCN, ammonium nitrate, sodium nitrite, sodium hydroxide.
Aluminum base	Air, seawater, salt and chemical combination
Magnesium base	Nitric acid, caustic, HF solutions, salts, coastal atmospheres
Copper base	Primarily ammonia and ammonium hydroxide, amines, mercury
Martensitic and precipitation hardening stainless steels	Seawater, chlorides, H_2S solutions
Austenitic stainless steels	Chlorides (organic and inorganic), caustic solutions, sulfurous and polythiuric acids
Nickel base	Caustic above 600°F (315°C), fused caustic, hydrofluoric acid
Titanium	Seawater, salt atmosphere, fused salt
Zirconium	$FeCl_3$ or $CuCl_2$ solutions

plished by the use of biocides or by the selection of a more resistant material of construction. For some species of bacteria a change in pH will provide control.

Dezincification (dealloying)

When one element in a solid alloy is removed by corrosion, the process is known as dezincification or dealloying. The most common example is the removal of zinc from brass alloys. When the zinc corrodes preferentially, a porous residue of copper and corrosion products remain. The corroded part often retains its original shape and may appear undamaged except for surface tarnish. However, its tensile strength and particularly its ductility have been seriously reduced.

Parting

Parting is similar to dezincification in that one or more of the active components of the alloy corrode preferentially as in dezincification with the same results. Copper-base alloys containing aluminum are subject to this reaction, with the aluminum corroding preferentially.

Concentration Cells

Concentration cells is not a type of corrosion but the cause of pitting and crevice corrosion. The primary factors causing pitting are electrical contact between dissimilar materials or areas of the same metal where oxygen or conductive salts in water differ. These couples are sufficient to cause a difference of potential, causing an electric current to flow through the water, or across moist steel, from the metallic anode to a nearby cathode. The cathode may be mill scale or any other portion of the metal surface that is cathodic to the more active metal areas. Mill scale is cathodic to steel and is one of the more common causes of pitting. If the cathodic area is relatively large compared to the anodic area, the damage is spread out and usually negligible. When the anode area is relatively small, the metal loss is concentrated and may be serious.

Concentration cells are capable of causing severe corrosion, leading to pitting, when differences in dissolved oxygen concentrations occur. That portion of the metal in contact with water relatively low in dissolved oxygen concentration is anodic to adjoining areas with water higher in dissolved oxygen concentration. This lack of oxygen may be caused by exhaustion of dissolved oxygen in a crevice. The low-oxygen area is always anodic. This type of cell is responsible for corrosion at crevices formed at the interface of two coupled pipes or at threaded connections, since the oxygen concentration is lower within the crevice or at the threads than elsewhere. It also is responsible for pitting damage under rust or at the water line (air-water interface). These differential aeration cells are responsible for initiating pits in stainless steel, aluminum, nickel, and other so-called passive metals when they are exposed to aqueous environments such as seawater.

Embrittlement

Embrittlement is the severe loss of ductility or toughness in a material which may result in cracking. Some metals, when stressed, crack on exposure to corrosive environments, but corrosion is not necessarily a part of crack initiation or crack growth. This type of failure is not properly called stress-corrosion cracking.

The most frequent occurrence of this form of attack is in steel equipment handling solutions containing hydrogen sulfide. Under these conditions corrosion of the steel generates atomic hydrogen which penetrates the steel and at submicroscopic discontinuity of pressures high enough to cause cracking or blistering. Failures of this kind are called "hydrogen cracking" or "hydrogen stress cracking." Dissolved atomic (ionic) hydrogen in concentrations of a few ppm can also cause cracking by a more complex and less understood mechanism.

Filiform

Metals with semipermeable coatings or films may undergo a type of corrosion resulting in numerous meandering threadlike filaments of corrosion beneath the coatings or films. The essential conditions for this form of corrosion to develop are generally high humidity (65–95% relative humidity at room temperature), sufficient water permeability of the film, stimulation by impurities, and the presence of film defects (mechanical damage, pores, insufficient coverage of localized areas, air bubbles, salt crystals, or dust particles).

Corrosion Fatigue

When a metal cracks on being stressed alternately or repeatedly it is said to fail by fatigue. The greater the applied stress at each cycle, the shorter is the time to failure. In a corrosive environment, failure at a given stress level usually occurs within fewer cycles, and a true fatigue limit is no longer observed. Cracking of a metal resulting from the combined action of a corrosive environment and repeated or alternate stress is called corrosion fatigue.

Fretting

Fretting corrosion also occurs because of mechanical stresses, and in the extreme may lead to failure by fatigue or corrosion fatigue. It is defined as damage occurring at the interface of two surfaces in contact, one or both of which are metals subject to slight relative slip.

Damage by fretting corrosion is characterized by discoloration of the metal surface and, in the case of oscillating motion, by formation of pits. It is at such pits that fatigue cracks eventually start.

Graphitization (Graphitic Corrosion)

Cast iron is composed of a mixture of ferrite phases (almost pure iron) and graphite flakes. When it corrodes it forms corrosion products, which in some soils or waters cement together the residual graphite flakes. The resulting structure (e.g., water pipe), although corroded completely, may have sufficient remaining strength, even with low ductility, to continue to operate under the required pressures and stresses.

This form of corrosion occurs only with gray cast iron or ductile cast iron containing spheroidal graphite but not with white cast iron which contains cementite and ferrite.

2.2 Stainless Steels

Probably the most widely known and most widely used metallic material for corrosion resistance is stainless steel. For many years this was the only material available.

Stainless steel is not a singular material as its name might imply, but a broad group of alloys each of which exhibits its own physical and corrosion-resistant properties. Stainless steels are alloys of iron to which a minimum of 11% chromium has been added to provide a passive film to resist "rusting" when the material is exposed to weather. This film is self-forming and self-healing in environments where the stainless steel is resistant. As more chromium is added to the alloy, improved corrosion resistance results. Consequently, there are stainless steels with chromium contents of 15%, 17%, 20%, and even higher. Chromium provides resistance to oxidizing environments such as nitric acid and also provides resistance to pitting and crevice attack.

Other alloying ingredients are added to further improve the corrosion resistance and mechanical strength. Molybdenum is extremely effective in improving pitting and crevice-corrosion resistance.

By the addition of copper, improved resistance to general corrosion in sulfuric acid is obtained. This will also strengthen some precipitation-hardenable grades. In sufficient amounts, though, copper reduces the pitting resistance of some alloys.

The addition of nickel improves resistance to reducing environments and stress-corrosion cracking. Nitrogen can also be added to improve corrosion resistance to pitting and crevice attack and to improve strength.

Columbium and titanium are added to stabilize carbon. They form carbides and reduce the amount of carbon available to form chromium carbides, which can be deleterious to corrosion resistance.

As a result of these alloying possibilities, more than 70 stainless steels are available. They can be divided into four major categories depending upon their microstructure:

Austenitic
Ferritic
Martensitic
Duplex

Austenitic Stainless Steels

Austenitic stainless steels are the most widely used for corrosion resistance of any of the stainless steels, and therefore this series is the most important to the process industry. The addition of substantial quantities of nickel to high-chromium alloys stabilizes the austenite at room temperature. This group of alloys contains 16–26% chromium and 6–22% nickel. The carbon content is kept low (0.08%) to minimize carbide precipitation. The types of stainless comprising the austenitic group are as follows:

201	304N	316SF	20Cb3
202	305N	316L	904L
205	308	316N	20Mo4
301	309	317L	20Mo6
302	309S	317LN	Al-6XN
302B	310N	321	25-6Mo
303	310S	330	Alloy 31
304	314S	347	Alloy 654
304L	316	348	

The corrosion resistance of the austenitic stainless steels is the result of the formation of an oxide film on the surface of the metal. Consequently they perform best under oxidizing conditions, since reducing conditions and chloride ions destroy the film, causing rapid attack. Chloride ions, combined with high tensile stresses, cause stress-corrosion cracking.

Types 304 and 304L. These grades of stainless, although possessing a wide range of corrosion resistance, have the least overall resistance of any of the austenitic grades we are considering. Type 304 is subject to intergranular corrosion as a result of carbide precipitation. Welding can cause this phenomenon, but competent welders using good welding techniques can control the problem. Depending on the corrodent being handled, the effect of the carbide precipitation may or may not present a problem. If the corrodent being handled will attack through intergranular corrosion, another alloy should be used.

If the carbon content in the alloy is not allowed to exceed 0.03%, carbide precipitation can be controlled. The letter L is added as a suffix to the alloy designation indicating this low carbon content. Type 304L is such an alloy. It can be used for welded sections without the danger of carbide precipitation.

Type 304N has nitrogen added to the alloy, which improves its resistance to pitting and crevice corrosion.

Type 304 and 304L stainless steel exhibit good overall corrosion resistance. They are used extensively in handling nitric acid and most organic acids. Refer to Appendix A for more detailed compatibility with various corrodents.

Types 316 and 316L. These chromium-nickel grades of stainless steel have 2–3% molybdenum added. The molybdenum substantially increases resistance to pitting and crevice corrosion in systems containing chlorides and improves overall resistance to most types of corrosion in chemically reducing and neutral solutions. They may be susceptible to chloride stress-corrosion cracking.

In general, these stainless steels are more corrosion resistant than the type 304 stainless steels. With the exception of oxidizing acids such as nitric acid, type 316 alloys provide satisfactory resistance to corrodents handled by type 304 with the added ability to handle some corrodents that type 304 alloys cannot handle. Refer to Appendix A for more detailed compatibility with various components.

Type 316L stainless is the low-carbon version of type 316 and offers the additional feature of preventing excessive intergranular precipitation of chromium carbides during welding and stress relieving.

Nitrogen is another element that improves resistance to pitting and crevice corrosion, and it is considerably less expensive than molybdenum. AOD and other processing methods make possible the addition of substantial amounts of nitrogen to steels. As a result, nitrogen-bearing types 316LN and 316L (Hi)N are now available.

Types 317 and 317L. Containing a greater amount of molybdenum, chromium, and nickel, type 317 stainless steels offer higher resistance to pitting and crevice corrosion than type 316 in various process environments encountered in the chemical process industry. However, they may still be susceptible to chloride stress-corrosion cracking.

Type 317L, like 316L, is a low-carbon variation of the basic alloy, which offers the additional advantage of preventing intergranular precipitation of chromium carbide during welding and stress relieving.

The newest of the high-molybdenum-containing type 300 stainless steels is type 317LN, also known as type 317MN and 317LXN. In addition to molybdenum it contains nitrogen as a critical alloying ingredient. Overall improvements in its corrosion resistance has not yet been proven, but pitting resistance appears to be better than for the other 317 alloys.

In general, these alloys can be used whenever the type 316 series is suitable and in many instances with improved performance.

Type 321. By alloying austenitic alloys with a small amount of an element having a higher affinity for carbon than does chromium, carbon is restrained from diffusing to the grain boundaries, and any which reaches the boundary reacts with the element instead of with the chromium. These are known as stabilized grades. Type 321 is such an alloy. It has titanium added to stabilize it.

This alloy is particularly useful in high-temperature service in the carbide precipitation range. Even though it has improved resistance to carbide precipitation it may still be susceptible to chloride stress-corrosion cracking.

Type 347. This is a columbium-stabilized stainless steel. The intergranular corrosion resistance has been improved by the addition of columbium. Basically this alloy is equivalent to type 304 stainless steel, with the added protection against carbide precipitation, and can be used wherever type 304 stainless steel is suitable.

Type 304L also supplies protection from carbide precipitation but has a maximum operating temperature of 800°F (427°C), while type 347 can be operated as high as 1000°F (538°C).

Type 348. This is a tantalum-stabilized stainless steel to prevent carbide precipitation.

20Cb3 Stainless Steel. This alloy was originally developed to provide improved corrosion resistance to sulfuric acid. However, it has found wide application throughout the chemical processing industry. The alloy is stabilized with columbium and tantalum and has a high nickel content, approximately 33%. It exhibits excellent resistance to chloride stress-corrosion cracking, with minimum carbide precipitation due to welding.

In high concentrations of chlorides, alloy 20 is vulnerable to pitting and crevice attack. For improved resistance to these localized types of corrosion the 2% molybdenum content must be increased to 4% or 6%, as has been done in alloy 20Mo-4 and alloy 20Mo-6.

Alloy 20 finds wide application in the manufacture of synthetic rubber, high octane gasoline, solvents, explosives, plastics, synthetic fibers, heavy chemicals, organic chemicals, pharmaceuticals, and food processing equipment.

Refer to Appendix A for the compatibility of alloy 20Cb3 with specific corrodents.

20Mo-4 Stainless Steel. This alloy is similar to alloy 20Cb3 but with a 4% molybdenum content in place of the 2%, providing improved pitting and crevice corrosion resistance over alloy 20Cb3. Other corrosion-resistant properties remain the same as those of alloy 20Cb3.

20Mo-6 Stainless Steel. Of the three grades of alloy 20 this one offers the highest level of pitting and crevice-corrosion resistance. It also possesses the same overall corrosion-resistant properties as alloy 20Cb3.

25-6Mo Stainless Steel. This alloy, produced by Inco International, contains nickel, chromium, molybdenum, copper, nitrogen, carbon, manganese, phosphorus, sulfur, and silicon as alloying ingredients. One of the outstanding properties of alloy 25-6Mo is its resistance to environments containing chlorides or other halides. It is especially suited for application in high-chloride environments such as brackish water, seawater, caustic chlorides, and pulp mill bleach systems. The alloy offers excellent resistance to pitting and crevice corrosion.

In brackish and wastewater systems microbially influenced corrosion can occur, especially in systems where equipment has been idle for extended periods. A 6% molybdenum alloy offers protection from manganese-bearing, sulfur-bearing, and generally reducing types of bacteria. Because of this resistance to microbially influenced corrosion, alloy 25-6Mo is being used in wastewater piping systems of power plants.

In saturated sodium chloride environments and pH values of 6–8, alloy 25-6Mo exhibits a corrosion rate of less than 1 mpy. Even under more aggressive oxidizing conditions involving sodium chlorate, alloy 25-6Mo maintains a corrosion rate of less than 1 mpy and shows no pitting even at temperatures up to boiling.

904L Stainless Steel. The high nickel and chromium content make alloy 904L resistant to corrosion in a wide variety of oxidizing and reducing environments. Molybdenum and copper are included in the alloy for increased resistance to pitting and crevice corrosion and to general corrosion in reducing acids. Other advantages of the alloy's composition are sufficient nickel for resistance to chloride ion stress-corrosion cracking and a low carbon content for resistance to intergranular corrosion.

The alloy's outstanding attributes are resistance to nonoxidizing acids along with resistance to pitting, crevice corrosion, and stress-corrosion cracking in such media as stack gas condensate and brackish water. Alloy 904L is especially suited to handle hot solutions of sulfuric acid at moderate concentrations. It also has excellent resistance to phosphoric acid. Refer to Appendix A for the compatibility of alloy 904L with selected corrodents.

AL-6XN Stainless Steel. This alloy is a low-carbon, high-purity superaustenitic stainless steel containing chromium, nickel, molybdenum, and nitrogen. It is manufactured by Allegheny Ludlum Corporation. The presence of nitrogen provides the alloy with improved pitting and crevice corrosion resistance, greater resistance to localized corrosion in oxidizing chlorides and reducing solutions.

The corrosion-resistant properties of Al-6XN show exceptional resistance to pitting, crevice attack, and stress cracking in high chlorides, and general resistance in various acid, alkaline, and salt solutions. It also exhibits excellent corrosion resistance to oxidizing chlorides, reducing solutions, and seawater corrosion. Al-6XN can also handle various concentrations at a variety of temperatures of sulfuric, nitric, phosphoric, acetic, and formic acids.

Alloy 31. With a 6.5% molybdenum content alloy 31 exhibits excellent resistance to pitting and crevice corrosion in neutral and aqueous acid solutions. The high chromium content of 27% imparts superior resistance to corrosive attack by oxidizing media.

Alloy 654. This alloy has better resistance to localized corrosion than do other superaustenitics, and it is on the same level as nickel-base alloys. Indications are that alloy 654 is as corrosion resistant as alloy C-276, based on tests in filtered seawater, bleach plants, and other aggressive chloride environments. It is intended to compete with titanium in the handling of high-chloride environments.

Pitting Resistance Summary. Although the austenitic stainless steels have a wide range of corrosion resistance and are the most widely used family of alloys in the corrosion resistance field, their single largest drawback has been their tendency to pit. As a result, new alloys have been developed for the purpose of improving their pitting and crevice corrosion resistance. As explained previously, performance in these areas is often measured by using the pitting resistance equivalent number (PREN): as a general rule the higher the PREN the better the resistance. The following table gives the PREN for various austenitic stainless steel alloys:

Alloy	PREN	Alloy	PREN
654	63.09	316LN	31.80
31	54.45	316	27.90
25-6Mo	47.45	20Cb3	27.26
Al-6XN	46.96	348	25.60
20Mo-6	42.81	347	19.0
317LM	39.60		
904L	36.51	331	19.0
20Mo-4	36.20	304N	18.3
317	33.20	304	18.0

Ferritic Stainless Steels

The ferritic stainless steels contain 15–30% chromium with a low carbon content (0.1%). Corrosion resistance of these alloys is rated good. Mildly corrosive solutions and oxidizing media are handled satisfactorily. These alloys exhibit a high level of resistance to pitting and crevice attack in many chloride-containing environments. They are also essentially immune to stress-corrosion cracking in chloride environments. Ferritic stainless steels are especially resistant to organic acids. Their resistance to chloride attack makes these alloys ideal candidates for food processing applications and a wide range of chemical processing uses.

These alloys will not resist reducing media such as hydrochloric acid.

Type 430. This alloy exhibits resistance to chloride stress-corrosion cracking and to elevated temperature sulfide attack. It is subject to embrittlement at 885°F (475°C) and loss of ductility at subzero temperatures.

Type 444. As with all ferritic stainless steels, this alloy relies on a passive film to resist corrosion, but exhibits relatively high corrosion rates when activitated. This explains the abrupt change in corrosion rates that occurs at particular acid concentrations. For example, it is resistant to very dilute solutions of sulfuric acid at boiling temperature, but corrodes rapidly at higher concentrations.

The corrosion rates of type 444 in strongly concentrated sodium hydroxide solutions are higher than those for austenitic stainless steels.

Type XM27. This alloy provides excellent resistance to pitting and crevice corrosion in chloride-containing environments, excellent resistance to chloride stress-corrosion cracking, resistance to intergranular corrosion and to a wide variety of other corrosive environments. This alloy can also be used in direct contact with foods.

Refer to Table 3 for compatibility of ferritic stainless steels with selected corrodents.

Martensitic Stainless Steels

The martensitic stainless steels are normally 11–13% chromium. Martensite is a body-centered tetragonal structure which provides increased strength and hardness versus the annealed stainlesses with other lattice structures. Sufficient carbon is added to permit martensite formation with rapid cooling. Other elements, such as nickel and molybdenum, may be added for improved corrosion resistance.

The corrosion resistance of the martensitic stainless steels is inferior to that of the austenitic stainless steels. These alloys are generally used in mildly corrosive services such as atmospheric, freshwater, and organic exposures. Their greatest weakness is their susceptibility to the absorption of atomic hydrogen, resulting in hydrogen-assisted cracking, particularly in sulfide environments. Primary applications are cutlery, turbine blades, and high-temperature parts.

Alloy 350. This is an austenitic/martensitic alloy. It may be subject to intergranular attack unless cooled to subzero temperatures before aging. The corrosion resistance of alloy 350 is similar to that of type 304 stainless steel. Refer to Appendix A for the compatibility of this alloy with selected corrodents.

Custom 450 Stainless Steel.* This is a martensitic age-hardenable stainless steel with good corrosion resistance similar to that of type 304 stainless steel. It is used where type 304 stainless steel is not strong enough or where type 420 is insufficiently corrosion resistant.

Custom 455 Stainless Steel.† This is an age-hardenable stainless steel having high strength with a corrosion resistance superior to that of type 410 stainless steel. It resists chloride cracking as well as alkalies and nitric acid, but it may be subject to hydrogen embrittlement under some conditions.

Alloy 718. This alloy is an age-hardenable martensitic alloy possessing good corrosion resistance, being compatible with alkalies, organic acids, and sulfuric acid. Excellent oxidation resistance is demonstrated throughout its operating temperature range.

Alloy 17-7PH. This is a semiaustenitic stainless steel. In the annealed condition it is austenitic, but in the aged condition it is martensitic.

In rural and mild industrial atmospheres it has excellent resistance to general corrosion, being equivalent to type 304 stainless steel. In seacoast applications alloy 17-7PH will gradually develop a light rusting and pitting in all heat-treated conditions. It is almost equal to type 304 in these applications.

Alloy 17-7PH is quite susceptible to stress-corrosion cracking when in condition H900. When hardened at temperatures of 1025°F/552°C, and higher, the alloy is highly resistant to SCC.

Refer to Appendix A for the compatibility of this alloy with selected corrodents.

---

*Custom 450 Stainless Steel is a registered trademark of Carpenter Technology Corporation.
†Custom 455 Stainless Steel is a registered trademark of Carpenter Technology Corporation.

Table 3 Compatibility of Ferritic Stainless Steels with Selected Corrodents

The chemicals listed are in the pure state or in a saturated solution unless otherwise indicated. Compatibility is shown to the maximum allowable temperature for which data are available. Incompatibility is shown by a X. A blank space indicates that data are not available. When compatible the corrosion rate is less than 20 mpy.

	Alloy		
Chemical	430 (°F/°C)	444 (°F/°C)	XM-27 (°F/°C)
Acetic acid 10%	70/21	200/93	200/93
Acetic acid 50%	X	200/93	200/93
Acetic acid 80%	70/21	200/93	130/54
Acetic acid, glacial	70/21		140/60
Acetic anhydride 90%	150/66		300/149
Aluminum chloride, aqueous	X		110/43
Aluminum hydroxide	70/21		
Aluminum sulfate	X		
Ammonia gas	212/100		
Ammonium carbonate	70/21		
Ammonium chloride 10%			200/93
Ammonium hydroxide 25%	70/21		
Ammonium hydroxide, sat.	70/21		
Ammonium nitrate	210/100		
Ammonium persulfate 5%	70/21		
Ammonium phosphate	70/21		
Ammonium sulfate 10–40%	X		
Amyl acetate	70/21		
Amyl chloride	X		
Aniline	70/21		
Antimony trichloride	X		
Aqua regia 3:1			X
Barium carbonate	70/21		
Barium chloride	70/21[a]		
Barium sulfate	70/21		
Barium sulfide	70/21		
Benzaldehyde			210/99
Benzene	70/21		
Benzoic acid	70/21		
Borax 5%	200/93		
Boric acid	200/93[a]		
Bromine gas, dry	X		
Bromine gas, moist	X		
Bromine liquid	X		
Butyric acid	200/93		
Calcium carbonate	200/93		
Calcium chloride	X		
Calcium hypochlorite	X		
Calcium sulfate	70/21		
Carbon bisulfide	70/21		
Carbon dioxide, dry	70/21		

Table 3 (Continued)

Chemical	Alloy 430 (°F/°C)	444 (°F/°C)	XM-27 (°F/°C)
Carbon monoxide	1600/821		
Carbon tetrachloride, dry	212/100		
Carbonic acid	X		
Chloracetic acid, 50% water	X		
Chloracetic acid	X		
Chlorine gas, dry	X		
Chlorine gas, wet	X		
Chloroform, dry	70/21		
Chromic acid 10%	70/21		120/49
Chromic acid 50%	X		X
Citric acid 15%	70/21	200/93	200/93
Citric acid, concentrated	X		
Copper acetate	70/21		
Copper carbonate	70/21		
Copper chloride	X		X
Copper cyanide	212/100		
Copper sulfate	212/100		
Cupric chloride 5%	X		
Cupric chloride 50%	X		
Ethylene glycol	70/21		
Ferric chloride	X		80/27
Ferric chloride, 10% in water			75/25
Ferric nitrate 10–50%	70/21		
Ferrous chloride	X		
Fluorine gas, dry	X		
Fluorine gas, moist	X		
Hydrobromic acid, dilute	X		
Hydrobromic acid, 20%	X		
Hydrobromic acid, 50%	X		
Hydrochloric acid, 20%	X		
Hydrochloric acid, 38%	X		
Hydrocyanic acid, 10%	X		
Hydrofluoric acid, 30%	X		X
Hydrofluoric acid, 70%	X		X
Hydrofluoric acid, 100%	X		X
Iodine solution 10%	X		
Iactic acid 20%	X	200/93	200/93
Iactic acid, concentrated	X		
Magnesium chloride			200/93
Malic acid	200/93		
Muriatic acid	X		
Nitric acid 5%	70/21	200/93	320/160
Nitric acid 20%	200/93	200/93	320/160
Nitric acid 70%	70/21	X	210/99
Nitric acid, anhydrous	X	X	
Nitrous acid 5%	70/21		

Table 3 (Continued)

	Alloy		
Chemical	430 (°F/°C)	444 (°F/°C)	XM-27 (°F/°C)
Phenol	200/93		
Phosphoric acid 50–80%	X	200/93	200/93
Picric acid	X		
Silver bromide	X		
Sodium choride	70/21[a]		
Sodium hydroxide 10%	70/21	212/100	200/93
Sodium hydroxide 50%		X	180/82
Sodium hydroxide, concentrated		X	
Sodium hypochlorite			90/32
Sodium sulfide to 50%	X		
Stannic chloride	X		
Stannous chloride 10%			90/32
Sulfuric acid 10%	X	X	X
Sulfuric acid 50%	X	X	X
Sulfuric acid 70%		X	X
Sulfuric acid 93%		X	X
Sulfuric acid 98%	X		280/132
Sulfuric acid 100%	70/21	X	
Sulfuric acid, fuming		X	
Sulfurous acid	X		360/182
Toluene			212/99
Zinc chloride 20%	70/21[a]		200/93

[a]Pitting may occur.
Source: Philip A. Schweitzer, *Corrosion Resistance Tables, Fourth Edition*, vols. 1–3, Marcel Dekker, New York, 1995.

Duplex Stainless Steels

While the austenitic stainless steels have a wide range of corrosion resistance, they do exhibit some shortcomings, such as in chloride stress-corrosion cracking, pitting, and maintaining corrosion resistance after welding. The duplex stainless steels were produced to overcome these shortcomings.

The duplex stainless steels are alloys whose microstructures are a mixture of austenite and ferrite. The original duplex stainless steels did not have nitrogen added specifically as an alloying ingredient. Adding 0.15–0.25% nitrogen reduces the chromium partitioning between the two phases, thereby improving the pitting and crevice corrosion resistance of the austenite. This nitrogen addition also improves the weldability of the stainless steel without losing any of its corrosion resistance. It is not necessary to heat-treat these stainless steels after welding.

The high chromium and molybdenum contents of duplex stainless steels are particularly important in providing resistance in oxidizing environments and are also responsible for exceptionally good pitting and crevice corrosion resistance, especially in chloride environments. In general, these stainless steels have greater pitting resistance than type 316, and several have an even greater resistance than alloy 904L.

The resistance to crevice corrosion of the duplexes is superior to the resistance of the 300 series austenitics. They also provide a greater resistance to crevice corrosion cracking. The duplexes are resistant to chloride stress-corrosion cracking in chloride-containing environments. However, under very severe conditions, such as boiling magnesium chloride, the duplexes will crack.

The duplexes have less resistance in reducing environments than the austenitics as a result of the lower nickel content. This reduction in nickel content is necessary in order to achieve the desired microstructure. The high chromium and molybdenum contents particularly offset this loss, and consequently they can be used in some reducing environments of dilute concentrations and lower temperatures.

Type 329. This is one of the original duplexes. Because this alloy has a high carbon content there is a considerable loss in corrosion resistance as a result of welding. Postweld heat treatment is required to maintain corrosion resistance.

Alloy 2205. This duplex resists oxidizing mineral acids and most organic acids in addition to reducing acids and chloride environments.

7Mo Plus.* The general corrosion resistance of 7Mo Plus is superior to those of stainless steel such as type 304 and type 316 in many environments. The high chromium content provides good corrosion resistance in strongly oxidizing media such as nitric acid. The corrosion resistance of this alloy is extended into the less-oxidizing environments as a result of the molybdenum content. The combination of the chromium and molybdenum content imparts a high level of resistance to pitting and crevice corrosion. This alloy also provides excellent chloride stress-corrosion cracking resistance in applications where type 304 and type 316 stainless steels would be used for general corrosion resistance.

2.3 Nickel

Commercially pure nickel is available in two main forms: alloy 200 and alloy 201. Alloy 200 is limited in operating temperatures to 600°F (315°C), and alloy 201 is used in applications having temperatures in excess of 600°F (315°C).

The outstanding characteristic of nickel 200 and 201 is their resistance to caustic soda and other alkalies (ammonium hydroxide is an exception). These alloys are not attacked by anhydrous ammonia or ammonium hydroxide in concentrations of 1% or less. Stronger concentrations can cause rapid attack.

Nickel 200 shows excellent resistance to all concentrations of caustic soda at temperatures up to and including the molten state. Below 50% the corrosion rates are negligible, usually being less than 0.2 mpy even in boiling solutions. As concentrations and temperatures increase, corrosion rates increase very slowly.

Nickel 200 is not subject to stress-corrosion cracking in any of the chlorine salts, and it has excellent general resistance to nonoxidizing halides. The oxidizing acid chlorides, such as ferric, cupric, and mercuric, are very corrosive and should

---

*Registered trademark of Carpenter Technology Corp.

not be used. Nickel 201 is extremely useful in the handling of dry chlorine and hydrogen chloride at elevated temperatures.

If aeration is not high, nickel 200 has excellent resistance to most organic acids, particularly fatty acids such as stearic and oleic.

Nickel is also used in handling food and synthetic fibers for maintaining product purity. For food products, the presence of nickel ions is not detrimental to the flavor and is not toxic. For many organic chemicals, such as phenol and viscose rayon, nickel does not discolor the solutions, unlike iron and copper.

The corrosion resistance of nickel 200, is the same as that of nickel 201. Refer to Appendix A for the compatibility of these alloys with selected corrodents.

2.4 High-Nickel Alloys

High-nickel alloys have the widest range of applications of any class of alloys. The electrochemical properties of nickel and its crystallographic characteristics enable it to accommodate large amounts of alloying elements. The alloys used for corrosion-resistant purposes are primarily solid-solution-strengthened alloys supplied in the fully annealed condition.

The element nickel is nobler than iron but more active than copper in the electromotive series. In reducing environments, such as dilute sulfuric acid, nickel is more corrosion resistant than iron but not as resistant as copper or nickel-copper alloys. The nickel-molybdenum alloys are more corrosion resistant in reducing acids than nickel or nickel-copper alloys.

Nickel can form a protective passive film in some environments. However, this passive film is not particularly stable, and therefore nickel cannot generally be used in oxidizing environments such as nitric acid. If alloyed with chromium a much more stable film is formed, and corrosion resistance is exhibited to a variety of oxidizing environments. However, these alloys can corrode in environments containing appreciable amounts of chlorides or other halides, especially if oxidizing species are present. Alloying with molybdenum or tungsten is necessary to improve the resistance to corrosion by oxidizing chloride solutions.

Monel Alloy 400

The alloying of 30–33% copper with nickel, producing Monel 400, provides an alloy with many of the characteristics of chemically pure nickel, with improvements. A major area of application is in water handling, including brackish water and seawater. As with nickel 200, the alloy can pit in stagnant seawater, but the rate of attack is significantly reduced. The absence of chloride stress-corrosion cracking is also a factor in the selection of this alloy.

The general corrosion resistance of Monel 400 in the nonoxidizing acids, such as sulfuric, hydrochloric, and phosphoric, is improved over that of pure nickel. The alloy is not resistant to oxidizing media, such as nitric acid, ferric chloride, chromic acid, wet chlorine, sulfur dioxide, or ammonia.

Monel 400 exhibits excellent resistance to hydrofluoric acid solutions at various concentrations and temperatures. It is subject to stress-corrosion cracking in moist, aerated hydrofluoric or hydrofluorsilicic acids and vapor. The corrosion of Monel is negligible in all types of atmospheres. Indoor exposure produces a very light tarnish that is easily removed by occasional wiping. Outdoor surfaces exposed to rain produce a thin gray-green patina. In sulfurous atmospheres a smooth brown, adherent film forms.

Because of its high nickel content Monel 400 is practically as resistant as nickel 200 to caustic soda throughout most of the concentration range. It is also resistant to anhydrous ammonia and to ammonium hydroxide solutions of up to 3% concentration. The alloy is subject to stress-corrosion cracking in high temperatures, concentrated caustic, and in mercury.

Refer to Appendix A for the compatibility of Monel 400 with selected corrodents.

Inconel Alloy 600

Inconel alloy 600 is a nickel-chromium-iron alloy for use in environments requiring resistance to heat and corrosion. Alloy 600 is practically free from corrosion by freshwaters, including the most corrosive of natural waters containing free carbon dioxide, iron compounds, and dissolved air. It remains free from stress-corrosion cracking even in boiling magnesium chloride.

Alloy 600 has excellent resistance to dry halogens at elevated temperatures and has been used successfully for chlorination equipment at temperatures up to 1000°F (538°C). Alloy 600 is not subject to stress-corrosion cracking in any of the chloride salts and has excellent resistance to all of the nonoxidizing halides.

Resistance to stress-corrosion cracking is imparted to alloy 600 by virtue of its nickel base. Therefore it finds application in handling water environments where stainless steel fails by cracking.

Because of its chromium content, alloy 600 exhibits greater resistance to sulfuric acid under oxidizing conditions than either nickel 200 or Monel 400. The addition of oxidizing salts to sulfuric acid tends to passivate alloy 600, which makes it suitable for use with acid mine waters or brass pickling solutions, applications where Monel 400 cannot be used.

Refer to Appendix A for the compatibility of alloy 600 with selected corrodents.

Inconel Alloy 625

Inconel alloy 625 is a nickel-chromium-molybdenum-columbium alloy. It is resistant to oxidation, general corrosion, pitting, and crevice corrosion, and is virtually immune to chloride ion stress-corrosion cracking.

In fresh and distilled water, corrosion rates are essentially zero, and in seawater, under flowing and stagnant conditions, weight losses are extremely low.

Resistance to aqueous solutions is good in a variety of applications including organic acids, sulfuric acid, and hydrochloric acid at temperatures below 150°F (60°C). The alloy also exhibits satisfactory resistance to hydrofluoric acid. Although nickel-base alloys are not normally used in nitric acid service because of cost, alloy 625 is resistant to mixtures of nitric-hydrofluoric, where stainless steel loses its resistance.

Alloy 625 also has excellent resistance to phosphoric acid solutions, including commercial grades of acids that contain fluorides, sulfates, and chlorides in the production of superphosphoric acid (72% P_2O_8).

In general, alloy 625 can be used wherever alloy 600 is suitable, and in some applications where alloy 600 does not have a satisfactory resistance.

Refer to Appendix A for the compatibility of alloy 625 with selected corrodents.

Incoloy Alloy 800

Incoloy alloy 800 contains approximately 20% chromium, 32% nickel, and 46% iron as balance. It is used primarily for its oxidation resistance at elevated temperatures.

At moderate temperatures the general corrosion resistance of alloy 800 is similar to that of the other austenitic nickel-iron-chromium alloys. However, as the temperature increases, alloy 800 continues to exhibit good corrosion resistance, whereas other austenitic alloys are unsatisfactory for such service.

Alloy 800 has excellent resistance to nitric acid at concentrations up to about 70%. It resists a variety of oxidizing salts, but not halide salts. It also has good resistance to organic acids, such as formic, acetic, and propionic. Alloy 800 is particularly suited for the handling of hot corrosive gases, such as hydrogen sulfide.

In aqueous service, alloy 800 has general corrosion resistance that falls between types 304 and 316 stainless steels. Therefore, the alloy is not widely used for aqueous service. The stress-corrosion cracking resistance of alloy 800 is, while not immune, better than that of the 300 series of stainless steels and may be substituted on this basis.

Refer to Appendix A for the compatibility of Incoloy alloy 800 with selected corrodents.

Incoloy Alloy 825

Alloy 825 is very similar to alloy 800; however, the composition has been modified to provide for improved aqueous corrosion resistance.

The higher nickel content of alloy 825, compared to alloy 800, makes it resistant to chloride ion stress-corrosion cracking. Additions of molybdenum and copper give resistance to pitting and to corrosion in reducing acid environments such as sulfuric or phosphoric acid solutions. Alloy 825 is resistant to pure sulfuric acid solution up to 40% by weight at boiling temperatures and at all concentrations at a maximum temperature of 150°F (60°C). In dilute solutions the presence of oxidizing salts, such as cupric or ferric, actually reduce the corrosion rates. It has limited use in hydrochloric or hydrofluoric acids.

The chromium content of alloy 825 gives it resistance to various oxidizing environments such as nitrates, nitric acid solutions, and oxidizing salts. The alloy is not fully resistant to stress-corrosion cracking when tested in boiling magnesium chloride, but is has good resistance in neutral chloride environments.

If localized corrosion is a problem with the 300 series stainless steels, alloy 825 may be substituted. Alloy 825 also provides excellent resistance to corrosion by seawater.

Refer to Appendix A for the compatibility of alloy 825 with selected corrodents.

Alloy B-2

Within the nickel-molybdenum series one alloy—B-2—is a low-carbon and low-silicon version of alloy B. These alloys are uniquely different from other corrosion-resistant alloys because they do not contain chromium. Molybdenum is the primary alloying ingredient and provides significant corrosion resistance to reducing environments.

Alloy B-2 is recommended for the handling of all concentrations of hydrochloric acid from 158 to 212°F (70 to 100°C) and for the handling of wet hydrogen chloride gas.

Alloy B-2 has excellent resistance to pure sulfuric acid at all concentrations and temperatures below 60% acid concentration and good resistance to 212°F (100°C) above 60% acid concentration. It is also resistant to nonoxidizing environments such as hydrofluoric and phosphoric acids, and several organic acids such as acetic, formic, and crysilic acids. It is also resistant to many nonoxidizing chloride-bearing salts as aluminum chloride, magnesium chloride, and antimony chloride.

Because alloy B-2 is nickel rich (approximately 70%), it is resistant to chloride-induced stress-corrosion cracking. By virtue of the high molybdenum content it is highly resistant to pitting attack in acid chloride environments.

The major disadvantage of alloy B-2 is its lack of corrosion resistance in oxidizing environments. It has virtually no corrosion resistance to oxidizing acids such as nitric and chromic or to oxidizing salts such as ferric chloride or cupric chloride. The presence of oxidizing salts such as ferric chloride, ferric sulfate, or cupric chloride, even when present in the part per million range, can accelerate the attack in hydrochloric or sulfuric acids. Even dissolved oxygen has sufficient oxidizing power to affect the corrosion rates in hydrochloric acid.

Refer to Appendix A for the compatibility of alloy B-2 with selected corrodents.

Alloy C-276

Alloy C-276 is a nickel-molybdenum-chromium-tungsten alloy with outstanding corrosion resistance that is maintained even in the welded condition.

Alloy C-276 possesses good resistance in both oxidizing and reducing media, including conditions with halogen ion contamination. The pitting and crevice corrosion resistance of this alloy makes it an excellent choice when dealing with acid chloride salts.

Alloy C-276 has exceptional resistance to highly oxidizing neutral and acid chlorides, solvents, chlorine, formic and acetic acids, and acetic anhydride. It also resists highly corrosive agents such as wet chlorine gas, hypochlorites, and chlorine solutions.

Alloy C-276 has excellent resistance to phosphoric acid. At all temperatures below the boiling point of phosphoric acid, when concentrations are less than 65% by weight, tests have shown corrosion rates of less than 5 mpy. At acid concentrations above 65% by weight and up to 85%, alloy C-276 displays similar corrosive behavior, except at temperatures between 240°F (116°C) and the boiling point where corrosion rates may be erratic and may reach 25 mpy.

Refer to Appendix A for the compatibility of alloy C-276 with selected corrodents.

Alloy G and G-3

Alloys G and G-3 are highly resistant to pitting and stress-corrosion cracking in both acid and alkaline environments, including hot sulfuric and phosphoric acids, hydrofluoric and contaminated nitric acids, mixed acids, and sulfate compounds.

Refer to Appendix A for the compatibility of alloy G and G-3 with selected corrodents.

2.5 Aluminum Alloys

Wrought alloys of aluminum are of two types: non-heat-treatable of the 1XXX, 3XXX, 4XXX, and 5XXX series, and heat-treatable of the 2XXX, 6XXX, and 7XXX series. All non-heat-treatable alloys have a high degree of corrosion resistance.

The non-heat-treatable alloys, which do not contain copper as a major alloying ingredient, have a high resistance to corrosion by many chemicals. They are compatible with dry salts of most inorganic salts and, within their passive range of pH 4–9, with aqueous solutions, mostly halide salts, under conditions at which the alloys are polarized to their pitting potentials. In most other solutions where conditions are less likely to occur that will polarize the alloys to these potentials, pitting is not a problem.

Aluminum alloys are not compatible with most inorganic acids, bases and salts with pH outside the passive range of the alloys (pH 4–9).

Aluminum alloys are resistant to a wide variety of organic compounds, including most aldehydes, esters, ethers, hydrocarbons, ketones, mercaptans, other sulfur-containing compounds, and nitro compounds. They are also resistant to most organic acids, alcohols, and phenols, except when these compounds are nearly dry and near their boiling points. Carbon tetrachloride also exhibits this behavior.

Aluminum alloys are most resistant to organic compounds halogenated with chlorine, bromine, and iodine. They are also resistant to highly polymerized compounds.

Note that the compatibility of aluminum alloy with mixtures of organic compounds cannot always be predicted from their compatibility with each of the compounds. For example, some aluminum alloys are corroded severely in mixtures of carbon tetrachloride and methyl alcohol, even though they are resistant to each compound alone. Caution should also be exercised in using data for pure organic compounds to predict performances of the alloys with commercial grades that may contain contaminants. Ions of halides and reducible metals, commonly chloride and copper, frequently have been found to be the cause of excessive corrosion of aluminum alloys in commercial grades of organic chemicals that would not have been predicted from their resistance to pure compounds.

Refer to Appendix A for the compatibility of aluminum with selected corrodents.

2.6 Copper

Copper exhibits good corrosion resistance. Many of the corrosion products that form on the copper produce adherent, relatively impervious films that provide the corrosion protection. When exposed to the atmosphere over a long time, copper forms a coloration on the surface known as a patina, which in reality is a corrosion product that acts as a protective film against further corrosion.

Copper is resistant to urban, marine, and industrial atmospheres. Pure copper is immune to stress-corrosion cracking. Copper is widely used in the handling of potable water. However, aggressive well waters can cause pitting. Waters with a pH below 7.8 and having less than 42 ppm potassium, less than 25 ppm nitrate, or more than 26 ppm silicate will cause pitting of copper.

Sodium and potassium hydroxide solutions can be handled at room temperature in all concentrations. Copper is not corroded by perfectly dry ammonia but may be rapidly corroded by moist ammonia and ammonium hydroxide solutions. Alkaline salts such as sodium carbonate, sodium phosphate, or sodium silicate act like hydroxides but are less corrosive.

Copper is rapidly corroded by oxidizing acids such as nitric and chromic. Organic acids are generally less corrosive than the mineral acids.

Refer to Appendix A for the compatibility of copper with selected corrodents.

2.7 Copper Alloys

To increase copper's strength and engineering usefulness, it is alloyed with other elements, such as zinc, aluminum, nickel, tin, and so on. The various alloy families are as follows:

Principal alloying element	Alloy family
Zinc	Brasses
Tin, aluminum, or silicon	Bronzes
Nickel	Copper nickels

There are other alloy families, but they are of little importance in the field of corrosion resistance and will not be covered.

Brasses

Brasses contain zinc as their principal alloying ingredient. Other alloying additions are lead, tin, and aluminum. Lead is added to improve machinability and does not improve corrosive resistance. The addition of approximately 1% tin increases the dealloying resistance of the alloys. Aluminum is added to stabilize the protective surface film. Alloys containing in excess of 15% zinc are susceptible to dealloying in environments such as acids, both organic and inorganic, dilute and concentrated alkalies, neutral solutions of chlorides and sulfate, and mild oxidizing agents such as hydrogen peroxide. As stated earlier, the addition of small amounts of tin improves the dezincification resistance of these alloys. The following brasses are those most commonly used for corrosive-resistant applications:

Copper alloy UNS no.	
C-27000	C-46400
C-28000	C-46500
C-44300	C-46600
C-44400	C-46700
C-44500	C-68700

The admiralty brasses, copper alloys UNS 44300 through C-44500, and the naval brasses, C-46400 through C-46700, owe their dealloying resistance to the tin.

The high-zinc brasses, such as C-27000, C-28000, C-44300, and C-46400 resist sulfides better than do the low-zinc brasses. Dry hydrogen sulfide is well resisted.

Alloys containing 15% or less of zinc resist dealloying and are generally more corrosion resistant than high-zinc-bearing alloys. These alloys are resistant to many acids, alkalies, and salt solutions that cause dealloying in the high-zinc brasses. Dissolved air, oxidizing materials, such as chlorine and ferric salts, compounds that form soluble copper complexes (e.g., ammonia), and compounds that react directly with copper (e.g., sulfur and mercury) are corrosive to the low-zinc brasses. These alloys are more resistant to stress-corrosion cracking than are the high-zinc-containing alloys. Red brass is a typical alloy in this group containing 15% zinc. It has basically the same corrosion resistance as copper.

Refer to Appendix A for the compatibility of red brass with selected corrodents.

Aluminum Bronze

Aluminum bronzes are not affected by pitting, crevice corrosion, or stress corrosion. They are resistant to nonoxidizing mineral acids, such as sulfuric and phosphoric. Alkalies such as sodium and potassium hydroxides can be handled by aluminum bronzes. They also resist many organic acids, such as acetic, citric, formic, and lactic. In some instances their use may be limited in these services because of the possibility of copper pickup in the finished product. However, a polishing step such as charcoal filtration easily removes the color contaminant.

Refer to Appendix A for the compatibility of aluminum bronze with selected corrodents. Corrosion data for both wrought and cast copper alloys are also available from Copper Development Association, 260 Madison Avenue, New York, New York 10016.

2.8 Titanium

Of the four grades of titanium available, unalloyed titanium, represented by grade 2 is the most widely used for corrosion resistance. Its exceptional corrosion resistance is due to a stable, protective, strongly adherent oxide film that covers its surface. This film forms instantly when a fresh surface is exposed to air and moisture. The oxide film on titanium, although very thin, is very stable and is attacked by only a few substances, most notable of which is hydrofluoric acid. Because of its strong affinity for oxygen, titanium is capable of healing ruptures in the film almost instantly in any environment where a trace of moisture or oxygen is present. Therefore, we find that titanium is impervious to attack by moist chlorine gas. If the moisture content of dry chlorine gas falls below a critical level of 0.5%, rapid or even catastrophic attack can occur. Anhydrous conditions in the absence of a source of oxygen should be avoided with titanium because the protective film may not be regenerated if it is damaged.

Titanium offers excellent resistance to oxidizing acids such as nitric and chromic acids. It has limited resistance to reducing acids. Hydrofluoric acid in very small amounts will attack titanium.

In general, titanium is quite resistant to organic acids. Its behavior is dependent on whether the environment is reducing or is oxidizing. Only a few organic acids are known to attack titanium. Among those are hot nonaerated formic acid and solutions of sulfamic acid.

Refer to Appendix A for the compatibility of titanium with selected corrodents.

2.9 Zirconium

Hafnium occurs naturally in ores with zirconium. It has chemical and metallurgical properties similar to those of zirconium. Grade 702 (ASTM designation R-60702) is an unalloyed zirconium which contains approximately 4.5% hafnium. Since this grade has the best overall corrosion resistance, it is the most widely used.

Zirconium exhibits excellent resistance to many chemical solutions, even at elevated temperatures and pressures. When exposed to an oxygen-containing environment, an adherent protective oxide film forms on its surface. This film is formed spontaneously in air or water at ambient temperatures and below, is self-healing, and protects the base metal from chemical and mechanical attack at temperatures up to 572°F (300°C).

As a result, zirconium resists attack by mineral acids, alkaline solutions, most organic and salt solutions, and molten alkalies. It has excellent oxidation resistance up to 752°F (400°C) in air, steam, carbon dioxide, sulfur dioxide, nitrogen, and oxygen. It will not be attacked by oxidizing media unless halides are present.

It is attacked by hydrofluoric acid, wet chlorine, concentrated sulfuric acid, aqua regia, ferric chloride, and cupric chloride.

Refer to Appendix A for the compatibility of zirconium with selected corrodents.

2.10 Tantalum

Because of the relatively high cost of tantalum, applications are limited to extremely corrosive conditions. Tantalum is inert to practically all organic and inorganic compounds at temperatures below 320°F (150°C). The only exceptions to this are hydrofluoric acid and fuming sulfuric acid. At temperatures below 302°F (150°C), it is inert to all concentrations of nitric acid (except fuming), to 98% sulfuric acid, to 85% phosphoric acid and to aqua regia.

Fuming sulfuric acid attacks tantalum even at room temperature. Similarly, hydrofluoric acid, anhydrous hydrogen fluoride, or any acid medium containing fluoride ion will rapidly attack the metal. Commercial phosphoric acid may attack tantalum if small amounts of fluoride impurity are present. Hot oxalic acid is the only organic acid known to attack tantalum. It is also attacked by concentrated alkaline solutions at room temperature, although it is fairly resistant to dilute solutions.

Refer to Appendix A for the compatibility of tantalum with selected corrodents.

SUGGESTED READING

Schweitzer, P. A., ed., *Corrosion and Corrosion Protection Handbook*, 2nd ed., Marcel Dekker, New York, 1989.

Schweitzer, P. A., *Corrosion Resistance Tables*, Fourth ed., vols. 1–3, Marcel Dekker, New York, 1995.

Schweitzer, P. A., *Corrosion Resistant Pipe Systems*, Marcel Dekker, New York, 1994.

Mansfield, F., ed., *Corrosion Mechanisms*, Marcel Dekker, New York, 1986.

3 POLYMERS

As discussed previously, metallic materials undergo a specific corrosion rate as a result of an electrochemical reaction. Because of this it is possible to predict the life of a metal when in contact with a specific corrodent under a given set of conditions. This is not the case with polymeric materials. Plastic materials do not experience a specific corrosion rate. They are usually completely resistant to chemical attack or they deteriorate rapidly. They are attacked either by chemical reaction or by solvation. Solvation is the penetration of the plastic by a corrodent, which causes softening, swelling, and ultimate failure. Corrosion of plastics can be classified in the following ways as to the attack mechanism:

1. Disintegration or degradation of a physical nature due to absorption, permeation, solvent action, or other factors
2. Oxidation, where chemical bonds are attacked
3. Hydrolysis, where ester linkages are attacked
4. Radiation
5. Thermal degradation involving depolymerization and possibly repolymerization
6. Dehydration (rather uncommon)
7. Any combination of the above

Results of such attacks will appear in the form of softening, charring, crazing, delamination, embrittlement, discoloration, dissolving, or swelling.

The corrosion resistance of polymer matrix composites is also affected by two other factors: the nature of the laminate and, in the case of the thermoset resins, the cure. Improper or insufficient cure time will adversely affect the corrosion resistance, whereas proper cure time and procedures will generally improve the corrosion resistance.

All of the polymers are compounded. The final product is produced to certain specific properties for a specific application. When the corrosion resistance of a polymer is discussed, the data referred to are that of the pure polymer. In many instances other ingredients are blended with the polymer to enhance certain properties, which in many cases reduce the ability of the polymer to resist the attack of some media. Therefore it is essential to know the makeup of any polymer prior to its use.

3.1 Thermoplasts

A general rule as to the differences in the corrosion resistance of the thermoplasts may be derived from the periodic table. In the periodic table the basic elements of nature are organized by atomic structure as well as by chemical nature. The elements are placed into classes with similar properties, i.e., elements and compounds that exhibit similar behavior. These classes are the alkali metals, alkaline earth metals, transition metals, rare earth series, other metals, nonmetals, and noble (inert) gases.

The category known as halogens is of particular importance and interest in the case of thermoplasts. These elements include fluorine, chlorine, bromine, and iodine. They are the most electronegative elements in the periodic table, making them the most likely to attract an electron from another element and become a stable structure. Of all the halogens, fluorine is the most electronegative, permitting it to bond strongly with carbon and hydrogen atoms but not well with itself. The carbon-fluorine bond is predominant in PVDF and is responsible for the important properties of these materials. These are among the strongest known organic compounds. The fluorine acts like a protective shield for other bonds of lesser strength within the main chain of the polymer. The carbon-hydrogen bond, of which such plastics as PE and PP are composed, is considerably weaker. The carbon-chlorine bond, a key bond in PVC, is still weaker.

The arrangement of the elements in the molecule, the symmetry of the structure, and the degree of branching of the polymer chains are as important as the specific elements contained in the molecule. Plastics containing the carbon-hydrogen bonds, such as PP and PE, and carbon-chlorine bonds such as PVC, ECTFE, and CTFE are different in the important property of chemical resistance from a fully fluorinated plastic such as PTFE.

The fluoroplastic materials are divided into two groups: (1) fully fluorinated fluorocarbon polymers, such as PTFE, FEP, and PFA, called perfluoropolymers, and (2) partially fluorinated polymers, such as ETFE, PVDF, and ECTFE, called fluoropolymers. The polymeric characteristics within each group are similar, but there are important differences between the groups.

Polyvinyl Chloride (PVC)

There are two basic types of PVC produced: Type 1, which is a rigid unplasticized PVC that has optimum chemical resistance, and Type 2, which has optimum impact resistance but reduced chemical resistance. Unplasticized PVC resists attack by most acids and strong alkalies as well as gasoline, kerosene, aliphatic alcohols, and hydrocarbons. It is particularly useful in the handling of inorganic materials such as hydrochloric acid. It has been approved by the National Sanitation Foundation for the handling of potable water. Type 2 PVC's resistance to oxidizing and highly alkaline mediums is reduced.

PVC may be attacked by aromatics, chlorinated organic compounds, and lacquer solvents.

Refer to Appendix A for the compatibility of PVC with selected corrodents.

Chlorinated Polyvinyl Chloride (CPVC)

Although the corrosion resistance of CPVC is similar to that of PVC, there are enough differences that prevent CPVC from being used in all environments where PVC is used. In general, CPVC cannot be used in the presence of most polar organic materials, including chlorinated or aromatic hydrocarbons, esters, and ketones. It is compatible for use with most acids, alkalies, salts, halogens, and many corrosive wastes.

Refer to Appendix A for the compatibility of CPVC with selected corrodents.

Polypropylene (PP)

Polypropylene is available as a homopolymer or a copolymer. The homopolymers are generally long-chain, high-molecular-weight molecules with a minimum of random molecular orientation, thus optimizing their chemical, thermal, and physical properties. For maximum corrosion resistance homopolymers should be used.

PP is resistant to saltwater, crude oil, sulfur-bearing compounds, caustic, solvents, acids, and other organic chemicals. It is not recommended for use with oxidizing-type acids, detergents, low-boiling hydrocarbons, alcohols, aromatics, and some chlorinated organic materials. Unpigmented PP is degraded by ultraviolet light.

Refer to Appendix A for the compatibility of PP with selected corrodents.

Polyethylene (PE)

Polyethylene material varies from type to type depending upon the molecular structure, its crystallinity, molecular weight, and molecular weight distribution. The terms low, high, and medium density refer to the ASTM designations based on the unmodified polyethylene. The densities, being related to the molecular structure, are indicators of the properties of the final product. High molecular weight (HMW) and ultrahigh molecular weight (UHMW) are the two forms most often used for corrosion resistance. When PE is exposed to ultraviolet radiation, usually from the sun, photo-, or light, oxidation will occur. To protect against this, carbon black must be incorporated into the resin to stabilize it. Other types of stabilizers will not provide complete protection.

PE exhibits a wide range of corrosion resistance ranging from potable water to corrosive wastes. It is resistant to most mineral acids, including sulfuric up to 70% concentration, inorganic salts, including chlorides, alkalies, and many organic acids. It is not resistant to bromine, aromatics, or chlorinated hydrocarbons.

Refer to Appendix A for the compatibility of HMW and UHMW with selected corrodents.

Polybutylene

The combination of stress-cracking resistance, chemical resistance, and abrasion resistance makes this polymer extremely useful. It is resistant to acids, bases, soaps, and detergents up to 200°F (93°C). It is not resistant to aliphatic solvents at room temperatures and is partially soluble in aromatic and chlorinated hydrocarbons. Chlorinated water will cause pitting attack.

Polyphenylene Sulfide (PPS) (Ryton)

Ryton exhibits good resistance to aqueous inorganic salts and bases and is inert to many organic solvents. It is also used in oxidizing environments. Refer to Appendix A for the compatibility of Ryton with selected corrodents.

Polycarbonate

Polycarbonate has exceptional weatherability and good corrosion resistance to mineral acids. Organic solvents will attack the polymer. Strong alkalies will decompose it. It is sold under the trade name Lexan.

Polyetheretherketone (PEEK)

PEEK exhibits excellent corrosion resistance to a wide range of organic and inorganic chemicals. It is resistant to acetic, nitric, hydrochloric, phosphoric, and sulfuric acids, among others. Refer to Appendix A for the compatibility of PEEK with selected corrodents.

Polyethersulfone (PES)

PES resists most inorganic chemicals, but is attacked by strong oxidizing acids. It has excellent resistance to aliphatic hydrocarbons and aromatics, is soluble in highly polar solvents, and is subject to stress cracking in certain solvents, notably ketones and esters.

Hydrocarbons and mineral oils, greases, and transmission fluids have no effect on PES. Refer to Appendix A for the compatibility of PES with selected corrodents.

Phenolics

Phenolics are relatively inert to acids but have little alkaline or bleach resistance. They exhibit a wider range of corrosion resistance as a composite material with a glass filling. Refer to Appendix A for the compatibility of phenolics with selected corrodents.

ABS

The thermoplastic resin ABS is resistant to aliphatic hydrocarbons but not to aromatic and chlorinated hydrocarbons. Refer to Appendix A for the compatibility of ABS with selected corrodents.

Vinylidene Fluoride (PVDF)

Vinylidene fluoride is chemically resistant to most acids, bases, and organic solvents. It also has the ability to handle wet or dry chlorine, bromine, and other halogens.

PVDF is not suitable for use with strong alkalies, fuming acids, polar solvents, amines, ketones, and esters. When used with strong alkalies, it is subject to stress cracking.

Refer to Appendix A for the compatibility of PVDF with selected corrodents.

Ethylene Chlorotrifluoroethylene (ECTFE)

The chemical resistance of ECTFE is outstanding. It is resistant to strong mineral and oxidizing acids, alkalies, metal etchants, liquid oxygen, and practically all organic solvents except hot amines, such as aniline, dimethylamine, etc. Severe stress tests have shown that ECTFE is not subject to chemically induced stress cracking from strong acids, bases, or solvents. Some halogenated solvents can cause ECTFE to become slightly plasticized when it contacts them. After removing the solvent and drying the ECTFE, its mechanical properties return to their original values, indicating that no chemical attack has taken place.

ECTFE will be attacked by metallic sodium and potassium and fluorine. Refer to Appendix A for the compatibility of ECTFE with selected corrodents.

Ethylene Tetrafluoroethylene (ETFE)

ETFE is inert to strong mineral acids, inorganic bases, halogens, and metal salt solutions. Even carboxylic acids, anhydrides, aromatic and aliphatic hydrocarbons, alcohols, aldehydes, ketones, ethers, esters, chlorocarbons, and classic polymer solvents have little effect on ETFE.

Very strong oxidizing acids, such as nitric, near their boiling points at high concentration will attack EFTE in varying degrees, as will organic bases, such as amines, and sulfonic acids.

Refer to Appendix A for the compatibility of ETFE with selected corrodents.

Polytetrafluoroethylene (PTFE)

PTFE is unique in its corrosion-resistant properties. Very few chemicals will attack PTFE within normal-use temperature. Elemental sodium in intimate contact removes fluorine from the polymer molecule. The other alkali metals (potassium, lithium, etc.) react in a similar manner.

Fluorine and related compounds (e.g., chlorine trifluoride) are absorbed into the PTFE resin with such intimate contact that the mixture becomes sensitive to a source of ignition such as impact. These potent oxidizers should only be handled with great care and a recognition of the potential hazards.

The handling of 80% sodium hydroxide, aluminum chloride, ammonia, and certain amines at high temperatures may produce the same effect as elemental sodium. Also, slow oxidative attack can be produced by 70% nitric acid under pressure and at 480°F (250°C).

Refer to Appendix A for the compatibility of PTFE with selected corrodents.

Fluorinated Ethylene Propylene (FEP)

FEP basically exhibits the same corrosion resistance as PTFE, with a few exceptions, but at a lower operating temperature. It is resistant to practically all chemicals, exceptions being the extremely potent oxidizing agents such as chlorine trifluoride and related compounds. Some chemicals will attack FEP when present in high concentrations at or near the service temperature limit of 400°F (204°C).

Refer to Appendix A for the compatibility of FEP with selected corrodents.

Perfluoroalkoxy (PFA)

PFA is inert to strong mineral acids, inorganic bases, inorganic oxidizers, aromatics, some aliphatic hydrocarbons, alcohols, aldehydes, ketones, ethers, esters, chlorocarbons, fluorocarbons, and mixtures of the aforementioned.

PFA will be attacked by certain halogenated complexes containing fluorine. These include chlorine trifluoride, bromine trifluoride, iodine pentachloride, and fluorine. It can also be attacked by such metals as sodium or potassium, particularly in their molten states.

Refer to Appendix A for the compatibility of PFA with selected corrodents.

3.2 Elastomers

Elastomeric materials fail in the same manner as other polymeric materials. When an elastomer is used as a lining material for a tank or piping, additional consideration must be given to the problems of permeation and absorption.

Physical properties of the elastomer determine the reaction of the elastomer to such physical actions as permeation and absorption. If a lining material is subject to permeation by a corrosive chemical, it is possible for the base metal to be attacked and corroded even though the lining material itself is unaffected. Because of this, permeation and absorption must be taken into account when specifying a lining material.

Permeation

All materials are somewhat permeable to chemical molecules, but plastic materials tend to be an order of magnitude greater in their permeability rates than metals. Gases, vapors, or liquids will permeate polymers.

Permeation is a molecular migration either through microvoids in the polymer (if the polymer is more or less porous) or between polymer molecules. In neither case is there any attack on the polymer. This action is strictly a physical phenomenon.

Permeation can result in

1. Failure of the substrate from corrosive attack.
2. Bond failure and blistering, resulting from accumulation of fluids at the bond when the substrate is less permeable than the liner, or from corrosion/reaction products if the substrate is attacked by the permeant.
3. Loss of contents through substrate and liner as a result of the eventual failure of the substrate. In unbonded linings it is important that the space between the liner and support member be vented to the atmosphere to allow minute quantities of permeant vapors to escape and to prevent expansion of entrapped air from collapsing the liner.

Permeation is a function of two variables, one relating to diffusion between molecular chains and the other to the solubility of the permeant in the polymer. The driving forces of diffusion are the partial pressure gradient for gases and the concentration gradient for liquids. Solubility is a function of the affinity of the permeant for the polymer.

There is no relation between permeation and the passage of materials through cracks and voids, even though in both cases migrating chemicals travel through the polymer from one side to the other.

The user has some control over permeation, which is affected by

1. Temperature and pressure
2. Permeant concentration
3. Polymer thickness

Increasing temperature increases permeation rate, since the permeant solubility in the polymer increases. As temperature rises, polymer chain movement is stimulated, permitting more permeants to diffuse among the chain more easily. For many gases the permeation rates increase linearly with the partial pressure gradient, and the same effect is experienced with concentration gradients of liquids. If the permeant is highly soluble in the polymer, the permeability increase may be nonlinear. Permeation is generally decreased by the square of the thickness. For general corrosion resistance, thicknesses of 0.010–0.020 in. are usually satisfactory.

Polymer density, as well as thickness, affect permeation rate: The greater the density of the polymer, the fewer the voids through which permeation can take place. A comparison of the density of sheets produced from different polymers does not provide any indication of the relative permeation rates. However, a comparison of the density of sheets produced from the same polymer indicates the relative permeation rates: the denser the sheet, the lower the permeation rate.

Other factors affecting permeation consisting of chemical and physiochemical properties are

1. Ease of condensation of the permeant: chemicals that condense readily permeate at higher rates.
2. The higher the intermolecular chain forces (e.g., van der Waals hydrogen bonding) of the polymer, the lower the permeation rate.
3. The higher the level of crystallinity in the polymer, the lower the permeation rate.
4. The greater the degree of cross-linking within the polymer, the lower the permeation rate.
5. Chemical similarity between the polymer and permeant. When the polymer and permeant have similar functional groups, the permeation rate increases.
6. The smaller the molecule of the permeant, the greater the permeation rate.

Absorption

Unlike metals, polymers will absorb varying quantities of the corrodents they contact, especially organic liquids. This can result in swelling, cracking, and penetration to the substrate. Swelling can cause softening of the polymer. If the polymer has a high absorption rate, permeation will probably occur. An approximation of the expected permeation and/or absorption of a polymer can be based on the absorption of water. These data are usually available. Table 4 provides the water absorption rates of the more common polymers used for linings.

Environmental Stress Cracking

When a tough polymer is stressed for a period of time under loads that are small relative to the polymers yield point, stress cracks develop. Crystallinity is an important factor affecting stress-corrosion cracking. The less crystallization that occurs, the less the likelihood of stress cracking; unfortunately, the lower the crystallinity, the greater the likelihood of permeation.

Contaminants in the fluid may accelerate stress-corrosion cracking. For example, polypropylene can safely handle sulfuric and hydrochloric acids. However, iron or copper contamination in concentrated sulfuric or hydrochloric acid can result in stress cracking of polypropylene.

Table 4 Water Absorption Rates of Polymers

Polymer	Water absorption, 24 hr at 73°F (23°C) (%)
PVC	0.05
CPVC	0.03
PP (homo)	0.02
PP (co)	0.03
EHMW PE	<0.01
ECTFE	<0.1
PVDF	<0.04
PFA	<0.03
ETFE	0.029
PTFE	<0.01
FEP	<0.01

Outdoor Use

Elastomers in outdoor use can degrade as a result of the weathering action of ozone, oxygen, and sunlight. Surface cracking, discoloration of colored stocks, and serious loss of tensile strength, elongation, and other rubberlike properties are the result.

Natural Rubber (NR)

Cold water preserves natural rubber, but if exposed to air, particularly in sunlight, rubber tends to become hard and brittle. It has only fair resistance to ozone. In general, it has poor weathering and aging properties.

Natural rubber offers excellent resistance to most inorganic salt solutions, alkalies, and nonoxidizing acids. Hydrochloric acid will react with rubber, forming rubber hydochloride. Strong oxidizing media such as nitric acid, concentrated sulfuric acid, permanganates, dichromates, chlorine dioxide, and sodium hypochlorite will severely attack rubber. Mineral and vegetable oils, gasoline, benzene, toluene, and chlorinated hydrocarbons also affect rubber. Natural rubber offers good resistance to radiation and alcohols.

Refer to Appendix A for the compatibility of natural rubber with selected corrodents.

Isoprene Rubber (IR)

This is the synthetic form of natural rubber and as such can be used in the same applications as natural rubber.

Neoprene (CR)

Neoprene possesses excellent resistance to sun, weather, and ozone. Because of its low rate of oxidation, neoprene has a high resistance to both outdoor and indoor aging. If severe ozone is to be expected, as for example around electrical equipment, neoprene can be compounded to resist thousands of parts per million of ozone for hours without surface cracking. Natural rubber will crack within minutes when exposed to ozone concentrations of only 50 ppm.

Neoprene provides excellent resistance to attack from solvents, fats, waxes, oils, greases, and many other petroleum-based products. A minimum amount of swelling and relatively little loss of strength are experienced when in contact with aliphatic compounds (methyl and ethyl alcohols, ethylene glycols, etc.), aliphatic hydrocarbons, and most freon refrigerants. Neoprene is also resistant to dilute mineral acids, inorganic salt solution, and alkalies.

Neoprene has only limited serviceability when exposed to chlorinated and aromatic hydrocarbons, organic esters, aromatic hydroxy compounds, and certain ketones. Highly oxidizing acid and salt solutions cause surface deterioration and loss of strength. This includes such materials as nitric acid and concentrated sulfuric acid.

Refer to Appendix A for the compatibility of neoprene with selected corrodents.

Butadiene-Styrene Rubber (SBR, Buna-S, GR-S)

Buna-S has poor weathering and aging properties. Sunlight causes it to deteriorate. It does have better water resistance than natural rubber.

The chemical resistance of Buna-S is similar to that of natural rubber. It is resistant to water and exhibits fair to good resistance to dilute acids, alkalies, and alcohols. It is not resistant to oils, gasoline, hydrocarbons, or oxidizing agents.

Refer to Appendix A for the compatibility of nitrile rubbers with selected corrodents.

Butyl Rubber (IIR) and Chlorobutyl Rubber (CIIR)

Butyl rubber has excellent resistance to sun, weather, and ozone. Its resistance to water absorption and its weathering qualities are outstanding. Butyl rubber is resistant to dilute mineral acids, alkalies, phosphate ester oils, acetone, ethylene, ethylene glycol, and water. It is resistant to swelling by vegetable and animal oils. However, it is not resistant to concentrated nitric and sulfuric acids, petroleum oils, gasoline, and most solvents, except oxygenated solvents.

Chlorobutyl rubber exhibits the same general resistance as natural rubber but can be used at a higher temperature. It cannot be used with hydrochloric acid even though butyl rubber is suitable.

Refer to Appendix A for the compatibility of butyl rubber and chlorobutyl rubber with selected corrodents.

Chlorosulfonated Polyethylene Rubber (Hypalon)

Hypalon is one of the most weather-resistant elastomers available. Sunlight and ultraviolet light have little if any adverse effect on its physical properties. Many elastomers are degraded by ozone concentrations of 1 ppm in air, while Hypalon is unaffected by concentrations as high as 1 part per 100 parts of air.

Hypalon is capable of resisting attack by hydrocarbon oils and fuels, to such oxidizing chemicals as sodium hypochlorite, sodium peroxide, ferric chloride, and sulfuric, chromic, and hydrofluoric acids. Concentrated hydrochloric acid (37%) at temperatures above 158°F (70°C) will attack Hypalon, but can be handled in all concentrations below this temperature. Nitric acid at room temperature and up to 60% concentration can also be handled without adverse effects. Hypalon is also resistant to salt solutions, alcohols, and weak and concentrated alkalies, and is generally unaffected by soil chemicals, moisture, and other deteriorating factors associated with burial in the earth.

Hypalon has poor resistance to aliphatic, aromatic and chlorinated hydrocarbons, aldehydes, and ketones.

Refer to Appendix A for the compatibility of Hypalon with selected corrodents.

Polybutadiene Rubber (BR)

Polybutadiene has good weather resistance but will deteriorate when exposed to sunlight for extended periods of time. It also has poor resistance to ozone. In general, the chemical resistance of BR is similar to that of natural rubber.

Refer to Appendix A for the compatibility of BR with selected corrodents.

Ethylene-Acrylic Rubber (EA)

The EA elastomers have extremely good resistance to sun, weather, and ozone. Its resistance to water absorption is very good.

The EA elastomers exhibit good resistance to hot oils, hydrocarbon or glycol-based lubricants, and to transmission and power steering fluids. Good resistance is also displayed to dilute acids, aliphatic hydrocarbons, gasoline, and animal and vegetable oils.

The EA elastomers will be attacked by esters, ketones, highly aromatic hydrocarbons, and concentrated acids.

Refer to Appendix A for the compatibility of the EA elastomers with selected corrodents.

Acrylate-Butadiene Rubber (ABR) and Acrylic-Ester-Acrylic-Halide Rubbers (ACM)

The ABR and ACM rubbers exhibit good resistance to sun, weather, and ozone. They have excellent resistance to aliphatic hydrocarbons (gasoline, kerosene) and offer good resistance to water, acids, synthetic lubricants, and silicate hydraulic fluids.

These rubbers will be attacked when exposed to alkalies, aromatic hydrocarbons (benzene, toluene), halogenated hydrocarbons, alcohol, and phosphate hydraulic fluids.

Refer to Appendix A for the compatibility of the ABR and ACM with selected corrodents.

Ethylene-Propylene Rubbers (EPDM and EPT)

Ethylene-propylene rubbers are particularly resistant to sun, weather, and ozone attack. Ozone resistance is inherent in the polymer, and for all practical purposes it can be considered immune to ozone attack. It is not necessary to add any compounding ingredients to produce the immunity.

Ethylene-propylene rubbers are resistant to oxygenated solvents, such as acetone, methyl ethyl ketone, ethyl acetate, weak acids and alkalies, detergents, phosphate esters, alcohols, and glycols.

The elastomer will be attacked by hydrocarbon solvents and oils, chlorinated hydrocarbons, and turpentine. EPT rubbers, in general, are resistant to most of the same corrodents as EPDM.

Refer to Appendix A for the compatibility of EPT and EPDM with selected corrodents.

Styrene-Butadiene-Styrene Rubber (SBS)

The SBS rubbers are not resistant to sun, weather, or ozone. Their chemical resistance is similar to that of natural rubber. They have excellent resistance to water, acids, and bases.

Refer to Appendix A for the compatibility of SBS with with selected corrodents.

Styrene-Ethylene-Butylene-Styrene Rubber (SEBS)

The SEBS rubbers possess excellent resistance to ozone. For prolonged outdoor exposure the addition of an ultraviolet light absorber or carbon black pigment or both is required. The chemical resistance of the SEBS rubbers is similar to that of natural rubber, and they possess excellent resistance to water, acids, and bases.

Refer to Appendix A for the compatibility of SEBS rubbers with selected corrodents.

Polysulfide Rubbers (ST and FA)

FA polysulfide rubbers possess excellent resistance to ozone, weathering, and ultraviolet light. ST polysulfide rubber, compounded with carbon black, is resistant to ultraviolet light and sunlight. It also has satisfactory weather resistance.

The polysulfide rubbers exhibit excellent resistance to oils, gasoline, aliphatic and aromatic hydrocarbon solvents, good water and alkali resistance, and fair acid resistance. The FA polysulfide rubbers are more resistant to solvents than the ST polysulfide rubbers. The ST rubbers exhibit better resistance to chlorinated organics than the FA rubbers.

The polysulfide rubbers are not resistant to strong concentrated inorganic acids, such as sulfuric, nitric, and hydrochloric.

Refer to Appendix A for the compatibility of the polysulfide rubbers with selected corrodents.

Urethane Rubbers (AU)

The urethane rubbers exhibit excellent resistance to ozone attack and have good resistance to weathering. Extended exposure to ultraviolet light will cause the rubbers to darken and reduce their physical properties. The addition of pigments or ultraviolet-screening agents will prevent this.

The urethane rubbers are resistant to most mineral and vegetable oils, grease and fuels, and aliphatic, aromatic and chlorinated hydrocarbons.

Aromatic hydrocarbons, polar, solvents, esters, ethers, and ketones will attack the urethane rubbers. The urethane rubbers have limited service in weak acid solutions and cannot be used in concentrated acids. Neither are they resistant to caustic or steam.

Refer to Appendix A for the compatibility of urethane rubbers with selected corrodents.

Polyamides

Of the many varieties of polyamides (Nylons) produced, only grades 11 and 12 find application as elastomeric materials. They are resistant to sun, weather, and ozone.

The polyamides are resistant to most inorganic alkalies, particularly ammonium hydroxide and ammonia at elevated temperatures, and sodium and potassium hydroxide at ambient temperatures. They are also resistant to almost all inorganic salts and almost all hydrocarbons and petroleum-based fuels. At normal temperatures they are also resistant to organic acids (citric, lactic, oleic, oxalic, stearic, tartaric, and uric) and most aldehydes and ketones.

The polyamides have limited resistance to hydrochloric, sulfonic, and phosphoric acids at ambient temperatures.

Refer to Appendix A for the compatibility of the polyamides with selected corrodents.

Polyester Elastomer (PE)

Polyesters exhibit excellent resistance to ozone and good resistance to weathering. When formulated with proper additives they are capable of exhibiting very good resistance to sunlight aging.

Polyester elastomers have excellent resistance to nonpolar materials such as oils and hydraulic fluids, even at elevated temperatures. At room temperature they are resistant to most polar fluids, such as acids, bases, amines, and glycols. Resistance is very poor at temperatures of 158°F (70°C) or higher.

Refer to Appendix A for the compatibility of the polyester elastomers with selected corrodents.

Thermoplastic Elastomers, Olefinic Type (TPE)

The TPEs exhibit good resistance to sun, weather, and ozone. Their water resistance is excellent, showing essentially no property changes after prolonged exposure to water at elevated temperatures. The TPEs display reasonably good resistance to oils and automotive fluids, comparable to that of neoprene. However, they do not have the outstanding oil resistance of the polyester elastomers.

Refer to Appendix A for the compatibility of the TPEs with selected corrodents.

Silicone (SI) and Fluorosilicone (FSI) Rubbers

The SI and FSI rubbers display excellent resistance to sun, weathering, and ozone, even after long-term exposure. Silicone rubbers are resistant to dilute acids and alkalies, alcohols, animal and vegetable oils, lubricating oils, and aliphatic hydrocarbons. Aromatic solvents such as benzene, toluene, gasoline, and chlorinated solvents, and high-temperature steam will attack the SI rubber.

The FSI rubbers have better chemical resistance than the SI rubbers. They possess excellent resistance to aliphatic hydrocarbons, and good resistance to aromatic hydrocarbons, oil and gasoline, animal and vegetable oils, dilute acids and alkalies, and alcohols.

Refer to Appendix A for the compatibility of the SI and FSI rubbers with selected corrodents.

Vinylidene Fluoride (PVDF)

PVDF is highly resistant to the chlorinated solvents, aliphatic solvents, weak bases and salts, strong acids, halogens, strong oxidants, and aromatic solvents. Strong bases will attack PVDF. Sodium hydroxide can cause stress cracking.

Refer to Appendix A for the compatibility of PVDF with selected corrodents.

Fluoroelastomers (FKM)

Fluoroelastomers possess excellent weathering resistance to sunlight and ozone. They possess excellent resistance to oils, fuels, lubricants, most mineral acids, many aliphatic and aromatic hydrocarbons (carbon tetrachloride, benzene, toluene, xylene, that act as solvents for other rubbers), gasoline, naphtha, chlorinated solvents, and pesticides.

The FKM elastomers will be attacked by low-molecular-weight esters and ethers, ketones, certain amines, and hot anhydrous hydrofluoric or chlorosulfonic acids.

Refer to Appendix A for the compatibility of fluoroelastomers with selected corrodents.

Ethylene-Tetrafluoroethylene Elastomer (ETFE)

Because of ETFE's outstanding resistance to sunlight, ozone, and weather, coupled with its wide range of corrosion resistance, it is ideally suited for outdoor applications subject to atmospheric corrosion.

ETFE is inert to strong mineral acids, inorganic bases, halogens, and metal-salt solutions. Carboxylic acids, anhydrides, aromatic and aliphatic hydrocarbons, alcohols, aldehydes, ketones, esters, ethers, chlorocarbons, and classic polymer solvent have little effect on ETFE.

Strong oxidizing acids near their boiling points, such as nitric acid at high concentrations, organic bases, such as amines, and sulfonic acids have a deleterious effect on ETFE.

Refer to Appendix A for the compatibility of ETFE with selected corrodents.

Ethylene-chlorotrifluoroethylene Elastomer (ECTFE)

ECTFE is extremely resistant to sun, weather, and ozone attack. It is resistant to strong mineral and oxidizing acids, alkalies, metal etchants, liquid oxygen, and practically all organic solvents except hot amines (aniline, dimethylamine, etc.). ECTFE will be attacked by metallic sodium and potassium.

Refer to Appendix A for the compatibility of ECTFE with selected corrodents.

Perfluoroelastomers (FPM)

The FPM elastomers have excellent resistance to sun, weather, and ozone, even after long-term exposure. They are resistant to polar solvents (ketones, esters, ethers), strong organic solvents (benzene, dimethyl formamide), inorganic and organic acids (hydrochloric, nitric, sulfuric) and bases, strong oxidizing agents (fuming nitric acid), metal halides, chlorine, wet and dry, inorganic salt solutions, hydraulic fluids, and heat-transfer fluids.

Molten or gaseous alkali metals such as sodium will attack the FPM elastomers.

Refer to Appendix A for the compatibility of FPM elastomers with selected corrodents.

3.3 Thermosets

Thermoset resins are "families" of compounds rather than unique individual compounds. Similar to the thermoplast resins, they can be formulated to improve certain specific properties but often at the expense of another property. In the chemical corrosion field there are four important families of thermoset resins: polyesters, epoxies, vinyl esters, and furans.

Polyester Resins

The members of the polyester family of resins that are more important for their corrosion resistance are

Isophthalic resins
Bisphenol A–fumarate resins
Hydrogenated–bisphenol A–bisphenol A resins
Halogenated resins
Terephthalate resins

Isophthalic Resins

Isophthalic resins have a relatively wide range of corrosion resistance. They are satisfactory for use up to 125°F (52°C) in such acids as 10% acetic, benzoic, boric, citric, oleic, 25% phosphoric, tartaric, and 10–25% sulfuric and fatty acids. Most inorganic salts are also compatible with the isophthalic resins. Solvents such as amyl alcohols, ethylene glycol, formaldehyde, gasoline, kerosene, and naphtha are also compatible.

Isophthalic resins are not resistant to acetone, amyl acetone, benzene, carbon disulfides, solutions of alkaline salts of potassium and sodium, hot distilled water, and higher concentrations of oxidizing acids.

Refer to Appendix A for the compatibility of the isophthalic resins with selected corrodents.

Bisphenol Polyesters

Bisphenol polyesters are superior in their corrosion-resistant properties to isophthalic polyesters. They show good performance with moderate alkaline solutions and excellent resistance to the various categories of bleaching agents. Bisphenol polyesters break down under highly concentrated acids or alkalies. These resins can be used in the handling of the following materials:

	Acids (to 200°F/93°C)	
acetic	fatty acids	stearic
benzoic	hydrochloric (10%)	sulfonic (50%)
boric	lactic	tannic
butyric	maleic	tartaric
chloroacetic (15%)	oleic	trichloroacetic (50%)
chromic (5%)	oxalic	rayon spin bath
citric	phosphoric (80%)	
	Salts (solutions to 200°F/93°C)	
all aluminum salts		copper salts
most ammonium salts		iron salts
calcium salts		zinc salts
		most plating solutions
	Solvents (all solvents shown are for the isophthalic resins)	
sour crude oil		linseed oil
alcohols at ambient temperatue		glycerine
	Alkalies	
ammonium hydroxide 5%		potassium hydroxide 25%
calcium hydroxide 25%		sodium hydroxide 25%
calcium hypochlorite 20%		chlorite
chlorine dioxide 15%		hydrosulfite

Solvents such as benzene, carbon disulfide, ether, methyl ethyl ketone, toluene, xylene, trichloroethylene, and trichloroethane will attack the resin. Sulfuric acid above 70% concentration, 73% sodium hydroxide, and 30% chromic acid will also attack the resin.

Refer to Appendix A for the compatibility of the bisphenyl esters with selected corrodents.

Halogenated Resins

Halogenated resins consist of chlorinated or brominated polyesters. Excellent resistance is exhibited in contact with oxidizing acids and solutions, such as 35% nitric acid at elevated temperature and 70% nitric acid at room temperature, 40% chromic acid, chlorine water, wet chlorine, and 15% hypochlorites. They also resist neutral and acid salts, nonoxidizing acids, organic acids, mercaptans, ketones, aldehydes, alcohols, glycols, organic esters, and fats and oils.

These polyesters are not resistant to highly alkaline solutions of sodium hydroxide, concentrated sulfuric acid, alkaline solutions with pH greater than 10, aliphatic, primary, and aromatic amines, amides and other alkaline organics, phenol, and acid halides.

Refer to Appendix A for the compatibility of the halogenated polyesters with selected corrodents.

Vinyl Ester Resins

The family of vinyl ester resins includes a wide variety. As a result there can be a difference in the compatibility of formulations between manufacturers. When one checks compatibility in a table, one must keep in mind that all formulations may not act as shown. An indication that vinyl ester is compatible generally means that at least one formulation is compatible. The resin manufacturer must be consulted to verify the resistance.

In general, vinyl esters can be used to handle most hot, highly chlorinated, and acidic mixtures at elevated temperatures. They also provide excellent resistance to strong mineral acids and bleaching solutions. Vinyl esters excel in alkaline and bleach environments and are used extensively in the very corrosive conditions in the pulp and paper industry.

Refer to Appendix A for the compatibility of vinyl ester resins with selected corrodents.

Epoxies

The epoxy resin family exhibits good resistance to alkalies, nonoxidizing acids, and many solvents. Specifically these resins are compatible to acids such as 10% acetic, benzoic, butyric, 10% hydrochloric, 20% sulfuric, oxalic, and fatty. On the alkaline side they are compatible with 50% sodium hydroxide, 10% sodium sulfite, calcium hydroxide, trisodium phosphate, magnesium hydroxide, aluminum, barium, calcium, iron, magnesium, potassium, and sodium. Solvents such as methanol, ethanol, isopropanol, benzene, ethyl acetate, naphtha, toluene, and xylene can also be handled safely.

Bromine water, chromic acid, bleaches, fluorine, methylene chloride, hydrogen peroxide, sulfuric acid above 80%, wet chlorine gas, and wet sulfur dioxide will attack the epoxies.

Refer to Appendix A for the compatibility of the epoxy resins with selected corrodents.

Furans

Since there are different formulations of the furan resins, the supplier should be consulted as to the compatibility of a resin with the corrodents to be encountered. Corrosion charts indicate the compatibility of at least one formulation.

Furans' strong point is their excellent resistance to solvents in combination with acids and alkalies. They are compatible with the following corrodents:

Solvents	
acetone	methyl ethyl ketone
benzene	perchlorethylene
carbon disulfide	styrene
chlorobenzene	toluene
ethanol	trichloroethylene
methanol	
Acids	
acetic	phosphoric
hydrochloric	60% sulfuric
5% nitric	
Bases	
dimethylamine	sodium sulfide
sodium carbonate	50% sodium hydroxide

Furans are not resistant to bleaches, such as peroxides and hypochlorites, concentrated sulfuric acid, phenol, and free chlorine, or to higher concentrations of chromic or nitric acid.

Refer to Appendix A for the compatibility of the furans with selected corrodents.

SUGGESTED READING

Mallinson, J. H., *Corrosion-Resistant Plastic Composites in Chemical Plant Design*, Marcel Dekker, New York, 1988.

Schweitzer, P. A., ed., *Corrosion and Corrosion Protection Handbook*, 2nd ed., Marcel Dekker, New York, 1989.

Schweitzer, P. A., *Corrosion Resistance Tables*, 4th ed., vols. 1–3, Marcel Dekker, New York, 1995.

Schweitzer, P.A., *Corrosion Resistant Piping Systems*, Marcel Dekker, New York, 1994.

4 CERAMIC MATERIALS

Corrosion of ceramic materials is a complicated process. It can take place by one or a combination of mechanisms. In general, the environment will attack the ceramic forming a reaction product, which may be a gas, liquid, solid, or combination. When the reaction product formed is a solid, it may form a protective layer preventing further corrosion. If the reaction product formed is a combination of a solid and a liquid, the protective layer formed may be removed by the process of erosion.

The following fundamental concepts of chemistry help us understand the corrosion of a ceramic section: a ceramic with acidic character tends to be attacked by an environment with a basic character, and vice versa; ionic materials tend to be soluble in polar solvents and covalent materials tend to be soluble in nonpolar solvents; and the solubility of solids in liquids generally increases with increasing temperatures.

4.1 Crystalline Materials

Polycrystalline materials consist of several components. Corrosive attack of these materials starts with the least corrosion-resistant component, which normally is the ingredient used for bonding, or more generally the minor component of the material.

The corrosion of a solid crystalline material by a liquid can result from indirect or direct dissolution. In the former, an interface or reaction product is formed between the solid crystalline material and the solvent. This reaction product, being less soluble than the bulk solid, may or may not form an attached surface layer. In direct dissolution the solid crystalline material dissolves directly into the solvent.

When a silicate is leached by an aqueous solution, an ion is removed from a site within the crystal structure and is placed into the aqueous phase. Whether or not leaching occurs will depend upon the ease with which the ions may be removed from the crystal structure.

The corrosion of polycrystalline ceramic by a vapor can be more serious than attack by either liquids or solids. Porosity or permeability of the ceramic is one of the most important properties related to its corrosion by a vapor or gas. If the vapor can penetrate the material, the surface area exposed to attack is greatly increased and corrosion proceeds rapidly. A combined attack of vapor and liquid may also take place. In this situation the vapor may penetrate the material under a thermal gradient to a lower temperature, condense, and then dissolve material by liquid solution. The liquid solution can then penetrate further along temperature gradients until it freezes. If the thermal gradient of the material is changed, it is possible for the solid reaction products to melt, causing excessive corrosion and spalling at the point of melting.

If two dissimilar solid materials react when in contact with each other, corrosion can take place. Common types of reactions involve the formation of a solid, liquid, or gas. Solid-solid reactions are predominantly reactions involving diffusion.

Porosity plays an important role in the corrosion resistance of ceramics. The greater the porosity, the greater will be the corrosion. The fact that one material may yield a better corrosion resistance than another does not necessarily make it a better material, if the two materials have different porosities.

Ceramics which have an acid or base characteristic similar to the corrodent will tend to resist corrosion the best. In some cases the minor components of a ceramic, such as the bonding agent, may have a different acid/base character than the major component. In this instance the acid/base character of the corrodent will determine which phase corrodes first.

In corrosion resistance of ceramics we will be dealing with the so-called traditional ceramics, which include brick-type products and the various mortars used for joining.

All "common" brick, though hard to the touch, has a high water adsorption (anywhere from 8% to 15%) and is leached or destroyed by exposure to strong acid or alkali. "Acid brick," whether of shale or fireclay body, is made from selected clays containing little acid-soluble components. These bricks are fired for longer periods of time at higher temperatures than the same clay used to make "common" brick. Firing eliminates any organics that may be present and produces a brick with a much lower absorption rate, under 1% for best-quality red shale and under 5% for high-quality fireclay.

Zirconia-containing materials will be attacked by alkali solutions containing lithium, potassium, or sodium hydroxide, and potassium carbonate.

The transition-metal carbides and nitrides are chemically stable at room temperature but exhibit some attack by concentrated acid solutions. The normally protective layer of silicon oxide that forms on the surface of silicon carbide and silicon nitride can exhibit accelerated corrosion when various molten salts are present. None of the carbides or nitrides are stable in oxygen-containing environments. Under certain conditions some carbides and nitrides form a protective metal oxide layer that allows them to exhibit reasonably good oxidation resistance. Silicon carbide and nitride are relatively inert to most silicate liquors as long as they do not contain significant amounts of iron oxide.

Quartz (silica) is not attacked by hydrochloric, nitric, or sulfuric acids at room temperature but will be slowly attacked by alkaline solutions. At elevated temperatures, quartz is readily attacked by sodium hydroxide, potassium hydroxide, sodium carbonate, and sodium borate. Organics dissolved in the water increase the solubility of silica. Fused silica is attacked by molten sodium sulfate.

Table 5 shows the compatibility of various ceramic materials with selected corrodents.

Mortars

No universal mortar meets all conditions. The corrodent being handled and the temperatures to be expected must be taken into account when selecting mortar.

Furan mortars resist attack by all acids and alkalies except those of a strongly oxidizing nature. They are also compatible with solvents and other organic materials except materials such as aniline. Oxidizing materials such as nitric acid over 5%, chromic acid, hypochlorous and perchloric acids, sodium hypochlorite, and other bleaches will attack furan mortars.

Polyester mortars provide excellent service in most acid conditions, particularly the oxidizing inorganic acids in many ranges. However, they have limited service in neutral and alkaline mediums.

Epoxy mortars generally have better organic and solvent resistance than polyester mortars, except for acetic acid, which in concentration above 10% can cause damage.

Sodium silicate mortars are useful for pH's of 0–6, except where sulfuric acid exposures exist in vapor phase, wet-dry exposures, or in concentrations above 93%.

Silica mortars operate well for a pH of 0–7, being attacked only by hydrofluoric acid and acid fluorides.

Table 5 Chemical Resistance of Various Brick Materials

	Resistance to																						
	Slags		Molten metals						Gases									Heated acids					
Material	Acid	Basic	Al	Fe	Na	Pb	Zn	Mg	CO_2	CO	Steam	Cl_2[a]	H_2[b]	HCl	NH_3	SO_2	S[c]	Nitric	sulfuric	Hydrochloric	Hydrofluoric	Phosphoric	Hydrocarbons
Zircon	G	P	G	G	P	G	G	F	A	A	A	B–D	E	A	A	A	B–D	A	A	A	B–C	A–B	A–D
Bonded 99% alumina	G	G	G	G	P	G	G	G	A	A	A	A	A	A	A	A	A	A	A	A	A–C	A	A
Fused cast alumina	EX	G	G	G	F	G	G	EX	A	A	A	A–D	A	A	A	A	A–C	A	A	A	A–C	A	A
Zirconia stabilized	G	P	G	G	P	G	G	F	A	A	B–C	B	E	A	E	A	A–B	A	A	A	B–C	A	A
Silicon carbide	G	F	G	P	P		EX	G	B	B	B	D		A	A	A	D	A	A	A	B–D	A–B	A–D
Silicon nitride bonded																							
Silicon carbide	EX	F	EX	P	P		EX	G	B	B	B	D		A	A	A	A–C	A	A	A	C–D	A	A–C
Magnesite	P	G	F	G	P	G	F	F	A	B–C	A	D		D	A	C–D	B–D	D	D	D	D	D	B–C
Chrome	P	F	P	G	P	G	F	F	A	B–C	A	C–D		C–D	A	C–D	B–C	B–C	B–D	B–D	B–E	B–D	B–C
Fosterite	P	F	F	G	P	G	F	F	A	A	A	C–D		C–D	A	C–D	B–C	B–D	B–D	B–D	B–E	B–E	B–C
Synthetic mullite	G	G	F	G	P	G	G	F	A	A	A	B–D		A	A–C	A	B	A	A	A	C–E	A–B	A–D
Converted mullite	G	F	F	G	P	G	G	F	A	B–C	A	B–E		A	A–C	A–C	A–C	A	A	A	C–E	A–B	A–D
Silica	G	F	P	G	P	F	G	P	A	A	A	C–E		A	A–C	B–C	B–C	A–C	A–C	A–C	E	A–C	B–D
Fireclay	G	F	F	G	P	F	G	P	A	C	A	C–E		A	A–C	B–C	B–C	A–C	A–C	A–C	E	A–C	B–D

EX = excellent, G = good, F = fair, P = poor, A = no reaction, material stable, B = slight reaction material suitable, C = reaction, material suitable under certain conditions, D = reaction, material not suited unless tested under operating conditions, E = rapid reaction, material not suitable.

[a]Chlorine attacks silicates above 1300°F (704°C).

[b]Nascent or atomic hydrogen attacks silica and iron.

[c]Sulfur in strong concentrations reacts with silica above 1700°F (927°C).

Refer to Table 6 for the compatibility of mortars with selected corrodents.

4.2 Glassy Materials

Glassy materials corrode primarily through the action of aqueous media. In general, the very high silica glasses (>96% SiO_2), such as aluminosilicate and borosilicate compositions, have excellent resistance to a variety of corrodents.

Borosilicate glass is the primary composition used in the corrosion resistance field. It is resistant to all chemicals except hydrofluoric acid, fluorides, and such strong caustics as sodium or potassium hydroxide. However, caustics of even up to 50% concentration at room temperature will not be detrimental to borosilicate glass.

Because of its inertness, borosilicate glass is widely used in contact with high-purity products. The glass will not contaminate the material.

There are many glass compositions. A list of about 30 with their resistance to weathering, water, and acid may be found on page 572 of the *Encyclopedia of Glass Technology*, 2nd ed., vol. 10, published by Wiley.

Refer to Table 7 for the compatibility of borosilicate glass with selected corrodents.

SUGGESTED READING

Schweitzer, P. A., ed., *Corrosion and Corrosion Protection Handbook*, 2nd ed., Marcel Dekker, New York, 1989.

Schweitzer, P. A., *Corrosion Resistance Tables*, 4th ed., vols. 1–3, Marcel Dekker, New York, 1995.

Sheppard, E. L., Jr., *Chemically Resistant Masonry*, 2nd ed., Marcel Dekker, New York, 1982.

5 COMPOSITES

There are many composite materials. Corrosion resistance of duplex stainless steels is covered in Section 2.2 along with the family of stainless steels. Our concern here is with polymer composites. In Section 3 the corrosion resistance of polymers was discussed. Included were thermoset resins. These resins are generally applied as composites incorporating some type of reinforcing material to provide additional strength. For such a construction to be effective, the reinforcing material as well as the base resin must be compatible with the environment.

5.1 Glass Fiber Reinforcement

The three grades of glass used for reinforcements are E glass, C glass, and S glass. E glass is a boroaluminosilicate glass, and C glass is a calcium aluminosilicate glass. These two glasses find wide application in the corrosion industry. S glass is not normally used because of its high cost.

These glasses are resistant to all acids except hydrofluoric. They will also be attacked by fluorides, strong alkalies, and commercial-grade phosphoric acid containing fluorides.

(*text continued on p. 420*)

Table 6 Compatibility of Various Mortars with Selected Corrodents

The table is arranged alphabetically according to corrodent. Unless otherwise noted, the corrodent is considered pure, in the case of liquids, and a saturated aqueous solution in the case of solids. All percentages shown are weight percents. Corrosion is a function of temperature. When using the tables, note that the vertical lines refer to temperatures midway between the temperatures cited. An R entry indicates that the material is resistant to the maximum temperature shown. A U entry indicates that the material is unsatisfactory. A blank indicates that no data are available.

Mortars	Acetic acid 10%
Silicate	U
Sodium silicate	R
Potassium silicate	R
Silica	R
Sulfur	R
Furan resin	R
Polyester	R
Epoxy	R

°F	60	80	100	120	140	160	180	200	220	240	260	280	300	320	340	360	380	400	420	440	460
°C	15	26	38	49	60	71	82	93	104	116	127	138	149	160	171	182	193	204	216	227	238

	Acetic acid 50%
Silicate	U
Sodium silicate	R
Potassium silicate	R
Silica	R
Sulfur	U
Furan resin	R
Polyester	U
Epoxy	U

°F	60	80	100	120	140	160	180	200	220	240	260	280	300	320	340	360	380	400	420	440	460
°C	15	26	38	49	60	71	82	93	104	116	127	138	149	160	171	182	193	204	216	227	238

	Acetic acid 80%
Silicate	U
Sodium silicate	R
Potassium silicate	R
Silica	R
Sulfur	U
Furan resin	R
Polyester	U
Epoxy	U

°F	60	80	100	120	140	160	180	200	220	240	260	280	300	320	340	360	380	400	420	440	460
°C	15	26	38	49	60	71	82	93	104	116	127	138	149	160	171	182	193	204	216	227	238

Table 6 Continued

Mortars	Acetic acid glacial
Silicate	U
Sodium silicate	R
Potassium silicate	R
Silica	R
Sulfur	U
Furan resin	R
Polyester	U
Epoxy	U

°F	60	80	100	120	140	160	180	200	220	240	260	280	300	320	340	360	380	400	420	440	460
°C	15	26	38	49	60	71	82	93	104	116	127	138	149	160	171	182	193	204	216	227	238

	Acetic anhydride
Silicate	
Sodium silicate	R
Potassium silicate	R
Silica	R
Sulfur	U
Furan resin	U
Polyester	U
Epoxy	U

°F	60	80	100	120	140	160	180	200	220	240	260	280	300	320	340	360	380	400	420	440	460
°C	15	26	38	49	60	71	82	93	104	116	127	138	149	160	171	182	193	204	216	227	238

	Aluminum chloride, aqueous
Silicate	R
Sodium silicate	R
Potassium silicate	R
Silica	R
Sulfur	R
Furan resin	R
Polyester	R
Epoxy	R

°F	60	80	100	120	140	160	180	200	220	240	260	280	300	320	340	360	380	400	420	440	460
°C	15	26	38	49	60	71	82	93	104	116	127	138	149	160	171	182	193	204	216	227	238

Table 6 Continued

Mortars	Aluminum fluoride
Silicate	U
Sodium silicate	U
Potassium silicate	U
Silica	U
Sulfur	R
Furan resin	R
Polyester	R
Epoxy	R

°F	60	80	100	120	140	160	180	200	220	240	260	280	300	320	340	360	380	400	420	440	460
°C	15	26	38	49	60	71	82	93	104	116	127	138	149	160	171	182	193	204	216	227	238

	Ammonium chloride 10%
Silicate	R
Sodium silicate	R
Potassium silicate	R
Silica	R
Sulfur	R
Furan resin	R
Polyester	R
Epoxy	R

°F	60	80	100	120	140	160	180	200	220	240	260	280	300	320	340	360	380	400	420	440	460
°C	15	26	38	49	60	71	82	93	104	116	127	138	149	160	171	182	193	204	216	227	238

	Ammonium chloride 50%
Silicate	R
Sodium silicate	R
Potassium silicate	R
Silica	R
Sulfur	R
Furan resin	R
Polyester	R
Epoxy	R

°F	60	80	100	120	140	160	180	200	220	240	260	280	300	320	340	360	380	400	420	440	460
°C	15	26	38	49	60	71	82	93	104	116	127	138	149	160	171	182	193	204	216	227	238

Table 6 Continued

Mortars	Ammonium chloride, saturated																			
Silicate	R	—	—	—	—	—	—	—	—	—	—	—	—	—	—	—	—	—	—	—
Sodium silicate	R	—	—																	
Potassium silicate	R	—	—	—	—	—	—	—	—	—	—	—	—	—	—	—	—	—	—	—
Silica	R	—	—	—	—	—	—	—	—	—	—	—	—	—	—	—	—	—	—	—
Sulfur	R	—	—	—	—	—	—	—												
Furan resin	R	—	—	—	—	—	—	—	—	—	—	—	—	—	—	—					
Polyester	R	—	—	—	—	—	—	—	—	—											
Epoxy	R	—	—	—	—	—	—	—	—	—											
°F	60	80	100	120	140	160	180	200	220	240	260	280	300	320	340	360	380	400	420	440	460
°C	15	26	38	49	60	71	82	93	104	116	127	138	149	160	171	182	193	204	216	227	238

	Ammonium fluoride 10%																			
Silicate	U																			
Sodium silicate	U																			
Potassium silicate	U																			
Silica	U																			
Sulfur	U																			
Furan resin	R	—	—	—	—	—	—	—	—	—	—	—	—	—	—	—					
Polyester	R	—	—	—	—																
Epoxy	R	—	—	—																	
°F	60	80	100	120	140	160	180	200	220	240	260	280	300	320	340	360	380	400	420	440	460
°C	15	26	38	49	60	71	82	93	104	116	127	138	149	160	171	182	193	204	216	227	238

	Ammonium fluoride 25%																			
Silicate	U																			
Sodium silicate	U																			
Potassium silicate	U																			
Silica	U																			
Sulfur	U																			
Furan resin	R	—	—	—	—	—	—	—	—	—	—	—	—	—	—	—					
Polyester	R	—	—	—	—																
Epoxy	R	—	—																		
°F	60	80	100	120	140	160	180	200	220	240	260	280	300	320	340	360	380	400	420	440	460
°C	15	26	38	49	60	71	82	93	104	116	127	138	149	160	171	182	193	204	216	227	238

Table 6 Continued

Mortars	Ammonium hydroxide 25%																				
°F	60	80	100	120	140	160	180	200	220	240	260	280	300	320	340	360	380	400	420	440	460
°C	15	26	38	49	60	71	82	93	104	116	127	138	149	160	171	182	193	204	216	227	238
Silicate	U																				
Sodium silicate	U																				
Potassium silicate	U																				
Silica	U																				
Sulfur	U																				
Furan resin	R	—	—	—	—	—	—	—	—	—	—	—	—	—	—	—	—				
Polyester	R	—	—	—	—	—	—	—	—	—											
Epoxy	R	—	—	—	—	—	—	—	—	—											

	Ammonium hydroxide, saturated																				
°F	60	80	100	120	140	160	180	200	220	240	260	280	300	320	340	360	380	400	420	440	460
°C	15	26	38	49	60	71	82	93	104	116	127	138	149	160	171	182	193	204	216	227	238
Silicate	U																				
Sodium silicate																					
Potassium silicate	U																				
Silica	U																				
Sulfur	U																				
Furan resin	R	—	—	—	—	—	—	—	—	—	—	—	—	—							
Polyester	R	—	—																		
Epoxy	R	—	—																		

	Aqua regia 3:1																				
°F	60	80	100	120	140	160	180	200	220	240	260	280	300	320	340	360	380	400	420	440	460
°C	15	26	38	49	60	71	82	93	104	116	127	138	149	160	171	182	193	204	216	227	238
Silicate	R	—	—	—	—	—	—	—	—	—	—	—	—	—	—	—	—	—	—	—	—
Sodium silicate	R	—	—	—	—	—	—	—	—	—	—	—	—	—	—	—	—	—	—	—	—
Potassium silicate	R	—	—	—	—	—	—	—	—	—	—	—	—	—	—	—	—	—	—	—	—
Silica	R	—	—	—	—	—	—	—	—	—	—	—	—	—	—	—	—	—	—	—	—
Sulfur	U																				
Furan resin	U																				
Polyester	U																				
Epoxy	U																				

Table 6 Continued

Mortars	Bromine gas, dry																				
°F	60	80	100	120	140	160	180	200	220	240	260	280	300	320	340	360	380	400	420	440	460
°C	15	26	38	49	60	71	82	93	104	116	127	138	149	160	171	182	193	204	216	227	238
Silicate																					
Sodium silicate	R	—	—	—	—	—	—	—	—	—	—	—	—	—	—	—	—	—	—	—	—
Potassium silicate																					
Silica																					
Sulfur	U																				
Furan resin	U																				
Polyester	U																				
Epoxy	U																				

	Bromine gas, moist																				
°F	60	80	100	120	140	160	180	200	220	240	260	280	300	320	340	360	380	400	420	440	460
°C	15	26	38	49	60	71	82	93	104	116	127	138	149	160	171	182	193	204	216	227	238
Silicate																					
Sodium silicate	R	—	—	—	—	—	—	—	—	—	—	—	—	—	—	—	—	—	—	—	—
Potassium silicate																					
Silica																					
Sulfur	U																				
Furan resin	U																				
Polyester	U																				
Epoxy	U																				

	Bromine liquid																				
°F	60	80	100	120	140	160	180	200	220	240	260	280	300	320	340	360	380	400	420	440	460
°C	15	26	38	49	60	71	82	93	104	116	127	138	149	160	171	182	193	204	216	227	238
Silicate	R	—	—	—	—	—	—	—	—	—	—	—	—	—	—	—	—	—	—	—	—
Sodium silicate																					
Potassium silicate	R	—	—	—	—	—	—	—	—	—	—	—	—	—	—	—	—	—	—	—	—
Silica	R	—	—	—	—	—	—	—	—	—	—	—	—	—	—	—	—	—	—	—	—
Sulfur	R	—	—	—	—	—	—	—													
Furan resin	U																				
Polyester	U																				
Epoxy	U																				

Table 6 Continued

Mortars		Calcium hypochlorite																				
Silicate																						
Sodium silicate		U																				
Potassium silicate		R																				
Silica		R																				
Sulfur		U																				
Furan resin		U																				
Polyester																						
Epoxy		U																				
	°F	60	80	100	120	140	160	180	200	220	240	260	280	300	320	340	360	380	400	420	440	460
	°C	15	26	38	49	60	71	82	93	104	116	127	138	149	160	171	182	193	204	216	227	238

		Carbon tetrachloride																				
Silicate		R																				
Sodium silicate		R																				
Potassium silicate		R																				
Silica		R																				
Sulfur		U																				
Furan resin		R																				
Polyester		R																				
Epoxy		R																				
	°F	60	80	100	120	140	160	180	200	220	240	260	280	300	320	340	360	380	400	420	440	460
	°C	15	26	38	49	60	71	82	93	104	116	127	138	149	160	171	182	193	204	216	227	238

		Chlorine gas, dry																				
Silicate		R																				
Sodium silicate		R																				
Potassium silicate		R																				
Silica		R																				
Sulfur		U																				
Furan resin		U																				
Polyester		U																				
Epoxy		U																				
	°F	60	80	100	120	140	160	180	200	220	240	260	280	300	320	340	360	380	400	420	440	460
	°C	15	26	38	49	60	71	82	93	104	116	127	138	149	160	171	182	193	204	216	227	238

Table 6 Continued

Mortars	Chlorine gas, wet
Silicate	R
Sodium silicate	R
Potassium silicate	R
Silica	R
Sulfur	U
Furan resin	U
Polyester	U
Epoxy	U

°F	60	80	100	120	140	160	180	200	220	240	260	280	300	320	340	360	380	400	420	440	460
°C	15	26	38	49	60	71	82	93	104	116	127	138	149	160	171	182	193	204	216	227	238

	Chlorine liquid
Silicate	R
Sodium silicate	
Potassium silicate	R
Silica	R
Sulfur	U
Furan resin	U
Polyester	U
Epoxy	U

°F	60	80	100	120	140	160	180	200	220	240	260	280	300	320	340	360	380	400	420	440	460
°C	15	26	38	49	60	71	82	93	104	116	127	138	149	160	171	182	193	204	216	227	238

	Chromic acid 10%
Silicate	U
Sodium silicate	R
Potassium silicate	R
Silica	R
Sulfur	U
Furan resin	U
Polyester	R
Epoxy	U

°F	60	80	100	120	140	160	180	200	220	240	260	280	300	320	340	360	380	400	420	440	460
°C	15	26	38	49	60	71	82	93	104	116	127	138	149	160	171	182	193	204	216	227	238

Table 6 Continued

Mortars	Chromic acid 50%																				
°F	60	80	100	120	140	160	180	200	220	240	260	280	300	320	340	360	380	400	420	440	460
°C	15	26	38	49	60	71	82	93	104	116	127	138	149	160	171	182	193	204	216	227	238
Silicate	R	—	—	—	—	—	—	—	—	—	—	—	—	—	—	—	—	—	—	—	—
Sodium silicate																					
Potassium silicate																					
Silica	R	—	—	—	—	—	—	—	—	—	—	—	—	—	—	—	—	—	—	—	—
Sulfur	U																				
Furan resin	U																				
Polyester 30%	R	—	—																		
Epoxy	U																				

	Ferric chloride																				
°F	60	80	100	120	140	160	180	200	220	240	260	280	300	320	340	360	380	400	420	440	460
°C	15	26	38	49	60	71	82	93	104	116	127	138	149	160	171	182	193	204	216	227	238
Silicate	R	—	—	—	—	—	—	—	—	—	—	—	—	—	—	—	—	—	—	—	—
Sodium silicate	U																				
Potassium silicate	R	—	—	—	—	—	—	—	—	—	—	—	—	—	—	—	—	—	—	—	—
Silica	R	—	—	—	—	—	—	—	—	—	—	—	—	—	—	—	—	—	—	—	—
Sulfur	R	—	—	—	—	—	—	—													
Furan resin	R	—	—	—	—	—	—	—	—	—	—	—	—	—	—	—					
Polyester	R	—	—	—	—	—	—	—	—	—											
Epoxy	R	—	—	—	—	—	—	—	—	—											

	Ferric chloride, 50% in water																				
°F	60	80	100	120	140	160	180	200	220	240	260	280	300	320	340	360	380	400	420	440	460
°C	15	26	38	49	60	71	82	93	104	116	127	138	149	160	171	182	193	204	216	227	238
Silicate	R	—	—	—	—	—	—	—	—	—	—	—	—	—	—	—	—	—	—	—	—
Sodium silicate	U																				
Potassium silicate	R	—	—	—	—	—	—	—	—	—	—	—	—	—	—	—	—	—	—	—	—
Silica	R	—	—	—	—	—	—	—	—	—	—	—	—	—	—	—	—	—	—	—	—
Sulfur	R	—	—	—	—	—	—	—													
Furan resin	R	—	—	—	—	—	—	—	—	—	—	—	—	—	—	—					
Polyester	R	—	—	—	—	—	—	—	—	—											
Epoxy	R	—	—	—	—	—	—	—	—	—											

Table 6 Continued

Mortars	Hydrobromic acid 20%
Silicate	R
Sodium silicate	R
Potassium silicate	R
Silica	R
Sulfur	R
Furan resin	R
Polyester	R
Epoxy	U

°F	60	80	100	120	140	160	180	200	220	240	260	280	300	320	340	360	380	400	420	440	460
°C	15	26	38	49	60	71	82	93	104	116	127	138	149	160	171	182	193	204	216	227	238

	Hydrobromic acid 50%
Silicate	R
Sodium silicate	R
Potassium silicate	R
Silica	R
Sulfur	R
Furan resin	R
Polyester	R
Epoxy	U

°F	60	80	100	120	140	160	180	200	220	240	260	280	300	320	340	360	380	400	420	440	460
°C	15	26	38	49	60	71	82	93	104	116	127	138	149	160	171	182	193	204	216	227	238

	Hydrochloric acid 20%
Silicate	R
Sodium silicate	R
Potassium silicate	R
Silica	R
Sulfur	R
Furan resin	R
Polyester	R
Epoxy	U

°F	60	80	100	120	140	160	180	200	220	240	260	280	300	320	340	360	380	400	420	440	460
°C	15	26	38	49	60	71	82	93	104	116	127	138	149	160	171	182	193	204	216	227	238

Table 6 Continued

Mortars	Hydrochloric acid 38%
Silicate	R
Sodium silicate	R
Potassium silicate	R
Silica	R
Sulfur	R
Furan resin	R
Polyester	R
Epoxy	U

°F	60	80	100	120	140	160	180	200	220	240	260	280	300	320	340	360	380	400	420	440	460
°C	15	26	38	49	60	71	82	93	104	116	127	138	149	160	171	182	193	204	216	227	238

	Hydrofluoric acid 30%
Silicate	U
Sodium silicate	U
Potassium silicate	U
Silica	U
Sulfur	R
Furan resin	R
Polyester	R
Epoxy	U

°F	60	80	100	120	140	160	180	200	220	240	260	280	300	320	340	360	380	400	420	440	460
°C	15	26	38	49	60	71	82	93	104	116	127	138	149	160	171	182	193	204	216	227	238

	Hydrofluoric acid 70%
Silicate	U
Sodium silicate	U
Potassium silicate	U
Silica	U
Sulfur	U
Furan resin	U
Polyester	
Epoxy	U

°F	60	80	100	120	140	160	180	200	220	240	260	280	300	320	340	360	380	400	420	440	460
°C	15	26	38	49	60	71	82	93	104	116	127	138	149	160	171	182	193	204	216	227	238

Table 6 Continued

Mortars	Hydrofluoric acid 100%																				
Silicate	U																				
Sodium silicate	U																				
Potassium silicate	U																				
Silica	U																				
Sulfur	U																				
Furan resin	U																				
Polyester																					
Epoxy	U																				
°F	60	80	100	120	140	160	180	200	220	240	260	280	300	320	340	360	380	400	420	440	460
°C	15	26	38	49	60	71	82	93	104	116	127	138	149	160	171	182	193	204	216	227	238

	Magnesium chloride																				
Silicate	R	—	—	—	—	—	—	—	—	—	—	—	—	—	—	—	—	—	—	—	—
Sodium silicate																					
Potassium silicate	R	—	—	—	—	—	—	—	—	—	—	—	—	—	—	—	—	—	—	—	—
Silica	R	—	—	—	—	—	—	—	—	—	—	—	—	—	—	—	—	—	—	—	—
Sulfur	R	—	—	—	—	—	—	—													
Furan resin	R	—	—	—	—	—	—	—	—	—	—	—	—	—	—	—					
Polyester	R	—	—	—	—	—	—	—	—												
Epoxy	R	—	—	—	—	—	—	—													
°F	60	80	100	120	140	160	180	200	220	240	260	280	300	320	340	360	380	400	420	440	460
°C	15	26	38	49	60	71	82	93	104	116	127	138	149	160	171	182	193	204	216	227	238

	Nitric acid 5%																				
Silicate	R	—	—	—	—	—	—	—	—	—	—	—	—	—	—	—	—	—	—	—	—
Sodium silicate	R	—	—	—	—	—	—	—	—	—	—	—	—	—	—	—	—	—	—	—	—
Potassium silicate	R	—	—	—	—	—	—	—	—	—	—	—	—	—	—	—	—	—	—	—	—
Silica	R	—	—	—	—	—	—	—	—	—	—	—	—	—	—	—	—	—	—	—	—
Sulfur	R	—	—	—	—																
Furan resin	U																				
Polyester	R	—	—	—	—																
Epoxy	U																				
°F	60	80	100	120	140	160	180	200	220	240	260	280	300	320	340	360	380	400	420	440	460
°C	15	26	38	49	60	71	82	93	104	116	127	138	149	160	171	182	193	204	216	227	238

Table 6 Continued

Mortars	Nitric acid 20%
Silicate	R
Sodium silicate	R
Potassium silicate	
Silica	
Sulfur	R
Furan resin	U
Polyester	R
Epoxy	U

°F	60	80	100	120	140	160	180	200	220	240	260	280	300	320	340	360	380	400	420	440	460
°C	15	26	38	49	60	71	82	93	104	116	127	138	149	160	171	182	193	204	216	227	238

	Nitric acid 70%
Silicate	R
Sodium silicate	R
Potassium silicate	
Silica	
Sulfur	U
Furan resin	U
Polyester	U
Epoxy	U

°F	60	80	100	120	140	160	180	200	220	240	260	280	300	320	340	360	380	400	420	440	460
°C	15	26	38	49	60	71	82	93	104	116	127	138	149	160	171	182	193	204	216	227	238

	Nitric acid, anhydrous
Silicate	R
Sodium silicate	R
Potassium silicate	
Silica	
Sulfur	U
Furan resin	U
Polyester	U
Epoxy	U

°F	60	80	100	120	140	160	180	200	220	240	260	280	300	320	340	360	380	400	420	440	460
°C	15	26	38	49	60	71	82	93	104	116	127	138	149	160	171	182	193	204	216	227	238

Table 6 Continued

Mortars	Oleum																			
Silicate																				
Sodium silicate	R																			
Potassium silicate																				
Silica																				
Sulfur	U																			
Furan resin	U																			
Polyester	U																			
Epoxy	U																			
°F	60	80	100	120	140	160	180	200	220	240	260	280	300	320	340	360	380	400	420	440	460
°C	15	26	38	49	60	71	82	93	104	116	127	138	149	160	171	182	193	204	216	227	238

	Phosphoric acid 50–80%																			
Silicate	R																			
Sodium silicate	U																			
Potassium silicate	R																			
Silica	R																			
Sulfur	R																			
Furan resin	R																			
Polyester	R																			
Epoxy	U																			
°F	60	80	100	120	140	160	180	200	220	240	260	280	300	320	340	360	380	400	420	440	460
°C	15	26	38	49	60	71	82	93	104	116	127	138	149	160	171	182	193	204	216	227	238

	Sodium chloride																			
Silicate	U																			
Sodium silicate	R																			
Potassium silicate	R																			
Silica	R																			
Sulfur	R																			
Furan resin	R																			
Polyester	R																			
Epoxy	R																			
°F	60	80	100	120	140	160	180	200	220	240	260	280	300	320	340	360	380	400	420	440	460
°C	15	26	38	49	60	71	82	93	104	116	127	138	149	160	171	182	193	204	216	227	238

Table 6 Continued

Mortars	Sodium hydroxide 10%																				
Silicate	U																				
Sodium silicate	U																				
Potassium silicate	U																				
Silica	U																				
Sulfur	U																				
Furan resin	R	—	—	—	—	—	—	—	—	—	—	—	—	—	—	—					
Polyester	R	—	—																		
Epoxy	R	—	—	—	—	—	—	—	—	—											
°F	60	80	100	120	140	160	180	200	220	240	260	280	300	320	340	360	380	400	420	440	460
°C	15	26	38	49	60	71	82	93	104	116	127	138	149	160	171	182	193	204	216	227	238

	Sodium hydroxide 50%																				
Silicate	U																				
Sodium silicate	U																				
Potassium silicate	U																				
Silica	U																				
Sulfur	U																				
Furan resin	R	—	—	—	—	—	—	—	—	—	—	—	—	—	—	—					
Polyester	R	—	—																		
Epoxy	R	—	—	—	—	—	—	—	—	—											
°F	60	80	100	120	140	160	180	200	220	240	260	280	300	320	340	360	380	400	420	440	460
°C	15	26	38	49	60	71	82	93	104	116	127	138	149	160	171	182	193	204	216	227	238

	Sodium hydroxide, concentrated																				
Silicate	U																				
Sodium silicate	U																				
Potassium silicate	U																				
Silica	U																				
Sulfur	U																				
Furan resin																					
Polyester																					
Epoxy	R	—																			
°F	60	80	100	120	140	160	180	200	220	240	260	280	300	320	340	360	380	400	420	440	460
°C	15	26	38	49	60	71	82	93	104	116	127	138	149	160	171	182	193	204	216	227	238

Table 6 Continued

Mortars	Sodium hypochlorite 20%																				
Silicate	U																				
Sodium silicate	U																				
Potassium silicate	U																				
Silica	U																				
Sulfur	U																				
Furan resin	U																				
Polyester	U																				
Epoxy	U																				
°F	60	80	100	120	140	160	180	200	220	240	260	280	300	320	340	360	380	400	420	440	460
°C	15	26	38	49	60	71	82	93	104	116	127	138	149	160	171	182	193	204	216	227	238

	Sodium hypochlorite, concentrated																				
Silicate	U																				
Sodium silicate	U																				
Potassium silicate	U																				
Silica	U																				
Sulfur	U																				
Furan resin	U																				
Polyester	U																				
Epoxy	U																				
°F	60	80	100	120	140	160	180	200	220	240	260	280	300	320	340	360	380	400	420	440	460
°C	15	26	38	49	60	71	82	93	104	116	127	138	149	160	171	182	193	204	216	227	238

	Sulfuric acid 10%																				
Silicate	U																				
Sodium silicate	R	—	—	—	—	—	—	—	—	—	—	—	—	—	—	—	—	—	—	—	—
Potassium silicate																					
Silica																					
Sulfur	R	—	—	—	—	—	—	—													
Furan resin	R	—	—	—	—	—	—	—	—	—	—	—	—	—	—	—					
Polyester	R	—	—	—	—	—	—	—	—	—											
Epoxy	R	—	—																		
°F	60	80	100	120	140	160	180	200	220	240	260	280	300	320	340	360	380	400	420	440	460
°C	15	26	38	49	60	71	82	93	104	116	127	138	149	160	171	182	193	204	216	227	238

Table 6 Continued

Mortars	Sulfuric acid 50%																			
°F	60	80	100	120	140	160	180	200	220	240	260	280	300	320	340	360	380	400	420	440	460
°C	15	26	38	49	60	71	82	93	104	116	127	138	149	160	171	182	193	204	216	227	238
Silicate	U																				
Sodium silicate	R	—	—	—	—	—	—	—	—	—	—	—	—	—	—	—	—	—	—	—	—
Potassium silicate																					
Silica																					
Sulfur	R	—	—	—	—																
Furan resin	R	—	—	—	—	—	—	—	—	—	—	—	—	—	—	—					
Polyester	R	—	—	—	—	—	—	—	—	—	—										
Epoxy	U																				

	Sulfuric acid 70%																			
°F	60	80	100	120	140	160	180	200	220	240	260	280	300	320	340	360	380	400	420	440	460
°C	15	26	38	49	60	71	82	93	104	116	127	138	149	160	171	182	193	204	216	227	238
Silicate	U																				
Sodium silicate	R	—	—	—	—																
Potassium silicate	R	—	—	—	—	—	—	—													
Silica	R	—	—	—	—	—	—	—													
Sulfur	R	—	—																		
Furan resin	R	—	—																		
Polyester	R	—	—																		
Epoxy	U																				

	Sulfuric acid 90%																			
°F	60	80	100	120	140	160	180	200	220	240	260	280	300	320	340	360	380	400	420	440	460
°C	15	26	38	49	60	71	82	93	104	116	127	138	149	160	171	182	193	204	216	227	238
Silicate	U																				
Sodium silicate	R	—	—																		
Potassium silicate																					
Silica																					
Sulfur	U																				
Furan resin	U																				
Polyester	U																				
Epoxy	U																				

Table 6 Continued

Mortars	Sulfuric acid 98%																				
Silicate	U																				
Sodium silicate	R	—																			
Potassium silicate																					
Silica																					
Sulfur	U																				
Furan resin	U																				
Polyester	U																				
Epoxy	U																				
°F	60	80	100	120	140	160	180	200	220	240	260	280	300	320	340	360	380	400	420	440	460
°C	15	26	38	49	60	71	82	93	104	116	127	138	149	160	171	182	193	204	216	227	238

Source: Philip A. Schweitzer, *Corrosion Resistance Tables, Fourth Edition* vols. 1–3, Marcel Dekker, New York, 1995.

5.2 Polyester

Polyester is used as a surfacing mat for the resin-rich inner surface of filament-wound or custom contact-molded structures. Nexus (registered trademark of Burlington Industries polyester) possesses excellent resistance to alcohols, bleaching agents, water, hydrocarbons, and aqueous solutions of most weak acids at boiling point. Polyesters are not resistant to strong acids such as 93% sulfuric acid.

5.3 Carbon Fiber

Carbon fibers are inert to most chemicals, including fluorides and medium-strength hydrofluoric acid.

5.4 Thermoplastic Fibers

Nylon and polypropylene fibers are used as reinforcing for thermoset resins to provide improved resistance to alkaline environments. For specific corrosion resistance of these materials refer to Section 3 and the tables located in Appendix A.

5.5 Composite Laminates

Many types of dual laminates are available. Typical combinations are ABS and polyester, bisphenol and isophthalic fibrous glass systems, vinyl ester, epoxy and polyester, glass and reinforced polyester, polypropylene-lined-reinforced polyester, and PVC and polyester composite. In each case it is necessary to evaluate each member of the composite as to its compatibility to the corrosive environment.

SUGGESTED READING

Mallinson, J. H., *Corrosion-Resistant Plastic Composites in Chemical Plant Design*, Marcel Dekker, New York, 1988.

Table 7 Compatibility of Borosilicate Glass with Selected Corrodents

The chemicals listed are in the pure state or in a saturated solution unless otherwise indicated. Compatibility is shown to the maximum allowable temperature for which data are available. Incompatibility is shown by an x. A blank space indicates that data are unavailable.

	Maximum temp.			Maximum temp.	
Chemical	°F	°C	Chemical	°F	°C
Acetaldehyde	450	232	Barium carbonate	250	121
Acetamide	270	132	Barium chloride	250	121
Acetic acid 10%	400	204	Barium hydroxide	250	121
Acetic acid 50%	400	204	Barium sulfate	250	121
Acetic acid 80%	400	204	Barium sulfide	250	121
Acetic acid, glacial	400	204	Benzaldehyde	200	93
Acetic anhydride	250	121	Benzene	200	93
Acetone	250	121	Benzene sulfonic acid 10%	200	93
Acetyl chloride			Benzoic acid	200	93
Acrylic acid			Benzyl alcohol	200	93
Acrylonitrile			Benzyl chloride	200	93
Adipic acid	210	99	Borax	250	121
Allyl alcohol	120	49	Boric acid	300	149
Allyl chloride	250	121	Bromine gas, dry		
Alum	250	121	Bromine gas, moist	250	121
Aluminum acetate			Bromine liquid	90	32
Aluminum chloride, aqueous	250	121	Butadiene	90	32
Aluminum chloride, dry	180	82	Butyl acetate	250	121
Aluminum fluoride	x	x	Butyl alcohol	200	93
Aluminum hydroxide	250	121	*n*-Butylamine		
Aluminum nitrate	100	38	Butyl phthalate		
Aluminum oxychloride	190	88	Butyric acid	200	93
Aluminum sulfate	250	121	Calcium bisulfide		
Ammonia gas			Calcium bisulfite	250	121
Ammonium bifluoride	x	x	Calcium carbonate	250	121
Ammonium carbonate	250	121	Calcium chlorate	200	93
Ammonium chloride 10%	250	121	Calcium chloride	200	93
Ammonium chloride 50%	250	121	Calcium hydroxide 10%	250	121
Ammonium chloride, sat.	250	121	Calcium hydroxide, sat.	x	x
Ammonium fluoride 10%	x	x	Calcium hypochlorite	200	93
Ammonium fluoride 25%	x	x	Calcium nitrate	100	38
Ammonium hydroxide 25%	250	121	Calcium oxide		
Ammonium hydroxide, sat.	250	121	Calcium sulfate		
Ammonium nitrate	200	93	Caprylic acid		
Ammonium persulfate	200	93	Carbon bisulfide	250	121
Ammonium phosphate	90	32	Carbon dioxide, dry	160	71
Ammonium sulfate 10–40%	200	93	Carbon dioxide, wet	160	71
Ammonium sulfide			Carbon disulfide	250	121
Ammonium sulfite			Carbon monoxide	450	232
Amyl acetate	200	93	Carbon tetrachloride	200	93
Amyl alcohol	250	121	Carbonic acid	200	93
Amyl chloride	250	121	Cellosolve	160	71
Aniline	200	93	Chloracetic acid, 50% water	250	121
Antimony trichloride	250	121	Chloracetic acid	250	121
Aqua regia 3:1	200	93	Chlorine gas, dry	450	232

Table 7 (Continued)

Chemical	Maximum temp. °F	Maximum temp. °C	Chemical	Maximum temp. °F	Maximum temp. °C
Chlorine gas, wet	400	204	Malic acid	160	72
Chlorine, liquid	140	60	Manganese chloride		
Chlorobenzene	200	93	Methyl chloride	200	93
Chloroform	200	93	Methyl ethyl ketone	200	93
Chlorosulfonic acid	200	93	Methyl isobutyl ketone	200	93
Chromic acid 10%	200	93	Muriatic acid		
Chromic acid 50%	200	93	Nitric acid 5%	400	204
Chromyl chloride			Nitric acid 20%	400	204
Citric acid 15%	200	93	Nitric acid 70%	400	204
Citric acid, concentrated	200	93	Nitric acid, anhydrous	250	121
Cooper acetate			Nitrous acid, concentrated		
Copper carbonate			Oleum	400	204
Copper chloride	250	121	Perchloric acid 10%	200	93
Copper cyanide			Perchloric acid 70%	200	93
Copper sulfate	200	93	Phenol	200	93
Cresol	200	93	Phosphoric acid 50–80%	300	149
Cupric chloride 5%	160	71	Picric acid	200	93
Cupric chloride 50%	160	71	Potassium bromide 30%	250	121
Cyclohexane	200	93	Salicylic acid		
Cyclohexanol			Silver bromide 10%		
Dichloroacetic acid	310	154	Sodium carbonate	250	121
Dichloroethane (ethylene dichloride)	250	121	Sodium chloride	250	121
Ethylene glycol	210	99	Sodium hydroxide 10%	x	x
			Sodium hydroxide 50%	x	x
Ferric chloride	290	143	Sodium hydroxide, concentrated	x	x
Ferric chloride 50% in water	280	138			
			Sodium hypochlorite 20%	150	66
Ferric nitrate 10–50%	180	138	Sodium hypochlorite, concentrated	150	66
Ferrous chloride	200	93	Sodium sulfide to 50%	x	x
Ferrous nitrate			Stannic chloride	210	99
Fluorine gas, dry	300	149	Stannous chloride	210	99
Fluorine gas, moist	x	x	Sulfuric acid 10%	400	204
Hydrobromic acid, dilute	200	93	Sulfuric acid 50%	400	204
Hydrobromic acid 20%	200	93	Sulfuric acid 70%	400	204
Hydrobromic acid 50%	200	93	Sulfuric acid 90%	400	204
Hydrobromic acid 20%	200	93	Sulfuric acid 98%	400	204
Hydrobromic acid 38%	200	93	Sulfuric acid 100%	400	204
Hydrocyanic acid 10%	200	93	Sulfuric acid, fuming		
Hydrofluoric acid 30%	x	x	Sulfurous acid	210	99
Hydrofluoric acid 70%	x	x			
Hydrofluoric acid 100%	x	x	Thionyl chloride	210	99
Hypochlorous acid	190	88	Toluene	250	121
Iodine solution 10%	200	93	Trichloroacetic acid	210	99
Ketones, general	200	93	White liquor	210	99
Lactic acid 25%	200	93	Zinc chloride	210	99
Magnesium chloride	250	121			

Source: Philip A. Schweitzer, *Corrosion Resistance Tables, Fourth Edition*, vols. 1–3, Marcel Dekker, New York, 1995.

Schweitzer, P. A., ed., *Corrosion and Corrosion Protection Handbook*, 2nd ed., Marcel Dekker, New York, 1989.
Schweitzer, P. A., *Corrosion Resistance Tables*, 4th ed., vols. 1–3, Marcel Dekker, New York, 1995.
Schweitzer, P. A., *Corrosion Resistant Piping Systems*, Marcel Dekker, New York, 1994.

16
Comparative Properties

G.T. Murray
California Polytechnic State University
San Luis Obispo, California

1 INTRODUCTION

In the materials selection process it helps to know in what ballpark the materials of interest for an application may be found. For example, if the materials are going to experience temperatures in excess of 500°C, should we waste our time and effort by considering polymers and polymer matrix composites? Let us narrow our field of consideration as quickly as possible. After all, we have many thousands of materials from which to choose. Our first objective is to reduce this number to a few hundred. This is sort of analogous to the block diagram approach used for detecting failures in electronic equipment. We must first isolate the block of interest. In this chapter we will attempt to establish some ballpark categories that should assist in narrowing the field of possible materials for any given application.

For a quick overall picture, Table 1 lists the range of properties stated earlier in this handbook. Some liberty has been taken in eliminating the extreme values in any property range. This helps to narrow the field of interest. However, the reader should be aware that extreme values do exist, and when the lack of a suitable material becomes apparent the detailed data in the specific chapters should be reexamined for materials that may possess these extreme values of interest. In the remainder of this section the specific properties of materials in each of the four major categories will be presented according to material category. This permits one to eliminate, or focus on, certain categories.

2 COMPARATIVE MECHANICAL PROPERTIES

The mechanical properties of paramount interest in most designs are strength, hardness, ductility, fatigue, and fracture toughness. For other than room-temperature applications these properties must also be known in the temperature range of the

Table 1 Materials Properties Summary

Materials	R.T.-Y. S. (MPA (ksi))	% Elong.	Modulus (GPa)	Toughness K_{1c} (MPa $\sqrt{m}$)	Thermal cond. (W/m · K @ R.T.)	CTE (ppm per °C @ R.T.)	Density (g/cc)
Metals							
Ultrahigh strength steels	1380–2070 (200–300)	10–40	200	28–154	—	—	7.7–7.9
Carbon, low-alloy & HSLA steels	310–1100 (45–159)	12–30	200	—	48.6–51.9	11.0–13.0	7.7–7.9
Austenitic stainless steels	172–262 (25–38)	40–50	193–197	—	14.2–16.3	15.7–17.2	7.8–8.0
Stainless steels, high strength	1000–1520 (145–220)	10–17	200	—	16–27	10–11	7.7–8.0
Titanium, high strength	700–1175 (101–170)	11–14	200	44–66	6.7–10	8.7–10.8	4.4–4.8
Bronzes	345–552 (50–88)	15–22	103–117	—	36–55	16.2–18.0	8.0–8.5
Al alloys aged	138–552 (20–80)	8–30	69–72	15–35	121–200	20–23	2.7–2.8
Ceramics	275–520 (40–75)	nil	120–550	1.5–7.0	6–160	3–7	2.3–6.0
CMCs	280–800 (41–116)	—	—	2.3–1.0	—	—	—
PMCs							
0°	965–1550 (140–225)	1.1–2.5	39–215	—	0.4–60	−0.9–7.13	1.4–1.9
Random	579 (84)	1.7	45–5	—	—	14.8–23.2	1.87
MMCs							
Particulate							
Al	124–607	0.1–4.0	94–265	13.0–1	119–220	6.2–20.7	2.8–3.0
	(18–14.5)	—	(14.2–34.8)		—		
Ti	1035 (150)	—	145 (21)	—	—	—	4.5
Continuous							
Al 0°	1550 (225)	—	193 (28)	—	—	7 (3.9)	2.8
Ti 0°	1725 (250)	—	193 (28)	—	—	—	3.9

intended application. For higher temperatures we must use creep and stress-rupture properties. In many room-temperature uses and in all elevated-temperature applications, the effect of environment must be considered. Environmental effects were presented in Chapter 15.

2.1 Comparison of Room-Temperature Mechanical Properties of Metals, Polymers, Ceramics, and Composites

In Fig. 1 the ultimate tensile strengths of the four categories are shown in ksi units. To obtain SI units MPa one can multiply the ksi figures by 6.9. Representative materials from each category have been selected. By choosing specific materials, at least some of the processing variables, such as heat treatment, porosity, and fabrication methods can be included. The ductilities and elastic moduli for similar materials are depicted in Figs. 2 and 3, respectively. It is evident that the highest ductilities can be found in polymers, whereas the lowest appear in ceramic materials. This situation is reversed for the moduli values. Figure 4 shows representative values of fracture toughness of materials in the four categories. Unfortunately, many polymers have been evaluated with respect to toughness by the impact test rather than by K_{1c} values. Likewise impact data have been used almost exclusively for

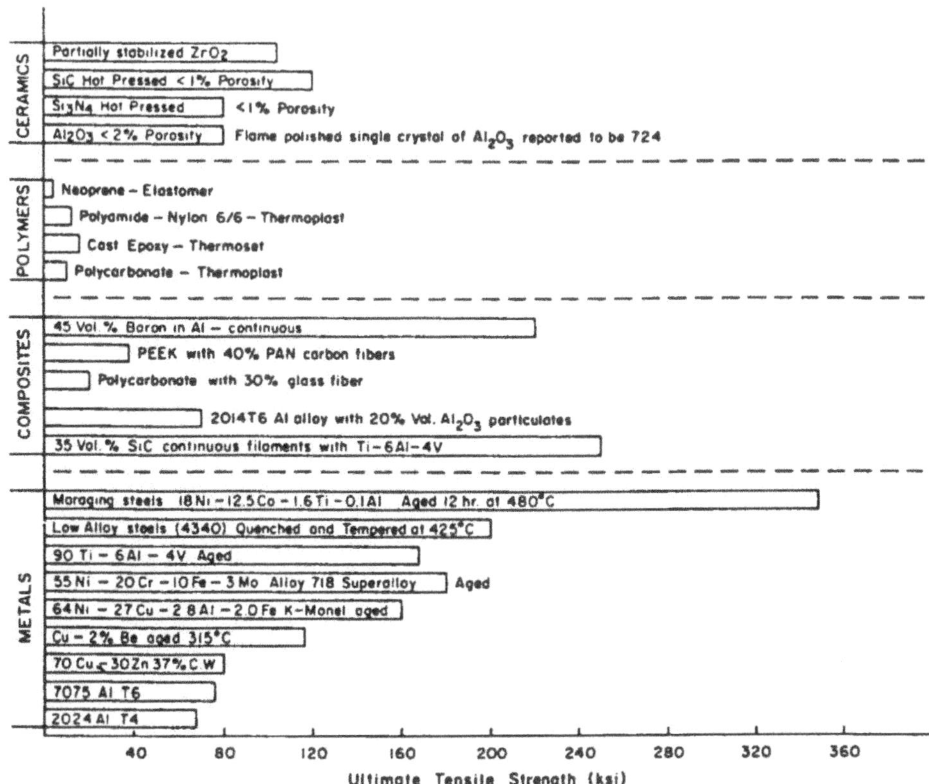

Figure 1 Ultimate tensile strength of selected metals, composites, polymers and ceramics.

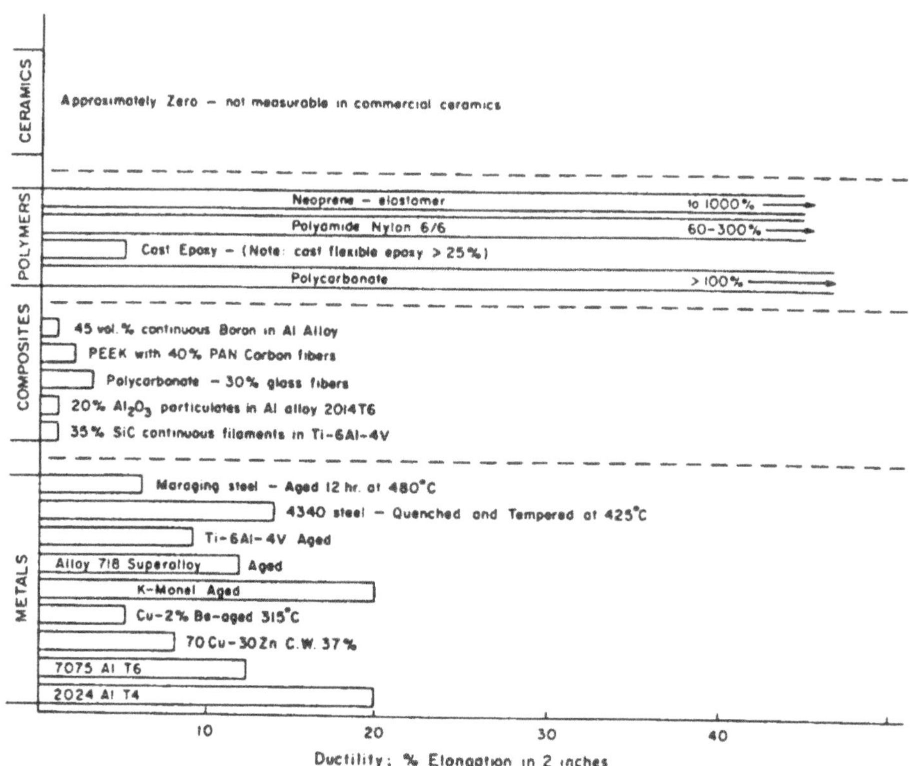

Figure 2 Ductilities of selected metals, composites, polymers, and ceramics.

the PMCs. Values generally range from 32 to 43 J/cm notch for PMCs. The SiC-Al alloy particulate composites are usually in the range of 10 to 30 MPa$\sqrt{m}$. The high value of 60 MPa$\sqrt{m}$ was reported on an aluminum sheet composite, which probably has some directional variation in its fracture toughness. There are many metals with higher toughness values than those shown in Fig. 4. Low-carbon steels in the annealed condition have very high values of fracture toughness. Likewise toughness is not reported for many of the ductile copper alloys since usually the toughness is sufficiently high to be of no concern. The continuous-fiber SiC-metal composites have good fracture toughness, and the particulate SiC-Al composites have more than adequate toughness in most cases. Many polymer composites and all ceramics and ceramic composites have very low fracture toughness and their usage is currently limited by this factor.

Fatigue data for some representative materials are shown in Fig. 5. Steels and the continuous-fiber composites are the best high-strength materials with respect to fatigue life. The fatigue life of these composites will, however, vary considerably with fiber and loading direction. The aluminum alloys that contain particulates have a somewhat better fatigue behavior than do the nonreinforced aluminum alloys.

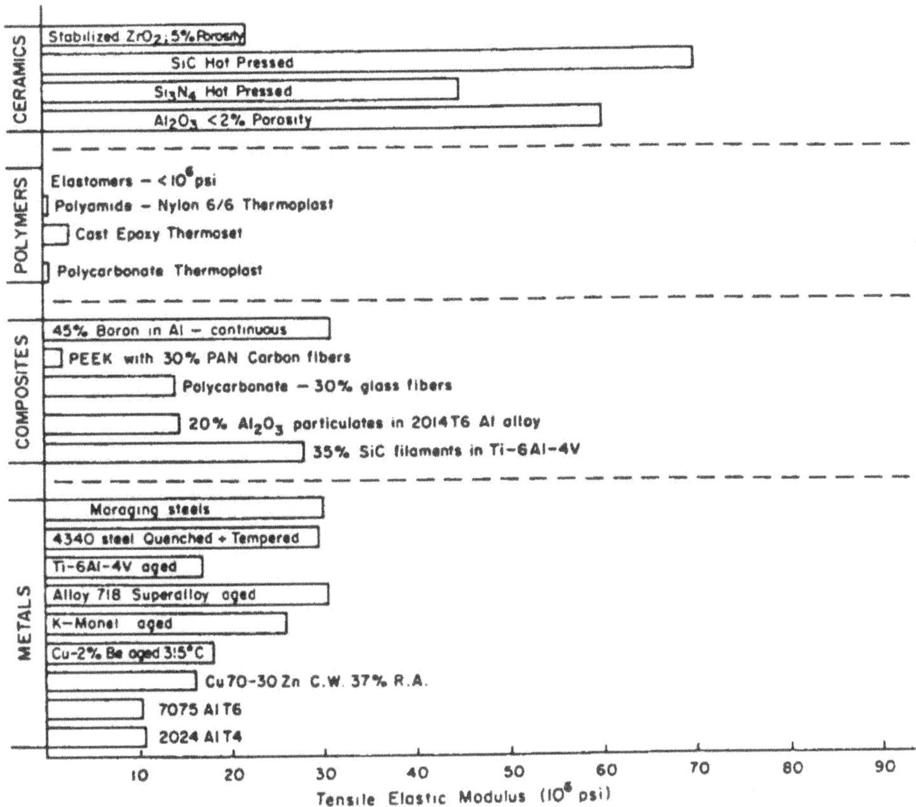

Figure 3 Tensile moduli of selected metals, composites, polymers, and ceramics.

2.2 Comparison of Elevated Temperature Properties of Metals, Ceramics, and Composites

Frequently, maximum-use temperatures are reported, particularly for polymers and metals. These temperatures must be less than the softening or decomposition temperatures for polymers and, of course, less than the melting point for metals. When stress is applied, these temperatures are markedly reduced due to creep. Thus, for elevated temperature applications under stress one must use creep and/or stress-rupture data for design purposes. Approximate maximum-use temperatures for some polymers, polymers with glass fibers, and some metals are listed in Table 2. Strength versus temperature data were used to obtain most of these values. Time was not a factor. These tests were conducted in a few minutes at the usual strain rates of 0.05 to 0.5 per minute. These temperatures may be used for narrowing the range of potential materials for use at elevated temperatures. Other strength versus temperature data for representative materials of all groups except the polymers are depicted in Fig. 6.

The time to rupture for a typical superalloy (Inconel 617) and two oxide dispersion-strengthened superalloys (Inconel MA 6000 and MA 956) and a ceramic material $(S_{13}N_4)$ are shown in Fig. 7.

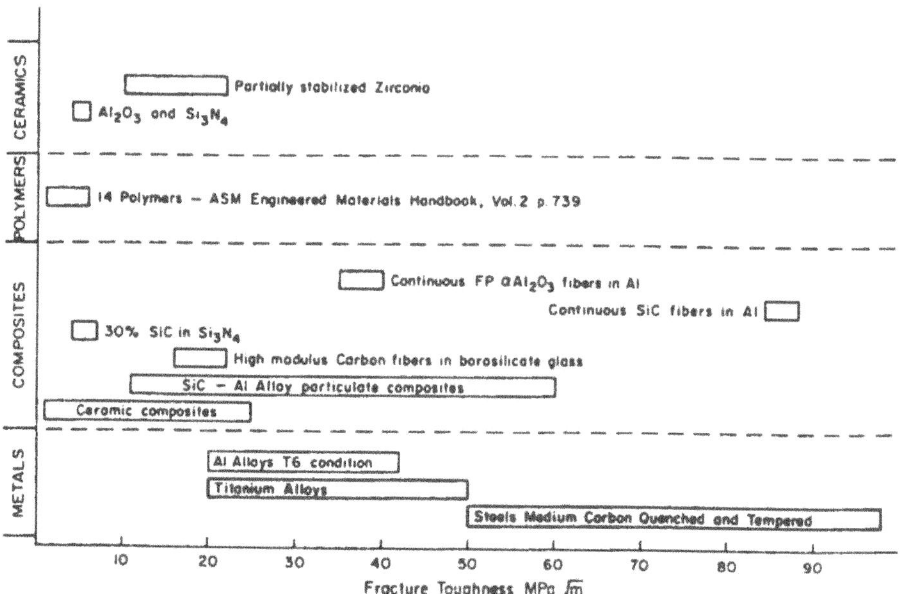

Figure 4 Fracture toughness of selected metals, composites, polymers, and ceramics.

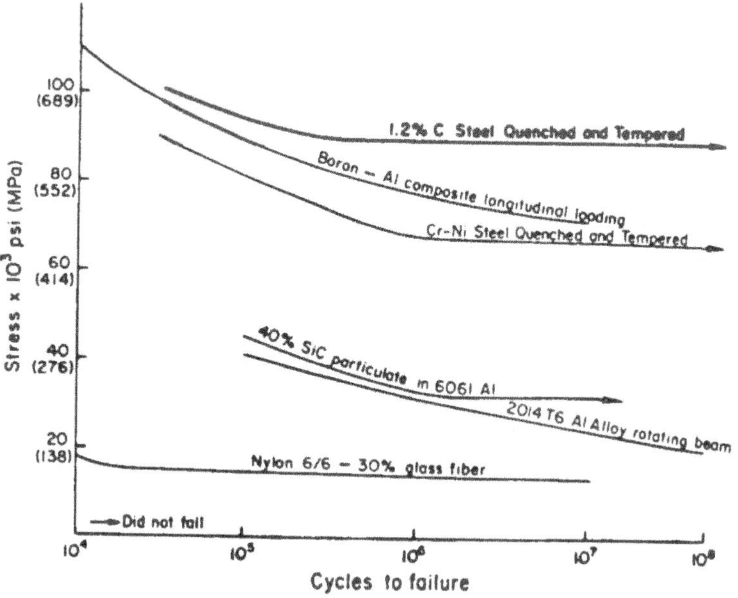

Figure 5 Fatigue behavior of selected composites and alloys.

Table 2 Approximate Use Temperatures of Selected Materials

Material	Approximate Maximum use temperature (°C)
Polymers	
Epoxies	180
Polyesters	130
Vinylesters	150
Phenolics	230
PEEK	250
PEEK + glass fibers	275
PPS	200
Nylons	140
Nylon 6/6 + glass fib.	200
Polypropylene	110
Polyimides, thermoset	310
Bismaelimides, therm.	230
Polyamide-imide (PAI)	260
Polyarl sulfone (PAS)	260
Polyetherimides	170
Metals	
Carbon steels	400
Low-alloy steels	500
2.25Cr–1Mo steels	600
Austenitic stainless	800
Nickel-base superalloys	1200
Other nickel alloys	800
Titanium alloys	500
Bronze (aluminum)	300
Aluminum alloys	275
Ceramics	
Si_3N_4 hot pressed	1100
Al_2O_3	1200
SiC	1500+
TiB_2	1400

3 COMPOSITE PHYSICAL PROPERTIES

Figure 8 shows the electrical resistivities of some of the more popular alloys. For a given group of alloys there is a wide variation in the resistivities, being higher as the alloy content increases, particularly when the alloying element is in solid solution, compared to compound form.

Probably the next most important and interesting physical property is thermal conductivity. These data for the various categories are summarized in Fig. 9. There are a few copper alloys with thermal conductivities well below the 225 W/m·K lowest value shown. However, most of the popular copper alloys have very high thermal conductivities. In aluminum alloys the wrought alloys have higher

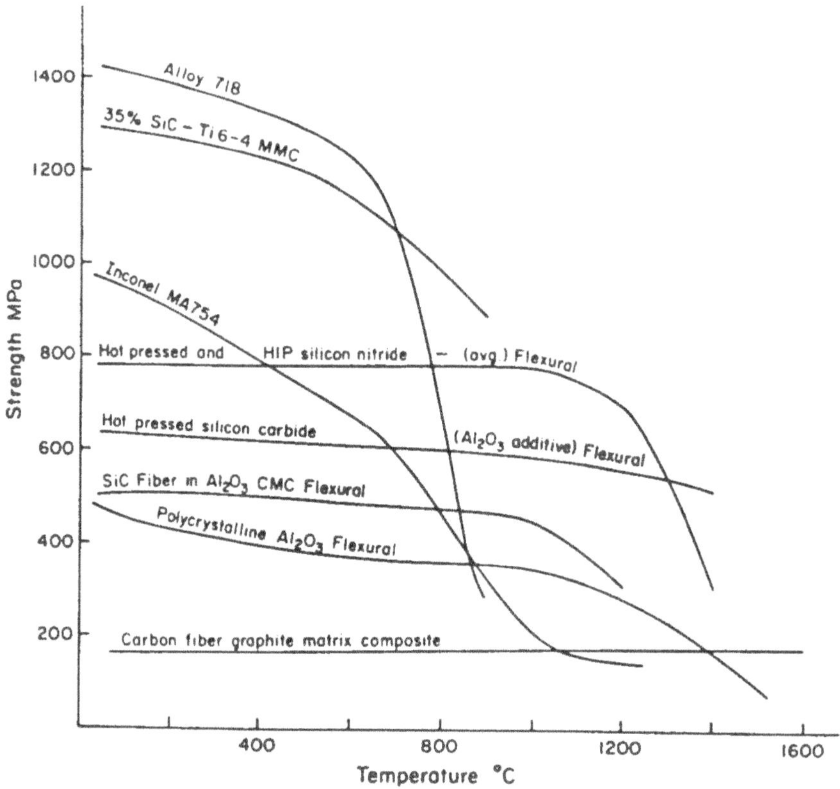

Figure 6 Ultimate strengths versus temperature for selected materials. Unless otherwise noted (e.g., flexural), the values were determined in tensile tests.

thermal conductivities than do the cast alloys, some of the latter being as low as 90 W/m·K. Also zinc and magnesium alloys (not shown) have thermal conductivities in the vicinity of 125 W/m·K. The values for composites vary considerably, as do all composite properties. For many composites physical property data are difficult to locate. Thermal conductivities of ceramics tend to decrease with increasing temperature, whereas those for metals remain reasonably constant. The metal matrix composites generally have good thermal conductivities and, as a result, are desirable for electronic packaging applications. Their low density and high moduli are additional benefits for this purpose.

The coefficient of thermal expansion must be considered in many designs where the materials experience large variations in temperature. Furthermore, dimensional changes created by thermal expansion can cause localized stress gradients due to nonuniform thermal conductivities. And if large members are restricted in movement during changes in temperature, stresses will be set up since the change in temperature and the corresponding desire to expand will place these members in compression. The coefficients of thermal expansion for a large number of materials are listed in Table 3. The polymers appear to have the largest thermal expansion coefficients, but these materials do not usually experience wide temperature

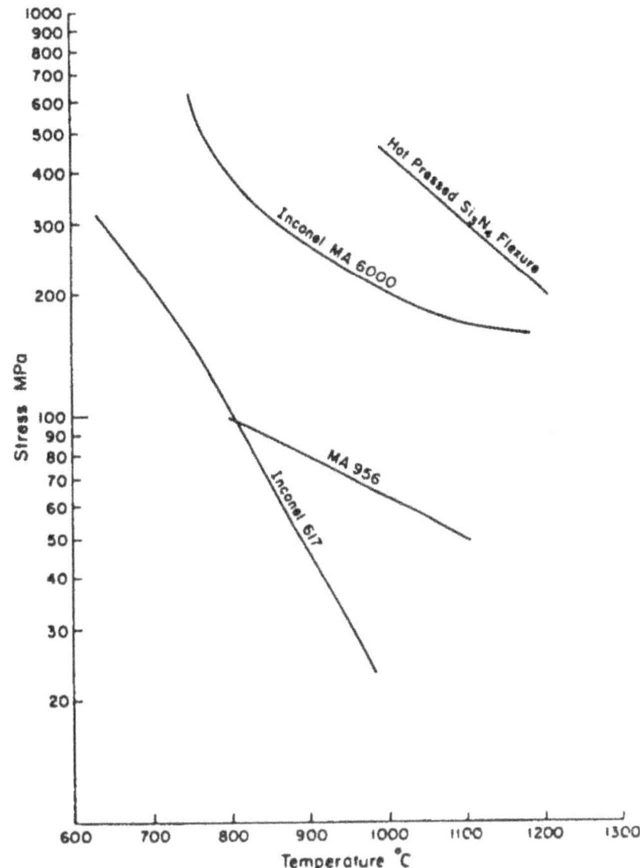

Figure 7 1000-hour stress-rupture curves for superalloys and hot-pressed Si_3N_4. (Superalloy data taken from Inco Alloys Int., Inc., and the Si_3N_4 from G. Quinn, *J. Am. Ceram. Soc.*, 73, 1990, p. 2374).

fluctuations since they are already temperature limited in their applications. Some ceramics have surprisingly high coefficients of thermal expansion. This factor along with their low thermal conductivity and high elastic moduli account for their susceptibility to thermal shock. Composites have sort of a mixed bag in the way of thermal expansion coefficients. Some of those with low thermal expansion coefficients and high thermal conductivities are desirable electronic device packaging materials.

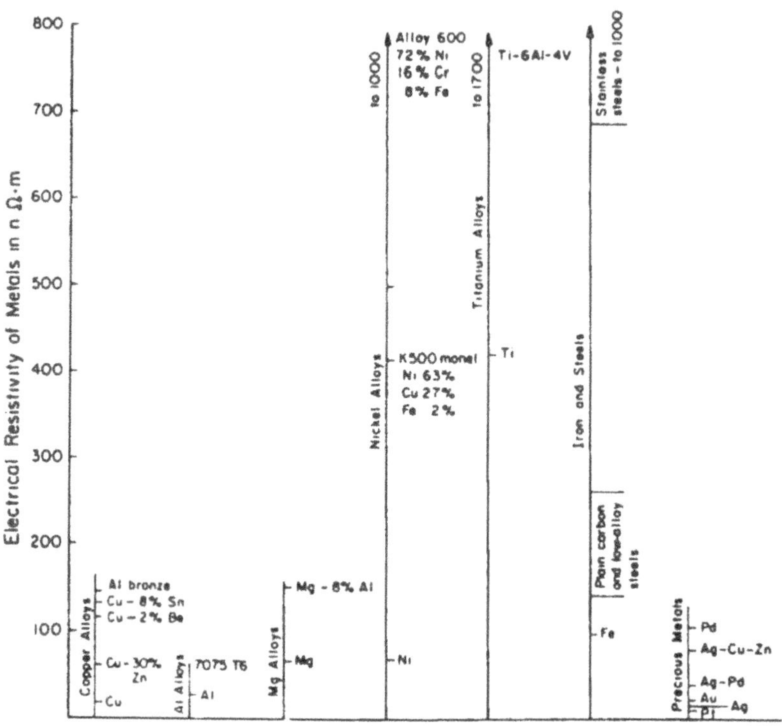

Figure 8 Electrical resistivities for selected metals.

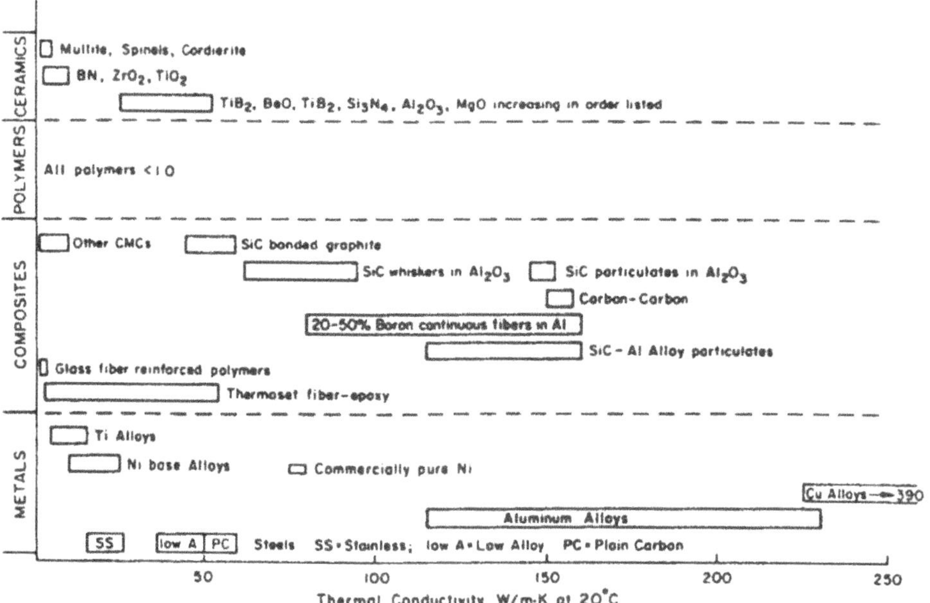

Figure 9 Thermal conductivities of selected materials.

Table 3 Coefficients of Thermal Expansion at 20°C (10^{-6} K^{-1})

Metals	CTE	Composites	CTE	Polymers	CTE	Ceramics	CTE
Al alloys	20–24	SiC whiskers in Al alloy	16	ABS	53	BeO	7.6
Cu alloys	17–20	SiC particulates 30% in Al alloy	12	Nylon 6/6	40	MgO	12.8
Plain C steels	11–14	Continuous boron fibers in Al	6	PEEK	26	SiO_2	22.2
Low-alloy steels	12–15	Al_2O_3	6.7	PET	15	ZrO_2	2.1
Stainless steels	11–19	Si whiskers 30% in Al_2O_3	6.7	PMMA	34	SiC	4.6
Ti alloys	7–8	Carbon-carbon	0	Bakelite	16	Al_2O_3	7.6
		Glass fibers in thermoplasts	34–63	PPS	30	Mullite	5.5
		Glass fibers in thermosets	9–32	Polysulfone	31		
		Grapite-epoxy thermosets	0.6				
		Boron-epoxy	4.5				

Appendix A
Corrosion Tables

Philip A. Schweitzer
Consultant
Fallston, Maryland

COMPARATIVE CORROSION RESISTANT PROPERTIES

The previous chapters have dealt with the corrosion resistant properties of individual materials. This appendix will permit a comparison of the relative resistance of various materials in the presence of selected corrodents.

The tables are arranged alphabetically according to corrodent. Unless otherwise noted, the corrodent is considered pure in the case of liquids and saturated aqueous solution in the case of solids. All percentages shown are weight percents. There are three pages for each corrodent; one for metals, one for plastics, and one for elastomers.

Corrosion is a function of temperature. Symbols used to designate specific corrosion rates are shown at the bottom of the metals pages. When using the tables, note that the vertical lines refer to temperatures midway between the temperatures cited.

TABLE NOTATIONS

1. Material is subject to pitting.
2. Material is subject to stress cracking.
3. Material is subject to crevice attack.
4. Material is subject to intergranular corrosion.
5. Synthetic veil or surfacing mat should be used.

GENERAL NOTES

1. Incoloy category is applicable to grades 800 and 825 only unless otherwise specified.
2. Inconel category is applicable to grade 600 unless otherwise specified.
3. There are many epoxy formulations. When epoxy is shown as being a satisfactory material of construction, it indicates that there is a suitable formulation. The supplier should be checked to ensure that the formulation is satisfactory.

ACETIC ACID 10%

METALS	60	80	100	120	140	160	180	200	220	240	260	280	300	320	340	360	380	400	420	440	460	480	500	520	540	560
°C	15	26	38	49	60	71	82	93	104	116	127	138	149	160	171	182	193	204	216	227	238	249	260	271	282	293
ADMIRALTY BRASS																										
ALUMINUM	G			S		U																				
ALUMINUM BRONZE																										
BRASS, RED	U																									
BRONZE	G																									
CARBON STEEL	U																									
COLUMBIUM (NIOBIUM)	G																									
COPPER	G																									
HASTELLOY B/B-2	E								G																	
HASTELLOY C/C-276	E																									
HASTELLOY D	G																									
HASTELLOY G/G-3																										
HIGH SILICON IRON	E																									
INCONEL	G																									
INCOLLOY 825	E																									
LEAD	U																									
MONEL	G																									
NAVAL BRONZE																										
NICKEL	G																									
NI-RESIST																										
STAINLESS STEELS																										
Type 304/347	E																									
Type 316	E																									
Type 410																										
Type 904-L																										
17-4 PH																										
20 Cb 3	E																									
E-Bright 26-1	E																									
SILICON BRONZE																										
SILICON COPPER																										
STELLITE																										
TANTALUM	E																									
TITANIUM	E																									
ZIRCONIUM	E																									
°F	60	80	100	120	140	160	180	200	220	240	260	280	300	320	340	360	380	400	420	440	460	480	500	520	540	560
°C	15	26	38	49	60	71	82	93	104	116	127	138	149	160	171	182	193	204	216	227	238	249	260	271	282	293

FOR METALS

E = < 2 Mils Penetration/Year; G = < 20 Mils Penetration /year

S = < 50 Mils Penetration/Year; U = > 50 Mils Penetration/year

FOR NONMETALLICS

R = Resistant

U = Unsatisfactory

ACETIC ACID 10%

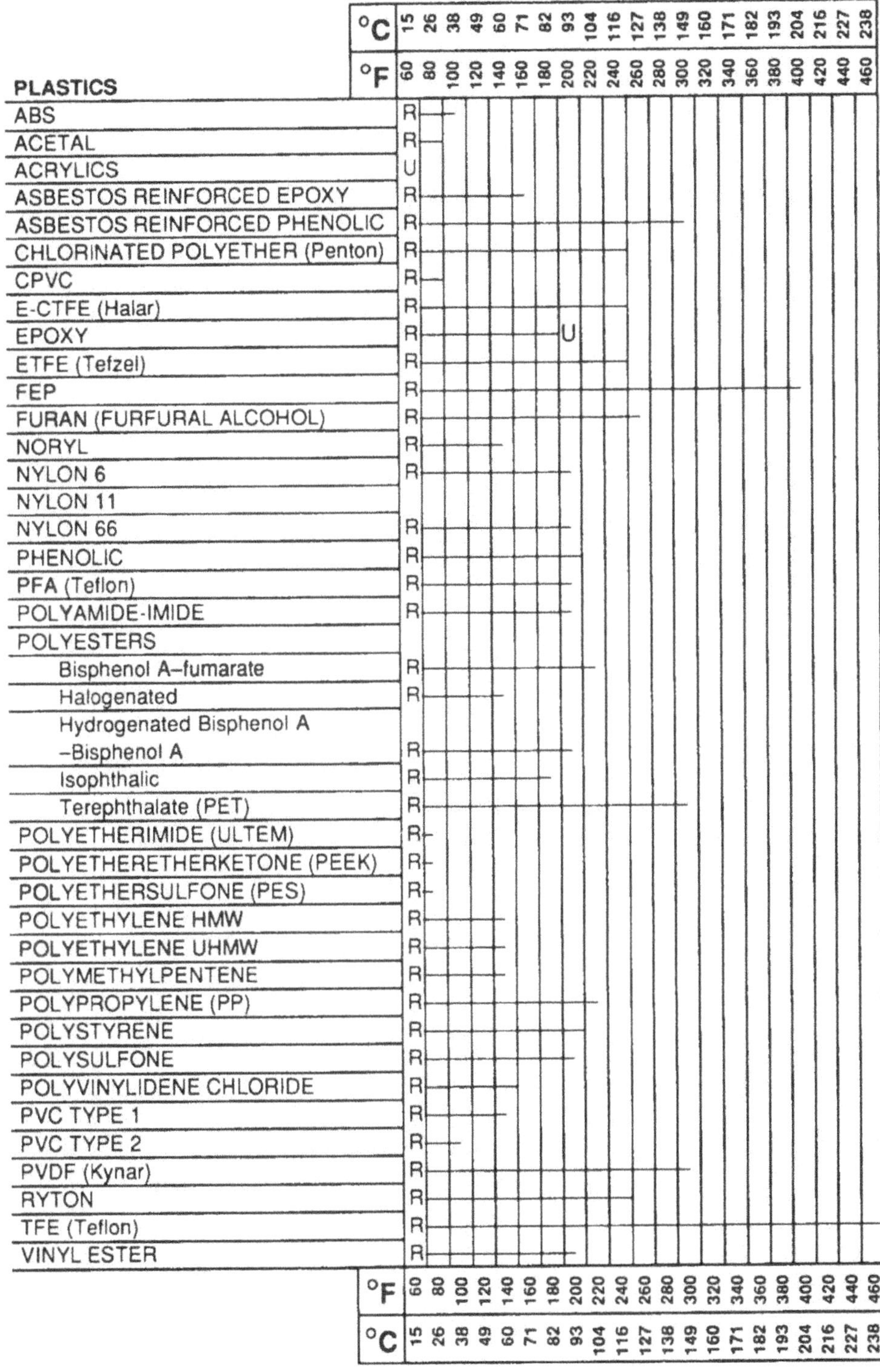
°C 15 26 38 49 60 71 82 93 104 116 127 138 149 160 171 182 193 204 216 227 238
°F 60 80 100 120 140 160 180 200 220 240 260 280 300 320 340 360 380 400 420 440 460
PLASTICS
ABS R
ACETAL R
ACRYLICS U
ASBESTOS REINFORCED EPOXY R
ASBESTOS REINFORCED PHENOLIC R
CHLORINATED POLYETHER (Penton) R
CPVC R
E-CTFE (Halar) R
EPOXY R U
ETFE (Tefzel) R
FEP R
FURAN (FURFURAL ALCOHOL) R
NORYL R
NYLON 6 R
NYLON 11
NYLON 66 R
PHENOLIC R
PFA (Teflon) R
POLYAMIDE-IMIDE R
POLYESTERS
Bisphenol A–fumarate R
Halogenated R
Hydrogenated Bisphenol A
–Bisphenol A R
Isophthalic R
Terephthalate (PET) R
POLYETHERIMIDE (ULTEM) R
POLYETHERETHERKETONE (PEEK) R
POLYETHERSULFONE (PES) R
POLYETHYLENE HMW R
POLYETHYLENE UHMW R
POLYMETHYLPENTENE R
POLYPROPYLENE (PP) R
POLYSTYRENE R
POLYSULFONE R
POLYVINYLIDENE CHLORIDE R
PVC TYPE 1 R
PVC TYPE 2 R
PVDF (Kynar) R
RYTON R
TFE (Teflon) R
VINYL ESTER R
°F 60 80 100 120 140 160 180 200 220 240 260 280 300 320 340 360 380 400 420 440 460
°C 15 26 38 49 60 71 82 93 104 116 127 138 149 160 171 182 193 204 216 227 238

ACETIC ACID 10%

°C	15	26	38	49	60	71	82	93	104	116	127	138	149	160	171	182	193	204	216	227	238
ELASTOMERS* AND LININGS / °F	60	80	100	120	140	160	180	200	220	240	260	280	300	320	340	360	380	400	420	440	460
ACRYLATE-BUTADIENE (ABR)																					
BUTYL GR-1 (IIR)	R																				
CARBOXYLIC-ACRYLONITRILE-BUTADIENE (XNBR)																					
CHEMRAZ (FPM)	R																				
CHLORO-ISOBUTENE-ISOPRENE (CIIR)	R																				
HYPALON (CHLORO-SULFONYL-POLYETHYLENE (CSM)	R																				
ETHYLENE-ACRYLIC (EA)																					
ETHYLENE-PROPYLENE (EPM)																					
ETHYLENE-PROPYLENE-DIENE (EPDM)	R																				
ETHYLENE-PROPYLENE TERPOLYMER (EPT)	U																				
FKM (Viton A)	R																				
HARD RUBBER	R																				
SOFT RUBBER																					
ISOPRENE (IR)	R																				
KALREZ (FPM)	R																				
KOROSEAL	R																				
NATURAL RUBBER (GRS)	R																				
NEOPRENE GR-M (CR)	R																				
NITRILE BUNA-N (NBR)	R																				
NORDEL (EPDM)	R																				
POLYBUTADIENE (BR)																					
POLYESTER (PE)	R																				
POLYETHER-URETHANE (EU)	U																				
POLYISOPRENE (IR)																					
POLYSULFIDES (T)	R																				
POLYURETHANE (AU)	U																				
SBR STYRENE (BUNA-S)	U																				
SILICONE RUBBERS	R																				
°F	60	80	100	120	140	160	180	200	220	240	260	280	300	320	340	360	380	400	420	440	460
°C	15	26	38	49	60	71	82	93	104	116	127	138	149	160	171	182	193	204	216	227	238

*See also Nylon 11 under PLASTICS

ACETIC ACID 50%

METALS	60 °F	80	100	120	140	160	180	200	220	240	260	280	300	320	340	360	380	400	420	440	460	480	500	520	540	560
°C	15	26	38	49	60	71	82	93	104	116	127	138	149	160	171	182	193	204	216	227	238	249	260	271	282	293
ADMIRALTY BRASS																										
ALUMINUM	G	—	—	—	S	—	U																			
ALUMINUM BRONZE	U																									
BRASS, RED	U																									
BRONZE	U																									
CARBON STEEL	U																									
COLUMBIUM (NIOBIUM)	G	—	—	—	—	—	—	—																		
COPPER	U																									
HASTELLOY B/B-2	E	—	—	—	—	—	—	—	—	—	—	—	—													
HASTELLOY C/C-276	E	—	—	—	—	—	—	—	G	—	—	—	—													
HASTELLOY D	G	—	—	—	—	—	—	—	—	—	—	—	—													
HASTELLOY G/G-3																										
HIGH SILICON IRON	E	—	—	—	—	—	—	—																		
INCONEL	S	—	—	—	U																					
INCOLLOY 825	E	—	—	—	—	—	—	—	—																	
LEAD	U																									
MONEL	G	—	—	—	—	—	—	—																		
NAVAL BRONZE																										
NICKEL	G	—	S	—	—																					
NI-RESIST																										
STAINLESS STEELS																										
Type 304/347	G	—	—	—	—	—	S	—	—																	
Type 316	E	—	—	—	—	—	—	—	—	—	—	—	—	—	—	—	—									
Type 410	G																									
Type 904-L																										
17-4 PH	G																									
20 Cb 3	E	—	—	—	—	—	—	—	G	—	—	—	—													
E-Bright 26-1	E	—	—	—	—	—	—	—																		
SILICON BRONZE	U																									
SILICON COPPER																										
STELLITE																										
TANTALUM	E	—	—	—	—	—	—	—	—	—	—	—	—	—	—	—	—	—	—	—	—	—	—	—	—	—
TITANIUM	E	—	—	—	—	—	—	—	—	—	—															
ZIRCONIUM	E	—	—	—	—	—	—	—	—																	

FOR METALS

E = < 2 Mils Penetration/Year; G = < 20 Mils Penetration /year

S = < 50 Mils Penetration/Year; U = > 50 Mils Penetration/year

FOR NONMETALLICS

R = Resistant

U = Unsatisfactory

ACETIC ACID 50%

PLASTICS	60	80	100	120	140	160	180	200	220	240	260	280	300	320	340	360	380	400	420	440	460
°C	15	26	38	49	60	71	82	93	104	116	127	138	149	160	171	182	193	204	216	227	238
ABS	R				U																
ACETAL	R																				
ACRYLICS	U																				
ASBESTOS REINFORCED EPOXY																					
ASBESTOS REINFORCED PHENOLIC	R																				
CHLORINATED POLYETHER (Penton)	R																				
CPVC	U																				
E-CTFE (Halar)	R																				
EPOXY	R			U																	
ETFE (Tefzel)	R																				
FEP	R																				
FURAN (FURFURAL ALCOHOL)	R																				
NORYL	R																				
NYLON 6	U																				
NYLON 11	U																				
NYLON 66	U																				
PHENOLIC																					
PFA (Teflon)	R																				
POLYAMIDE-IMIDE	R																				
POLYESTERS																					
Bisphenol A–fumarate	R																				
Halogenated	R																				
Hydrogenated Bisphenol A –Bisphenol A	R																				
Isophthalic	R			U																	
Terephthalate (PET)	R																				
POLYETHERIMIDE (ULTEM)	R																				
POLYETHERETHERKETONE (PEEK)	R																				
POLYETHERSULFONE (PES)	R																				
POLYETHYLENE HMW	R																				
POLYETHYLENE UHMW	R																				
POLYMETHYLPENTENE	U																				
POLYPROPYLENE (PP)	R																				
POLYSTYRENE	U																				
POLYSULFONE	R																				
POLYVINYLIDENE CHLORIDE	R																				
PVC TYPE 1	R																				
PVC TYPE 2	R																				
PVDF (Kynar)	R																				
RYTON	R																				
TFE (Teflon)	R																				
VINYL ESTER	R																				
°F	60	80	100	120	140	160	180	200	220	240	260	280	300	320	340	360	380	400	420	440	460
°C	15	26	38	49	60	71	82	93	104	116	127	138	149	160	171	182	193	204	216	227	238

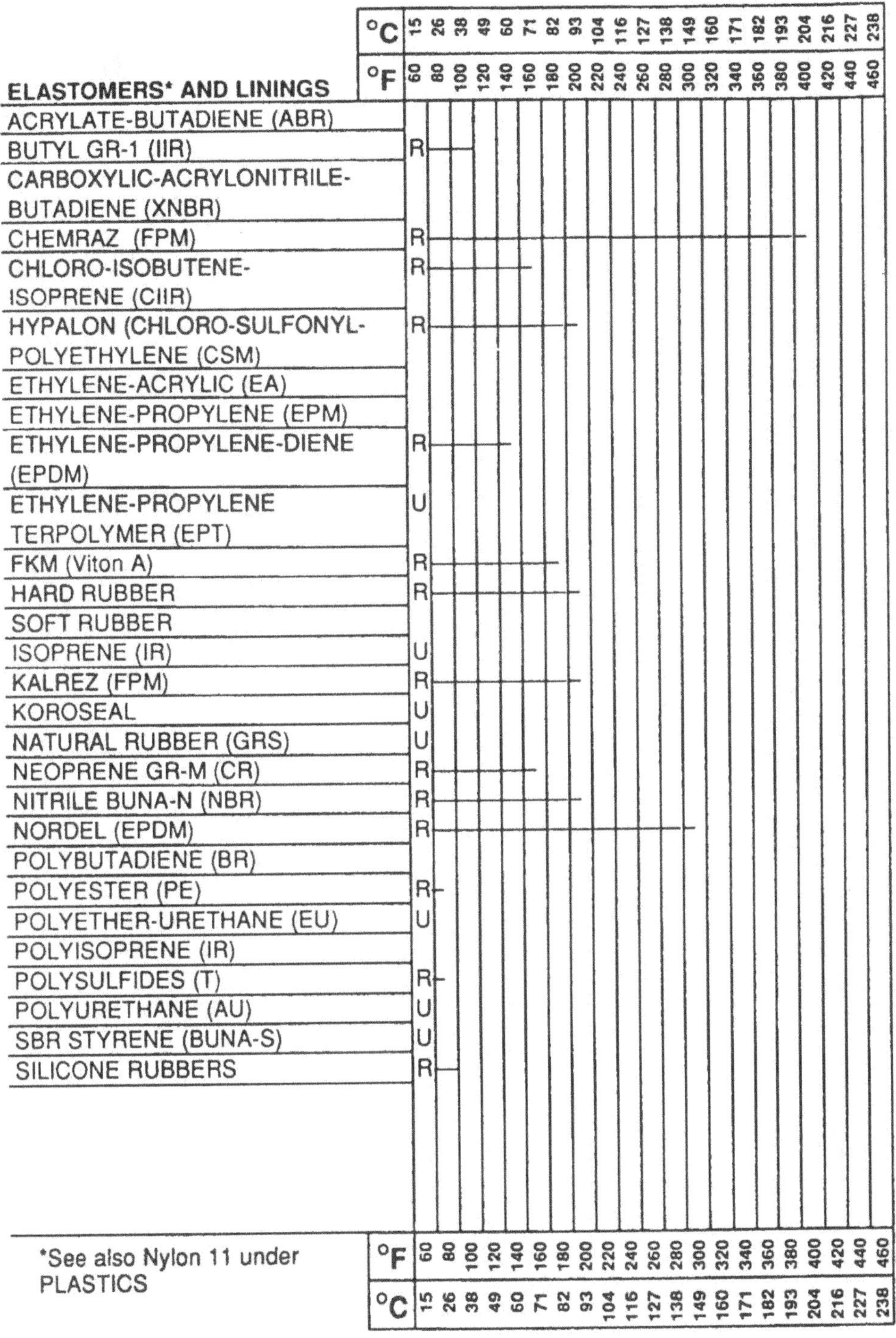

ELASTOMERS* AND LININGS	Rating	Max. temperature shown °F (°C)
ACRYLATE-BUTADIENE (ABR)		
BUTYL GR-1 (IIR)	R	100 (38)
CARBOXYLIC-ACRYLONITRILE-BUTADIENE (XNBR)		
CHEMRAZ (FPM)	R	400 (204)
CHLORO-ISOBUTENE-ISOPRENE (CIIR)	R	160 (71)
HYPALON (CHLORO-SULFONYL-POLYETHYLENE (CSM)	R	200 (93)
ETHYLENE-ACRYLIC (EA)		
ETHYLENE-PROPYLENE (EPM)		
ETHYLENE-PROPYLENE-DIENE (EPDM)	R	140 (60)
ETHYLENE-PROPYLENE TERPOLYMER (EPT)	U	
FKM (Viton A)	R	180 (82)
HARD RUBBER	R	200 (93)
SOFT RUBBER		
ISOPRENE (IR)	U	
KALREZ (FPM)	R	200 (93)
KOROSEAL	U	
NATURAL RUBBER (GRS)	U	
NEOPRENE GR-M (CR)	R	160 (71)
NITRILE BUNA-N (NBR)	R	200 (93)
NORDEL (EPDM)	R	300 (149)
POLYBUTADIENE (BR)		
POLYESTER (PE)	R	60 (15)
POLYETHER-URETHANE (EU)	U	
POLYISOPRENE (IR)		
POLYSULFIDES (T)	R	60 (15)
POLYURETHANE (AU)	U	
SBR STYRENE (BUNA-S)	U	
SILICONE RUBBERS	R	80 (26)

*See also Nylon 11 under PLASTICS

ACETIC ACID 80%

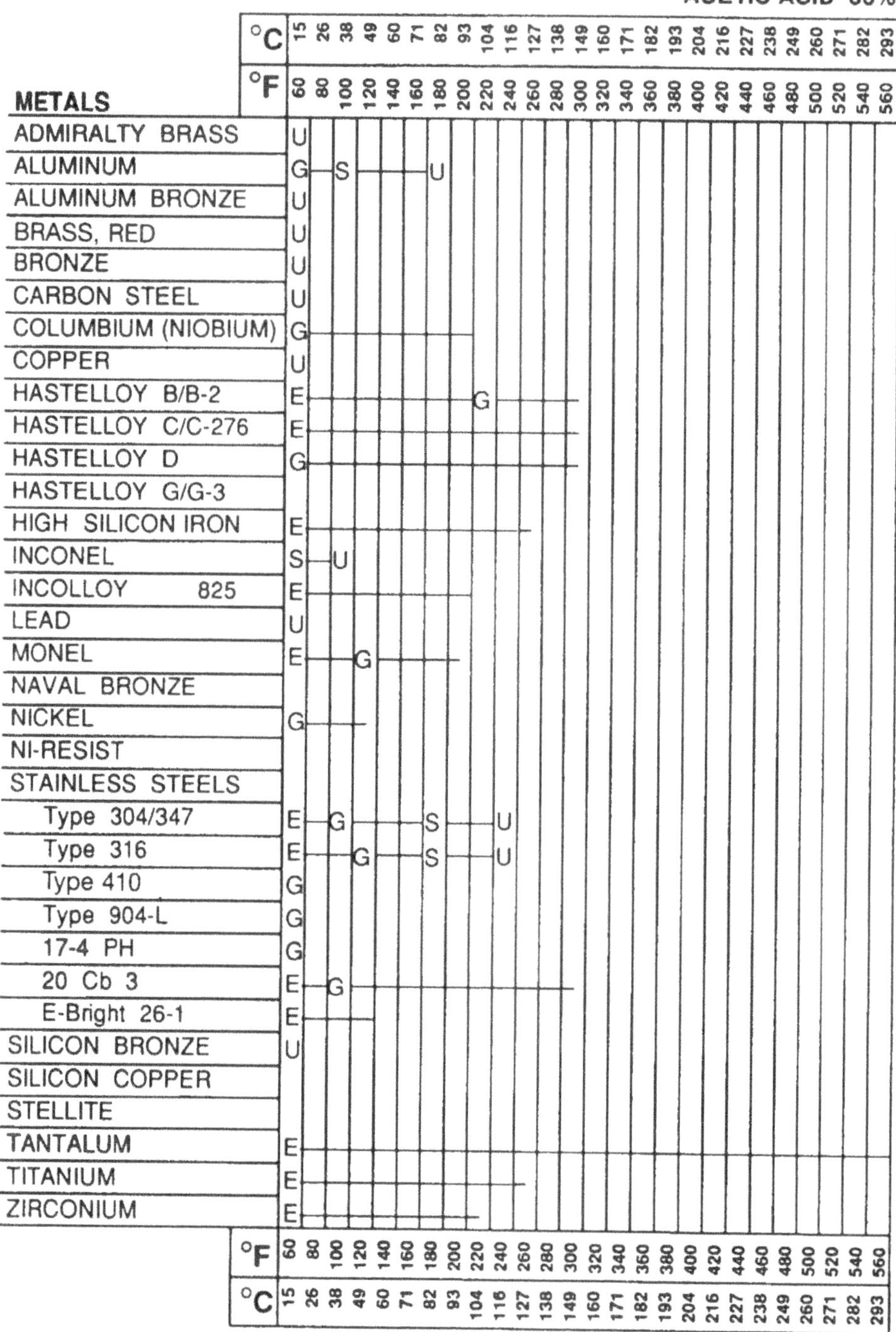

°C	15	26	38	49	60	71	82	93	104	116	127	138	149	160	171	182	193	204	216	227	238	249	260	271	282	293
METALS °F	60	80	100	120	140	160	180	200	220	240	260	280	300	320	340	360	380	400	420	440	460	480	500	520	540	560
ADMIRALTY BRASS	U																									
ALUMINUM	G		S				U																			
ALUMINUM BRONZE	U																									
BRASS, RED	U																									
BRONZE	U																									
CARBON STEEL	U																									
COLUMBIUM (NIOBIUM)	G																									
COPPER	U																									
HASTELLOY B/B-2	E								G																	
HASTELLOY C/C-276	E																									
HASTELLOY D	G																									
HASTELLOY G/G-3																										
HIGH SILICON IRON	E																									
INCONEL	S		U																							
INCOLLOY 825	E																									
LEAD	U																									
MONEL	E			G																						
NAVAL BRONZE																										
NICKEL	G																									
NI-RESIST																										
STAINLESS STEELS																										
Type 304/347	E		G				S			U																
Type 316	E			G			S			U																
Type 410	G																									
Type 904-L	G																									
17-4 PH	G																									
20 Cb 3	E		G																							
E-Bright 26-1	E																									
SILICON BRONZE	U																									
SILICON COPPER																										
STELLITE																										
TANTALUM	E																									
TITANIUM	E																									
ZIRCONIUM	E																									

FOR METALS

E = < 2 Mils Penetration/Year; G = < 20 Mils Penetration /year

S = < 50 Mils Penetration/Year; U = > 50 Mils Penetration/year

FOR NONMETALLICS

R = Resistant

U = Unsatisfactory

ACETIC ACID 80%

PLASTICS	60	80	100	120	140	160	180	200	220	240	260	280	300	320	340	360	380	400	420	440	460
°C	15	26	38	49	60	71	82	93	104	116	127	138	149	160	171	182	193	204	216	227	238
ABS	U																				
ACETAL	R																				
ACRYLICS	U																				
ASBESTOS REINFORCED EPOXY																					
ASBESTOS REINFORCED PHENOLIC	R	—	—	—	—	—	—	—	—	—	—	—	—								
CHLORINATED POLYETHER (Penton)	R	—	—	—	—	—	—	—	—	—											
CPVC	U																				
E-CTFE (Halar)	R	—	—	—	—																
EPOXY	R	—	—	U																	
ETFE (Tefzel)	R	—	—	—	—	—	—	—	—												
FEP	R	—	—	—	—	—	—	—	—	—	—	—	—	—	—	—	—	—			
FURAN (FURFURAL ALCOHOL)	R	—	—	—	—	—	—	—													
NORYL	R	—	—	—	—																
NYLON 6	U																				
NYLON 11	U																				
NYLON 66	U																				
PHENOLIC																					
PFA (Teflon)	R	—	—	—	—	—	—	—													
POLYAMIDE-IMIDE	R	—	—	—	—	—	—	—													
POLYESTERS																					
Bisphenol A–fumarate	R	—	—	—	—	—															
Halogenated																					
Hydrogenated Bisphenol A –Bisphenol A																					
Isophthalic	U																				
Terephthalate (PET)	R	—	—	—	—	—	—	—	—	—	—	—	—								
POLYETHERIMIDE (ULTEM)	R																				
POLYETHERETHERKETONE (PEEK)	R	—	—	—	—																
POLYETHERSULFONE (PES)	R	—	—	—	—																
POLYETHYLENE HMW	R																				
POLYETHYLENE UHMW	R																				
POLYMETHYLPENTENE	U																				
POLYPROPYLENE (PP)	R	—	—	—	—	—	—	—													
POLYSTYRENE	U																				
POLYSULFONE	R	—	—	—	—	—	—	—													
POLYVINYLIDENE CHLORIDE	R	—	—	—																	
PVC TYPE 1	R	—	—	—	—																
PVC TYPE 2	U																				
PVDF (Kynar)	R	—	—	—	—	—	—	U													
RYTON	R	—	—	—	—	—	—	—	—	—											
TFE (Teflon)	R	—	—	—	—	—	—	—	—	—	—	—	—	—	—	—	—	—	—	—	—
VINYL ESTER	R	—	—	—	—																
°F	60	80	100	120	140	160	180	200	220	240	260	280	300	320	340	360	380	400	420	440	460
°C	15	26	38	49	60	71	82	93	104	116	127	138	149	160	171	182	193	204	216	227	238

ACETIC ACID 80%

ELASTOMERS* AND LININGS	Rating	Max. °F (°C)
ACRYLATE-BUTADIENE (ABR)		
BUTYL GR-1 (IIR)	R	100 (38)
CARBOXYLIC-ACRYLONITRILE-BUTADIENE (XNBR)		
CHEMRAZ (FPM)	R	400 (204)
CHLORO-ISOBUTENE-ISOPRENE (CIIR)	R	160 (71)
HYPALON (CHLORO-SULFONYL-POLYETHYLENE (CSM)	R	200 (93)
ETHYLENE-ACRYLIC (EA)		
ETHYLENE-PROPYLENE (EPM)		
ETHYLENE-PROPYLENE-DIENE (EPDM)	R	140 (60)
ETHYLENE-PROPYLENE TERPOLYMER (EPT)	U	
FKM (Viton A)	R	180 (82)
HARD RUBBER	R	140 (60)
SOFT RUBBER		
ISOPRENE (IR)	U	
KALREZ (FPM)	R	80 (26)
KOROSEAL	U	
NATURAL RUBBER (GRS)	U	
NEOPRENE GR-M (CR)	R	160 (71)
NITRILE BUNA-N (NBR)	R	200 (93)
NORDEL (EPDM)	R	300 (149)
POLYBUTADIENE (BR)		
POLYESTER (PE)	R	80 (26)
POLYETHER-URETHANE (EU)	U	
POLYISOPRENE (IR)		
POLYSULFIDES (T)	R	80 (26)
POLYURETHANE (AU)	U	
SBR STYRENE (BUNA-S)	U	
SILICONE RUBBERS	R	80 (26)

Temperature scale: °F 60, 80, 100, 120, 140, 160, 180, 200, 220, 240, 260, 280, 300, 320, 340, 360, 380, 400, 420, 440, 460; °C 15, 26, 38, 49, 60, 71, 82, 93, 104, 116, 127, 138, 149, 160, 171, 182, 193, 204, 216, 227, 238

*See also Nylon 11 under PLASTICS

METALS	60	80	100	120	140	160	180	200	220	240	260	280	300	320	340	360	380	400	420	440	460	480	500	520	540	560
°C	15	26	38	49	60	71	82	93	104	116	127	138	149	160	171	182	193	204	216	227	238	249	260	271	282	293
ADMIRALTY BRASS	U																									
ALUMINUM	E	—	G	—	—	—	—	—	U																	
ALUMINUM BRONZE	S																									
BRASS, RED	U																									
BRONZE	U																									
CARBON STEEL	U																									
COLUMBIUM (NIOBIUM)	E	—	—	—	—	—	—	—																		
COPPER	U																									
HASTELLOY B/B-2	E	—	—	—	—	—	—	—	—	—	—	—	—	—	—	—	—	—	—	—	—	—	—	—	—	—
HASTELLOY C/C-276	E	—	—	—	—	—	—	—	—	—	—	—	—	—	—	—	—	—	—	—	—	—	—	—	—	—
HASTELLOY D	E	—	—	—	—	—	—	—	—																	
HASTELLOY G/G-3																										
HIGH SILICON IRON	E	—	—	—	—	—	—	—	—																	
INCONEL	G	—	—	—	—	—	—	—	—																	
INCOLLOY 825	E	—	—	—	—	—	—	—	—																	
LEAD	U																									
MONEL	E	—	—	G	—	—	—	—	—	—	—	—	U													
NAVAL BRONZE	U																									
NICKEL	U																									
NI-RESIST																										
STAINLESS STEELS																										
Type 304/347	E	—	G	—	—	—	—	—	U																	
Type 316	E	—	—	—	—	—	—	—	—	—	—	—	—	—	—	—	—	—								
Type 410	U																									
Type 904-L																										
17-4 PH	S																									
20 Cb 3	E	—	—	G	—	—	—	—	—	—	—	—	—													
E-Bright 26-1	E	—	—	—	—																					
SILICON BRONZE	S																									
SILICON COPPER	G																									
STELLITE																										
TANTALUM	E	—	—	—	—	—	—	—	—	—	—	—	—	—	—	—	—	—	—	—	—	—	—	—	—	—
TITANIUM	E	—	—	—	—	—	—	—	—	—	—															
ZIRCONIUM	G	—	—	—	—	—	—	—	—																	
°F	60	80	100	120	140	160	180	200	220	240	260	280	300	320	340	360	380	400	420	440	460	480	500	520	540	560
°C	15	26	38	49	60	71	82	93	104	116	127	138	149	160	171	182	193	204	216	227	238	249	260	271	282	293

FOR METALS

E = < 2 Mils Penetration/Year; G = < 20 Mils Penetration /year

S = < 50 Mils Penetration/Year; U = > 50 Mils Penetration/year

FOR NONMETALLICS

R = Resistant

U = Unsatisfactory

ACETIC ACID, GLACIAL

PLASTICS	60	80	100	120	140	160	180	200	220	240	260	280	300	320	340	360	380	400	420	440	460
°C	15	26	38	49	60	71	82	93	104	116	127	138	149	160	171	182	193	204	216	227	238
ABS	U																				
ACETAL																					
ACRYLICS	U																				
ASBESTOS REINFORCED EPOXY	U																				
ASBESTOS REINFORCED PHENOLIC	R																				
CHLORINATED POLYETHER (Penton)	R																				
CPVC	U																				
E-CTFE (Halar)	R																				
EPOXY																					
ETFE (Tefzel)	R																				
FEP	R																				
FURAN (FURFURAL ALCOHOL)	R											U									
NORYL	R																				
NYLON 6	U																				
NYLON 11	U																				
NYLON 66	U																				
PHENOLIC	R																				
PFA (Teflon)	R																				
POLYAMIDE-IMIDE	R																				
POLYESTERS																					
Bisphenol A–fumarate	U																				
Halogenated	R																				
Hydrogenated Bisphenol A –Bisphenol A																					
Isophthalic	U																				
Terephthalate (PET)	R																				
POLYETHERIMIDE (ULTEM)	R																				
POLYETHERETHERKETONE (PEEK)	R																				
POLYETHERSULFONE (PES)	R																				
POLYETHYLENE HMW	R																				
POLYETHYLENE UHMW 40%	R																				
POLYMETHYLPENTENE	U																				
POLYPROPYLENE (PP)	R							U													
POLYSTYRENE	U																				
POLYSULFONE	R																				
POLYVINYLIDENE CHLORIDE	R																				
PVC TYPE 1	R				U																
PVC TYPE 2	U																				
PVDF (Kynar)	R							U													
RYTON	R																				
TFE (Teflon)																					
VINYL ESTER	R																				

°F 60 80 100 120 140 160 180 200 220 240 260 280 300 320 340 360 380 400 420 440 460
°C 15 26 38 49 60 71 82 93 104 116 127 138 149 160 171 182 193 204 216 227 238

ACETIC ACID, GLACIAL

ELASTOMERS* AND LININGS	°F 60	80	100	120	140	160	180	200	220	240	260	280	300	320	340	360	380	400	420	440	460
	°C 15	26	38	49	60	71	82	93	104	116	127	138	149	160	171	182	193	204	216	227	238
ACRYLATE-BUTADIENE (ABR)																					
BUTYL GR-1 (IIR)	R		U																		
CARBOXYLIC-ACRYLONITRILE-BUTADIENE (XNBR)																					
CHEMRAZ (FPM)	R																				
CHLORO-ISOBUTENE-ISOPRENE (CIIR)	R																				
HYPALON (CHLORO-SULFONYL-POLYETHYLENE (CSM)	U																				
ETHYLENE-ACRYLIC (EA)																					
ETHYLENE-PROPYLENE (EPM)	U																				
ETHYLENE-PROPYLENE-DIENE (EPDM)	R																				
ETHYLENE-PROPYLENE TERPOLYMER (EPT)	U																				
FKM (Viton A)	U																				
HARD RUBBER	R																				
SOFT RUBBER	U																				
ISOPRENE (IR)	U																				
KALREZ (FPM)	R																				
KOROSEAL	U																				
NATURAL RUBBER (GRS)	U																				
NEOPRENE GR-M (CR)	U																				
NITRILE BUNA-N (NBR)	R			U																	
NORDEL (EPDM)	U																				
POLYBUTADIENE (BR)																					
POLYESTER (PE)	R																				
POLYETHER-URETHANE (EU)	U																				
POLYISOPRENE (IR)																					
POLYSULFIDES (T)	R																				
POLYURETHANE (AU)	U																				
SBR STYRENE (BUNA-S)	U																				
SILICONE RUBBERS	R																				

*See also Nylon 11 under PLASTICS

°F 60 80 100 120 140 160 180 200 220 240 260 280 300 320 340 360 380 400 420 440 460
°C 15 26 38 49 60 71 82 93 104 116 127 138 149 160 171 182 193 204 216 227 238

ACETIC ANHYDRIDE

METALS	60 °F	80	100	120	140	160	180	200	220	240	260	280	300	320	340	360	380	400	420	440	460	480	500	520	540	560
°C	15	26	38	49	60	71	82	93	104	116	127	138	149	160	171	182	193	204	216	227	238	249	260	271	282	293
ADMIRALTY BRASS	S																									
ALUMINUM	G																									
ALUMINUM BRONZE	G																									
BRASS, RED	S			U																						
BRONZE	G																									
CARBON STEEL	G																									
COLUMBIUM (NIOBIUM)	E																									
COPPER	G																									
HASTELLOY B/B-2	G																									
HASTELLOY C/C-276	E																									
HASTELLOY D	G																									
HASTELLOY G/G-3																										
HIGH SILICON IRON	E																									
INCONEL	G																									
INCOLLOY	G																									
LEAD	G																									
MONEL	G						S																			
NAVAL BRONZE	S																									
NICKEL	G						S																			
NI-RESIST	S																									
STAINLESS STEELS																										
Type 304/347	G																									
Type 316	G																									
Type 410	U																									
Type 904-L																										
17-4 PH	G																									
20 Cb 3	G																									
E-Bright 26-1	E		G																							
SILICON BRONZE	G																									
SILICON COPPER	G																									
STELLITE																										
TANTALUM	E		G																							
TITANIUM	E																									
ZIRCONIUM	E																									

FOR METALS
E = < 2 Mils Penetration/Year; G = < 20 Mils Penetration /year
S = < 50 Mils Penetration/Year; U = > 50 Mils Penetration/year

FOR NONMETALLICS
R = Resistant
U = Unsatisfactory

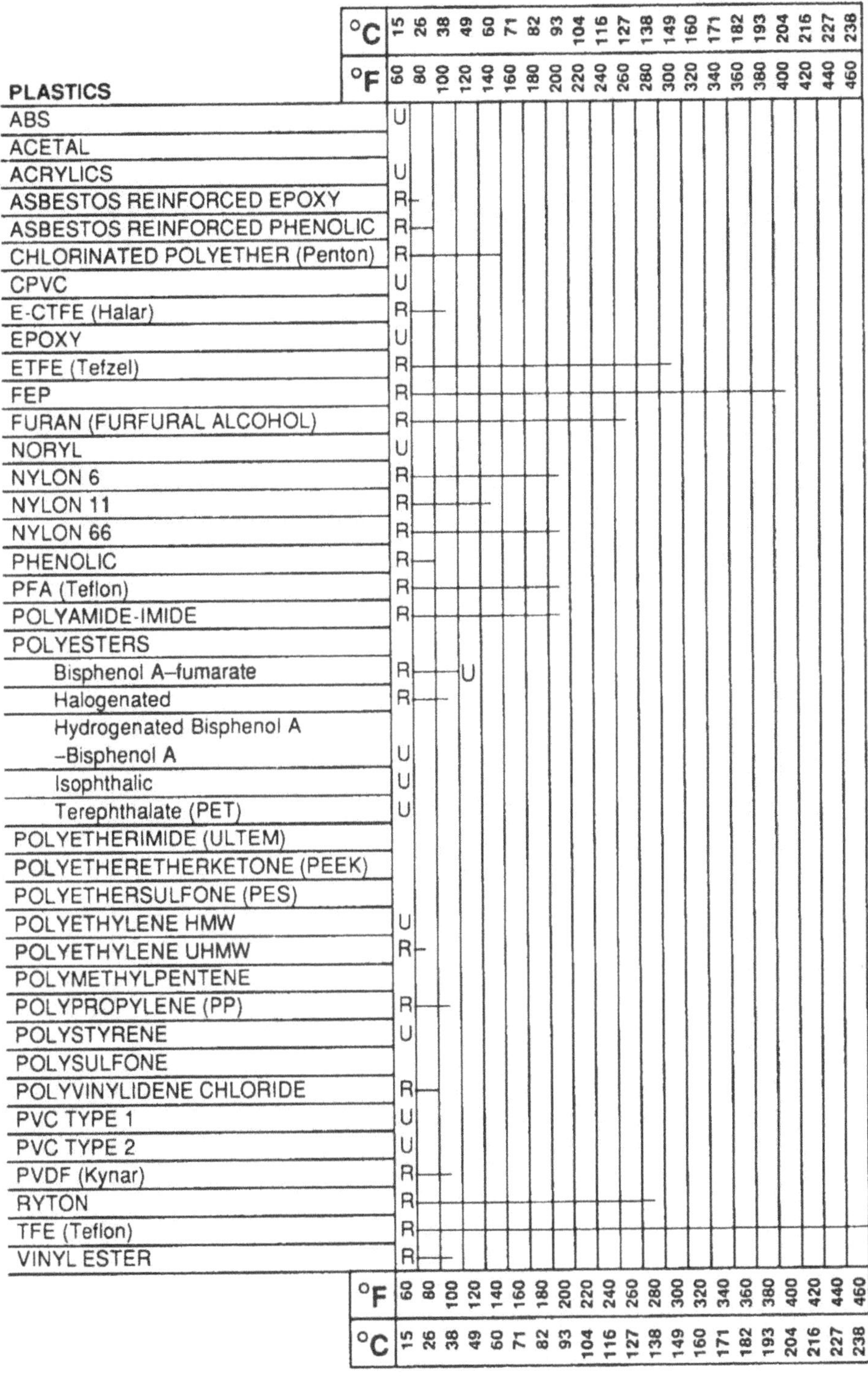

°F: 60, 80, 100, 120, 140, 160, 180, 200, 220, 240, 260, 280, 300, 320, 340, 360, 380, 400, 420, 440, 460

°C: 15, 26, 38, 49, 60, 71, 82, 93, 104, 116, 127, 138, 149, 160, 171, 182, 193, 204, 216, 227, 238

PLASTICS	60 °F (15 °C)	120 °F (49 °C)
ABS	U	
ACETAL		
ACRYLICS	U	
ASBESTOS REINFORCED EPOXY	R	
ASBESTOS REINFORCED PHENOLIC	R	
CHLORINATED POLYETHER (Penton)	R	
CPVC	U	
E-CTFE (Halar)	R	
EPOXY	U	
ETFE (Tefzel)	R	
FEP	R	
FURAN (FURFURAL ALCOHOL)	R	
NORYL	U	
NYLON 6	R	
NYLON 11	R	
NYLON 66	R	
PHENOLIC	R	
PFA (Teflon)	R	
POLYAMIDE-IMIDE	R	
POLYESTERS		
Bisphenol A–fumarate	R	U
Halogenated	R	
Hydrogenated Bisphenol A –Bisphenol A	U	
Isophthalic	U	
Terephthalate (PET)	U	
POLYETHERIMIDE (ULTEM)		
POLYETHERETHERKETONE (PEEK)		
POLYETHERSULFONE (PES)		
POLYETHYLENE HMW	U	
POLYETHYLENE UHMW	R	
POLYMETHYLPENTENE		
POLYPROPYLENE (PP)	R	
POLYSTYRENE	U	
POLYSULFONE		
POLYVINYLIDENE CHLORIDE	R	
PVC TYPE 1	U	
PVC TYPE 2	U	
PVDF (Kynar)	R	
RYTON	R	
TFE (Teflon)	R	
VINYL ESTER	R	

°C: 15 26 38 49 60 71 82 93 104 116 127 138 149 160 171 182 193 204 216 227 238

°F: 60 80 100 120 140 160 180 200 220 240 260 280 300 320 340 360 380 400 420 440 460

ELASTOMERS* AND LININGS	
ACRYLATE-BUTADIENE (ABR)	
BUTYL GR-1 (IIR)	R
CARBOXYLIC-ACRYLONITRILE-BUTADIENE (XNBR)	
CHEMRAZ (FPM)	R
CHLORO-ISOBUTENE-ISOPRENE (CIIR)	
HYPALON (CHLORO-SULFONYL-POLYETHYLENE (CSM)	R
ETHYLENE-ACRYLIC (EA)	
ETHYLENE-PROPYLENE (EPM)	U
ETHYLENE-PROPYLENE-DIENE (EPDM)	U
ETHYLENE-PROPYLENE TERPOLYMER (EPT)	U
FKM (Viton A)	U
HARD RUBBER	R
SOFT RUBBER	U
ISOPRENE (IR)	U
KALREZ (FPM)	R
KOROSEAL	
NATURAL RUBBER (GRS)	U
NEOPRENE GR-M (CR)	R
NITRILE BUNA-N (NBR)	R
NORDEL (EPDM)	
POLYBUTADIENE (BR)	
POLYESTER (PE)	
POLYETHER-URETHANE (EU)	U
POLYISOPRENE (IR)	
POLYSULFIDES (T)	
POLYURETHANE (AU)	U
SBR STYRENE (BUNA-S)	R
SILICONE RUBBERS	

*See also Nylon 11 under PLASTICS

°F: 60 80 100 120 140 160 180 200 220 240 260 280 300 320 340 360 380 400 420 440 460

°C: 15 26 38 49 60 71 82 93 104 116 127 138 149 160 171 182 193 204 216 227 238

ALUMINUM CHLORIDE AQUEOUS

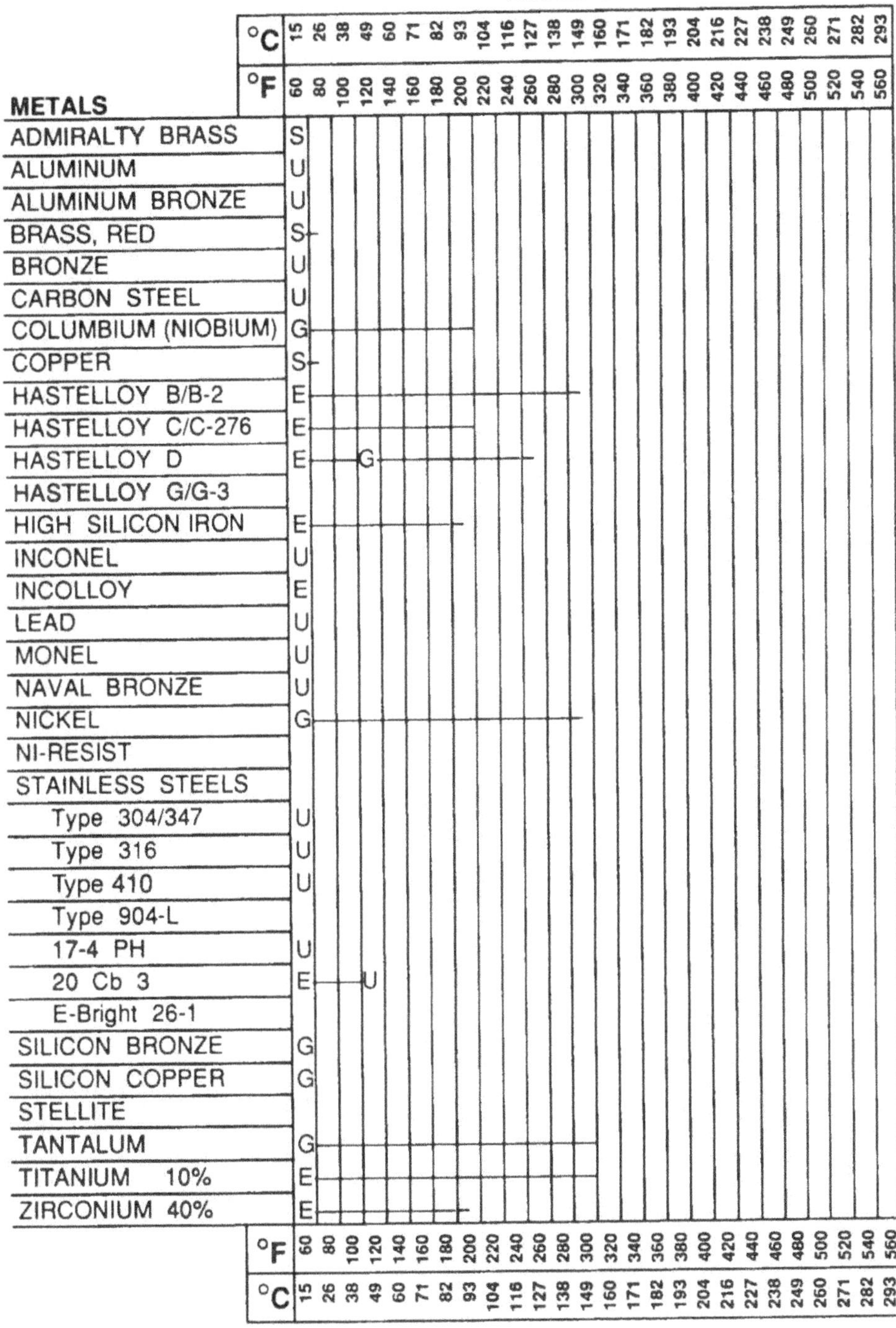

METALS °F	60	80	100	120	140	160	180	200	220	240	260	280	300	320	340	360	380	400	420	440	460	480	500	520	540	560
°C	15	26	38	49	60	71	82	93	104	116	127	138	149	160	171	182	193	204	216	227	238	249	260	271	282	293
ADMIRALTY BRASS	S																									
ALUMINUM	U																									
ALUMINUM BRONZE	U																									
BRASS, RED	S																									
BRONZE	U																									
CARBON STEEL	U																									
COLUMBIUM (NIOBIUM)	G																									
COPPER	S																									
HASTELLOY B/B-2	E																									
HASTELLOY C/C-276	E																									
HASTELLOY D	E			G																						
HASTELLOY G/G-3																										
HIGH SILICON IRON	E																									
INCONEL	U																									
INCOLLOY	E																									
LEAD	U																									
MONEL	U																									
NAVAL BRONZE	U																									
NICKEL	G																									
NI-RESIST																										
STAINLESS STEELS																										
Type 304/347	U																									
Type 316	U																									
Type 410	U																									
Type 904-L																										
17-4 PH	U																									
20 Cb 3	E			U																						
E-Bright 26-1																										
SILICON BRONZE	G																									
SILICON COPPER	G																									
STELLITE																										
TANTALUM	G																									
TITANIUM 10%	E																									
ZIRCONIUM 40%	E																									
°F	60	80	100	120	140	160	180	200	220	240	260	280	300	320	340	360	380	400	420	440	460	480	500	520	540	560
°C	15	26	38	49	60	71	82	93	104	116	127	138	149	160	171	182	193	204	216	227	238	249	260	271	282	293

FOR METALS

E = < 2 Mils Penetration/Year; G = < 20 Mils Penetration /year

S = < 50 Mils Penetration/Year; U = > 50 Mils Penetration/year

FOR NONMETALLICS

R = Resistant

U = Unsatisfactory

ALUMINUM CHLORIDE AQUEOUS

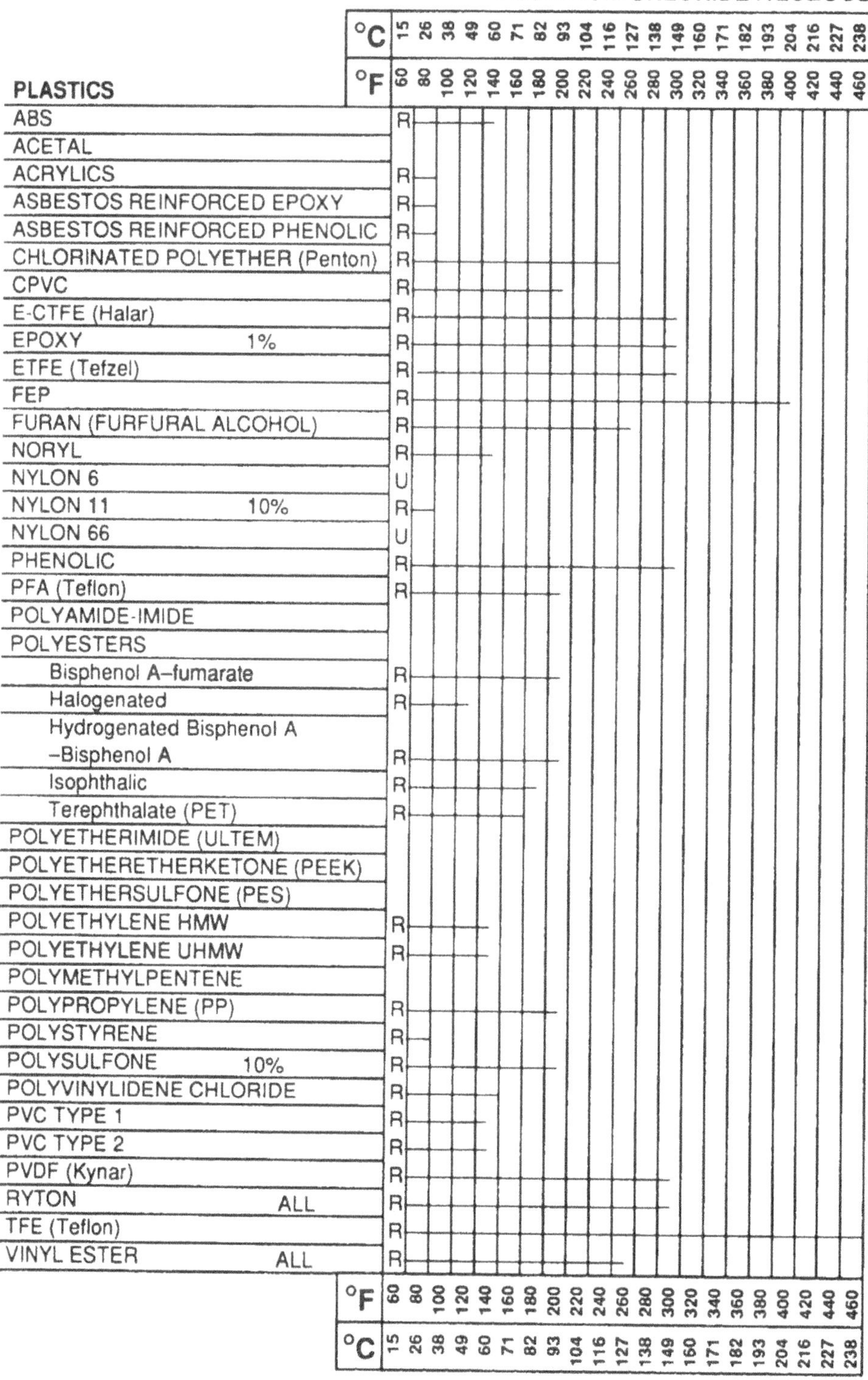

PLASTICS	Concentration	Rating	Max. °F (°C)
ABS		R	140 (60)
ACETAL			
ACRYLICS		R	80 (26)
ASBESTOS REINFORCED EPOXY		R	80 (26)
ASBESTOS REINFORCED PHENOLIC		R	80 (26)
CHLORINATED POLYETHER (Penton)		R	240 (116)
CPVC		R	200 (93)
E-CTFE (Halar)		R	300 (149)
EPOXY	1%	R	300 (149)
ETFE (Tefzel)		R	300 (149)
FEP		R	400 (204)
FURAN (FURFURAL ALCOHOL)		R	260 (127)
NORYL		R	140 (60)
NYLON 6		U	
NYLON 11	10%	R	80 (26)
NYLON 66		U	
PHENOLIC		R	300 (149)
PFA (Teflon)		R	200 (93)
POLYAMIDE-IMIDE			
POLYESTERS			
Bisphenol A–fumarate		R	200 (93)
Halogenated		R	120 (49)
Hydrogenated Bisphenol A –Bisphenol A		R	200 (93)
Isophthalic		R	180 (82)
Terephthalate (PET)		R	160 (71)
POLYETHERIMIDE (ULTEM)			
POLYETHERETHERKETONE (PEEK)			
POLYETHERSULFONE (PES)			
POLYETHYLENE HMW		R	140 (60)
POLYETHYLENE UHMW		R	140 (60)
POLYMETHYLPENTENE			
POLYPROPYLENE (PP)		R	200 (93)
POLYSTYRENE		R	80 (26)
POLYSULFONE	10%	R	200 (93)
POLYVINYLIDENE CHLORIDE		R	140 (60)
PVC TYPE 1		R	140 (60)
PVC TYPE 2		R	140 (60)
PVDF (Kynar)		R	300 (149)
RYTON	ALL	R	300 (149)
TFE (Teflon)		R	460 (238)
VINYL ESTER	ALL	R	260 (127)

ALUMINUM CHLORIDE AQUEOUS

ELASTOMERS* AND LININGS	Rating	Max. temp. (°F)
ACRYLATE-BUTADIENE (ABR)		
BUTYL GR-1 (IIR)	R	150
CARBOXYLIC-ACRYLONITRILE-BUTADIENE (XNBR)		
CHEMRAZ (FPM)	R	400
CHLORO-ISOBUTENE-ISOPRENE (CIIR)	R	200
HYPALON (CHLORO-SULFONYL-POLYETHYLENE (CSM)	R	250
ETHYLENE-ACRYLIC (EA)		
ETHYLENE-PROPYLENE (EPM)	R	180
ETHYLENE-PROPYLENE-DIENE (EPDM)	R	300
ETHYLENE-PROPYLENE TERPOLYMER (EPT)	R	180
FKM (Viton A)	R	400
HARD RUBBER	R	200
SOFT RUBBER	R	140
ISOPRENE (IR)	R	140
KALREZ (FPM)	R	210
KOROSEAL	R	140
NATURAL RUBBER (GRS)	R	140
NEOPRENE GR-M (CR)	R	200
NITRILE BUNA-N (NBR)	R	200
NORDEL (EPDM)	R	300
POLYBUTADIENE (BR)	R	90
POLYESTER (PE)		
POLYETHER-URETHANE (EU)	R	130
POLYISOPRENE (IR)		
POLYSULFIDES (T)		
POLYURETHANE (AU)		
SBR STYRENE (BUNA-S)		
SILICONE RUBBERS		

°F: 60 80 100 120 140 160 180 200 220 240 260 280 300 320 340 360 380 400 420 440 460

°C: 15 26 38 49 60 71 82 93 104 116 127 138 149 160 171 182 193 204 216 227 238

*See also Nylon 11 under PLASTICS

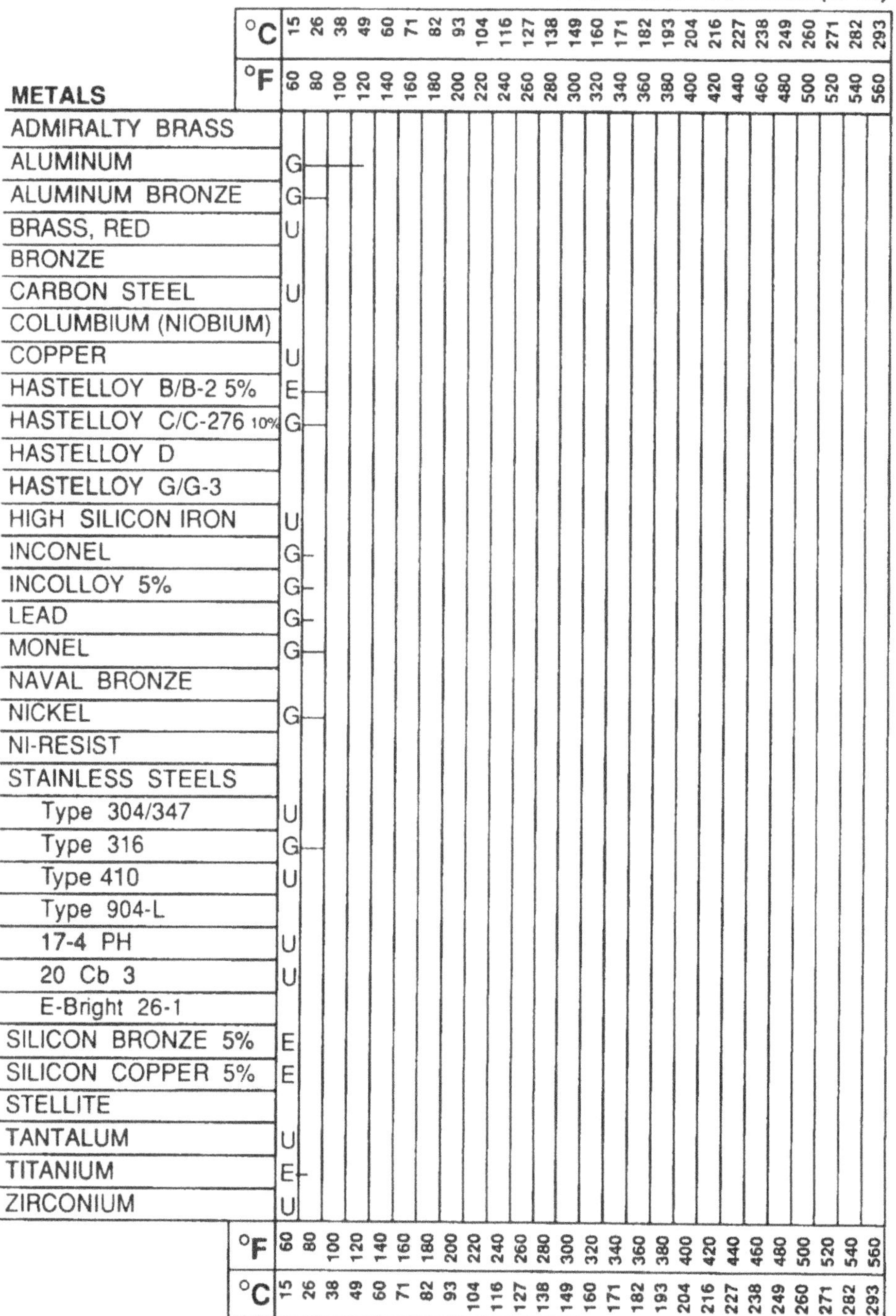

ALUMINUM FLUORIDE (SAT.)

METALS	60 °F / 15 °C	80 / 26	100 / 38	120 / 49	140 / 60	160 / 71	180 / 82	200 / 93	220 / 104	240 / 116	260 / 127	280 / 138	300 / 149	320 / 160	340 / 171	360 / 182	380 / 193	400 / 204	420 / 216	440 / 227	460 / 238	480 / 249	500 / 260	520 / 271	540 / 282	560 / 293
ADMIRALTY BRASS																										
ALUMINUM	G																									
ALUMINUM BRONZE	G																									
BRASS, RED	U																									
BRONZE																										
CARBON STEEL	U																									
COLUMBIUM (NIOBIUM)																										
COPPER	U																									
HASTELLOY B/B-2 5%	E																									
HASTELLOY C/C-276 10%	G																									
HASTELLOY D																										
HASTELLOY G/G-3																										
HIGH SILICON IRON	U																									
INCONEL	G																									
INCOLLOY 5%	G																									
LEAD	G																									
MONEL	G																									
NAVAL BRONZE																										
NICKEL	G																									
NI-RESIST																										
STAINLESS STEELS																										
Type 304/347	U																									
Type 316	G																									
Type 410	U																									
Type 904-L																										
17-4 PH	U																									
20 Cb 3	U																									
E-Bright 26-1																										
SILICON BRONZE 5%	E																									
SILICON COPPER 5%	E																									
STELLITE																										
TANTALUM	U																									
TITANIUM	E																									
ZIRCONIUM	U																									

FOR METALS
E = < 2 Mils Penetration/Year; G = < 20 Mils Penetration /year
S = < 50 Mils Penetration/Year; U = > 50 Mils Penetration/year

FOR NONMETALLICS
R = Resistant
U = Unsatisfactory

ALUMINUM FLUORIDE (SAT.)

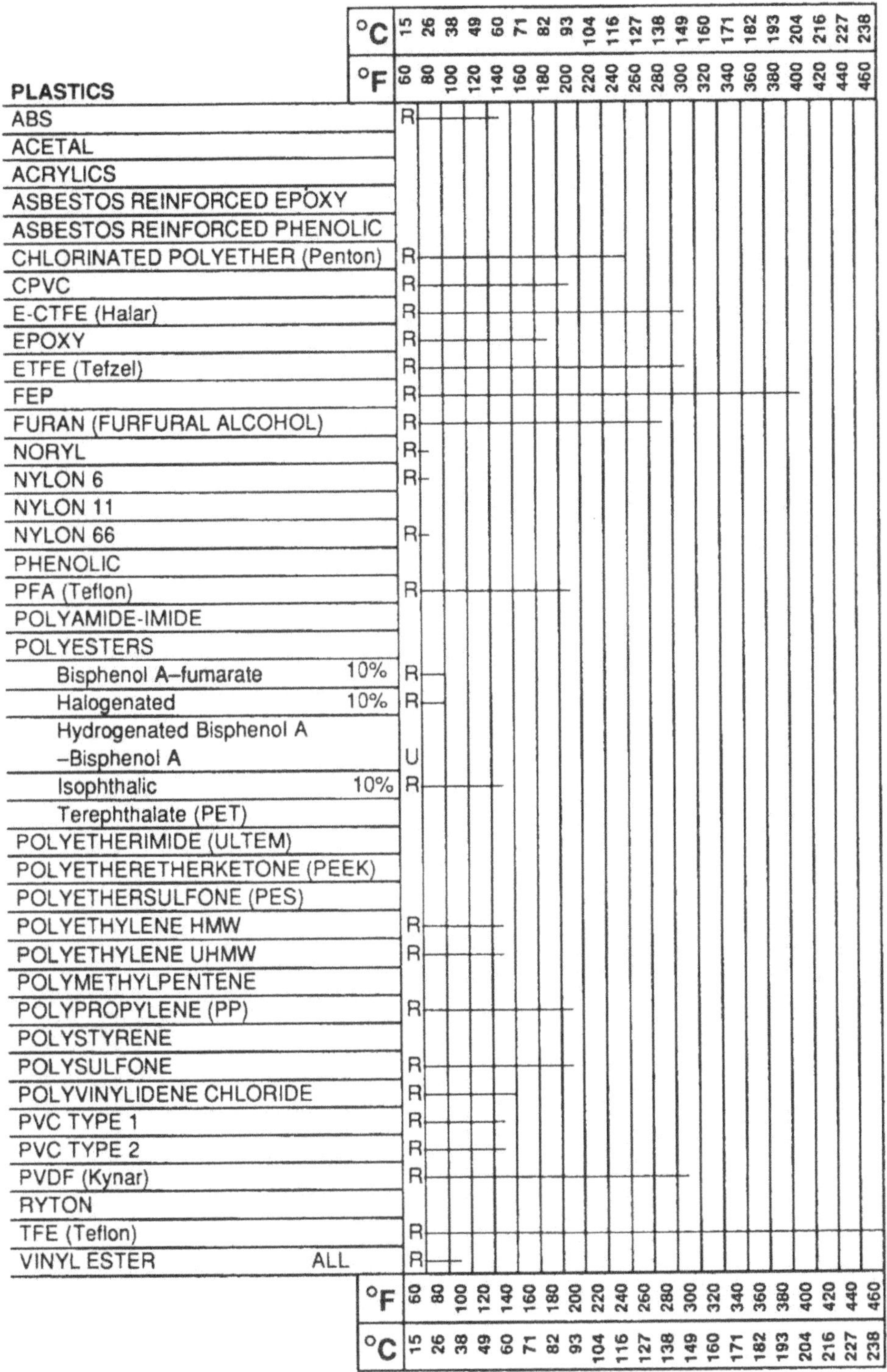

°C: 15, 26, 38, 49, 60, 71, 82, 93, 104, 116, 127, 138, 149, 160, 171, 182, 193, 204, 216, 227, 238

°F: 60, 80, 100, 120, 140, 160, 180, 200, 220, 240, 260, 280, 300, 320, 340, 360, 380, 400, 420, 440, 460

PLASTICS		Rating
ABS		R
ACETAL		
ACRYLICS		
ASBESTOS REINFORCED EPOXY		
ASBESTOS REINFORCED PHENOLIC		
CHLORINATED POLYETHER (Penton)		R
CPVC		R
E-CTFE (Halar)		R
EPOXY		R
ETFE (Tefzel)		R
FEP		R
FURAN (FURFURAL ALCOHOL)		R
NORYL		R
NYLON 6		R
NYLON 11		
NYLON 66		R
PHENOLIC		
PFA (Teflon)		R
POLYAMIDE-IMIDE		
POLYESTERS		
Bisphenol A–fumarate	10%	R
Halogenated	10%	R
Hydrogenated Bisphenol A –Bisphenol A		U
Isophthalic	10%	R
Terephthalate (PET)		
POLYETHERIMIDE (ULTEM)		
POLYETHERETHERKETONE (PEEK)		
POLYETHERSULFONE (PES)		
POLYETHYLENE HMW		R
POLYETHYLENE UHMW		R
POLYMETHYLPENTENE		
POLYPROPYLENE (PP)		R
POLYSTYRENE		
POLYSULFONE		R
POLYVINYLIDENE CHLORIDE		R
PVC TYPE 1		R
PVC TYPE 2		R
PVDF (Kynar)		R
RYTON		
TFE (Teflon)		R
VINYL ESTER	ALL	R

ALUMINUM FLUORIDE (SAT.)

°C: 15, 26, 38, 49, 60, 71, 82, 93, 104, 116, 127, 138, 149, 160, 171, 182, 193, 204, 216, 227, 238

°F: 60, 80, 100, 120, 140, 160, 180, 200, 220, 240, 260, 280, 300, 320, 340, 360, 380, 400, 420, 440, 460

ELASTOMERS* AND LININGS	
ACRYLATE-BUTADIENE (ABR)	
BUTYL GR-1 (IIR)	R
CARBOXYLIC-ACRYLONITRILE-BUTADIENE (XNBR)	
CHEMRAZ (FPM)	R
CHLORO-ISOBUTENE-ISOPRENE (CIIR)	
HYPALON (CHLORO-SULFONYL-POLYETHYLENE (CSM)	R
ETHYLENE-ACRYLIC (EA)	
ETHYLENE-PROPYLENE (EPM)	R
ETHYLENE-PROPYLENE-DIENE (EPDM)	R
ETHYLENE-PROPYLENE TERPOLYMER (EPT)	R
FKM (Viton A)	R
HARD RUBBER	R
SOFT RUBBER	U
ISOPRENE (IR)	R
KALREZ (FPM)	R
KOROSEAL	
NATURAL RUBBER (GRS)	R
NEOPRENE GR-M (CR)	R
NITRILE BUNA-N (NBR)	R
NORDEL (EPDM)	R
POLYBUTADIENE (BR)	
POLYESTER (PE)	
POLYETHER-URETHANE (EU)	
POLYISOPRENE (IR)	
POLYSULFIDES (T)	
POLYURETHANE (AU)	
SBR STYRENE (BUNA-S)	
SILICONE RUBBERS	

*See also Nylon 11 under PLASTICS

°F: 60, 80, 100, 120, 140, 160, 180, 200, 220, 240, 260, 280, 300, 320, 340, 360, 380, 400, 420, 440, 460

°C: 15, 26, 38, 49, 60, 71, 82, 93, 104, 116, 127, 138, 149, 160, 171, 182, 193, 204, 216, 227, 238

AMMONIUM CHLORIDE 10%

METALS	60	80	100	120	140	160	180	200	220	240	260	280	300	320	340	360	380	400	420	440	460	480	500	520	540	560
°C	15	26	38	49	60	71	82	93	104	116	127	138	149	160	171	182	193	204	216	227	238	249	260	271	282	293
ADMIRALTY BRASS																										
ALUMINUM	U																									
ALUMINUM BRONZE	U																									
BRASS, RED																										
BRONZE																										
CARBON STEEL	S																									
COLUMBIUM (NIOBIUM)																										
COPPER	U																									
HASTELLOY B/B-2	E																									
HASTELLOY C/C-276	E																									
HASTELLOY D																										
HASTELLOY G/G-3																										
HIGH SILICON IRON																										
INCONEL 2	E																									
INCOLLOY 825	E																									
LEAD																										
MONEL	E			G																						
NAVAL BRONZE																										
NICKEL	E			G																						
NI-RESIST																										
STAINLESS STEELS																										
Type 304/347	G																									
Type 316	G																									
Type 410 1	G																									
Type 904-L																										
17-4 PH																										
20 Cb 3	E																									
E-Bright 26-1																										
SILICON BRONZE	U																									
SILICON COPPER																										
STELLITE																										
TANTALUM	E																									
TITANIUM	E																									
ZIRCONIUM	E																									
°F	60	80	100	120	140	160	180	200	220	240	260	280	300	320	340	360	380	400	420	440	460	480	500	520	540	560
°C	15	26	38	49	60	71	82	93	104	116	127	138	149	160	171	182	193	204	216	227	238	249	260	271	282	293

FOR METALS
E = < 2 Mils Penetration/Year; G = < 20 Mils Penetration /year
S = < 50 Mils Penetration/Year; U = > 50 Mils Penetration/year

FOR NONMETALLICS
R = Resistant
U = Unsatisfactory

AMMONIUM CHLORIDE 10%

PLASTICS	Rating	Maximum temperature per chart (°F / °C)
ABS		
ACETAL		
ACRYLICS		
ASBESTOS REINFORCED EPOXY		
ASBESTOS REINFORCED PHENOLIC		
CHLORINATED POLYETHER (Penton)		
CPVC	R	180 / 82
E-CTFE (Halar)	R	between 280 / 138 and 300 / 149
EPOXY	R	200 / 93
ETFE (Tefzel)	R	300 / 149
FEP	R	400 / 204
FURAN (FURFURAL ALCOHOL)	R	220 / 104
NORYL		
NYLON 6	U	
NYLON 11	R	200 / 93
NYLON 66	U	
PHENOLIC		
PFA (Teflon)	R	200 / 93
POLYAMIDE-IMIDE	R	200 / 93
POLYESTERS		
Bisphenol A–fumarate	R	200 / 93
Halogenated	R	200 / 93
Hydrogenated Bisphenol A –Bisphenol A		
Isophthalic	R	160 / 71
Terephthalate (PET)	R	between 160 / 71 and 180 / 82
POLYETHERIMIDE (ULTEM)		
POLYETHERETHERKETONE (PEEK)		
POLYETHERSULFONE (PES)		
POLYETHYLENE HMW	R	140 / 60
POLYETHYLENE UHMW	R	140 / 60
POLYMETHYLPENTENE		
POLYPROPYLENE (PP)	R	180 / 82
POLYSTYRENE		
POLYSULFONE	R	200 / 93
POLYVINYLIDENE CHLORIDE		
PVC TYPE 1	R	140 / 60
PVC TYPE 2		
PVDF (Kynar)	R	280 / 138
RYTON	R	300 / 149
TFE (Teflon)	R	460 / 238
VINYL ESTER	R	200 / 93

°F	60	80	100	120	140	160	180	200	220	240	260	280	300	320	340	360	380	400	420	440	460
°C	15	26	38	49	60	71	82	93	104	116	127	138	149	160	171	182	193	204	216	227	238

AMMONIUM CHLORIDE 10%

Temperature scale: °F 60, 80, 100, 120, 140, 160, 180, 200, 220, 240, 260, 280, 300, 320, 340, 360, 380, 400, 420, 440, 460 / °C 15, 26, 38, 49, 60, 71, 82, 93, 104, 116, 127, 138, 149, 160, 171, 182, 193, 204, 216, 227, 238

ELASTOMERS* AND LININGS	Rating	Max °F	Max °C
ACRYLATE-BUTADIENE (ABR)			
BUTYL GR-1 (IIR)	R	180	82
CARBOXYLIC-ACRYLONITRILE-BUTADIENE (XNBR)			
CHEMRAZ (FPM)	R	400	204
CHLORO-ISOBUTENE-ISOPRENE (CIIR)	R	200	93
HYPALON (CHLORO-SULFONYL-POLYETHYLENE (CSM)	R	200	93
ETHYLENE-ACRYLIC (EA)			
ETHYLENE-PROPYLENE (EPM)	R	300	149
ETHYLENE-PROPYLENE-DIENE (EPDM)	R	200	93
ETHYLENE-PROPYLENE TERPOLYMER (EPT)	R	180	82
FKM (Viton A)	R	400	204
HARD RUBBER	R	200	93
SOFT RUBBER	R	140	60
ISOPRENE (IR)	R	140	60
KALREZ (FPM)	R	200	93
KOROSEAL	R	140	60
NATURAL RUBBER (GRS)	R	140	60
NEOPRENE GR-M (CR)	R	200	93
NITRILE BUNA-N (NBR)	R	200	93
NORDEL (EPDM)	R	300	149
POLYBUTADIENE (BR)	R	80	26
POLYESTER (PE)	R	80	26
POLYETHER-URETHANE (EU)	R	120	49
POLYISOPRENE (IR)			
POLYSULFIDES (T)	R	140	60
POLYURETHANE (AU)	R	80	26
SBR STYRENE (BUNA-S)	R	200	93
SILICONE RUBBERS	R	80	26

*See also Nylon 11 under PLASTICS

AMMONIUM CHLORIDE 50%

METALS	60	80	100	120	140	160	180	200	220	240	260	280	300	320	340	360	380	400	420	440	460	480	500	520	540	560
°C	15	26	38	49	60	71	82	93	104	116	127	138	149	160	171	182	193	204	216	227	238	249	260	271	282	293
ADMIRALTY BRASS																										
ALUMINUM	U																									
ALUMINUM BRONZE	U																									
BRASS, RED																										
BRONZE																										
CARBON STEEL																										
COLUMBIUM (NIOBIUM)																										
COPPER	U																									
HASTELLOY B/B-2	E																									
HASTELLOY C/C-276	E																									
HASTELLOY D																										
HASTELLOY G/G-3																										
HIGH SILICON IRON																										
INCONEL	E																									
INCOLLOY																										
LEAD																										
MONEL	E																									
NAVAL BRONZE																										
NICKEL	E																									
NI-RESIST	G																									
STAINLESS STEELS																										
Type 304/347	U																									
Type 316	U																									
Type 410	U																									
Type 904-L																										
17-4 PH																										
20 Cb 3	G																									
E-Bright 26-1																										
SILICON BRONZE	U																									
SILICON COPPER																										
STELLITE																										
TANTALUM	E																									
TITANIUM	E																									
ZIRCONIUM	E																									
°F	60	80	100	120	140	160	180	200	220	240	260	280	300	320	340	360	380	400	420	440	460	480	500	520	540	560
°C	15	26	38	49	60	71	82	93	104	116	127	138	149	160	171	182	193	204	216	227	238	249	260	271	282	293

FOR METALS

E = < 2 Mils Penetration/Year; G = < 20 Mils Penetration /year

S = < 50 Mils Penetration/Year; U = > 50 Mils Penetration/year

FOR NONMETALLICS

R = Resistant

U = Unsatisfactory

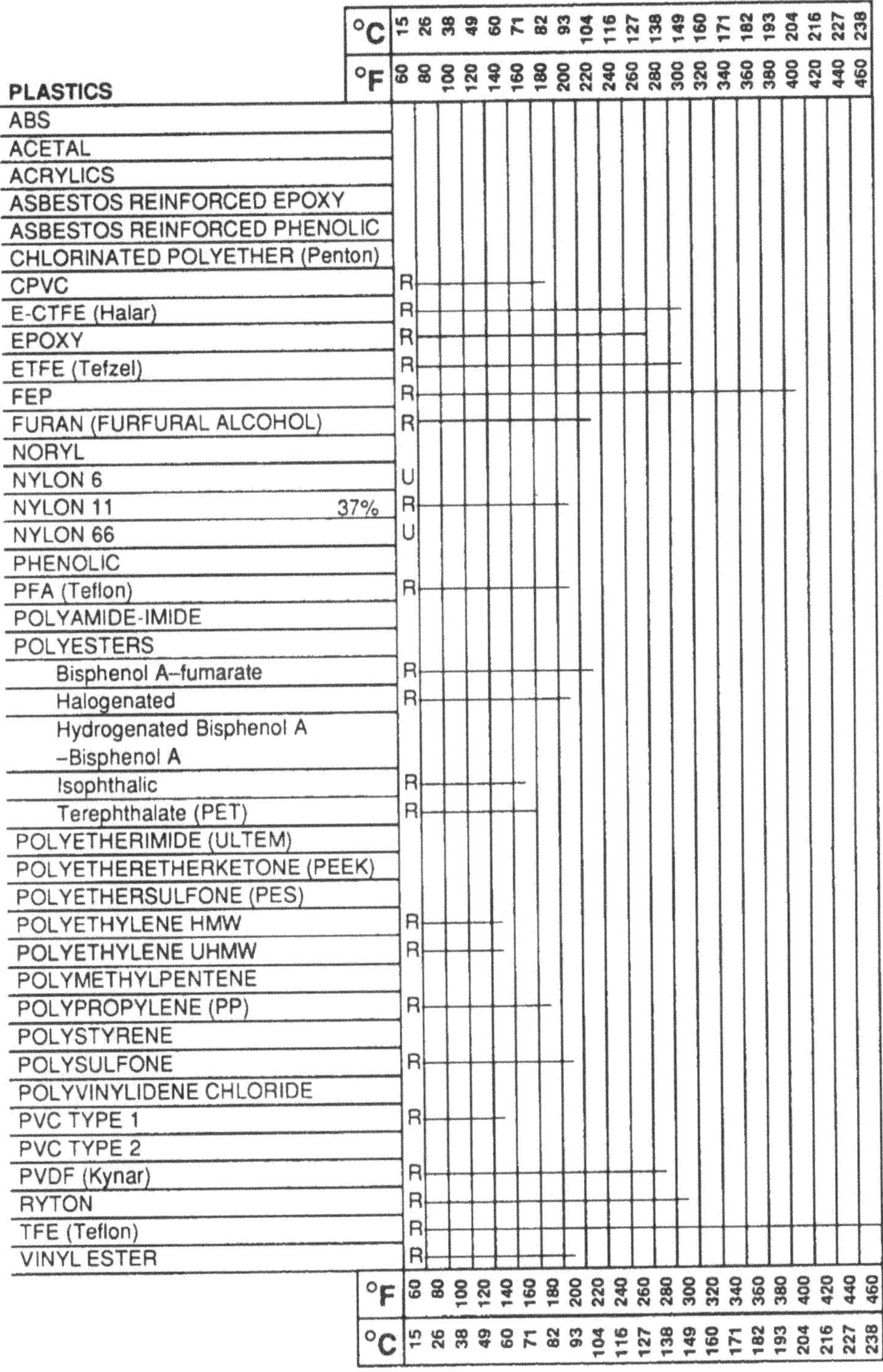

AMMONIUM CHLORIDE 50%

PLASTICS		Rating	Max. temp. °F	Max. temp. °C
ABS				
ACETAL				
ACRYLICS				
ASBESTOS REINFORCED EPOXY				
ASBESTOS REINFORCED PHENOLIC				
CHLORINATED POLYETHER (Penton)				
CPVC		R	180	82
E-CTFE (Halar)		R	300	149
EPOXY		R	280	138
ETFE (Tefzel)		R	300	149
FEP		R	400	204
FURAN (FURFURAL ALCOHOL)		R	220	104
NORYL				
NYLON 6		U		
NYLON 11	37%	R	200	93
NYLON 66		U		
PHENOLIC				
PFA (Teflon)		R	200	93
POLYAMIDE-IMIDE				
POLYESTERS				
Bisphenol A–fumarate		R	220	104
Halogenated		R	200	93
Hydrogenated Bisphenol A –Bisphenol A				
Isophthalic		R	160	71
Terephthalate (PET)		R	180	82
POLYETHERIMIDE (ULTEM)				
POLYETHERETHERKETONE (PEEK)				
POLYETHERSULFONE (PES)				
POLYETHYLENE HMW		R	140	60
POLYETHYLENE UHMW		R	140	60
POLYMETHYLPENTENE				
POLYPROPYLENE (PP)		R	180	82
POLYSTYRENE				
POLYSULFONE		R	200	93
POLYVINYLIDENE CHLORIDE				
PVC TYPE 1		R	140	60
PVC TYPE 2				
PVDF (Kynar)		R	280	138
RYTON		R	300	149
TFE (Teflon)		R	460	238
VINYL ESTER		R	200	93

AMMONIUM CHLORIDE 50%

°C	15	26	38	49	60	71	82	93	104	116	127	138	149	160	171	182	193	204	216	227	238
°F	60	80	100	120	140	160	180	200	220	240	260	280	300	320	340	360	380	400	420	440	460

ELASTOMERS* AND LININGS	
ACRYLATE-BUTADIENE (ABR)	
BUTYL GR-1 (IIR)	R
CARBOXYLIC-ACRYLONITRILE-BUTADIENE (XNBR)	
CHEMRAZ (FPM)	R
CHLORO-ISOBUTENE-ISOPRENE (CIIR)	R
HYPALON (CHLORO-SULFONYL-POLYETHYLENE (CSM)	R
ETHYLENE-ACRYLIC (EA)	
ETHYLENE-PROPYLENE (EPM)	
ETHYLENE-PROPYLENE-DIENE (EPDM)	R
ETHYLENE-PROPYLENE TERPOLYMER (EPT)	R
FKM (Viton A)	R
HARD RUBBER	R
SOFT RUBBER	R
ISOPRENE (IR)	R
KALREZ (FPM)	R
KOROSEAL	R
NATURAL RUBBER (GRS)	R
NEOPRENE GR-M (CR)	R
NITRILE BUNA-N (NBR)	R
NORDEL (EPDM)	R
POLYBUTADIENE (BR)	R
POLYESTER (PE)	R
POLYETHER-URETHANE (EU)	R
POLYISOPRENE (IR)	
POLYSULFIDES (T)	R
POLYURETHANE (AU)	R
SBR STYRENE (BUNA-S)	R
SILICONE RUBBERS	R

*See also Nylon 11 under PLASTICS

°F	60	80	100	120	140	160	180	200	220	240	260	280	300	320	340	360	380	400	420	440	460
°C	15	26	38	49	60	71	82	93	104	116	127	138	149	160	171	182	193	204	216	227	238

AMMONIUM CHLORIDE (SAT.)

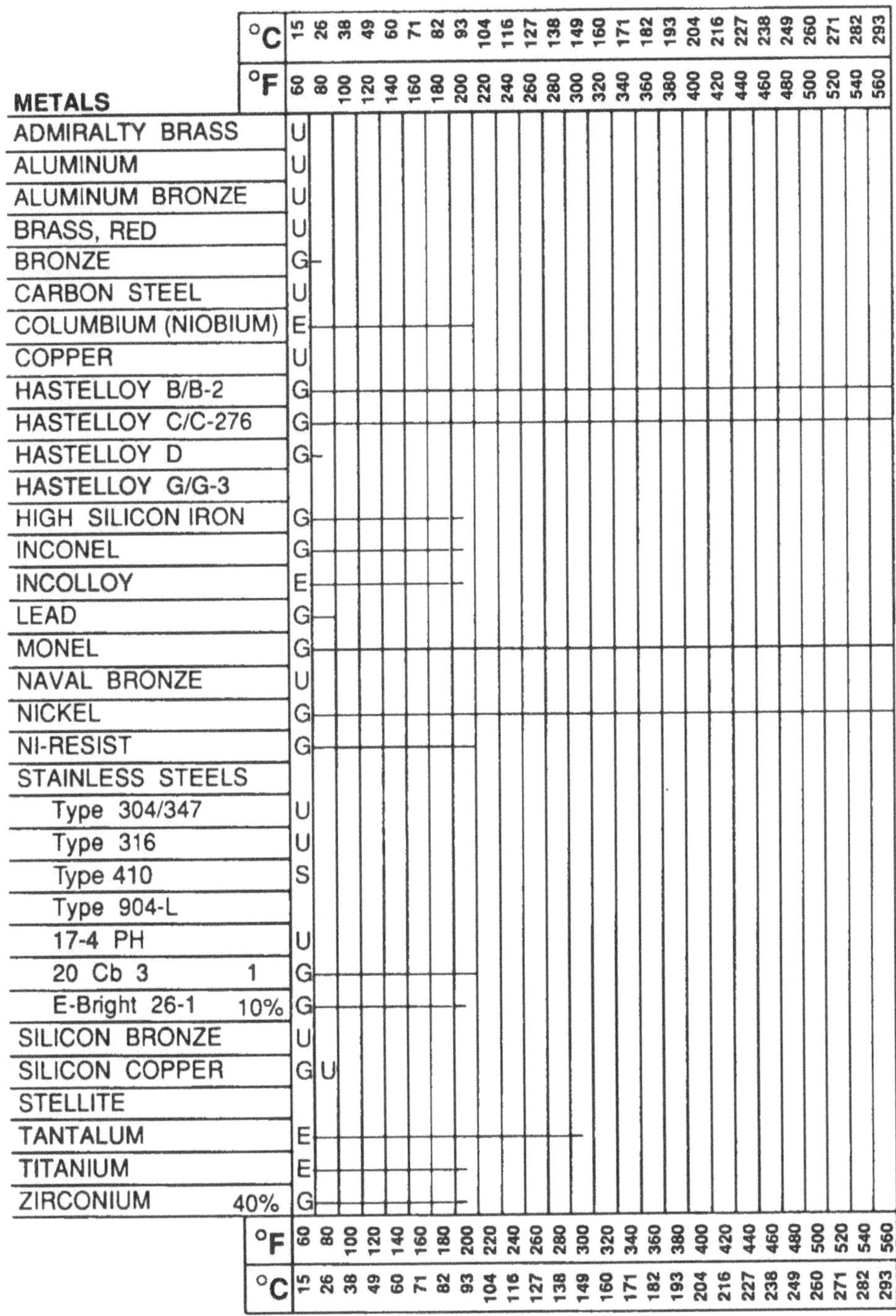

METALS	°F 60 / °C 15	80 / 26	100 / 38	120 / 49	140 / 60	160 / 71	180 / 82	200 / 93	220 / 104	240 / 116	260 / 127	280 / 138	300 / 149	320 / 160	340 / 171	360 / 182	380 / 193	400 / 204	420 / 216	440 / 227	460 / 238	480 / 249	500 / 260	520 / 271	540 / 282	560 / 293
ADMIRALTY BRASS	U																									
ALUMINUM	U																									
ALUMINUM BRONZE	U																									
BRASS, RED	U																									
BRONZE	G																									
CARBON STEEL	U																									
COLUMBIUM (NIOBIUM)	E																									
COPPER	U																									
HASTELLOY B/B-2	G																									
HASTELLOY C/C-276	G																									
HASTELLOY D	G																									
HASTELLOY G/G-3																										
HIGH SILICON IRON	G																									
INCONEL	G																									
INCOLLOY	E																									
LEAD	G																									
MONEL	G																									
NAVAL BRONZE	U																									
NICKEL	G																									
NI-RESIST	G																									
STAINLESS STEELS																										
Type 304/347	U																									
Type 316	U																									
Type 410	S																									
Type 904-L																										
17-4 PH	U																									
20 Cb 3 1	G																									
E-Bright 26-1 10%	G																									
SILICON BRONZE	U																									
SILICON COPPER	G	U																								
STELLITE																										
TANTALUM	E																									
TITANIUM	E																									
ZIRCONIUM 40%	G																									

°F 60 80 100 120 140 160 180 200 220 240 260 280 300 320 340 360 380 400 420 440 460 480 500 520 540 560
°C 15 26 38 49 60 71 82 93 104 116 127 138 149 160 171 182 193 204 216 227 238 249 260 271 282 293

FOR METALS

E = < 2 Mils Penetration/Year; G = < 20 Mils Penetration /year

S = < 50 Mils Penetration/Year; U = > 50 Mils Penetration/year

FOR NONMETALLICS

R = Resistant

U = Unsatisfactory

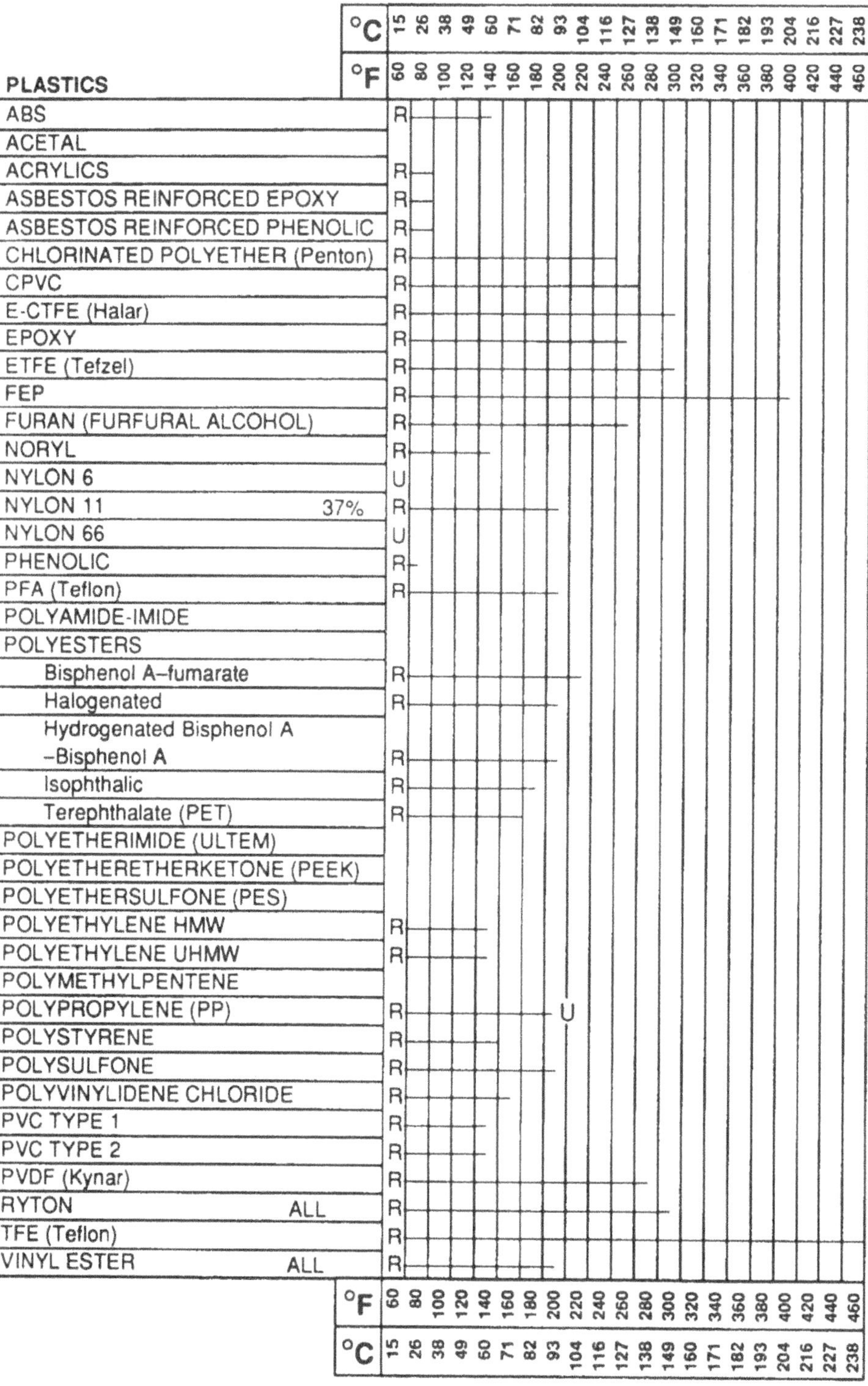

AMMONIUM CHLORIDE (SAT.)

°C	15	26	38	49	60	71	82	93	104	116	127	138	149	160	171	182	193	204	216	227	238
°F	60	80	100	120	140	160	180	200	220	240	260	280	300	320	340	360	380	400	420	440	460

PLASTICS		Rating
ABS		R
ACETAL		
ACRYLICS		R
ASBESTOS REINFORCED EPOXY		R
ASBESTOS REINFORCED PHENOLIC		R
CHLORINATED POLYETHER (Penton)		R
CPVC		R
E-CTFE (Halar)		R
EPOXY		R
ETFE (Tefzel)		R
FEP		R
FURAN (FURFURAL ALCOHOL)		R
NORYL		R
NYLON 6		U
NYLON 11	37%	R
NYLON 66		U
PHENOLIC		R
PFA (Teflon)		R
POLYAMIDE-IMIDE		
POLYESTERS		
Bisphenol A–fumarate		R
Halogenated		R
Hydrogenated Bisphenol A –Bisphenol A		R
Isophthalic		R
Terephthalate (PET)		R
POLYETHERIMIDE (ULTEM)		
POLYETHERETHERKETONE (PEEK)		
POLYETHERSULFONE (PES)		
POLYETHYLENE HMW		R
POLYETHYLENE UHMW		R
POLYMETHYLPENTENE		
POLYPROPYLENE (PP)		R … U
POLYSTYRENE		R
POLYSULFONE		R
POLYVINYLIDENE CHLORIDE		R
PVC TYPE 1		R
PVC TYPE 2		R
PVDF (Kynar)		R
RYTON	ALL	R
TFE (Teflon)		R
VINYL ESTER	ALL	R

AMMONIUM CHLORIDE (SAT.)

ELASTOMERS* AND LININGS	Rating	Maximum temperature °F	Maximum temperature °C
ACRYLATE-BUTADIENE (ABR)			
BUTYL GR-1 (IIR)	R	180	82
CARBOXYLIC-ACRYLONITRILE-BUTADIENE (XNBR)			
CHEMRAZ (FPM)	R	400	204
CHLORO-ISOBUTENE-ISOPRENE (CIIR)	R	200	93
HYPALON (CHLORO-SULFONYL-POLYETHYLENE (CSM)	R	200	93
ETHYLENE-ACRYLIC (EA)			
ETHYLENE-PROPYLENE (EPM)			
ETHYLENE-PROPYLENE-DIENE (EPDM)	R	300	149
ETHYLENE-PROPYLENE TERPOLYMER (EPT)	R	180	82
FKM (Viton A)	R	300	149
HARD RUBBER	R	200	93
SOFT RUBBER	R	140	60
ISOPRENE (IR)	R	140	60
KALREZ (FPM)	R	200	93
KOROSEAL	R	140	60
NATURAL RUBBER (GRS)	R	140	60
NEOPRENE GR-M (CR)	R	200	93
NITRILE BUNA-N (NBR)	R	200	93
NORDEL (EPDM)	R	300	149
POLYBUTADIENE (BR)	R	80	26
POLYESTER (PE)	R	80	26
POLYETHER-URETHANE (EU)	R	120	49
POLYISOPRENE (IR)			
POLYSULFIDES (T)	R	140	60
POLYURETHANE (AU)	R	80	26
SBR STYRENE (BUNA-S)	R	200	93
SILICONE RUBBERS	R	80	26

*See also Nylon 11 under PLASTICS

AMMONIUM FLUORIDE 10%

METALS	°C 15	26	38	49	60	71	82	93	104	116	127	138	149	160	171	182	193	204	216	227	238	249	260	271	282	293
	°F 60	80	100	120	140	160	180	200	220	240	260	280	300	320	340	360	380	400	420	440	460	480	500	520	540	560
ADMIRALTY BRASS																										
ALUMINUM	U																									
ALUMINUM BRONZE																										
BRASS, RED	U																									
BRONZE	U																									
CARBON STEEL	U																									
COLUMBIUM (NIOBIUM)																										
COPPER	U																									
HASTELLOY B/B-2	G	—	—	—	—	—	—	—																		
HASTELLOY C/C-276	E	—	—	—	—	—	—	—																		
HASTELLOY D																										
HASTELLOY G/G-3																										
HIGH SILICON IRON	U																									
INCONEL	G	—																								
INCOLLOY																										
LEAD	G	—																								
MONEL	G	—	—	—	—	—	—	—	—	—	—	—	—	—	—	—	—	—								
NAVAL BRONZE																										
NICKEL	G	—	—	—	—	—	—	—																		
NI-RESIST																										
STAINLESS STEELS																										
Type 304/347	U																									
Type 316	G	—	U																							
Type 410																										
Type 904-L																										
17-4 PH																										
20 Cb 3	E	—	U																							
E-Bright 26-1																										
SILICON BRONZE																										
SILICON COPPER																										
STELLITE																										
TANTALUM	U																									
TITANIUM	G	—																								
ZIRCONIUM	U																									
	°F 60	80	100	120	140	160	180	200	220	240	260	280	300	320	340	360	380	400	420	440	460	480	500	520	540	560
	°C 15	26	38	49	60	71	82	93	104	116	127	138	149	160	171	182	193	204	216	227	238	249	260	271	282	293

FOR METALS

E = < 2 Mils Penetration/Year; G = < 20 Mils Penetration /year

S = < 50 Mils Penetration/Year; U = > 50 Mils Penetration/year

FOR NONMETALLICS

R = Resistant

U = Unsatisfactory

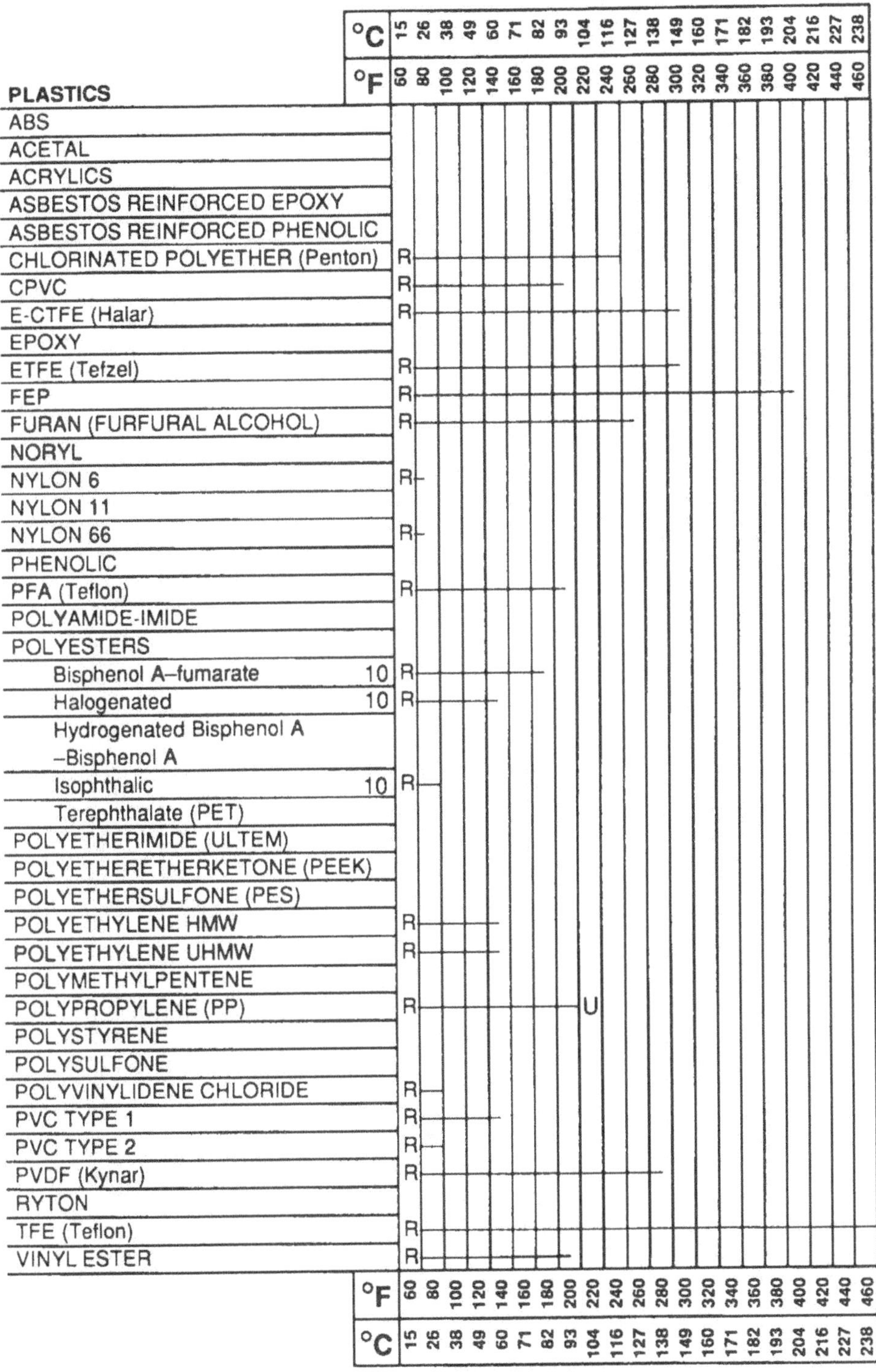

AMMONIUM FLUORIDE 10%

PLASTICS	Conc.	Rating	Max. temp. (°F)	Max. temp. (°C)
ABS				
ACETAL				
ACRYLICS				
ASBESTOS REINFORCED EPOXY				
ASBESTOS REINFORCED PHENOLIC				
CHLORINATED POLYETHER (Penton)		R	250	
CPVC		R	200	93
E-CTFE (Halar)		R	300	149
EPOXY				
ETFE (Tefzel)		R	300	149
FEP		R	400	204
FURAN (FURFURAL ALCOHOL)		R	260	127
NORYL				
NYLON 6		R	60	15
NYLON 11				
NYLON 66		R	60	15
PHENOLIC				
PFA (Teflon)		R	200	93
POLYAMIDE-IMIDE				
POLYESTERS				
Bisphenol A–fumarate	10	R	180	82
Halogenated	10	R	140	60
Hydrogenated Bisphenol A –Bisphenol A				
Isophthalic	10	R	70	
Terephthalate (PET)				
POLYETHERIMIDE (ULTEM)				
POLYETHERETHERKETONE (PEEK)				
POLYETHERSULFONE (PES)				
POLYETHYLENE HMW		R	140	60
POLYETHYLENE UHMW		R	140	60
POLYMETHYLPENTENE				
POLYPROPYLENE (PP)		R to 210; U at 220	210	
POLYSTYRENE				
POLYSULFONE				
POLYVINYLIDENE CHLORIDE		R	70	
PVC TYPE 1		R	140	60
PVC TYPE 2		R	70	
PVDF (Kynar)		R	280	138
RYTON				
TFE (Teflon)		R	460	238
VINYL ESTER		R	200	93

AMMONIUM FLUORIDE 10%

ELASTOMERS* AND LININGS	Rating	Max. temp. (°F)	Max. temp. (°C)
ACRYLATE-BUTADIENE (ABR)			
BUTYL GR-1 (IIR)	R	140	60
CARBOXYLIC-ACRYLONITRILE-BUTADIENE (XNBR)			
CHEMRAZ (FPM)	R	400	204
CHLORO-ISOBUTENE-ISOPRENE (CIIR)			
HYPALON (CHLORO-SULFONYL-POLYETHYLENE (CSM)	R	200	93
ETHYLENE-ACRYLIC (EA)			
ETHYLENE-PROPYLENE (EPM)			
ETHYLENE-PROPYLENE-DIENE (EPDM)	R	200	93
ETHYLENE-PROPYLENE TERPOLYMER (EPT)	R	200	93
FKM (Viton A)	R	140	60
HARD RUBBER	R	200	93
SOFT RUBBER	U		
ISOPRENE (IR)	R	160	71
KALREZ (FPM)	R	200	93
KOROSEAL			
NATURAL RUBBER (GRS)	R	160	71
NEOPRENE GR-M (CR)	R	160	71
NITRILE BUNA-N (NBR)	R	200	93
NORDEL (EPDM)	R	300	149
POLYBUTADIENE (BR)			
POLYESTER (PE)			
POLYETHER-URETHANE (EU)			
POLYISOPRENE (IR)			
POLYSULFIDES (T)			
POLYURETHANE (AU)			
SBR STYRENE (BUNA-S)			
SILICONE RUBBERS			

Temperature scale: °F 60, 80, 100, 120, 140, 160, 180, 200, 220, 240, 260, 280, 300, 320, 340, 360, 380, 400, 420, 440, 460; °C 15, 26, 38, 49, 60, 71, 82, 93, 104, 116, 127, 138, 149, 160, 171, 182, 193, 204, 216, 227, 238

*See also Nylon 11 under PLASTICS

AMMONIUM FLUORIDE 25%

METALS	°F: 60 / °C: 15
ADMIRALTY BRASS	
ALUMINUM	U
ALUMINUM BRONZE	
BRASS, RED	U
BRONZE	U
CARBON STEEL	U
COLUMBIUM (NIOBIUM)	
COPPER	U
HASTELLOY B/B-2	
HASTELLOY C/C-276	E
HASTELLOY D	
HASTELLOY G/G-3	
HIGH SILICON IRON	U
INCONEL	G
INCOLLOY	
LEAD	G
MONEL	G
NAVAL BRONZE	
NICKEL	G
NI-RESIST	
STAINLESS STEELS	
Type 304/347	U
Type 316	U
Type 410	
Type 904-L	
17-4 PH	
20 Cb 3	G
E-Bright 26-1	
SILICON BRONZE	
SILICON COPPER	
STELLITE	
TANTALUM	U
TITANIUM	E
ZIRCONIUM	U

°F: 60 80 100 120 140 160 180 200 220 240 260 280 300 320 340 360 380 400 420 440 460 480 500 520 540 560

°C: 15 26 38 49 60 71 82 93 104 116 127 138 149 160 171 182 193 204 216 227 238 249 260 271 282 293

FOR METALS
E = < 2 Mils Penetration/Year; G = < 20 Mils Penetration /year
S = < 50 Mils Penetration/Year; U = > 50 Mils Penetration/year

FOR NONMETALLICS
R = Resistant
U = Unsatisfactory

AMMONIUM FLUORIDE 25%

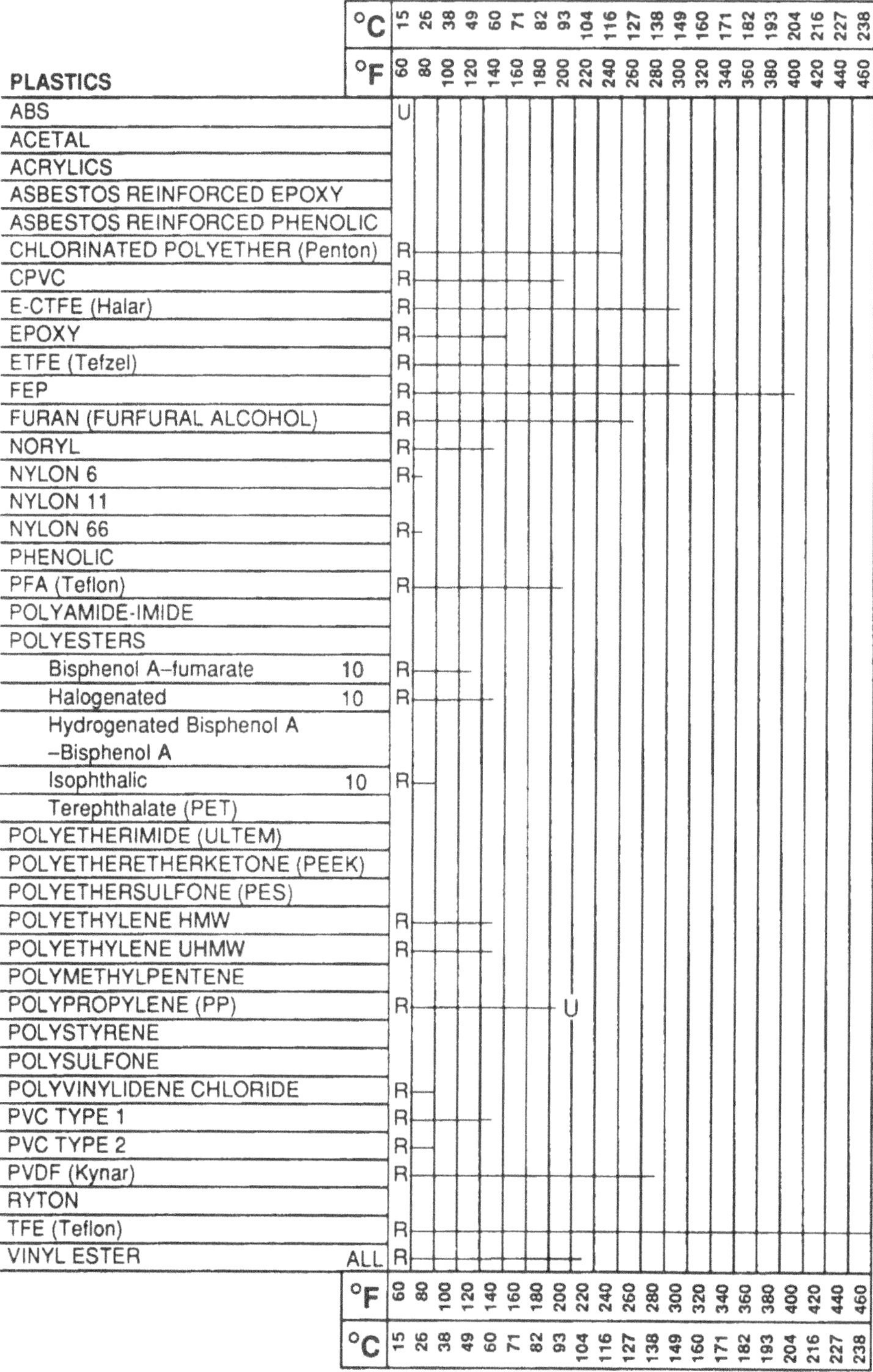

PLASTICS		°F: 60 80 100 120 140 160 180 200 220 240 260 280 300 320 340 360 380 400 420 440 460 / °C: 15 26 38 49 60 71 82 93 104 116 127 138 149 160 171 182 193 204 216 227 238
ABS		U
ACETAL		
ACRYLICS		
ASBESTOS REINFORCED EPOXY		
ASBESTOS REINFORCED PHENOLIC		
CHLORINATED POLYETHER (Penton)		R
CPVC		R
E-CTFE (Halar)		R
EPOXY		R
ETFE (Tefzel)		R
FEP		R
FURAN (FURFURAL ALCOHOL)		R
NORYL		R
NYLON 6		R
NYLON 11		
NYLON 66		R
PHENOLIC		
PFA (Teflon)		R
POLYAMIDE-IMIDE		
POLYESTERS		
Bisphenol A–fumarate	10	R
Halogenated	10	R
Hydrogenated Bisphenol A –Bisphenol A		
Isophthalic	10	R
Terephthalate (PET)		
POLYETHERIMIDE (ULTEM)		
POLYETHERETHERKETONE (PEEK)		
POLYETHERSULFONE (PES)		
POLYETHYLENE HMW		R
POLYETHYLENE UHMW		R
POLYMETHYLPENTENE		
POLYPROPYLENE (PP)		R U
POLYSTYRENE		
POLYSULFONE		
POLYVINYLIDENE CHLORIDE		R
PVC TYPE 1		R
PVC TYPE 2		R
PVDF (Kynar)		R
RYTON		
TFE (Teflon)		R
VINYL ESTER	ALL	R

AMMONIUM FLUORIDE 25%

ELASTOMERS* AND LININGS	15	26	38	49	60	71	82	93	104	116	127	138	149	160	171	182	193	204	216	227	238
°F	60	80	100	120	140	160	180	200	220	240	260	280	300	320	340	360	380	400	420	440	460
ACRYLATE-BUTADIENE (ABR)																					
BUTYL GR-1 (IIR)	R																				
CARBOXYLIC-ACRYLONITRILE-BUTADIENE (XNBR)																					
CHEMRAZ (FPM)	R																				
CHLORO-ISOBUTENE-ISOPRENE (CIIR)																					
HYPALON (CHLORO-SULFONYL-POLYETHYLENE (CSM)																					
ETHYLENE-ACRYLIC (EA)																					
ETHYLENE-PROPYLENE (EPM)																					
ETHYLENE-PROPYLENE-DIENE (EPDM)	R																				
ETHYLENE-PROPYLENE TERPOLYMER (EPT)	R																				
FKM (Viton A)	R																				
HARD RUBBER	R																				
SOFT RUBBER	U																				
ISOPRENE (IR)	R																				
KALREZ (FPM)	R																				
KOROSEAL																					
NATURAL RUBBER (GRS)	R																				
NEOPRENE GR-M (CR)	R																				
NITRILE BUNA-N (NBR)	R																				
NORDEL (EPDM)	R																				
POLYBUTADIENE (BR)																					
POLYESTER (PE)																					
POLYETHER-URETHANE (EU)																					
POLYISOPRENE (IR)																					
POLYSULFIDES (T)																					
POLYURETHANE (AU)																					
SBR STYRENE (BUNA-S)																					
SILICONE RUBBERS																					

°F 60 80 100 120 140 160 180 200 220 240 260 280 300 320 340 360 380 400 420 440 460
°C 15 26 38 49 60 71 82 93 104 116 127 138 149 160 171 182 193 204 216 227 238

*See also Nylon 11 under PLASTICS

METALS	60 °F / 15 °C	80 °F / 26 °C	100 °F / 38 °C	120 °F / 49 °C	140 °F / 60 °C	160 °F / 71 °C	180 °F / 82 °C	200 °F / 93 °C	220 °F / 104 °C	240 °F / 116 °C	260 °F / 127 °C	280 °F / 138 °C	300 °F / 149 °C	320 °F / 160 °C	340 °F / 171 °C	360 °F / 182 °C	380 °F / 193 °C	400 °F / 204 °C	420 °F / 216 °C	440 °F / 227 °C	460 °F / 238 °C	480 °F / 249 °C	500 °F / 260 °C	520 °F / 271 °C	540 °F / 282 °C	560 °F / 293 °C
ADMIRALTY BRASS	U																									
ALUMINUM	G																									
ALUMINUM BRONZE	U																									
BRASS, RED	U																									
BRONZE	U																									
CARBON STEEL	E		G																							
COLUMBIUM (NIOBIUM)																										
COPPER	U																									
HASTELLOY B/B-2	G																									
HASTELLOY C/C-276	G																									
HASTELLOY D	G																									
HASTELLOY G/G-3																										
HIGH SILICON IRON	G																									
INCONEL	G																									
INCOLLOY																										
LEAD	G																									
MONEL	U																									
NAVAL BRONZE	U																									
NICKEL	U																									
NI-RESIST																										
STAINLESS STEELS																										
Type 304/347	E		G																							
Type 316	E		G																							
Type 410																										
Type 904-L																										
17-4 PH																										
20 Cb 3	G																									
E-Bright 26-1																										
SILICON BRONZE	U																									
SILICON COPPER	U																									
STELLITE																										
TANTALUM	E																									
TITANIUM	E																									
ZIRCONIUM	E																									

FOR METALS

E = < 2 Mils Penetration/Year; G = < 20 Mils Penetration /year

S = < 50 Mils Penetration/Year; U = > 50 Mils Penetration/year

FOR NONMETALLICS

R = Resistant

U = Unsatisfactory

AMMONIUM HYDROXIDE 25%

°C	15	26	38	49	60	71	82	93	104	116	127	138	149	160	171	182	193	204	216	227	238
°F	60	80	100	120	140	160	180	200	220	240	260	280	300	320	340	360	380	400	420	440	460

PLASTICS	Rating
ABS	R
ACETAL	U
ACRYLICS	R
ASBESTOS REINFORCED EPOXY	R
ASBESTOS REINFORCED PHENOLIC	R U
CHLORINATED POLYETHER (Penton)	R
CPVC	R
E-CTFE (Halar)	R
EPOXY	R U
ETFE (Tefzel)	R
FEP	R
FURAN (FURFURAL ALCOHOL)	R
NORYL	R
NYLON 6	R
NYLON 11	R
NYLON 66	R
PHENOLIC	R
PFA (Teflon)	R
POLYAMIDE-IMIDE	R
POLYESTERS	
Bisphenol A–fumarate	R
Halogenated	R
Hydrogenated Bisphenol A –Bisphenol A	
Isophthalic	U
Terephthalate (PET)	R
POLYETHERIMIDE (ULTEM)	R
POLYETHERETHERKETONE (PEEK)	
POLYETHERSULFONE (PES)	
POLYETHYLENE HMW	R
POLYETHYLENE UHMW	R
POLYMETHYLPENTENE	
POLYPROPYLENE (PP)	R
POLYSTYRENE	R
POLYSULFONE	R
POLYVINYLIDENE CHLORIDE	U
PVC TYPE 1	R
PVC TYPE 2	R
PVDF (Kynar)	R
RYTON	R
TFE (Teflon)	R
VINYL ESTER	R

°F	60	80	100	120	140	160	180	200	220	240	260	280	300	320	340	360	380	400	420	440	460
°C	15	26	38	49	60	71	82	93	104	116	127	138	149	160	171	182	193	204	216	227	238

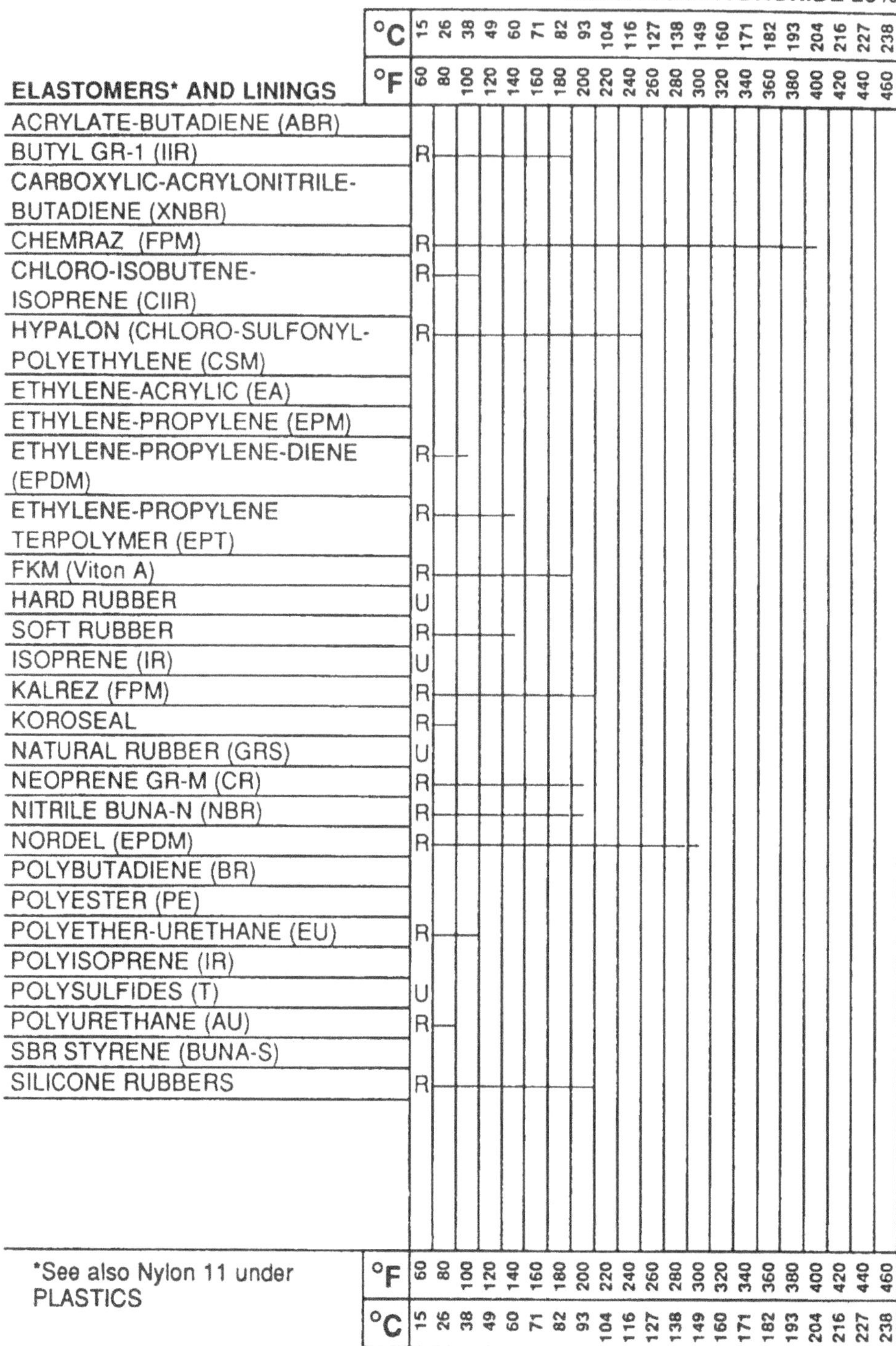

AMMONIUM HYDROXIDE 25%

°C: 15 26 38 49 60 71 82 93 104 116 127 138 149 160 171 182 193 204 216 227 238

°F: 60 80 100 120 140 160 180 200 220 240 260 280 300 320 340 360 380 400 420 440 460

ELASTOMERS* AND LININGS	
ACRYLATE-BUTADIENE (ABR)	
BUTYL GR-1 (IIR)	R
CARBOXYLIC-ACRYLONITRILE-BUTADIENE (XNBR)	
CHEMRAZ (FPM)	R
CHLORO-ISOBUTENE-ISOPRENE (CIIR)	R
HYPALON (CHLORO-SULFONYL-POLYETHYLENE (CSM)	R
ETHYLENE-ACRYLIC (EA)	
ETHYLENE-PROPYLENE (EPM)	
ETHYLENE-PROPYLENE-DIENE (EPDM)	R
ETHYLENE-PROPYLENE TERPOLYMER (EPT)	R
FKM (Viton A)	R
HARD RUBBER	U
SOFT RUBBER	R
ISOPRENE (IR)	U
KALREZ (FPM)	R
KOROSEAL	R
NATURAL RUBBER (GRS)	U
NEOPRENE GR-M (CR)	R
NITRILE BUNA-N (NBR)	R
NORDEL (EPDM)	R
POLYBUTADIENE (BR)	
POLYESTER (PE)	
POLYETHER-URETHANE (EU)	R
POLYISOPRENE (IR)	
POLYSULFIDES (T)	U
POLYURETHANE (AU)	R
SBR STYRENE (BUNA-S)	
SILICONE RUBBERS	R

*See also Nylon 11 under PLASTICS

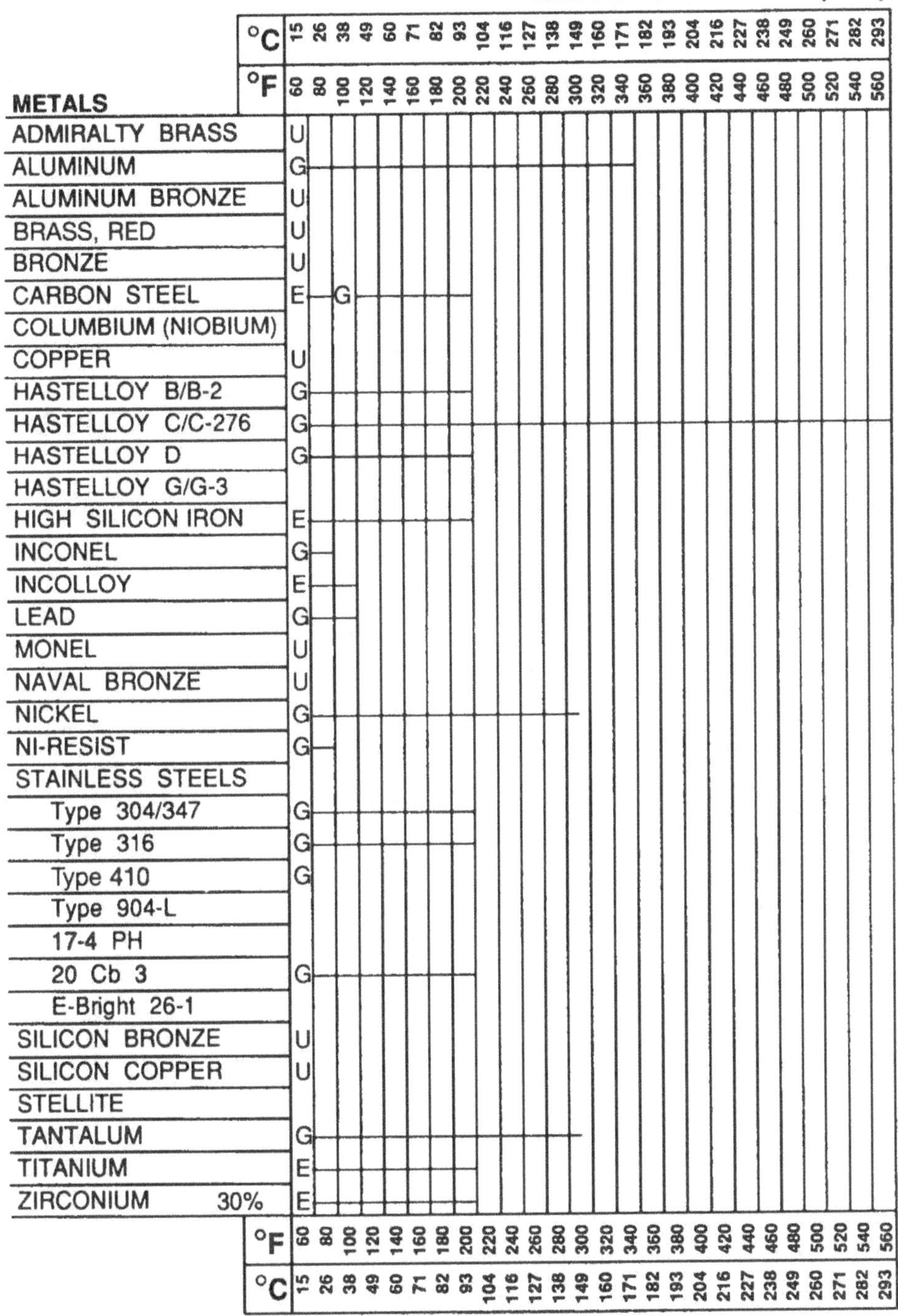

METALS	60 °F	80	100	120	140	160	180	200	220	240	260	280	300	320	340	360	380	400	420	440	460	480	500	520	540	560
°C	15	26	38	49	60	71	82	93	104	116	127	138	149	160	171	182	193	204	216	227	238	249	260	271	282	293
ADMIRALTY BRASS	U																									
ALUMINUM	G																									
ALUMINUM BRONZE	U																									
BRASS, RED	U																									
BRONZE	U																									
CARBON STEEL	E		G																							
COLUMBIUM (NIOBIUM)																										
COPPER	U																									
HASTELLOY B/B-2	G																									
HASTELLOY C/C-276	G																									
HASTELLOY D	G																									
HASTELLOY G/G-3																										
HIGH SILICON IRON	E																									
INCONEL	G																									
INCOLLOY	E																									
LEAD	G																									
MONEL	U																									
NAVAL BRONZE	U																									
NICKEL	G																									
NI-RESIST	G																									
STAINLESS STEELS																										
Type 304/347	G																									
Type 316	G																									
Type 410	G																									
Type 904-L																										
17-4 PH																										
20 Cb 3	G																									
E-Bright 26-1																										
SILICON BRONZE	U																									
SILICON COPPER	U																									
STELLITE																										
TANTALUM	G																									
TITANIUM	E																									
ZIRCONIUM 30%	E																									

FOR METALS

E = < 2 Mils Penetration/Year; G = < 20 Mils Penetration /year

S = < 50 Mils Penetration/Year; U = > 50 Mils Penetration/year

FOR NONMETALLICS

R = Resistant

U = Unsatisfactory

AMMONIUM HYDROXIDE (SAT.)

°C	15	26	38	49	60	71	82	93	104	116	127	138	149	160	171	182	193	204	216	227	238
°F	60	80	100	120	140	160	180	200	220	240	260	280	300	320	340	360	380	400	420	440	460

PLASTICS		Rating
ABS		R
ACETAL		U
ACRYLICS		R
ASBESTOS REINFORCED EPOXY		R
ASBESTOS REINFORCED PHENOLIC		R
CHLORINATED POLYETHER (Penton)		R
CPVC		R
E-CTFE (Halar)		R
EPOXY		R
ETFE (Tefzel)		R
FEP		R
FURAN (FURFURAL ALCOHOL)		R
NORYL		R
NYLON 6		R
NYLON 11		R
NYLON 66		R
PHENOLIC		U
PFA (Teflon)		R
POLYAMIDE-IMIDE		R
POLYESTERS		
Bisphenol A–fumarate	20%	R
Halogenated		R
Hydrogenated Bisphenol A –Bisphenol A		
Isophthalic		U
Terephthalate (PET)		R
POLYETHERIMIDE (ULTEM)		R
POLYETHERETHERKETONE (PEEK)		R
POLYETHERSULFONE (PES)		
POLYETHYLENE HMW		R
POLYETHYLENE UHMW		R
POLYMETHYLPENTENE		
POLYPROPYLENE (PP)		R
POLYSTYRENE		R
POLYSULFONE		R
POLYVINYLIDENE CHLORIDE		U
PVC TYPE 1		R
PVC TYPE 2		R
PVDF (Kynar)		R
RYTON		R
TFE (Teflon)		R
VINYL ESTER	10	R

°F	60	80	100	120	140	160	180	200	220	240	260	280	300	320	340	360	380	400	420	440	460
°C	15	26	38	49	60	71	82	93	104	116	127	138	149	160	171	182	193	204	216	227	238

AMMONIUM HYDROXIDE (SAT.)

ELASTOMERS* AND LININGS	°F: 60 80 100 120 140 160 180 200 220 240 260 280 300 320 340 360 380 400 420 440 460 / °C: 15 26 38 49 60 71 82 93 104 116 127 138 149 160 171 182 193 204 216 227 238
ACRYLATE-BUTADIENE (ABR)	
BUTYL GR-1 (IIR)	R
CARBOXYLIC-ACRYLONITRILE-BUTADIENE (XNBR)	
CHEMRAZ (FPM)	R
CHLORO-ISOBUTENE-ISOPRENE (CIIR)	R
HYPALON (CHLORO-SULFONYL-POLYETHYLENE (CSM)	R
ETHYLENE-ACRYLIC (EA)	
ETHYLENE-PROPYLENE (EPM)	R
ETHYLENE-PROPYLENE-DIENE (EPDM)	R
ETHYLENE-PROPYLENE TERPOLYMER (EPT)	R
FKM (Viton A)	R
HARD RUBBER	U
SOFT RUBBER	R
ISOPRENE (IR)	R
KALREZ (FPM)	R
KOROSEAL	R
NATURAL RUBBER (GRS)	R
NEOPRENE GR-M (CR)	R
NITRILE BUNA-N (NBR)	R
NORDEL (EPDM)	R
POLYBUTADIENE (BR)	
POLYESTER (PE)	
POLYETHER-URETHANE (EU)	R
POLYISOPRENE (IR)	
POLYSULFIDES (T)	U
POLYURETHANE (AU)	R
SBR STYRENE (BUNA-S)	
SILICONE RUBBERS	R

*See also Nylon 11 under PLASTICS

METALS	60	80	100	120	140	160	180	200	220	240	260	280	300	320	340	360	380	400	420	440	460	480	500	520	540	560
°C	15	26	38	49	60	71	82	93	104	116	127	138	149	160	171	182	193	204	216	227	238	249	260	271	282	293
ADMIRALTY BRASS																										
ALUMINUM	U																									
ALUMINUM BRONZE	U																									
BRASS, RED	U																									
BRONZE	U																									
CARBON STEEL	U																									
COLUMBIUM (NIOBIUM)	E	U																								
COPPER	U																									
HASTELLOY B/B-2	U																									
HASTELLOY C/C-276	U																									
HASTELLOY D																										
HASTELLOY G/G-3																										
HIGH SILICON IRON	U																									
INCONEL	U																									
INCOLLOY																										
LEAD																										
MONEL	U																									
NAVAL BRONZE	U																									
NICKEL	U																									
NI-RESIST																										
STAINLESS STEELS																										
Type 304/347	U																									
Type 316	U																									
Type 410																										
Type 904-L																										
17-4 PH																										
20 Cb 3	U																									
E-Bright 26-1	U																									
SILICON BRONZE	U																									
SILICON COPPER																										
STELLITE																										
TANTALUM	E																									
TITANIUM	E	S																								
ZIRCONIUM	U																									
°F	60	80	100	120	140	160	180	200	220	240	260	280	300	320	340	360	380	400	420	440	460	480	500	520	540	560
°C	15	26	38	49	60	71	82	93	104	116	127	138	149	160	171	182	193	204	216	227	238	249	260	271	282	293

FOR METALS
E = < 2 Mils Penetration/Year; G = < 20 Mils Penetration /year
S = < 50 Mils Penetration/Year; U = > 50 Mils Penetration/year

FOR NONMETALLICS
R = Resistant
U = Unsatisfactory

PLASTICS °C	15	26	38	49	60	71	82	93	104	116	127	138	149	160	171	182	193	204	216	227	238
°F	60	80	100	120	140	160	180	200	220	240	260	280	300	320	340	360	380	400	420	440	460
ABS	U																				
ACETAL	U																				
ACRYLICS																					
ASBESTOS REINFORCED EPOXY																					
ASBESTOS REINFORCED PHENOLIC																					
CHLORINATED POLYETHER (Penton)	R–																				
CPVC	R–	U																			
E-CTFE (Halar)	R	—	—	—	—	—	—	—	—	—											
EPOXY	U																				
ETFE (Tefzel)	R	—	—	—	—	—	—	—													
FEP	R	—	—	—	—	—	—	—	—	—	—	—	—	—	—	—	—	—			
FURAN (FURFURAL ALCOHOL)	U																				
NORYL	U																				
NYLON 6	U																				
NYLON 11	U																				
NYLON 66	U																				
PHENOLIC	U																				
PFA (Teflon)	R	—	—	—	—	—	—	—	—	—											
POLYAMIDE-IMIDE																					
POLYESTERS																					
Bisphenol A–fumarate	U																				
Halogenated	U																				
Hydrogenated Bisphenol A –Bisphenol A																					
Isophthalic	U																				
Terephthalate (PET)	R–																				
POLYETHERIMIDE (ULTEM)																					
POLYETHERETHERKETONE (PEEK)	U																				
POLYETHERSULFONE (PES)																					
POLYETHYLENE HMW	R	—	—	—	U																
POLYETHYLENE UHMW	R	—	—	—	U																
POLYMETHYLPENTENE																					
POLYPROPYLENE (PP)	U																				
POLYSTYRENE																					
POLYSULFONE	U																				
POLYVINYLIDENE CHLORIDE	R	—	—	—																	
PVC TYPE 1	R	U																			
PVC TYPE 2	U																				
PVDF (Kynar)	R	—	—	—	—	—															
RYTON																					
TFE (Teflon)	R	—	—	—	—	—	—	—	—	—	—	—	—	—	—	—	—	—	—	—	—
VINYL ESTER	U																				
°F	60	80	100	120	140	160	180	200	220	240	260	280	300	320	340	360	380	400	420	440	460
°C	15	26	38	49	60	71	82	93	104	116	127	138	149	160	171	182	193	204	216	227	238

ELASTOMERS* AND LININGS	Rating at 60°F (15°C)	Line extends to °F (°C)
ACRYLATE-BUTADIENE (ABR)		
BUTYL GR-1 (IIR)		
CARBOXYLIC-ACRYLONITRILE-BUTADIENE (XNBR)		
CHEMRAZ (FPM)	R	400 (204)
CHLORO-ISOBUTENE-ISOPRENE (CIIR)		
HYPALON (CHLORO-SULFONYL-POLYETHYLENE (CSM)		
ETHYLENE-ACRYLIC (EA)		
ETHYLENE-PROPYLENE (EPM)		
ETHYLENE-PROPYLENE-DIENE (EPDM)	U	
ETHYLENE-PROPYLENE TERPOLYMER (EPT)	U	
FKM (Viton A)	R	180 (82)
HARD RUBBER	U	
SOFT RUBBER		
ISOPRENE (IR)	U	
KALREZ (FPM)	R	200 (93)
KOROSEAL	U	
NATURAL RUBBER (GRS)	U	
NEOPRENE GR-M (CR)	U	
NITRILE BUNA-N (NBR)	U	
NORDEL (EPDM)		
POLYBUTADIENE (BR)		
POLYESTER (PE)		
POLYETHER-URETHANE (EU)		
POLYISOPRENE (IR)		
POLYSULFIDES (T)		
POLYURETHANE (AU)	U	
SBR STYRENE (BUNA-S)		
SILICONE RUBBERS	U	

Temperature scale: °F 60, 80, 100, 120, 140, 160, 180, 200, 220, 240, 260, 280, 300, 320, 340, 360, 380, 400, 420, 440, 460; °C 15, 26, 38, 49, 60, 71, 82, 93, 104, 116, 127, 138, 149, 160, 171, 182, 193, 204, 216, 227, 238

*See also Nylon 11 under PLASTICS

BROMINE GAS DRY

METALS °F	60	80	100	120	140	160	180	200	220	240	260	280	300	320	340	360	380	400	420	440	460	480	500	520	540	560
°C	15	26	38	49	60	71	82	93	104	116	127	138	149	160	171	182	193	204	216	227	238	249	260	271	282	293
ADMIRALTY BRASS	G																									
ALUMINUM	G	U																								
ALUMINUM BRONZE	U																									
BRASS, RED																										
BRONZE	U																									
CARBON STEEL	U																									
COLUMBIUM (NIOBIUM)	E																									
COPPER	E																									
HASTELLOY B/B-2	E																									
HASTELLOY C/C-276	E					G																				
HASTELLOY D																										
HASTELLOY G/G-3																										
HIGH SILICON IRON																										
INCONEL	G																									
INCOLLOY 825	E																									
LEAD																										
MONEL	E																									
NAVAL BRONZE	G																									
NICKEL	E																									
NI-RESIST	U																									
STAINLESS STEELS																										
Type 304/347	U																									
Type 316	U																									
Type 410	U																									
Type 904-L																										
17-4 PH	U																									
20 Cb 3	E																									
E-Bright 26-1																										
SILICON BRONZE	U																									
SILICON COPPER	E																									
STELLITE																										
TANTALUM	E																									
TITANIUM	U																									
ZIRCONIUM	U																									

FOR METALS

E = < 2 Mils Penetration/Year; G = < 20 Mils Penetration /year

S = < 50 Mils Penetration/Year; U = > 50 Mils Penetration/year

FOR NONMETALLICS

R = Resistant

U = Unsatisfactory

PLASTICS	°F 60 / °C 15
ABS	
ACETAL	
ACRYLICS	
ASBESTOS REINFORCED EPOXY	
ASBESTOS REINFORCED PHENOLIC	
CHLORINATED POLYETHER (Penton)	U
CPVC	U
E-CTFE (Halar)	U
EPOXY	U
ETFE (Tefzel)	R
FEP	R
FURAN (FURFURAL ALCOHOL)	U
NORYL	
NYLON 6	U
NYLON 11	U
NYLON 66	U
PHENOLIC	
PFA (Teflon)	R
POLYAMIDE-IMIDE	
POLYESTERS	
Bisphenol A–fumarate	R
Halogenated	R
Hydrogenated Bisphenol A –Bisphenol A	
Isophthalic	U
Terephthalate (PET)	
POLYETHERIMIDE (ULTEM)	
POLYETHERETHERKETONE (PEEK)	U
POLYETHERSULFONE (PES)	
POLYETHYLENE HMW	U
POLYETHYLENE UHMW	
POLYMETHYLPENTENE	
POLYPROPYLENE (PP)	U
POLYSTYRENE	
POLYSULFONE	
POLYVINYLIDENE CHLORIDE	
PVC TYPE 1	U
PVC TYPE 2	U
PVDF (Kynar)	R
RYTON	U
TFE (Teflon)	R
VINYL ESTER	R

°F: 60, 80, 100, 120, 140, 160, 180, 200, 220, 240, 260, 280, 300, 320, 340, 360, 380, 400, 420, 440, 460

°C: 15, 26, 38, 49, 60, 71, 82, 93, 104, 116, 127, 138, 149, 160, 171, 182, 193, 204, 216, 227, 238

ELASTOMERS* AND LININGS		60	80	100	120	140	160	180	200	220	240	260	280	300	320	340	360	380	400	420	440	460
°C		15	26	38	49	60	71	82	93	104	116	127	138	149	160	171	182	193	204	216	227	238
ACRYLATE-BUTADIENE (ABR)																						
BUTYL GR-1 (IIR)																						
CARBOXYLIC-ACRYLONITRILE-BUTADIENE (XNBR)																						
CHEMRAZ (FPM)		R																				
CHLORO-ISOBUTENE-ISOPRENE (CIIR)																						
HYPALON (CHLORO-SULFONYL-POLYETHYLENE (CSM)		R																				
ETHYLENE-ACRYLIC (EA)																						
ETHYLENE-PROPYLENE (EPM)																						
ETHYLENE-PROPYLENE-DIENE (EPDM)		U																				
ETHYLENE-PROPYLENE TERPOLYMER (EPT)		U																				
FKM (Viton A)	25%	R																				
HARD RUBBER																						
SOFT RUBBER																						
ISOPRENE (IR)																						
KALREZ (FPM)		R																				
KOROSEAL																						
NATURAL RUBBER (GRS)																						
NEOPRENE GR-M (CR)		U																				
NITRILE BUNA-N (NBR)		U																				
NORDEL (EPDM)		U																				
POLYBUTADIENE (BR)																						
POLYESTER (PE)																						
POLYETHER-URETHANE (EU)																						
POLYISOPRENE (IR)																						
POLYSULFIDES (T)																						
POLYURETHANE (AU)																						
SBR STYRENE (BUNA-S)																						
SILICONE RUBBERS																						
°F		60	80	100	120	140	160	180	200	220	240	260	280	300	320	340	360	380	400	420	440	460
°C		15	26	38	49	60	71	82	93	104	116	127	138	149	160	171	182	193	204	216	227	238

*See also Nylon 11 under PLASTICS

°C	15	26	38	49	60	71	82	93	104	116	127	138	149	160	171	182	193	204	216	227	238	249	260	271	282	293
METALS °F	60	80	100	120	140	160	180	200	220	240	260	280	300	320	340	360	380	400	420	440	460	480	500	520	540	560
ADMIRALTY BRASS	U																									
ALUMINUM	U																									
ALUMINUM BRONZE	U																									
BRASS, RED																										
BRONZE																										
CARBON STEEL	U																									
COLUMBIUM (NIOBIUM)	E																									
COPPER	U																									
HASTELLOY B/B-2																										
HASTELLOY C/C-276	E																									
HASTELLOY D																										
HASTELLOY G/G-3																										
HIGH SILICON IRON	U																									
INCONEL	U																									
INCOLLOY																										
LEAD																										
MONEL	U																									
NAVAL BRONZE	U																									
NICKEL	U																									
NI-RESIST																										
STAINLESS STEELS																										
Type 304/347	U																									
Type 316	U																									
Type 410	U																									
Type 904-L																										
17-4 PH	U																									
20 Cb 3	U																									
E-Bright 26-1																										
SILICON BRONZE	U																									
SILICON COPPER	U																									
STELLITE																										
TANTALUM	E																									
TITANIUM	E																									
ZIRCONIUM	E																									
°F	60	80	100	120	140	160	180	200	220	240	260	280	300	320	340	360	380	400	420	440	460	480	500	520	540	560
°C	15	26	38	49	60	71	82	93	104	116	127	138	149	160	171	182	193	204	216	227	238	249	260	271	282	293

FOR METALS

E = < 2 Mils Penetration/Year; G = < 20 Mils Penetration /year

S = < 50 Mils Penetration/Year; U = > 50 Mils Penetration/year

FOR NONMETALLICS

R = Resistant

U = Unsatisfactory

BROMINE GAS MOIST

PLASTICS	Rating	Max. temp. °F	Max. temp. °C
ABS			
ACETAL			
ACRYLICS			
ASBESTOS REINFORCED EPOXY			
ASBESTOS REINFORCED PHENOLIC			
CHLORINATED POLYETHER (Penton)	U		
CPVC	U		
E-CTFE (Halar)			
EPOXY	U		
ETFE (Tefzel)			
FEP	R	200	93
FURAN (FURFURAL ALCOHOL)	U		
NORYL			
NYLON 6	U		
NYLON 11	U		
NYLON 66	U		
PHENOLIC			
PFA (Teflon)	R	200	93
POLYAMIDE-IMIDE	R	120	49
POLYESTERS			
Bisphenol A–fumarate	R	100	38
Halogenated	R	100	38
Hydrogenated Bisphenol A –Bisphenol A			
Isophthalic	U		
Terephthalate (PET)			
POLYETHERIMIDE (ULTEM)			
POLYETHERETHERKETONE (PEEK)	U		
POLYETHERSULFONE (PES)			
POLYETHYLENE HMW	U		
POLYETHYLENE UHMW			
POLYMETHYLPENTENE			
POLYPROPYLENE (PP)	U		
POLYSTYRENE			
POLYSULFONE	R	200	93
POLYVINYLIDENE CHLORIDE			
PVC TYPE 1	U		
PVC TYPE 2	U		
PVDF (Kynar)	R	200	93
RYTON	U		
TFE (Teflon)	R	240	116
VINYL ESTER	R	100	38

Temperature scale: °F 60, 80, 100, 120, 140, 160, 180, 200, 220, 240, 260, 280, 300, 320, 340, 360, 380, 400, 420, 440, 460; °C 15, 26, 38, 49, 60, 71, 82, 93, 104, 116, 127, 138, 149, 160, 171, 182, 193, 204, 216, 227, 238

BROMINE GAS MOIST

ELASTOMERS* AND LININGS		°C 15	26	38	49	60	71	82	93	104	116	127	138	149	160	171	182	193	204	216	227	238
		°F 60	80	100	120	140	160	180	200	220	240	260	280	300	320	340	360	380	400	420	440	460
ACRYLATE-BUTADIENE (ABR)																						
BUTYL GR-1 (IIR)																						
CARBOXYLIC-ACRYLONITRILE-BUTADIENE (XNBR)																						
CHEMRAZ (FPM)		U																				
CHLORO-ISOBUTENE-ISOPRENE (CIIR)																						
HYPALON (CHLORO-SULFONYL-POLYETHYLENE (CSM)		R																				
ETHYLENE-ACRYLIC (EA)																						
ETHYLENE-PROPYLENE (EPM)																						
ETHYLENE-PROPYLENE-DIENE (EPDM)		U																				
ETHYLENE-PROPYLENE TERPOLYMER (EPT)																						
FKM (Viton A)	25%	R																				
HARD RUBBER																						
SOFT RUBBER																						
ISOPRENE (IR)																						
KALREZ (FPM)																						
KOROSEAL																						
NATURAL RUBBER (GRS)																						
NEOPRENE GR-M (CR)		U																				
NITRILE BUNA-N (NBR)		U																				
NORDEL (EPDM)		U																				
POLYBUTADIENE (BR)																						
POLYESTER (PE)																						
POLYETHER-URETHANE (EU)																						
POLYISOPRENE (IR)																						
POLYSULFIDES (T)																						
POLYURETHANE (AU)																						
SBR STYRENE (BUNA-S)																						
SILICONE RUBBERS																						
		°F 60	80	100	120	140	160	180	200	220	240	260	280	300	320	340	360	380	400	420	440	460
		°C 15	26	38	49	60	71	82	93	104	116	127	138	149	160	171	182	193	204	216	227	238

*See also Nylon 11 under PLASTICS

BROMINE LIQUID

METALS °C	15	26	38	49	60	71	82	93	104	116	127	138	149	160	171	182	193	204	216	227	238	249	260	271	282	293
°F	60	80	100	120	140	160	180	200	220	240	260	280	300	320	340	360	380	400	420	440	460	480	500	520	540	560
ADMIRALTY BRASS																										
ALUMINUM	G																									
ALUMINUM BRONZE	U																									
BRASS, RED																										
BRONZE																										
CARBON STEEL																										
COLUMBIUM (NIOBIUM)	E																									
COPPER																										
HASTELLOY B/B-2																										
HASTELLOY C/C-276																										
HASTELLOY D																										
HASTELLOY G/G-3																										
HIGH SILICON IRON	U																									
INCONEL																										
INCOLLOY																										
LEAD																										
MONEL																										
NAVAL BRONZE																										
NICKEL																										
NI-RESIST																										
STAINLESS STEELS																										
Type 304/347	U																									
Type 316	U																									
Type 410	U																									
Type 904-L																										
17-4 PH	U																									
20 Cb 3																										
E-Bright 26-1																										
SILICON BRONZE	U																									
SILICON COPPER																										
STELLITE																										
TANTALUM	E																									
TITANIUM	U																									
ZIRCONIUM 1	G																									
°F	60	80	100	120	140	160	180	200	220	240	260	280	300	320	340	360	380	400	420	440	460	480	500	520	540	560
°C	15	26	38	49	60	71	82	93	104	116	127	138	149	160	171	182	193	204	216	227	238	249	260	271	282	293

FOR METALS
E = < 2 Mils Penetration/Year; G = < 20 Mils Penetration /year
S = < 50 Mils Penetration/Year; U = > 50 Mils Penetration/year

FOR NONMETALLICS
R = Resistant
U = Unsatisfactory

PLASTICS																					
°C	15	26	38	49	60	71	82	93	104	116	127	138	149	160	171	182	193	204	216	227	238
°F	60	80	100	120	140	160	180	200	220	240	260	280	300	320	340	360	380	400	420	440	460
ABS	U																				
ACETAL																					
ACRYLICS	U																				
ASBESTOS REINFORCED EPOXY																					
ASBESTOS REINFORCED PHENOLIC																					
CHLORINATED POLYETHER (Penton)	U																				
CPVC	U																				
E-CTFE (Halar)	R																				
EPOXY	U																				
ETFE (Tefzel)																					
FEP	R																				
FURAN (FURFURAL ALCOHOL)	R	U																			
NORYL																					
NYLON 6	U																				
NYLON 11	U																				
NYLON 66	U																				
PHENOLIC																					
PFA (Teflon)																					
POLYAMIDE-IMIDE																					
POLYESTERS																					
Bisphenol A–fumarate	U																				
Halogenated	U																				
Hydrogenated Bisphenol A –Bisphenol A	U																				
Isophthalic	U																				
Terephthalate (PET)	R																				
POLYETHERIMIDE (ULTEM)																					
POLYETHERETHERKETONE (PEEK)																					
POLYETHERSULFONE (PES)																					
POLYETHYLENE HMW	U																				
POLYETHYLENE UHMW	U																				
POLYMETHYLPENTENE	U																				
POLYPROPYLENE (PP)	U																				
POLYSTYRENE	U																				
POLYSULFONE																					
POLYVINYLIDENE CHLORIDE	U																				
PVC TYPE 1	U																				
PVC TYPE 2	U																				
PVDF (Kynar)	R																				
RYTON	U																				
TFE (Teflon)	R																				
VINYL ESTER	U																				
°F	60	80	100	120	140	160	180	200	220	240	260	280	300	320	340	360	380	400	420	440	460
°C	15	26	38	49	60	71	82	93	104	116	127	138	149	160	171	182	193	204	216	227	238

BROMINE LIQUID

ELASTOMERS* AND LININGS	60	80	100	120	140	160	180	200	220	240	260	280	300	320	340	360	380	400	420	440	460
°C	15	26	38	49	60	71	82	93	104	116	127	138	149	160	171	182	193	204	216	227	238
ACRYLATE-BUTADIENE (ABR)																					
BUTYL GR-1 (IIR)																					
CARBOXYLIC-ACRYLONITRILE-BUTADIENE (XNBR)																					
CHEMRAZ (FPM)	R																				
CHLORO-ISOBUTENE-ISOPRENE (CIIR)																					
HYPALON (CHLORO-SULFONYL-POLYETHYLENE (CSM)	R																				
ETHYLENE-ACRYLIC (EA)																					
ETHYLENE-PROPYLENE (EPM)																					
ETHYLENE-PROPYLENE-DIENE (EPDM)	U																				
ETHYLENE-PROPYLENE TERPOLYMER (EPT)	U																				
FKM (Viton A)	R																				
HARD RUBBER																					
SOFT RUBBER																					
ISOPRENE (IR)																					
KALREZ (FPM)	R																				
KOROSEAL																					
NATURAL RUBBER (GRS)																					
NEOPRENE GR-M (CR)	U																				
NITRILE BUNA-N (NBR)	U																				
NORDEL (EPDM)	U																				
POLYBUTADIENE (BR)																					
POLYESTER (PE)	U																				
POLYETHER-URETHANE (EU)																					
POLYISOPRENE (IR)																					
POLYSULFIDES (T)																					
POLYURETHANE (AU)	U																				
SBR STYRENE (BUNA-S)																					
SILICONE RUBBERS																					
°F	60	80	100	120	140	160	180	200	220	240	260	280	300	320	340	360	380	400	420	440	460
°C	15	26	38	49	60	71	82	93	104	116	127	138	149	160	171	182	193	204	216	227	238

*See also Nylon 11 under PLASTICS

CALCIUM HYPOCHLORITE (SAT.)

METALS	°F 60	80	100	120	140	160	180	200	220	240	260	280	300	320	340	360	380	400	420	440	460	480	500	520	540	560
	°C 15	26	38	49	60	71	82	93	104	116	127	138	149	160	171	182	193	204	216	227	238	249	260	271	282	293
ADMIRALTY BRASS	U																									
ALUMINUM	U																									
ALUMINUM BRONZE	U																									
BRASS, RED	U																									
BRONZE	U																									
CARBON STEEL	U																									
COLUMBIUM (NIOBIUM)	E																									
COPPER	U																									
HASTELLOY B/B-2	U																									
HASTELLOY C/C-276 50%	E	—	G	—	—	—	—	—																		
HASTELLOY D	U																									
HASTELLOY G/G-3																										
HIGH SILICON IRON	G	—																								
INCONEL	U																									
INCOLLOY	U																									
LEAD	U																									
MONEL	U																									
NAVAL BRONZE	S																									
NICKEL	U																									
NI-RESIST	G																									
STAINLESS STEELS																										
Type 304/347	U																									
Type 316	G	—																								
Type 410	U																									
Type 904-L																										
17-4 PH	U																									
20 Cb 3	G	—																								
E-Bright 26-1																										
SILICON BRONZE	U																									
SILICON COPPER	S																									
STELLITE																										
TANTALUM	G	—	—	—	—	—	—	—	—	—	—	—	—													
TITANIUM	E	—	—	—	—	—	—	—																		
ZIRCONIUM	S	—	—	—	—	—	—	—																		
	°F 60	80	100	120	140	160	180	200	220	240	260	280	300	320	340	360	380	400	420	440	460	480	500	520	540	560
	°C 15	26	38	49	60	71	82	93	104	116	127	138	149	160	171	182	193	204	216	227	238	249	260	271	282	293

FOR METALS

E = < 2 Mils Penetration/Year; G = < 20 Mils Penetration /year

S = < 50 Mils Penetration/Year; U = > 50 Mils Penetration/year

FOR NONMETALLICS

R = Resistant

U = Unsatisfactory

CALCIUM HYPOCHLORITE (SAT.)

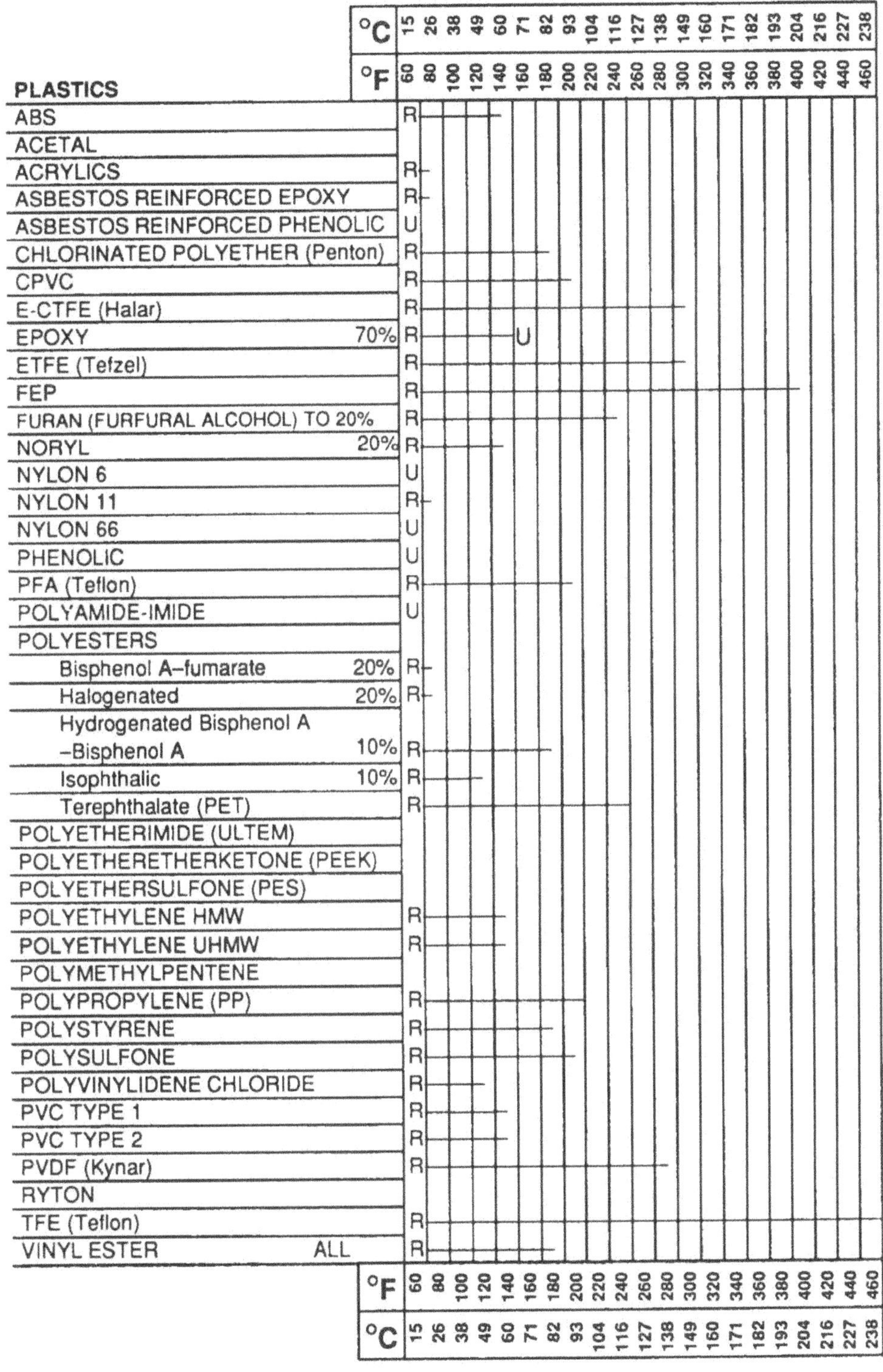

PLASTICS	Conc.	Rating
ABS		R
ACETAL		
ACRYLICS		R
ASBESTOS REINFORCED EPOXY		R
ASBESTOS REINFORCED PHENOLIC		U
CHLORINATED POLYETHER (Penton)		R
CPVC		R
E-CTFE (Halar)		R
EPOXY	70%	R; U (160°F / 71°C)
ETFE (Tefzel)		R
FEP		R
FURAN (FURFURAL ALCOHOL) TO 20%		R
NORYL	20%	R
NYLON 6		U
NYLON 11		R
NYLON 66		U
PHENOLIC		U
PFA (Teflon)		R
POLYAMIDE-IMIDE		U
POLYESTERS		
Bisphenol A–fumarate	20%	R
Halogenated	20%	R
Hydrogenated Bisphenol A –Bisphenol A	10%	R
Isophthalic	10%	R
Terephthalate (PET)		R
POLYETHERIMIDE (ULTEM)		
POLYETHERETHERKETONE (PEEK)		
POLYETHERSULFONE (PES)		
POLYETHYLENE HMW		R
POLYETHYLENE UHMW		R
POLYMETHYLPENTENE		
POLYPROPYLENE (PP)		R
POLYSTYRENE		R
POLYSULFONE		R
POLYVINYLIDENE CHLORIDE		R
PVC TYPE 1		R
PVC TYPE 2		R
PVDF (Kynar)		R
RYTON		
TFE (Teflon)		R
VINYL ESTER	ALL	R

CALCIUM HYPOCHLORITE (SAT.)

ELASTOMERS* AND LININGS		°F / °C	60 / 15	80 / 26	100 / 38	120 / 49	140 / 60	160 / 71	180 / 82	200 / 93	220 / 104	240 / 116	260 / 127	280 / 138	300 / 149	320 / 160	340 / 171	360 / 182	380 / 193	400 / 204	420 / 216	440 / 227	460 / 238
ACRYLATE-BUTADIENE (ABR)																							
BUTYL GR-1 (IIR)			R																				
CARBOXYLIC-ACRYLONITRILE-BUTADIENE (XNBR)																							
CHEMRAZ (FPM)			R																				
CHLORO-ISOBUTENE-ISOPRENE (CIIR)	TO 25%		R																				
HYPALON (CHLORO-SULFONYL-POLYETHYLENE (CSM)			R																				
ETHYLENE-ACRYLIC (EA)																							
ETHYLENE-PROPYLENE (EPM)																							
ETHYLENE-PROPYLENE-DIENE (EPDM)			R																				
ETHYLENE-PROPYLENE TERPOLYMER (EPT)			U																				
FKM (Viton A)			R																				
HARD RUBBER			R																				
SOFT RUBBER			U																				
ISOPRENE (IR)			R																				
KALREZ (FPM)			R																				
KOROSEAL																							
NATURAL RUBBER (GRS)			R																				
NEOPRENE GR-M (CR)			R																				
NITRILE BUNA-N (NBR)			U																				
NORDEL (EPDM)			R																				
POLYBUTADIENE (BR)			R																				
POLYESTER (PE)	5%		R																				
POLYETHER-URETHANE (EU)																							
POLYISOPRENE (IR)																							
POLYSULFIDES (T)																							
POLYURETHANE (AU)			U																				
SBR STYRENE (BUNA-S)			U																				
SILICONE RUBBERS																							

*See also Nylon 11 under PLASTICS

°F 60 80 100 120 140 160 180 200 220 240 260 280 300 320 340 360 380 400 420 440 460
°C 15 26 38 49 60 71 82 93 104 116 127 138 149 160 171 182 193 204 216 227 238

CARBON TETRACHLORIDE

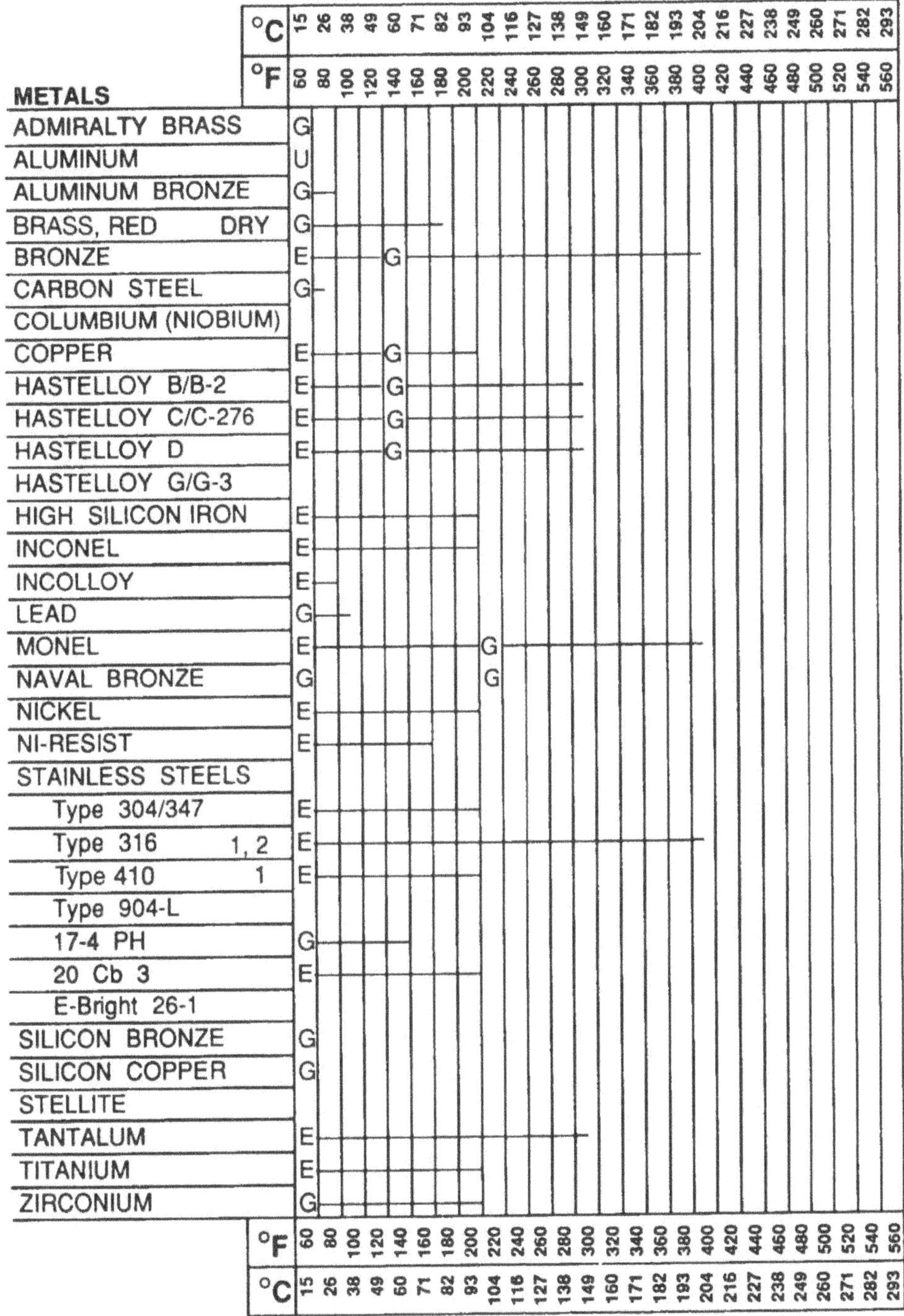

METALS		60 °F (15 °C)	140 °F (60 °C)	220 °F (104 °C)	Range extends to (°F)
ADMIRALTY BRASS		G			
ALUMINUM		U			
ALUMINUM BRONZE		G			80
BRASS, RED	DRY	G			180
BRONZE		E	G		400
CARBON STEEL		G			60
COLUMBIUM (NIOBIUM)					
COPPER		E	G		200
HASTELLOY B/B-2		E	G		300
HASTELLOY C/C-276		E	G		300
HASTELLOY D		E	G		300
HASTELLOY G/G-3					
HIGH SILICON IRON		E			200
INCONEL		E			200
INCOLLOY		E			80
LEAD		G			100
MONEL		E		G	400
NAVAL BRONZE		G		G	
NICKEL		E			200
NI-RESIST		E			160
STAINLESS STEELS					
Type 304/347		E			200
Type 316	1, 2	E			400
Type 410	1	E			200
Type 904-L					
17-4 PH		G			140
20 Cb 3		E			200
E-Bright 26-1					
SILICON BRONZE		G			
SILICON COPPER		G			
STELLITE					
TANTALUM		E			300
TITANIUM		E			200
ZIRCONIUM		G			200

°F: 60, 80, 100, 120, 140, 160, 180, 200, 220, 240, 260, 280, 300, 320, 340, 360, 380, 400, 420, 440, 460, 480, 500, 520, 540, 560

°C: 15, 26, 38, 49, 60, 71, 82, 93, 104, 116, 127, 138, 149, 160, 171, 182, 193, 204, 216, 227, 238, 249, 260, 271, 282, 293

FOR METALS
E = < 2 Mils Penetration/Year; G = < 20 Mils Penetration /year
S = < 50 Mils Penetration/Year; U = > 50 Mils Penetration/year

FOR NONMETALLICS
R = Resistant
U = Unsatisfactory

	°C	15	26	38	49	60	71	82	93	104	116	127	138	149	160	171	182	193	204	216	227	238
PLASTICS	°F	60	80	100	120	140	160	180	200	220	240	260	280	300	320	340	360	380	400	420	440	460
ABS		U																				
ACETAL		R																				
ACRYLICS		U																				
ASBESTOS REINFORCED EPOXY		U																				
ASBESTOS REINFORCED PHENOLIC		R																				
CHLORINATED POLYETHER (Penton)		R																				
CPVC		R	U																			
E-CTFE (Halar)		R																				
EPOXY		R						U														
ETFE (Tefzel)		R																				
FEP		R																				
FURAN (FURFURAL ALCOHOL)		R																				
NORYL		U																				
NYLON 6		R																				
NYLON 11		R																				
NYLON 66		R																				
PHENOLIC		R																				
PFA (Teflon)		R																				
POLYAMIDE-IMIDE																						
POLYESTERS																						
Bisphenol A–fumarate		R			U																	
Halogenated		R																				
Hydrogenated Bisphenol A –Bisphenol A		U																				
Isophthalic		U																				
Terephthalate (PET)		R																				
POLYETHERIMIDE (ULTEM)		R																				
POLYETHERETHERKETONE (PEEK)		R																				
POLYETHERSULFONE (PES)		R																				
POLYETHYLENE HMW		U																				
POLYETHYLENE UHMW		U																				
POLYMETHYLPENTENE		U																				
POLYPROPYLENE (PP)		U																				
POLYSTYRENE		U																				
POLYSULFONE		U																				
POLYVINYLIDENE CHLORIDE		R																				
PVC TYPE 1		R	U																			
PVC TYPE 2		U																				
PVDF (Kynar)		R																				
RYTON		R																				
TFE (Teflon)		R																				
VINYL ESTER		R																				
	°F	60	80	100	120	140	160	180	200	220	240	260	280	300	320	340	360	380	400	420	440	460
	°C	15	26	38	49	60	71	82	93	104	116	127	138	149	160	171	182	193	204	216	227	238

ELASTOMERS* AND LININGS	60	80	100	120	140	160	180	200	220	240	260	280	300	320	340	360	380	400	420	440	460
°C	15	26	38	49	60	71	82	93	104	116	127	138	149	160	171	182	193	204	216	227	238
°F	60	80	100	120	140	160	180	200	220	240	260	280	300	320	340	360	380	400	420	440	460
ACRYLATE-BUTADIENE (ABR)																					
BUTYL GR-1 (IIR)	U																				
CARBOXYLIC-ACRYLONITRILE-BUTADIENE (XNBR)																					
CHEMRAZ (FPM)	R																				
CHLORO-ISOBUTENE-ISOPRENE (CIIR)	U																				
HYPALON (CHLORO-SULFONYL-POLYETHYLENE (CSM)	U																				
ETHYLENE-ACRYLIC (EA)																					
ETHYLENE-PROPYLENE (EPM)	U																				
ETHYLENE-PROPYLENE-DIENE (EPDM)	U																				
ETHYLENE-PROPYLENE TERPOLYMER (EPT)	U																				
FKM (Viton A)	R																				
HARD RUBBER	U																				
SOFT RUBBER	U																				
ISOPRENE (IR)	U																				
KALREZ (FPM)	R																				
KOROSEAL	U																				
NATURAL RUBBER (GRS)	U																				
NEOPRENE GR-M (CR)	U																				
NITRILE BUNA-N (NBR)	U																				
NORDEL (EPDM)	U																				
POLYBUTADIENE (BR)																					
POLYESTER (PE)	U																				
POLYETHER-URETHANE (EU)	U																				
POLYISOPRENE (IR)																					
POLYSULFIDES (T)	U																				
POLYURETHANE (AU)	U																				
SBR STYRENE (BUNA-S)	U																				
SILICONE RUBBERS	U																				
°F	60	80	100	120	140	160	180	200	220	240	260	280	300	320	340	360	380	400	420	440	460
°C	15	26	38	49	60	71	82	93	104	116	127	138	149	160	171	182	193	204	216	227	238

*See also Nylon 11 under PLASTICS

CHLORINE GAS DRY

METALS	60	80	100	120	140	160	180	200	220	240	260	280	300	320	340	360	380	400	420	440	460	480	500	520	540	560
°C	15	26	38	49	60	71	82	93	104	116	127	138	149	160	171	182	193	204	216	227	238	249	260	271	282	293
ADMIRALTY BRASS	E																									
ALUMINUM	G								U																	
ALUMINUM BRONZE	E																									
BRASS, RED	G																									
BRONZE	G																									
CARBON STEEL	G																									
COLUMBIUM (NIOBIUM)	G																									
COPPER	G																									
HASTELLOY B/B-2	G																									
HASTELLOY C/C-276	G																									
HASTELLOY D	G																									
HASTELLOY G/G-3																										
HIGH SILICON IRON																										
INCONEL	G																									
INCOLLOY 825	E																									
LEAD																										
MONEL	E								G																	
NAVAL BRONZE	E																									
NICKEL	G																									
NI-RESIST	G																									
STAINLESS STEELS																										
Type 304/347	U																									
Type 316	G																									
Type 410	U																									
Type 904-L																										
17-4 PH																										
20 Cb 3	E								G																	
E-Bright 26-1																										
SILICON BRONZE	E																									
SILICON COPPER	E																									
STELLITE																										
TANTALUM	E																					U				
TITANIUM	U																									
ZIRCONIUM	G																									
°F	60	80	100	120	140	160	180	200	220	240	260	280	300	320	340	360	380	400	420	440	460	480	500	520	540	560
°C	15	26	38	49	60	71	82	93	104	116	127	138	149	160	171	182	193	204	216	227	238	249	260	271	282	293

FOR METALS
E = < 2 Mils Penetration/Year; G = < 20 Mils Penetration /year
S = < 50 Mils Penetration/Year; U = > 50 Mils Penetration/year

FOR NONMETALLICS
R = Resistant
U = Unsatisfactory

PLASTICS	60	80	100	120	140	160	180	200	220	240	260	280	300	320	340	360	380	400	420	440	460
°C	15	26	38	49	60	71	82	93	104	116	127	138	149	160	171	182	193	204	216	227	238
ABS	R																				
ACETAL	U																				
ACRYLICS																					
ASBESTOS REINFORCED EPOXY																					
ASBESTOS REINFORCED PHENOLIC																					
CHLORINATED POLYETHER (Penton)	R			U																	
CPVC	R																				
E-CTFE (Halar)	R					U															
EPOXY	R																				
ETFE (Tefzel)	R																				
FEP	R																				
FURAN (FURFURAL ALCOHOL)	R																				
NORYL	U																				
NYLON 6	U																				
NYLON 11	U																				
NYLON 66	U																				
PHENOLIC																					
PFA (Teflon)	R																				
POLYAMIDE-IMIDE																					
POLYESTERS																					
Bisphenol A–fumarate	R																				
Halogenated	R																				
Hydrogenated Bisphenol A –Bisphenol A	R																				
Isophthalic	R																				
Terephthalate (PET)	R																				
POLYETHERIMIDE (ULTEM)																					
POLYETHERETHERKETONE (PEEK)	R																				
POLYETHERSULFONE (PES)																					
POLYETHYLENE HMW 10%	R																				
POLYETHYLENE UHMW	R																				
POLYMETHYLPENTENE																					
POLYPROPYLENE (PP)	U																				
POLYSTYRENE																					
POLYSULFONE																					
POLYVINYLIDENE CHLORIDE	R																				
PVC TYPE 1	R																				
PVC TYPE 2	R																				
PVDF (Kynar)	R																				
RYTON	U																				
TFE (Teflon)	R																				
VINYL ESTER	R																				

°F 60 80 100 120 140 160 180 200 220 240 260 280 300 320 340 360 380 400 420 440 460
°C 15 26 38 49 60 71 82 93 104 116 127 138 149 160 171 182 193 204 216 227 238

CHLORINE GAS DRY

ELASTOMERS* AND LININGS	60	80	100	120	140	160	180	200	220	240	260	280	300	320	340	360	380	400	420	440	460
°C	15	26	38	49	60	71	82	93	104	116	127	138	149	160	171	182	193	204	216	227	238
ACRYLATE-BUTADIENE (ABR)																					
BUTYL GR-1 (IIR)	U																				
CARBOXYLIC-ACRYLONITRILE-BUTADIENE (XNBR)																					
CHEMRAZ (FPM)	R	—	—	—	—	—	—	—	—	—	—	—	—	—	—	—	—	—			
CHLORO-ISOBUTENE-ISOPRENE (CIIR)																					
HYPALON (CHLORO-SULFONYL-POLYETHYLENE (CSM)	U																				
ETHYLENE-ACRYLIC (EA)																					
ETHYLENE-PROPYLENE (EPM)																					
ETHYLENE-PROPYLENE-DIENE (EPDM)	U																				
ETHYLENE-PROPYLENE TERPOLYMER (EPT)	U																				
FKM (Viton A)	R	—	—	—	—	—	—														
HARD RUBBER	U																				
SOFT RUBBER																					
ISOPRENE (IR)	U																				
KALREZ (FPM)	R	—	—	—	—	—	—	—													
KOROSEAL	R	—																			
NATURAL RUBBER (GRS)	U																				
NEOPRENE GR-M (CR)	U																				
NITRILE BUNA-N (NBR)	U																				
NORDEL (EPDM)	U																				
POLYBUTADIENE (BR)																					
POLYESTER (PE)	U																				
POLYETHER-URETHANE (EU)																					
POLYISOPRENE (IR)																					
POLYSULFIDES (T)																					
POLYURETHANE (AU)	U																				
SBR STYRENE (BUNA-S)	U																				
SILICONE RUBBERS																					
°F	60	80	100	120	140	160	180	200	220	240	260	280	300	320	340	360	380	400	420	440	460
°C	15	26	38	49	60	71	82	93	104	116	127	138	149	160	171	182	193	204	216	227	238

*See also Nylon 11 under PLASTICS

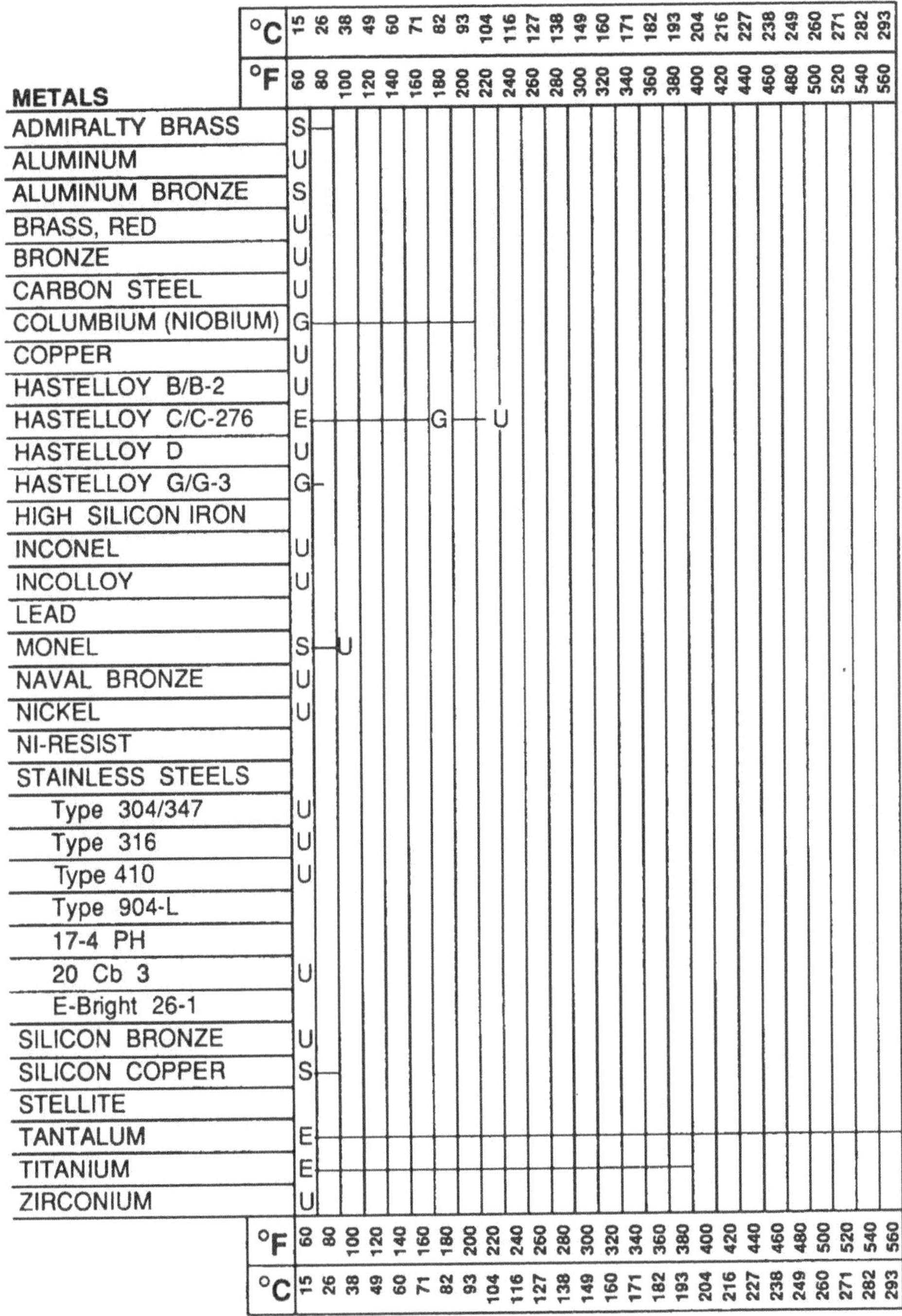

CHLORINE GAS WET

METALS	°F 60	80	100	120	140	160	180	200	220	240	260	280	300	320	340	360	380	400	420	440	460	480	500	520	540	560
°C	15	26	38	49	60	71	82	93	104	116	127	138	149	160	171	182	193	204	216	227	238	249	260	271	282	293
ADMIRALTY BRASS	S																									
ALUMINUM	U																									
ALUMINUM BRONZE	S																									
BRASS, RED	U																									
BRONZE	U																									
CARBON STEEL	U																									
COLUMBIUM (NIOBIUM)	G																									
COPPER	U																									
HASTELLOY B/B-2	U																									
HASTELLOY C/C-276	E						G			U																
HASTELLOY D	U																									
HASTELLOY G/G-3	G																									
HIGH SILICON IRON																										
INCONEL	U																									
INCOLLOY	U																									
LEAD																										
MONEL	S		U																							
NAVAL BRONZE	U																									
NICKEL	U																									
NI-RESIST																										
STAINLESS STEELS																										
Type 304/347	U																									
Type 316	U																									
Type 410	U																									
Type 904-L																										
17-4 PH																										
20 Cb 3	U																									
E-Bright 26-1																										
SILICON BRONZE	U																									
SILICON COPPER	S																									
STELLITE																										
TANTALUM	E																									
TITANIUM	E																									
ZIRCONIUM	U																									
°F	60	80	100	120	140	160	180	200	220	240	260	280	300	320	340	360	380	400	420	440	460	480	500	520	540	560
°C	15	26	38	49	60	71	82	93	104	116	127	138	149	160	171	182	193	204	216	227	238	249	260	271	282	293

FOR METALS
E = < 2 Mils Penetration/Year; G = < 20 Mils Penetration /year
S = < 50 Mils Penetration/Year; U = > 50 Mils Penetration/year

FOR NONMETALLICS
R = Resistant
U = Unsatisfactory

°C	15	26	38	49	60	71	82	93	104	116	127	138	149	160	171	182	193	204	216	227	238
°F	60	80	100	120	140	160	180	200	220	240	260	280	300	320	340	360	380	400	420	440	460

PLASTICS	
ABS	R
ACETAL	R
ACRYLICS	
ASBESTOS REINFORCED EPOXY	R
ASBESTOS REINFORCED PHENOLIC	U
CHLORINATED POLYETHER (Penton)	R – U
CPVC	U
E-CTFE (Halar)	R
EPOXY	U
ETFE (Tefzel)	R
FEP	R
FURAN (FURFURAL ALCOHOL)	R
NORYL	R – U
NYLON 6	U
NYLON 11	U
NYLON 66	U
PHENOLIC	U
PFA (Teflon)	R
POLYAMIDE-IMIDE	
POLYESTERS	
Bisphenol A–fumarate	R
Halogenated	R
Hydrogenated Bisphenol A –Bisphenol A	R
Isophthalic	R
Terephthalate (PET)	R
POLYETHERIMIDE (ULTEM)	
POLYETHERETHERKETONE (PEEK)	
POLYETHERSULFONE (PES)	
POLYETHYLENE HMW	U
POLYETHYLENE UHMW 10%	R
POLYMETHYLPENTENE	
POLYPROPYLENE (PP)	U
POLYSTYRENE	R
POLYSULFONE	
POLYVINYLIDENE CHLORIDE	R
PVC TYPE 1	R – U
PVC TYPE 2	U
PVDF (Kynar)	R
RYTON	U
TFE (Teflon)	R
VINYL ESTER	R

°F	60	80	100	120	140	160	180	200	220	240	260	280	300	320	340	360	380	400	420	440	460
°C	15	26	38	49	60	71	82	93	104	116	127	138	149	160	171	182	193	204	216	227	238

CHLORINE GAS WET

ELASTOMERS* AND LININGS	15	26	38	49	60	71	82	93	104	116	127	138	149	160	171	182	193	204	216	227	238
°F	60	80	100	120	140	160	180	200	220	240	260	280	300	320	340	360	380	400	420	440	460
ACRYLATE-BUTADIENE (ABR)																					
BUTYL GR-1 (IIR)																					
CARBOXYLIC-ACRYLONITRILE-BUTADIENE (XNBR)																					
CHEMRAZ (FPM)	R	—	—	—	—	—	—	—	—	—	—	—	—	—	—	—	—	—			
CHLORO-ISOBUTENE-ISOPRENE (CIIR)																					
HYPALON (CHLORO-SULFONYL-POLYETHYLENE (CSM)	R	—																			
ETHYLENE-ACRYLIC (EA)																					
ETHYLENE-PROPYLENE (EPM)																					
ETHYLENE-PROPYLENE-DIENE (EPDM)	U																				
ETHYLENE-PROPYLENE TERPOLYMER (EPT)	U																				
FKM (Viton A)	R	—	—	—	—	—	—	—													
HARD RUBBER	R	—	—	—	—	—	—	—													
SOFT RUBBER	U																				
ISOPRENE (IR)	R	—																			
KALREZ (FPM)																					
KOROSEAL	R	—	—	—	—	—	—	—	—												
NATURAL RUBBER (GRS)	R	—																			
NEOPRENE GR-M (CR)	U																				
NITRILE BUNA-N (NBR)	U																				
NORDEL (EPDM)	R	—	—	—	—	—	—	—	—	—	—	—	—								
POLYBUTADIENE (BR)	U																				
POLYESTER (PE)	U																				
POLYETHER-URETHANE (EU)																					
POLYISOPRENE (IR)																					
POLYSULFIDES (T)																					
POLYURETHANE (AU)	U																				
SBR STYRENE (BUNA-S)	U																				
SILICONE RUBBERS																					

°F	60	80	100	120	140	160	180	200	220	240	260	280	300	320	340	360	380	400	420	440	460
°C	15	26	38	49	60	71	82	93	104	116	127	138	149	160	171	182	193	204	216	227	238

*See also Nylon 11 under PLASTICS

METALS	°F: 60 (°C: 15)
ADMIRALTY BRASS	G
ALUMINUM	
ALUMINUM BRONZE	U
BRASS, RED	
BRONZE	
CARBON STEEL	G
COLUMBIUM (NIOBIUM)	E
COPPER	
HASTELLOY B/B-2	
HASTELLOY C/C-276	E
HASTELLOY D	
HASTELLOY G/G-3	
HIGH SILICON IRON	
INCONEL	
INCOLLOY	
LEAD	G
MONEL	G
NAVAL BRONZE	U
NICKEL	
NI-RESIST	
STAINLESS STEELS	
Type 304/347 1	G
Type 316	S
Type 410	U
Type 904-L	
17-4 PH	U
20 Cb 3	
E-Bright 26-1	
SILICON BRONZE	S
SILICON COPPER	S
STELLITE	
TANTALUM	G
TITANIUM	
ZIRCONIUM	U

°F: 60 80 100 120 140 160 180 200 220 240 260 280 300 320 340 360 380 400 420 440 460 480 500 520 540 560

°C: 15 26 38 49 60 71 82 93 104 116 127 138 149 160 171 182 193 204 216 227 238 249 260 271 282 293

FOR METALS
E = < 2 Mils Penetration/Year; G = < 20 Mils Penetration /year
S = < 50 Mils Penetration/Year; U = > 50 Mils Penetration/year

FOR NONMETALLICS
R = Resistant
U = Unsatisfactory

PLASTICS	60 °F / 15 °C	80 / 26	100 / 38	120 / 49	140 / 60	160 / 71	180 / 82	200 / 93	220 / 104	240 / 116	260 / 127	280 / 138	300 / 149	320 / 160	340 / 171	360 / 182	380 / 193	400 / 204	420 / 216	440 / 227	460 / 238
ABS	U																				
ACETAL																					
ACRYLICS																					
ASBESTOS REINFORCED EPOXY																					
ASBESTOS REINFORCED PHENOLIC																					
CHLORINATED POLYETHER (Penton)	U																				
CPVC	U																				
E-CTFE (Halar)	R	—	—	—	—	—	—	—	—	—											
EPOXY																					
ETFE (Tefzel)																					
FEP	R	—	—	—	—	—	—	—	—	—	—	—	—	—	—	—	—	—			
FURAN (FURFURAL ALCOHOL)	U																				
NORYL	U																				
NYLON 6	U																				
NYLON 11	U																				
NYLON 66	U																				
PHENOLIC																					
PFA (Teflon)																					
POLYAMIDE-IMIDE																					
POLYESTERS																					
Bisphenol A–fumarate	U																				
Halogenated	U																				
Hydrogenated Bisphenol A –Bisphenol A																					
Isophthalic	U																				
Terephthalate (PET)																					
POLYETHERIMIDE (ULTEM)																					
POLYETHERETHERKETONE (PEEK)	U																				
POLYETHERSULFONE (PES)																					
POLYETHYLENE HMW	U																				
POLYETHYLENE UHMW	U																				
POLYMETHYLPENTENE	U																				
POLYPROPYLENE (PP)	U																				
POLYSTYRENE																					
POLYSULFONE	U																				
POLYVINYLIDENE CHLORIDE	U																				
PVC TYPE 1	U																				
PVC TYPE 2	U																				
PVDF (Kynar)	R	—	—	—	—	—	—	—													
RYTON																					
TFE (Teflon)	R	—	—	—	—	—	—	—	—	—	—	—	—	—	—	—	—	—	—	—	—
VINYL ESTER	U																				

°F 60 80 100 120 140 160 180 200 220 240 260 280 300 320 340 360 380 400 420 440 460
°C 15 26 38 49 60 71 82 93 104 116 127 138 149 160 171 182 193 204 216 227 238

ELASTOMERS* AND LININGS	60°F / 15°C	80 / 26	100 / 38	120 / 49	140 / 60	160 / 71	180 / 82	200 / 93	220 / 104	240 / 116	260 / 127	280 / 138	300 / 149	320 / 160	340 / 171	360 / 182	380 / 193	400 / 204	420 / 216	440 / 227	460 / 238
ACRYLATE-BUTADIENE (ABR)																					
BUTYL GR-1 (IIR)	U																				
CARBOXYLIC-ACRYLONITRILE-BUTADIENE (XNBR)																					
CHEMRAZ (FPM)	R	—	—	—	—	—	—	—	—	—	—	—	—	—	—	—	—	—			
CHLORO-ISOBUTENE-ISOPRENE (CIIR)																					
HYPALON (CHLORO-SULFONYL-POLYETHYLENE (CSM)																					
ETHYLENE-ACRYLIC (EA)																					
ETHYLENE-PROPYLENE (EPM)																					
ETHYLENE-PROPYLENE-DIENE (EPDM)	U																				
ETHYLENE-PROPYLENE TERPOLYMER (EPT)	U																				
FKM (Viton A)	R	—	—	—	—	—	—														
HARD RUBBER	U																				
SOFT RUBBER																					
ISOPRENE (IR)	U																				
KALREZ (FPM)	R	—	—	—	—	—	—	—													
KOROSEAL																					
NATURAL RUBBER (GRS)	U																				
NEOPRENE GR-M (CR)	U																				
NITRILE BUNA-N (NBR)	U																				
NORDEL (EPDM)	U																				
POLYBUTADIENE (BR)																					
POLYESTER (PE)																					
POLYETHER-URETHANE (EU)																					
POLYISOPRENE (IR)																					
POLYSULFIDES (T)																					
POLYURETHANE (AU)																					
SBR STYRENE (BUNA-S)																					
SILICONE RUBBERS																					

*See also Nylon 11 under PLASTICS

CHROMIC ACID 10%

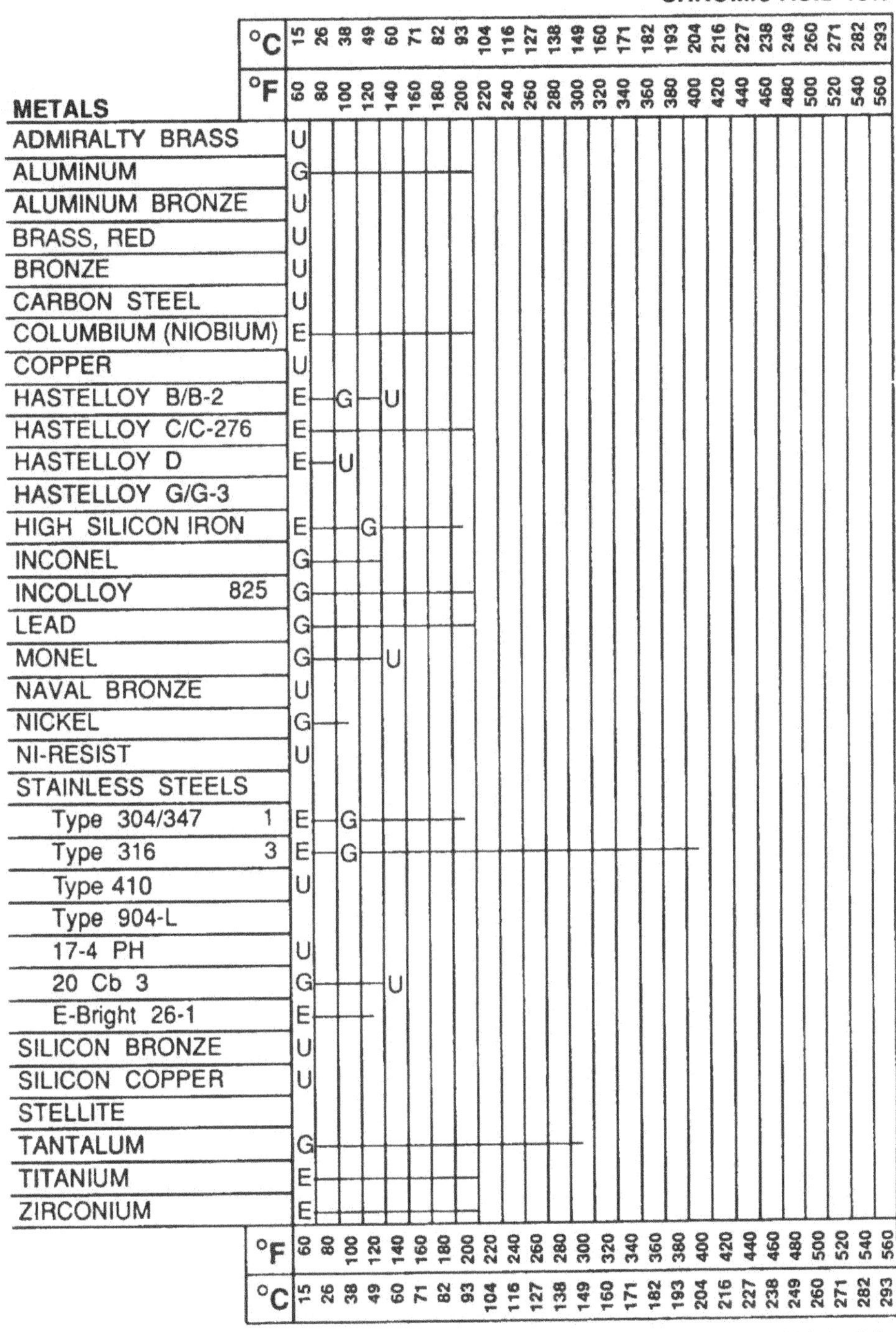

°C	15	26	38	49	60	71	82	93	104	116	127	138	149	160	171	182	193	204	216	227	238	249	260	271	282	293
METALS °F	60	80	100	120	140	160	180	200	220	240	260	280	300	320	340	360	380	400	420	440	460	480	500	520	540	560
ADMIRALTY BRASS	U																									
ALUMINUM	G	—	—	—	—	—	—	—																		
ALUMINUM BRONZE	U																									
BRASS, RED	U																									
BRONZE	U																									
CARBON STEEL	U																									
COLUMBIUM (NIOBIUM)	E	—	—	—	—	—	—	—																		
COPPER	U																									
HASTELLOY B/B-2	E	—	G	—	U																					
HASTELLOY C/C-276	E	—	—	—	—	—	—	—																		
HASTELLOY D	E	—	U																							
HASTELLOY G/G-3																										
HIGH SILICON IRON	E	—	—	G	—	—	—	—																		
INCONEL	G	—	—	—																						
INCOLLOY 825	G	—	—	—	—	—	—	—																		
LEAD	G	—	—	—	—	—	—	—																		
MONEL	G	—	—	—	U																					
NAVAL BRONZE	U																									
NICKEL	G	—	—																							
NI-RESIST	U																									
STAINLESS STEELS																										
Type 304/347 1	E	—	G	—	—	—	—	—																		
Type 316 3	E	—	G	—	—	—	—	—	—	—	—	—	—	—	—	—	—	—								
Type 410	U																									
Type 904-L																										
17-4 PH	U																									
20 Cb 3	G	—	—	—	U																					
E-Bright 26-1	E	—	—	—																						
SILICON BRONZE	U																									
SILICON COPPER	U																									
STELLITE																										
TANTALUM	G	—	—	—	—	—	—	—	—	—	—	—	—													
TITANIUM	E	—	—	—	—	—	—	—																		
ZIRCONIUM	E	—	—	—	—	—	—	—																		
°F	60	80	100	120	140	160	180	200	220	240	260	280	300	320	340	360	380	400	420	440	460	480	500	520	540	560
°C	15	26	38	49	60	71	82	93	104	116	127	138	149	160	171	182	193	204	216	227	238	249	260	271	282	293

FOR METALS

E = < 2 Mils Penetration/Year; G = < 20 Mils Penetration /year

S = < 50 Mils Penetration/Year; U = > 50 Mils Penetration/year

FOR NONMETALLICS

R = Resistant

U = Unsatisfactory

CHROMIC ACID 10%

PLASTICS	60 °F (15 °C)	Notes
ABS	R	
ACETAL	R	
ACRYLICS	U	
ASBESTOS REINFORCED EPOXY	U	
ASBESTOS REINFORCED PHENOLIC	U	
CHLORINATED POLYETHER (Penton)	R	
CPVC	R	
E-CTFE (Halar)	R	
EPOXY	R	U at 120 °F (49 °C)
ETFE (Tefzel)	R	
FEP	R	
FURAN (FURFURAL ALCOHOL)	R	
NORYL	U	
NYLON 6	U	
NYLON 11	U	
NYLON 66	U	
PHENOLIC	U	
PFA (Teflon)	R	
POLYAMIDE-IMIDE	R	
POLYESTERS		
Bisphenol A–fumarate	U	
Halogenated	R	
Hydrogenated Bisphenol A –Bisphenol A		
Isophthalic	U	
Terephthalate (PET)	R	
POLYETHERIMIDE (ULTEM)		
POLYETHERETHERKETONE (PEEK)	R	
POLYETHERSULFONE (PES)	U	
POLYETHYLENE HMW	R	
POLYETHYLENE UHMW	R	
POLYMETHYLPENTENE	R	
POLYPROPYLENE (PP)	R	
POLYSTYRENE	U	
POLYSULFONE 12%	R	
POLYVINYLIDENE CHLORIDE	R	
PVC TYPE 1	R	
PVC TYPE 2	R	
PVDF (Kynar)	R	
RYTON	R	
TFE (Teflon)	R	
VINYL ESTER	R	

Temperature scale: °F 60, 80, 100, 120, 140, 160, 180, 200, 220, 240, 260, 280, 300, 320, 340, 360, 380, 400, 420, 440, 460; °C 15, 26, 38, 49, 60, 71, 82, 93, 104, 116, 127, 138, 149, 160, 171, 182, 193, 204, 216, 227, 238

CHROMIC ACID 10%

°C: 15, 26, 38, 49, 60, 71, 82, 93, 104, 116, 127, 138, 149, 160, 171, 182, 193, 204, 216, 227, 238

°F: 60, 80, 100, 120, 140, 160, 180, 200, 220, 240, 260, 280, 300, 320, 340, 360, 380, 400, 420, 440, 460

ELASTOMERS* AND LININGS	
ACRYLATE-BUTADIENE (ABR)	
BUTYL GR-1 (IIR)	R
CARBOXYLIC-ACRYLONITRILE-BUTADIENE (XNBR)	U
CHEMRAZ (FPM)	R
CHLORO-ISOBUTENE-ISOPRENE (CIIR)	U
HYPALON (CHLORO-SULFONYL-POLYETHYLENE (CSM)	R
ETHYLENE-ACRYLIC (EA)	
ETHYLENE-PROPYLENE (EPM)	
ETHYLENE-PROPYLENE-DIENE (EPDM)	R
ETHYLENE-PROPYLENE TERPOLYMER (EPT)	U
FKM (Viton A)	R
HARD RUBBER	U
SOFT RUBBER	U
ISOPRENE (IR)	U
KALREZ (FPM)	R
KOROSEAL	R
NATURAL RUBBER (GRS)	U
NEOPRENE GR-M (CR)	R
NITRILE BUNA-N (NBR)	R
NORDEL (EPDM)	
POLYBUTADIENE (BR)	U
POLYESTER (PE)	
POLYETHER-URETHANE (EU)	U
POLYISOPRENE (IR)	
POLYSULFIDES (T)	U
POLYURETHANE (AU)	U
SBR STYRENE (BUNA-S)	U
SILICONE RUBBERS	R

°F: 60, 80, 100, 120, 140, 160, 180, 200, 220, 240, 260, 280, 300, 320, 340, 360, 380, 400, 420, 440, 460

°C: 15, 26, 38, 49, 60, 71, 82, 93, 104, 116, 127, 138, 149, 160, 171, 182, 193, 204, 216, 227, 238

*See also Nylon 11 under PLASTICS

CHROMIC ACID 50%

METALS	60	80	100	120	140	160	180	200	220	240	260	280	300	320	340	360	380	400	420	440	460	480	500	520	540	560
°C	15	26	38	49	60	71	82	93	104	116	127	138	149	160	171	182	193	204	216	227	238	249	260	271	282	293
ADMIRALTY BRASS	U																									
ALUMINUM	G																									
ALUMINUM BRONZE	U																									
BRASS, RED	U																									
BRONZE	U																									
CARBON STEEL	U																									
COLUMBIUM (NIOBIUM)	E																									
COPPER	U																									
HASTELLOY B/B-2	U																									
HASTELLOY C/C-276	G																									
HASTELLOY D	U																									
HASTELLOY G/G-3																										
HIGH SILICON IRON	E			G																						
INCONEL	G																									
INCOLLOY	U																									
LEAD	G																									
MONEL	U																									
NAVAL BRONZE	U																									
NICKEL	U																									
NI-RESIST	U																									
STAINLESS STEELS																										
Type 304/347	G		U																							
Type 316 3	G					S																				
Type 410	U																									
Type 904-L																										
17-4 PH	U																									
20 Cb 3	G																									
E-Bright 26-1	U																									
SILICON BRONZE	U																									
SILICON COPPER	U																									
STELLITE																										
TANTALUM	E																									
TITANIUM	E																									
ZIRCONIUM	E																									

FOR METALS
E = < 2 Mils Penetration/Year; G = < 20 Mils Penetration /year
S = < 50 Mils Penetration/Year; U = > 50 Mils Penetration/year

FOR NONMETALLICS
R = Resistant
U = Unsatisfactory

CHROMIC ACID 50%

°C: 15, 26, 38, 49, 60, 71, 82, 93, 104, 116, 127, 138, 149, 160, 171, 182, 193, 204, 216, 227, 238

°F: 60, 80, 100, 120, 140, 160, 180, 200, 220, 240, 260, 280, 300, 320, 340, 360, 380, 400, 420, 440, 460

PLASTICS	
ABS	U
ACETAL	R
ACRYLICS	U
ASBESTOS REINFORCED EPOXY	U
ASBESTOS REINFORCED PHENOLIC	U
CHLORINATED POLYETHER (Penton)	R
CPVC	R
E-CTFE (Halar)	R
EPOXY	U
ETFE (Tefzel)	R
FEP	R
FURAN (FURFURAL ALCOHOL)	U
NORYL	U
NYLON 6	U
NYLON 11	U
NYLON 66	U
PHENOLIC	U
PFA (Teflon)	R
POLYAMIDE-IMIDE	
POLYESTERS	
Bisphenol A–fumarate	U
Halogenated	R
Hydrogenated Bisphenol A –Bisphenol A	U
Isophthalic	U
Terephthalate (PET)	R
POLYETHERIMIDE (ULTEM)	
POLYETHERETHERKETONE (PEEK)	R
POLYETHERSULFONE (PES)	U
POLYETHYLENE HMW	R
POLYETHYLENE UHMW	R
POLYMETHYLPENTENE	
POLYPROPYLENE (PP)	R
POLYSTYRENE	U
POLYSULFONE	U
POLYVINYLIDENE CHLORIDE	R
PVC TYPE 1	U
PVC TYPE 2	U
PVDF (Kynar)	R
RYTON	R
TFE (Teflon)	R
VINYL ESTER	U

°F: 60, 80, 100, 120, 140, 160, 180, 200, 220, 240, 260, 280, 300, 320, 340, 360, 380, 400, 420, 440, 460

°C: 15, 26, 38, 49, 60, 71, 82, 93, 104, 116, 127, 138, 149, 160, 171, 182, 193, 204, 216, 227, 238

CHROMIC ACID 50%

ELASTOMERS* AND LININGS	Rating
ACRYLATE-BUTADIENE (ABR)	
BUTYL GR-1 (IIR)	U
CARBOXYLIC-ACRYLONITRILE-BUTADIENE (XNBR)	U
CHEMRAZ (FPM)	R
CHLORO-ISOBUTENE-ISOPRENE (CIIR)	U
HYPALON (CHLORO-SULFONYL-POLYETHYLENE (CSM)	R
ETHYLENE-ACRYLIC (EA)	
ETHYLENE-PROPYLENE (EPM)	U
ETHYLENE-PROPYLENE-DIENE (EPDM)	U
ETHYLENE-PROPYLENE TERPOLYMER (EPT)	U
FKM (Viton A)	R
HARD RUBBER	U
SOFT RUBBER	U
ISOPRENE (IR)	U
KALREZ (FPM)	R
KOROSEAL	U
NATURAL RUBBER (GRS)	U
NEOPRENE GR-M (CR)	R
NITRILE BUNA-N (NBR)	R
NORDEL (EPDM)	
POLYBUTADIENE (BR)	U
POLYESTER (PE)	
POLYETHER-URETHANE (EU)	U
POLYISOPRENE (IR)	
POLYSULFIDES (T)	U
POLYURETHANE (AU)	U
SBR STYRENE (BUNA-S)	U
SILICONE RUBBERS	

°F	60	80	100	120	140	160	180	200	220	240	260	280	300	320	340	360	380	400	420	440	460
°C	15	26	38	49	60	71	82	93	104	116	127	138	149	160	171	182	193	204	216	227	238

*See also Nylon 11 under PLASTICS

FERRIC CHLORIDE

METALS	60	80	100	120	140	160	180	200	220	240	260	280	300	320	340	360	380	400	420	440	460	480	500	520	540	560
°C	15	26	38	49	60	71	82	93	104	116	127	138	149	160	171	182	193	204	216	227	238	249	260	271	282	293
ADMIRALTY BRASS	U																									
ALUMINUM	U																									
ALUMINUM BRONZE	U																									
BRASS, RED	U																									
BRONZE	G																									
CARBON STEEL	U																									
COLUMBIUM (NIOBIUM)	E																									
COPPER	G																									
HASTELLOY B/B-2	G		U																							
HASTELLOY C/C-276	G		U																							
HASTELLOY D	G		U																							
HASTELLOY G/G-3																										
HIGH SILICON IRON	U																									
INCONEL	U																									
INCOLLOY	U																									
LEAD	U																									
MONEL	U																									
NAVAL BRONZE	U																									
NICKEL	U																									
NI-RESIST																										
STAINLESS STEELS																										
Type 304/347	U																									
Type 316	U																									
Type 410	U																									
Type 904-L																										
17-4 PH	U																									
20 Cb 3	U																									
E-Bright 26-1	E																									
SILICON BRONZE	U																									
SILICON COPPER	U																									
STELLITE	G																									
TANTALUM	E																									
TITANIUM	E																									
ZIRCONIUM	U																									
°F	60	80	100	120	140	160	180	200	220	240	260	280	300	320	340	360	380	400	420	440	460	480	500	520	540	560
°C	15	26	38	49	60	71	82	93	104	116	127	138	149	160	171	182	193	204	216	227	238	249	260	271	282	293

FOR METALS

E = < 2 Mils Penetration/Year; G = < 20 Mils Penetration /year

S = < 50 Mils Penetration/Year; U = > 50 Mils Penetration/year

FOR NONMETALLICS

R = Resistant

U = Unsatisfactory

PLASTICS	Rating	Bar extends to °F (°C)
ABS	R	140 (60)
ACETAL	U	
ACRYLICS	R	~70
ASBESTOS REINFORCED EPOXY	R	~250
ASBESTOS REINFORCED PHENOLIC	R	300 (149)
CHLORINATED POLYETHER (Penton)	R	~250
CPVC	R	~210
E-CTFE (Halar)	R	300 (149)
EPOXY	R	300 (149)
ETFE (Tefzel)		
FEP	R	400 (204)
FURAN (FURFURAL ALCOHOL)	R	260 (127)
NORYL	R	140 (60)
NYLON 6	U	
NYLON 11	U	
NYLON 66	U	
PHENOLIC	R	300 (149)
PFA (Teflon)	R	200 (93)
POLYAMIDE-IMIDE		
POLYESTERS		
Bisphenol A–fumarate	R	220 (104)
Halogenated	R	~250
Hydrogenated Bisphenol A –Bisphenol A	R	~210
Isophthalic	R	180 (82)
Terephthalate (PET)	R	~250
POLYETHERIMIDE (ULTEM)		
POLYETHERETHERKETONE (PEEK)		
POLYETHERSULFONE (PES)		
POLYETHYLENE HMW	R	140 (60)
POLYETHYLENE UHMW	R	140 (60)
POLYMETHYLPENTENE		
POLYPROPYLENE (PP)	R; U	R to ~210; U at 220 (104)
POLYSTYRENE	R	160 (71)
POLYSULFONE	R	200 (93)
POLYVINYLIDENE CHLORIDE	R	140 (60)
PVC TYPE 1	R	140 (60)
PVC TYPE 2	R	140 (60)
PVDF (Kynar)	R	280 (138)
RYTON	R	~210
TFE (Teflon)	R	460 (238)
VINYL ESTER	R	~210

FERRIC CHLORIDE

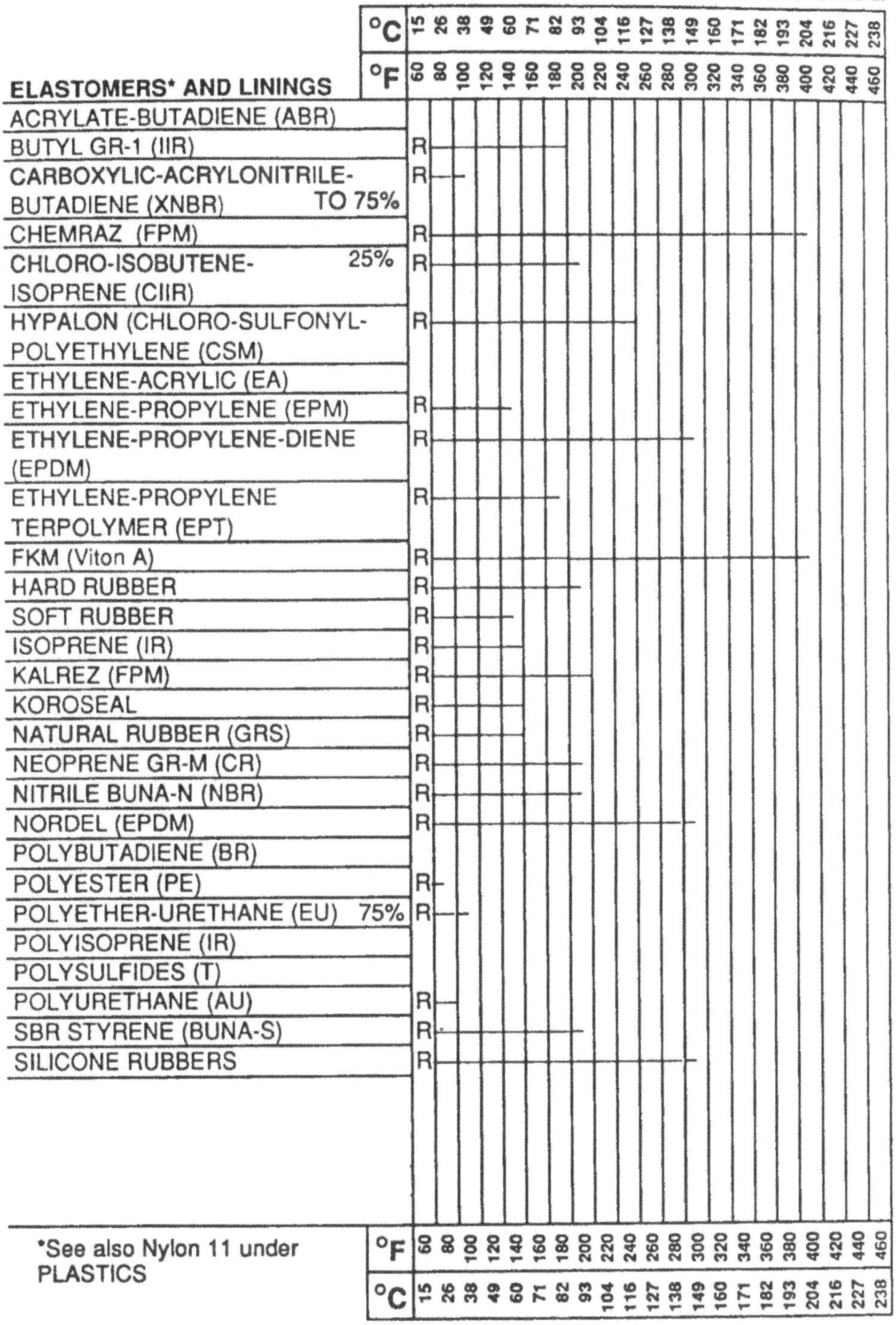

ELASTOMERS* AND LININGS	Concentration	Rating	Max. temp. (°F)
ACRYLATE-BUTADIENE (ABR)			
BUTYL GR-1 (IIR)		R	190
CARBOXYLIC-ACRYLONITRILE-BUTADIENE (XNBR)	TO 75%	R	100
CHEMRAZ (FPM)		R	400
CHLORO-ISOBUTENE-ISOPRENE (CIIR)	25%	R	200
HYPALON (CHLORO-SULFONYL-POLYETHYLENE (CSM)		R	250
ETHYLENE-ACRYLIC (EA)			
ETHYLENE-PROPYLENE (EPM)		R	140
ETHYLENE-PROPYLENE-DIENE (EPDM)		R	300
ETHYLENE-PROPYLENE TERPOLYMER (EPT)		R	180
FKM (Viton A)		R	400
HARD RUBBER		R	200
SOFT RUBBER		R	140
ISOPRENE (IR)		R	150
KALREZ (FPM)		R	210
KOROSEAL		R	150
NATURAL RUBBER (GRS)		R	150
NEOPRENE GR-M (CR)		R	200
NITRILE BUNA-N (NBR)		R	200
NORDEL (EPDM)		R	300
POLYBUTADIENE (BR)			
POLYESTER (PE)		R	80
POLYETHER-URETHANE (EU)	75%	R	100
POLYISOPRENE (IR)			
POLYSULFIDES (T)			
POLYURETHANE (AU)		R	90
SBR STYRENE (BUNA-S)		R	200
SILICONE RUBBERS		R	300

Temperature scale: °F 60, 80, 100, 120, 140, 160, 180, 200, 220, 240, 260, 280, 300, 320, 340, 360, 380, 400, 420, 440, 460; °C 15, 26, 38, 49, 60, 71, 82, 93, 104, 116, 127, 138, 149, 160, 171, 182, 193, 204, 216, 227, 238

*See also Nylon 11 under PLASTICS

FERRIC CHLORIDE 50% IN WATER

METALS																										
°C	15	26	38	49	60	71	82	93	104	116	127	138	149	160	171	182	193	204	216	227	238	249	260	271	282	293
°F	60	80	100	120	140	160	180	200	220	240	260	280	300	320	340	360	380	400	420	440	460	480	500	520	540	560
ADMIRALTY BRASS	U																									
ALUMINUM	U																									
ALUMINUM BRONZE	U																									
BRASS, RED	U																									
BRONZE	U																									
CARBON STEEL	U																									
COLUMBIUM (NIOBIUM)																										
COPPER	U																									
HASTELLOY B/B-2	U																									
HASTELLOY C/C-276 10%	G																									
HASTELLOY D																										
HASTELLOY G/G-3																										
HIGH SILICON IRON																										
INCONEL	S																									
INCOLLOY 5%	U																									
LEAD	U																									
MONEL	U																									
NAVAL BRONZE	U																									
NICKEL	U																									
NI-RESIST																										
STAINLESS STEELS																										
Type 304/347	U																									
Type 316	U																									
Type 410	U																									
Type 904-L 10%	G																									
17-4 PH																										
20 Cb 3	U																									
E-Bright 26-1																										
SILICON BRONZE	U																									
SILICON COPPER	U																									
STELLITE																										
TANTALUM	E																									
TITANIUM	E																									
ZIRCONIUM	U																									
°F	60	80	100	120	140	160	180	200	220	240	260	280	300	320	340	360	380	400	420	440	460	480	500	520	540	560
°C	15	26	38	49	60	71	82	93	104	116	127	138	149	160	171	182	193	204	216	227	238	249	260	271	282	293

FOR METALS

E = < 2 Mils Penetration/Year; G = < 20 Mils Penetration /year

S = < 50 Mils Penetration/Year; U = > 50 Mils Penetration/year

FOR NONMETALLICS

R = Resistant

U = Unsatisfactory

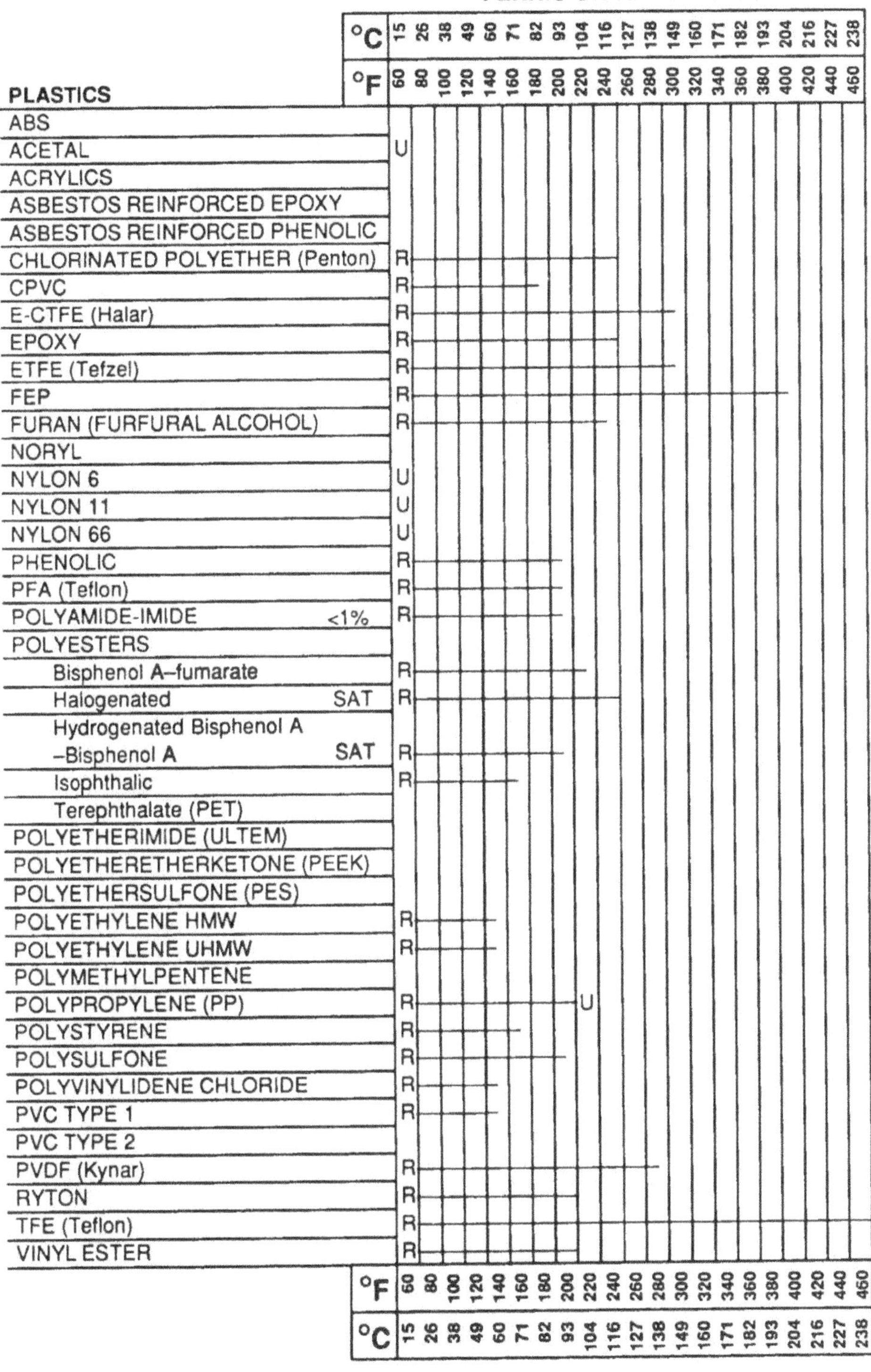

FERRIC CHLORIDE 50% IN WATER

°F	60	80	100	120	140	160	180	200	220	240	260	280	300	320	340	360	380	400	420	440	460
°C	15	26	38	49	60	71	82	93	104	116	127	138	149	160	171	182	193	204	216	227	238

PLASTICS		Rating	Line extends to (°F)
ABS			
ACETAL		U	
ACRYLICS			
ASBESTOS REINFORCED EPOXY			
ASBESTOS REINFORCED PHENOLIC			
CHLORINATED POLYETHER (Penton)		R	250
CPVC		R	180
E-CTFE (Halar)		R	300
EPOXY		R	250
ETFE (Tefzel)		R	300
FEP		R	400
FURAN (FURFURAL ALCOHOL)		R	240
NORYL			
NYLON 6		U	
NYLON 11		U	
NYLON 66		U	
PHENOLIC		R	200
PFA (Teflon)		R	200
POLYAMIDE-IMIDE	<1%	R	200
POLYESTERS			
Bisphenol A–fumarate		R	220
Halogenated	SAT	R	250
Hydrogenated Bisphenol A –Bisphenol A	SAT	R	200
Isophthalic		R	160
Terephthalate (PET)			
POLYETHERIMIDE (ULTEM)			
POLYETHERETHERKETONE (PEEK)			
POLYETHERSULFONE (PES)			
POLYETHYLENE HMW		R	140
POLYETHYLENE UHMW		R	140
POLYMETHYLPENTENE			
POLYPROPYLENE (PP)		R	210, then U at 220
POLYSTYRENE		R	160
POLYSULFONE		R	200
POLYVINYLIDENE CHLORIDE		R	140
PVC TYPE 1		R	140
PVC TYPE 2			
PVDF (Kynar)		R	280
RYTON		R	210
TFE (Teflon)		R	450
VINYL ESTER		R	210

FERRIC CHLORIDE 50% IN WATER

Temperature scale (°C): 15, 26, 38, 49, 60, 71, 82, 93, 104, 116, 127, 138, 149, 160, 171, 182, 193, 204, 216, 227, 238

Temperature scale (°F): 60, 80, 100, 120, 140, 160, 180, 200, 220, 240, 260, 280, 300, 320, 340, 360, 380, 400, 420, 440, 460

ELASTOMERS* AND LININGS	Rating	Rating line extends to (approx. °F)
ACRYLATE-BUTADIENE (ABR)		
BUTYL GR-1 (IIR)	R	160
CARBOXYLIC-ACRYLONITRILE-BUTADIENE (XNBR)	R	100
CHEMRAZ (FPM)	R	400
CHLORO-ISOBUTENE-ISOPRENE (CIIR) TO 25%	R	200
HYPALON (CHLORO-SULFONYL-POLYETHYLENE (CSM)	R	250
ETHYLENE-ACRYLIC (EA)		
ETHYLENE-PROPYLENE (EPM)		
ETHYLENE-PROPYLENE-DIENE (EPDM)	R	300
ETHYLENE-PROPYLENE TERPOLYMER (EPT)	R	180
FKM (Viton A)	R	400
HARD RUBBER	R	200
SOFT RUBBER	R	140
ISOPRENE (IR)	R	150
KALREZ (FPM)	R	210
KOROSEAL	R	150
NATURAL RUBBER (GRS)	R	150
NEOPRENE GR-M (CR)	R	160
NITRILE BUNA-N (NBR)	R	180
NORDEL (EPDM)	R	300
POLYBUTADIENE (BR)		
POLYESTER (PE)	R	80
POLYETHER-URETHANE (EU)	R	100
POLYISOPRENE (IR)		
POLYSULFIDES (T)		
POLYURETHANE (AU)	R	80
SBR STYRENE (BUNA-S)	R	200
SILICONE RUBBERS	R	300

*See also Nylon 11 under PLASTICS

HYDROBROMIC ACID 20%

METALS °F	60	80	100	120	140	160	180	200	220	240	260	280	300	320	340	360	380	400	420	440	460	480	500	520	540	560
°C	15	26	38	49	60	71	82	93	104	116	127	138	149	160	171	182	193	204	216	227	238	249	260	271	282	293
ADMIRALTY BRASS	U																									
ALUMINUM	U																									
ALUMINUM BRONZE	U																									
BRASS, RED	U																									
BRONZE	U																									
CARBON STEEL	U																									
COLUMBIUM (NIOBIUM)																										
COPPER	U																									
HASTELLOY B/B-2	G	—	—	—	—	—	—	—																		
HASTELLOY C/C-276	E	—																								
HASTELLOY D																										
HASTELLOY G/G-3																										
HIGH SILICON IRON	U																									
INCONEL	G	U																								
INCOLLOY	U																									
LEAD	U																									
MONEL	U																									
NAVAL BRONZE	U																									
NICKEL	U																									
NI-RESIST																										
STAINLESS STEELS																										
Type 304/347	U																									
Type 316	U																									
Type 410	U																									
Type 904-L																										
17-4 PH	U																									
20 Cb 3	U																									
E-Bright 26-1																										
SILICON BRONZE	U																									
SILICON COPPER	U																									
STELLITE	U																									
TANTALUM	E	—	—	—	—	—	—	—	—	—	—	—	—													
TITANIUM	E	—	—	—	—	—	—	—																		
ZIRCONIUM	U																									

FOR METALS
E = < 2 Mils Penetration/Year; G = < 20 Mils Penetration /year
S = < 50 Mils Penetration/Year; U = > 50 Mils Penetration/year

FOR NONMETALLICS
R = Resistant
U = Unsatisfactory

PLASTICS	
ABS	R
ACETAL	
ACRYLICS	
ASBESTOS REINFORCED EPOXY	
ASBESTOS REINFORCED PHENOLIC	
CHLORINATED POLYETHER (Penton)	R
CPVC	R
E-CTFE (Halar)	R
EPOXY	R
ETFE (Tefzel)	R
FEP	R
FURAN (FURFURAL ALCOHOL)	R
NORYL	R
NYLON 6	U
NYLON 11	U
NYLON 66	U
PHENOLIC	R
PFA (Teflon)	R
POLYAMIDE-IMIDE	
POLYESTERS	
Bisphenol A–fumarate	R
Halogenated	R
Hydrogenated Bisphenol A –Bisphenol A	R
Isophthalic	R
Terephthalate (PET)	R
POLYETHERIMIDE (ULTEM)	
POLYETHERETHERKETONE (PEEK)	U
POLYETHERSULFONE (PES)	
POLYETHYLENE HMW	R
POLYETHYLENE UHMW	R
POLYMETHYLPENTENE	
POLYPROPYLENE (PP)	R
POLYSTYRENE	
POLYSULFONE	R
POLYVINYLIDENE CHLORIDE	R
PVC TYPE 1	R
PVC TYPE 2	R
PVDF (Kynar)	R
RYTON	R
TFE (Teflon)	R
VINYL ESTER	R

°C: 15, 26, 38, 49, 60, 71, 82, 93, 104, 116, 127, 138, 149, 160, 171, 182, 193, 204, 216, 227, 238

°F: 60, 80, 100, 120, 140, 160, 180, 200, 220, 240, 260, 280, 300, 320, 340, 360, 380, 400, 420, 440, 460

HYDROBROMIC ACID 20%

Temperature scale: °F 60, 80, 100, 120, 140, 160, 180, 200, 220, 240, 260, 280, 300, 320, 340, 360, 380, 400, 420, 440, 460 / °C 15, 26, 38, 49, 60, 71, 82, 93, 104, 116, 127, 138, 149, 160, 171, 182, 193, 204, 216, 227, 238

ELASTOMERS* AND LININGS	Rating	Chart line extends to (°F / °C)
ACRYLATE-BUTADIENE (ABR)		
BUTYL GR-1 (IIR)	R	160 / 71
CARBOXYLIC-ACRYLONITRILE-BUTADIENE (XNBR)		
CHEMRAZ (FPM)	R	400 / 204
CHLORO-ISOBUTENE-ISOPRENE (CIIR)	R	120 / 49
HYPALON (CHLORO-SULFONYL-POLYETHYLENE (CSM)	R	100 / 38
ETHYLENE-ACRYLIC (EA)		
ETHYLENE-PROPYLENE (EPM)	R	140 / 60
ETHYLENE-PROPYLENE-DIENE (EPDM)	R	300 / 149
ETHYLENE-PROPYLENE TERPOLYMER (EPT)	R	140 / 60
FKM (Viton A)	R	400 / 204
HARD RUBBER	R	200 / 93
SOFT RUBBER	R	140 / 60
ISOPRENE (IR)	R	between 100 and 120
KALREZ (FPM)	R	between 200 and 220
KOROSEAL	R	between 80 and 100
NATURAL RUBBER (GRS)	R	between 100 and 120
NEOPRENE GR-M (CR)	U	
NITRILE BUNA-N (NBR)	U	
NORDEL (EPDM)		
POLYBUTADIENE (BR)		
POLYESTER (PE)		
POLYETHER-URETHANE (EU)	U	
POLYISOPRENE (IR)		
POLYSULFIDES (T)		
POLYURETHANE (AU)		
SBR STYRENE (BUNA-S)		
SILICONE RUBBERS	U	

*See also Nylon 11 under PLASTICS

HYDROBROMIC ACID 50%

METALS	60	80	100	120	140	160	180	200	220	240	260	280	300	320	340	360	380	400	420	440	460	480	500	520	540	560
°C	15	26	38	49	60	71	82	93	104	116	127	138	149	160	171	182	193	204	216	227	238	249	260	271	282	293
ADMIRALTY BRASS	U																									
ALUMINUM	U																									
ALUMINUM BRONZE	U																									
BRASS, RED	U																									
BRONZE	U																									
CARBON STEEL	U																									
COLUMBIUM (NIOBIUM)																										
COPPER	U																									
HASTELLOY B/B-2	G	—	—	—	—	—	—	—	—	—	—															
HASTELLOY C/C-276	G	—																								
HASTELLOY D																										
HASTELLOY G/G-3																										
HIGH SILICON IRON	U																									
INCONEL	U																									
INCOLLOY	U																									
LEAD	U																									
MONEL	U																									
NAVAL BRONZE	U																									
NICKEL	U																									
NI-RESIST																										
STAINLESS STEELS																										
Type 304/347	U																									
Type 316	U																									
Type 410	U																									
Type 904-L																										
17-4 PH	U																									
20 Cb 3	U																									
E-Bright 26-1																										
SILICON BRONZE	U																									
SILICON COPPER	U																									
STELLITE	U																									
TANTALUM	E	—	—	—	—	—	—	—	—	—	—	—	—													
TITANIUM	E	—	—	—	—	—	—	—																		
ZIRCONIUM	U																									
°F	60	80	100	120	140	160	180	200	220	240	260	280	300	320	340	360	380	400	420	440	460	480	500	520	540	560
°C	15	26	38	49	60	71	82	93	104	116	127	138	149	160	171	182	193	204	216	227	238	249	260	271	282	293

FOR METALS
E = < 2 Mils Penetration/Year; G = < 20 Mils Penetration /year
S = < 50 Mils Penetration/Year; U = > 50 Mils Penetration/year

FOR NONMETALLICS
R = Resistant
U = Unsatisfactory

HYDROBROMIC ACID 50%

PLASTICS	°C 15	26	38	49	60	71	82	93	104	116	127	138	149	160	171	182	193	204	216	227	238
	°F 60	80	100	120	140	160	180	200	220	240	260	280	300	320	340	360	380	400	420	440	460
ABS																					
ACETAL																					
ACRYLICS																					
ASBESTOS REINFORCED EPOXY																					
ASBESTOS REINFORCED PHENOLIC																					
CHLORINATED POLYETHER (Penton)	R	—	—	—	—	—	—	—	—	—											
CPVC	R	—	—	—	—	—	—														
E-CTFE (Halar)	R	—	—	—	—	—	—	—	—	—	—	—	—								
EPOXY	R	—	—	U																	
ETFE (Tefzel)	R	—	—	—	—	—	—	—	—	—	—	—	—								
FEP	R	—	—	—	—	—	—	—	—	—	—	—	—	—	—	—	—	—			
FURAN (FURFURAL ALCOHOL)	R	—	—	—	—	—	—	—	—	—	—										
NORYL	R	—																			
NYLON 6	U																				
NYLON 11	U																				
NYLON 66	U																				
PHENOLIC	R	—	—	—	—	—	—	—													
PFA (Teflon)	R	—	—	—	—	—	—	—													
POLYAMIDE-IMIDE																					
POLYESTERS																					
Bisphenol A–fumarate	R	—	—	—	—	—															
Halogenated	R	—	—	—	—	—	—	—													
Hydrogenated Bisphenol A –Bisphenol A	R	—																			
Isophthalic	R	—	—	—	—																
Terephthalate (PET)	R	—	—	—	—	—	—	—	—	—											
POLYETHERIMIDE (ULTEM)																					
POLYETHERETHERKETONE (PEEK)	U																				
POLYETHERSULFONE (PES)																					
POLYETHYLENE HMW	R	—	—	—	—																
POLYETHYLENE UHMW	R	—	—	—	—																
POLYMETHYLPENTENE																					
POLYPROPYLENE (PP)	R	—	—	—	—	—	—	U													
POLYSTYRENE																					
POLYSULFONE																					
POLYVINYLIDENE CHLORIDE	R	—	—	—																	
PVC TYPE 1	R	—	—	—	—																
PVC TYPE 2	R	—	—	—	—																
PVDF (Kynar)	R	—	—	—	—	—	—	—	—	—	—	—									
RYTON	R	—	—	—	—	—	—	—													
TFE (Teflon)	R	—	—	—	—	—	—	—	—	—	—	—	—	—	—	—	—	—	—	—	—
VINYL ESTER	R	—	—	—	—	—	—	—													

HYDROBROMIC ACID 50%

ELASTOMERS* AND LININGS	Rating	Max. temp. °F (°C)
ACRYLATE-BUTADIENE (ABR)		
BUTYL GR-1 (IIR)	R	110
CARBOXYLIC-ACRYLONITRILE-BUTADIENE (XNBR)		
CHEMRAZ (FPM)	R	400 (204)
CHLORO-ISOBUTENE-ISOPRENE (CIIR)	R	120 (49)
HYPALON (CHLORO-SULFONYL-POLYETHYLENE (CSM)	R	100 (38)
ETHYLENE-ACRYLIC (EA)		
ETHYLENE-PROPYLENE (EPM)		
ETHYLENE-PROPYLENE-DIENE (EPDM)	R	300 (149)
ETHYLENE-PROPYLENE TERPOLYMER (EPT)	R	140 (60)
FKM (Viton A)	R	400 (204)
HARD RUBBER	R	200 (93)
SOFT RUBBER	R	140 (60)
ISOPRENE (IR)	R	150
KALREZ (FPM)	R	210
KOROSEAL	R	90
NATURAL RUBBER (GRS)	R	150
NEOPRENE GR-M (CR)	U	
NITRILE BUNA-N (NBR)	U	
NORDEL (EPDM)		
POLYBUTADIENE (BR)		
POLYESTER (PE)		
POLYETHER-URETHANE (EU)	U	
POLYISOPRENE (IR)		
POLYSULFIDES (T)		
POLYURETHANE (AU)		
SBR STYRENE (BUNA-S)		
SILICONE RUBBERS	U	

Temperature scale: °F 60, 80, 100, 120, 140, 160, 180, 200, 220, 240, 260, 280, 300, 320, 340, 360, 380, 400, 420, 440, 460; °C 15, 26, 38, 49, 60, 71, 82, 93, 104, 116, 127, 138, 149, 160, 171, 182, 193, 204, 216, 227, 238

*See also Nylon 11 under PLASTICS

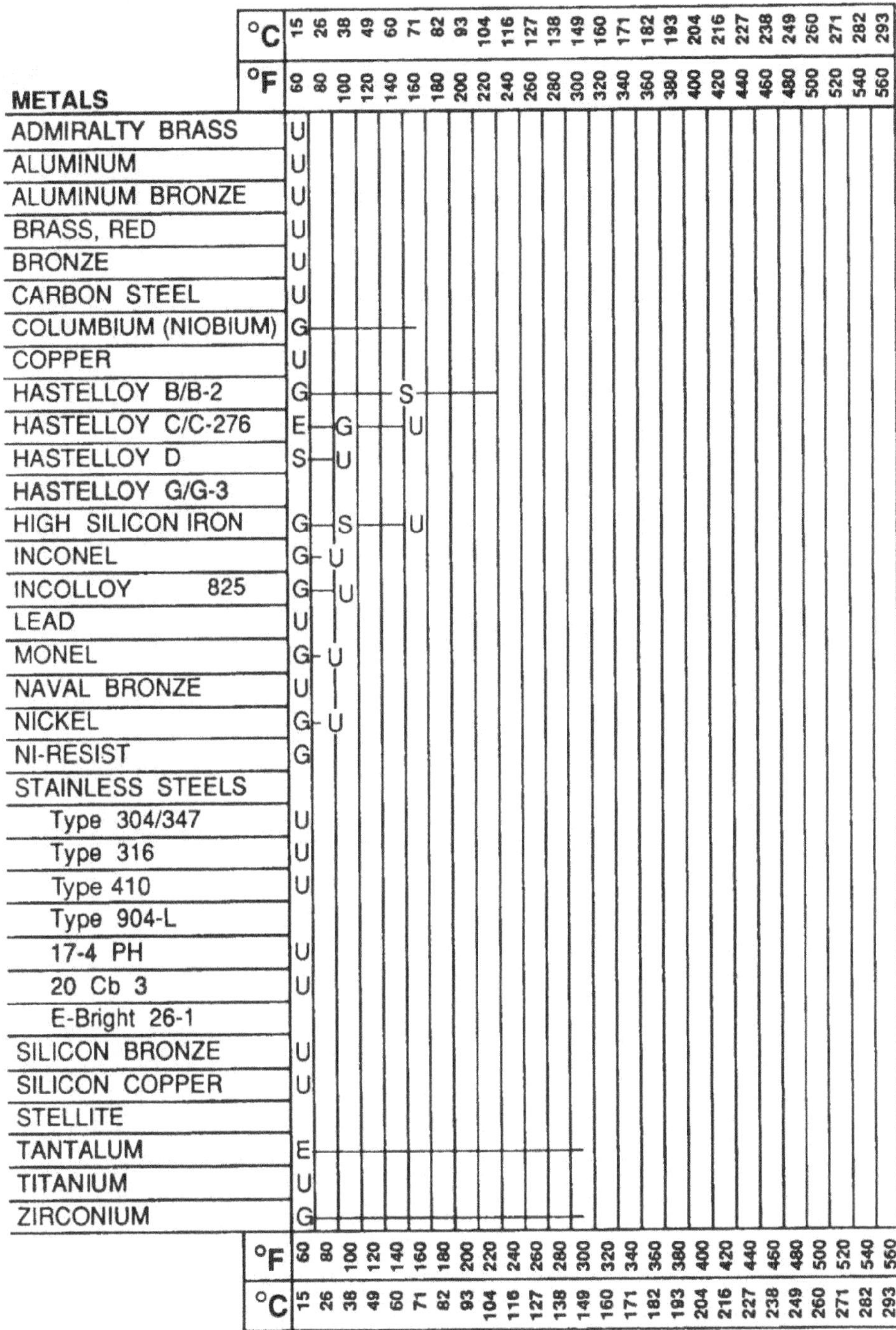

HYDROCHLORIC ACID 20%

METALS	60	80	100	120	140	160	180	200	220	240	260	280	300	320	340	360	380	400	420	440	460	480	500	520	540	560
°F / °C	15	26	38	49	60	71	82	93	104	116	127	138	149	160	171	182	193	204	216	227	238	249	260	271	282	293
ADMIRALTY BRASS	U																									
ALUMINUM	U																									
ALUMINUM BRONZE	U																									
BRASS, RED	U																									
BRONZE	U																									
CARBON STEEL	U																									
COLUMBIUM (NIOBIUM)	G																									
COPPER	U																									
HASTELLOY B/B-2	G					S																				
HASTELLOY C/C-276	E		G			U																				
HASTELLOY D	S		U																							
HASTELLOY G/G-3																										
HIGH SILICON IRON	G		S			U																				
INCONEL	G		U																							
INCOLLOY 825	G		U																							
LEAD	U																									
MONEL	G		U																							
NAVAL BRONZE	U																									
NICKEL	G		U																							
NI-RESIST	G																									
STAINLESS STEELS																										
Type 304/347	U																									
Type 316	U																									
Type 410	U																									
Type 904-L																										
17-4 PH	U																									
20 Cb 3	U																									
E-Bright 26-1																										
SILICON BRONZE	U																									
SILICON COPPER	U																									
STELLITE																										
TANTALUM	E																									
TITANIUM	U																									
ZIRCONIUM	G																									

FOR METALS

E = < 2 Mils Penetration/Year; G = < 20 Mils Penetration /year

S = < 50 Mils Penetration/Year; U = > 50 Mils Penetration/year

FOR NONMETALLICS

R = Resistant

U = Unsatisfactory

HYDROCHLORIC ACID 20%

°C: 15, 26, 38, 49, 60, 71, 82, 93, 104, 116, 127, 138, 149, 160, 171, 182, 193, 204, 216, 227, 238

°F: 60, 80, 100, 120, 140, 160, 180, 200, 220, 240, 260, 280, 300, 320, 340, 360, 380, 400, 420, 440, 460

PLASTICS	
ABS	R
ACETAL	U
ACRYLICS	R
ASBESTOS REINFORCED EPOXY	R
ASBESTOS REINFORCED PHENOLIC	R
CHLORINATED POLYETHER (Penton)	R
CPVC	R
E-CTFE (Halar)	R
EPOXY	R
ETFE (Tefzel)	R
FEP	R
FURAN (FURFURAL ALCOHOL)	R
NORYL	R
NYLON 6	U
NYLON 11	U
NYLON 66	U
PHENOLIC	R
PFA (Teflon)	R
POLYAMIDE-IMIDE	R
POLYESTERS	
Bisphenol A–fumarate	R … U
Halogenated	R
Hydrogenated Bisphenol A –Bisphenol A	R
Isophthalic	R
Terephthalate (PET)	R
POLYETHERIMIDE (ULTEM)	R
POLYETHERETHERKETONE (PEEK)	R
POLYETHERSULFONE (PES)	R
POLYETHYLENE HMW	R
POLYETHYLENE UHMW	R
POLYMETHYLPENTENE	R
POLYPROPYLENE (PP)	R
POLYSTYRENE	R
POLYSULFONE	R
POLYVINYLIDENE CHLORIDE	R
PVC TYPE 1	R
PVC TYPE 2	R
PVDF (Kynar)	R
RYTON	R
TFE (Teflon)	R
VINYL ESTER	R

°F: 60, 80, 100, 120, 140, 160, 180, 200, 220, 240, 260, 280, 300, 320, 340, 360, 380, 400, 420, 440, 460

°C: 15, 26, 38, 49, 60, 71, 82, 93, 104, 116, 127, 138, 149, 160, 171, 182, 193, 204, 216, 227, 238

HYDROCHLORIC ACID 20%

°C: 15, 26, 38, 49, 60, 71, 82, 93, 104, 116, 127, 138, 149, 160, 171, 182, 193, 204, 216, 227, 238

°F: 60, 80, 100, 120, 140, 160, 180, 200, 220, 240, 260, 280, 300, 320, 340, 360, 380, 400, 420, 440, 460

ELASTOMERS* AND LININGS	
ACRYLATE-BUTADIENE (ABR)	
BUTYL GR-1 (IIR)	U
CARBOXYLIC-ACRYLONITRILE-BUTADIENE (XNBR)	
CHEMRAZ (FPM)	R
CHLORO-ISOBUTENE-ISOPRENE (CIIR)	U
HYPALON (CHLORO-SULFONYL-POLYETHYLENE (CSM)	R
ETHYLENE-ACRYLIC (EA)	
ETHYLENE-PROPYLENE (EPM)	R
ETHYLENE-PROPYLENE-DIENE (EPDM)	R
ETHYLENE-PROPYLENE TERPOLYMER (EPT)	U
FKM (Viton A)	R
HARD RUBBER	R
SOFT RUBBER	R
ISOPRENE (IR)	R
KALREZ (FPM)	R
KOROSEAL	R
NATURAL RUBBER (GRS)	R
NEOPRENE GR-M (CR)	R
NITRILE BUNA-N (NBR)	R
NORDEL (EPDM)	R
POLYBUTADIENE (BR)	R
POLYESTER (PE)	U
POLYETHER-URETHANE (EU)	U
POLYISOPRENE (IR)	
POLYSULFIDES (T)	U
POLYURETHANE (AU)	R
SBR STYRENE (BUNA-S)	
SILICONE RUBBERS	R

°F: 60, 80, 100, 120, 140, 160, 180, 200, 220, 240, 260, 280, 300, 320, 340, 360, 380, 400, 420, 440, 460

°C: 15, 26, 38, 49, 60, 71, 82, 93, 104, 116, 127, 138, 149, 160, 171, 182, 193, 204, 216, 227, 238

*See also Nylon 11 under PLASTICS

HYDROCHLORIC ACID 38%

METALS °F	60	80	100	120	140	160	180	200	220	240	260	280	300	320	340	360	380	400	420	440	460	480	500	520	540	560
°C	15	26	38	49	60	71	82	93	104	116	127	138	149	160	171	182	193	204	216	227	238	249	260	271	282	293
ADMIRALTY BRASS	U																									
ALUMINUM	U																									
ALUMINUM BRONZE	U																									
BRASS, RED	U																									
BRONZE	U																									
CARBON STEEL	U																									
COLUMBIUM (NIOBIUM)	G																									
COPPER	U																									
HASTELLOY B/B-2	E				U																					
HASTELLOY C/C-276	E		S		U																					
HASTELLOY D																										
HASTELLOY G/G-3																										
HIGH SILICON IRON	U																									
INCONEL	U																									
INCOLLOY	U																									
LEAD	U																									
MONEL	U																									
NAVAL BRONZE	U																									
NICKEL	U																									
NI-RESIST	U																									
STAINLESS STEELS																										
Type 304/347	U																									
Type 316	U																									
Type 410	U																									
Type 904-L																										
17-4 PH	U																									
20 Cb 3	U																									
E-Bright 26-1																										
SILICON BRONZE	U																									
SILICON COPPER	U																									
STELLITE	U																									
TANTALUM	E																									
TITANIUM	U																									
ZIRCONIUM	E				U																					
°F	60	80	100	120	140	160	180	200	220	240	260	280	300	320	340	360	380	400	420	440	460	480	500	520	540	560
°C	15	26	38	49	60	71	82	93	104	116	127	138	149	160	171	182	193	204	216	227	238	249	260	271	282	293

FOR METALS

E = < 2 Mils Penetration/Year; G = < 20 Mils Penetration /year

S = < 50 Mils Penetration/Year; U = > 50 Mils Penetration/year

FOR NONMETALLICS

R = Resistant

U = Unsatisfactory

HYDROCHLORIC ACID 38%

°C: 15, 26, 38, 49, 60, 71, 82, 93, 104, 116, 127, 138, 149, 160, 171, 182, 193, 204, 216, 227, 238
°F: 60, 80, 100, 120, 140, 160, 180, 200, 220, 240, 260, 280, 300, 320, 340, 360, 380, 400, 420, 440, 460

PLASTICS	Rating	Bar extends to (°F)
ABS	R	140
ACETAL	U	
ACRYLICS	R	90
ASBESTOS REINFORCED EPOXY	R	160
ASBESTOS REINFORCED PHENOLIC	R	300
CHLORINATED POLYETHER (Penton)	R	230
CPVC	R	170; U at 180
E-CTFE (Halar)	R	300
EPOXY	R	140
ETFE (Tefzel)	R	300
FEP	R	400
FURAN (FURFURAL ALCOHOL)	R	230
NORYL	R	190
NYLON 6	U	
NYLON 11	U	
NYLON 66	U	
PHENOLIC	R	300
PFA (Teflon)	R	200
POLYAMIDE-IMIDE	R	200
POLYESTERS		
Bisphenol A–fumarate	U	
Halogenated	R	180
Hydrogenated Bisphenol A –Bisphenol A	R	170
Isophthalic	R	160
Terephthalate (PET)	R	90
POLYETHERIMIDE (ULTEM)	R	90
POLYETHERETHERKETONE (PEEK)	R	80
POLYETHERSULFONE (PES)	R	140
POLYETHYLENE HMW	R	140
POLYETHYLENE UHMW	R	140
POLYMETHYLPENTENE	R	80
POLYPROPYLENE (PP)	R	200
POLYSTYRENE	R	90
POLYSULFONE	R	140
POLYVINYLIDENE CHLORIDE	R	180
PVC TYPE 1	R	140
PVC TYPE 2	R	140
PVDF (Kynar)	R	280
RYTON	R	210
TFE (Teflon)	R	460
VINYL ESTER	R	180

ELASTOMERS* AND LININGS	60	80	100	120	140	160	180	200	220	240	260	280	300	320	340	360	380	400	420	440	460
°C	15	26	38	49	60	71	82	93	104	116	127	138	149	160	171	182	193	204	216	227	238
ACRYLATE-BUTADIENE (ABR)																					
BUTYL GR-1 (IIR)	U																				
CARBOXYLIC-ACRYLONITRILE-BUTADIENE (XNBR)																					
CHEMRAZ (FPM)	R																				
CHLORO-ISOBUTENE-ISOPRENE (CIIR)	U																				
HYPALON (CHLORO-SULFONYL-POLYETHYLENE (CSM)	R					U															
ETHYLENE-ACRYLIC (EA)																					
ETHYLENE-PROPYLENE (EPM)	R																				
ETHYLENE-PROPYLENE-DIENE (EPDM)	R																				
ETHYLENE-PROPYLENE TERPOLYMER (EPT)	U																				
FKM (Viton A)	R																				
HARD RUBBER	R																				
SOFT RUBBER	R																				
ISOPRENE (IR)	R																				
KALREZ (FPM)	R																				
KOROSEAL	R																				
NATURAL RUBBER (GRS)	R																				
NEOPRENE GR-M (CR)	R																				
NITRILE BUNA-N (NBR)	U																				
NORDEL (EPDM)	R																				
POLYBUTADIENE (BR)																					
POLYESTER (PE)	U																				
POLYETHER-URETHANE (EU)	U																				
POLYISOPRENE (IR)																					
POLYSULFIDES (T)	U																				
POLYURETHANE (AU)	U																				
SBR STYRENE (BUNA-S)	U																				
SILICONE RUBBERS	U																				

*See also Nylon 11 under PLASTICS

HYDROFLUORIC ACID 30%

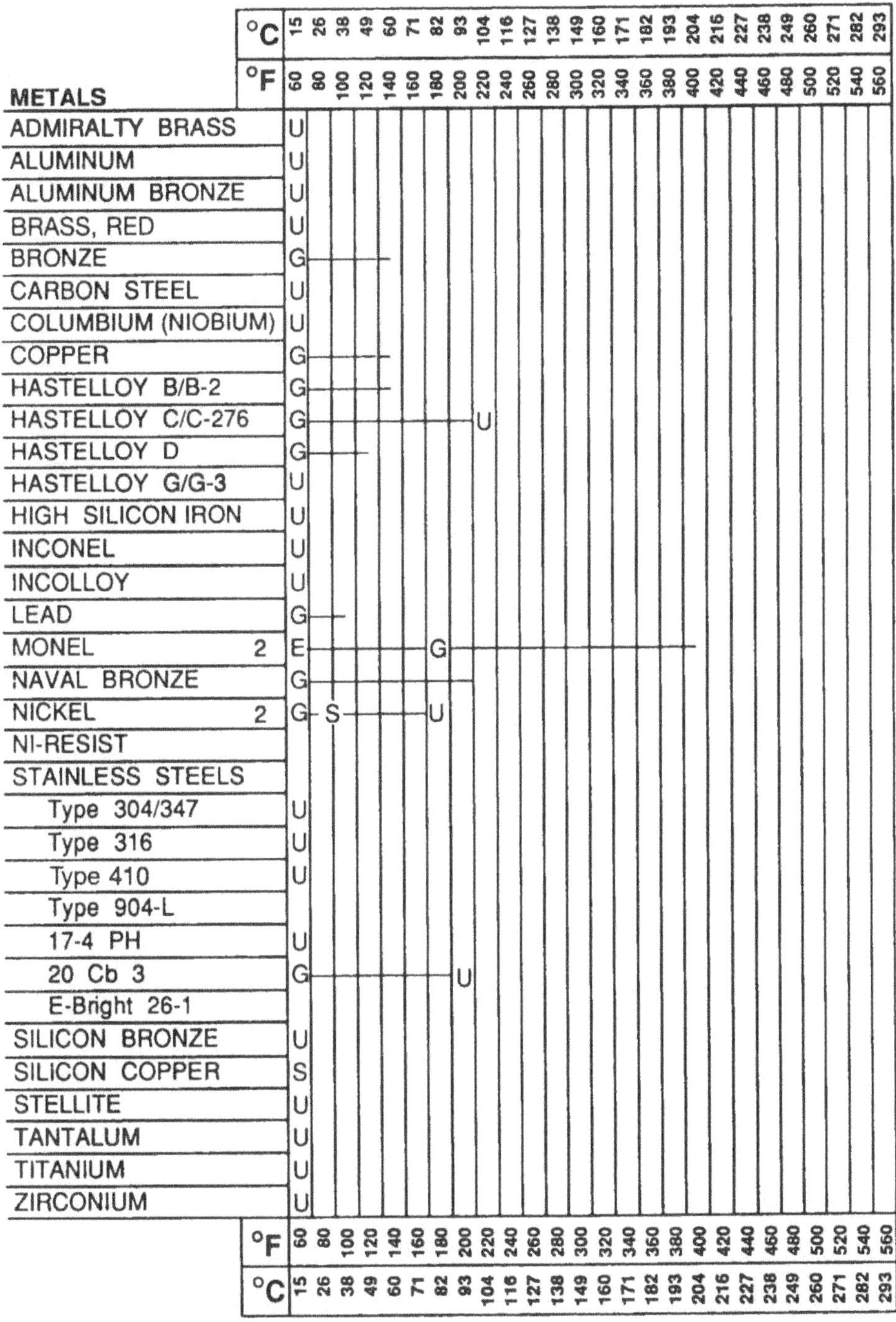

METALS °F	60	80	100	120	140	160	180	200	220	240	260	280	300	320	340	360	380	400	420	440	460	480	500	520	540	560
°C	15	26	38	49	60	71	82	93	104	116	127	138	149	160	171	182	193	204	216	227	238	249	260	271	282	293
ADMIRALTY BRASS	U																									
ALUMINUM	U																									
ALUMINUM BRONZE	U																									
BRASS, RED	U																									
BRONZE	G	—	—	—	—																					
CARBON STEEL	U																									
COLUMBIUM (NIOBIUM)	U																									
COPPER	G	—	—	—	—																					
HASTELLOY B/B-2	G	—	—	—	—																					
HASTELLOY C/C-276	G	—	—	—	—	—	—	—	U																	
HASTELLOY D	G	—	—	—																						
HASTELLOY G/G-3	U																									
HIGH SILICON IRON	U																									
INCONEL	U																									
INCOLLOY	U																									
LEAD	G	—	—																							
MONEL 2	E	—	—	—	—	—	G	—	—	—	—	—	—	—	—	—	—	—								
NAVAL BRONZE	G	—	—	—	—	—	—	—																		
NICKEL 2	G	S	—	—	—	—	U																			
NI-RESIST																										
STAINLESS STEELS																										
Type 304/347	U																									
Type 316	U																									
Type 410	U																									
Type 904-L																										
17-4 PH	U																									
20 Cb 3	G	—	—	—	—	—	—	U																		
E-Bright 26-1																										
SILICON BRONZE	U																									
SILICON COPPER	S																									
STELLITE	U																									
TANTALUM	U																									
TITANIUM	U																									
ZIRCONIUM	U																									
°F	60	80	100	120	140	160	180	200	220	240	260	280	300	320	340	360	380	400	420	440	460	480	500	520	540	560
°C	15	26	38	49	60	71	82	93	104	116	127	138	149	160	171	182	193	204	216	227	238	249	260	271	282	293

FOR METALS

E = < 2 Mils Penetration/Year; G = < 20 Mils Penetration /year

S = < 50 Mils Penetration/Year; U = > 50 Mils Penetration/year

FOR NONMETALLICS

R = Resistant

U = Unsatisfactory

HYDROFLUORIC ACID 30%

PLASTICS		°F 60 / °C 15
ABS		U
ACETAL		U
ACRYLICS		
ASBESTOS REINFORCED EPOXY		
ASBESTOS REINFORCED PHENOLIC		
CHLORINATED POLYETHER (Penton)		R
CPVC		U
E-CTFE (Halar)		R
EPOXY		U
ETFE (Tefzel)		R
FEP		R
FURAN (FURFURAL ALCOHOL)		U
NORYL		R
NYLON 6		U
NYLON 11		U
NYLON 66		U
PHENOLIC		U
PFA (Teflon)		R
POLYAMIDE-IMIDE		
POLYESTERS		
Bisphenol A–fumarate	10	R
Halogenated	10	R
Hydrogenated Bisphenol A –Bisphenol A		U
Isophthalic		U
Terephthalate (PET)		U
POLYETHERIMIDE (ULTEM)		R
POLYETHERETHERKETONE (PEEK)		U
POLYETHERSULFONE (PES)		
POLYETHYLENE HMW		R
POLYETHYLENE UHMW		R
POLYMETHYLPENTENE		R
POLYPROPYLENE (PP)		R
POLYSTYRENE		R
POLYSULFONE		R
POLYVINYLIDENE CHLORIDE		R
PVC TYPE 1		R (U at 140°F)
PVC TYPE 2		R (U at 140°F)
PVDF (Kynar)		R
RYTON		R
TFE (Teflon)		R
VINYL ESTER		U

°F: 60, 80, 100, 120, 140, 160, 180, 200, 220, 240, 260, 280, 300, 320, 340, 360, 380, 400, 420, 440, 460

°C: 15, 26, 38, 49, 60, 71, 82, 93, 104, 116, 127, 138, 149, 160, 171, 182, 193, 204, 216, 227, 238

HYDROFLUORIC ACID 30%

ELASTOMERS* AND LININGS	60 °F / 15 °C	80 / 26	100 / 38	120 / 49	140 / 60	160 / 71	180 / 82	200 / 93	220 / 104	240 / 116	260 / 127	280 / 138	300 / 149	320 / 160	340 / 171	360 / 182	380 / 193	400 / 204	420 / 216	440 / 227	460 / 238
ACRYLATE-BUTADIENE (ABR)																					
BUTYL GR-1 (IIR)	R	—	—	—	—	—	—	—	—	—	—	—	—	—	—						
CARBOXYLIC-ACRYLONITRILE-BUTADIENE (XNBR)																					
CHEMRAZ (FPM)	R	—	—	—	—	—	—	—	—	—	—	—	—								
CHLORO-ISOBUTENE-ISOPRENE (CIIR)	R	—	—	—	—	—															
HYPALON (CHLORO-SULFONYL-POLYETHYLENE (CSM)	R	—																			
ETHYLENE-ACRYLIC (EA)																					
ETHYLENE-PROPYLENE (EPM)	R	—	—	—	—																
ETHYLENE-PROPYLENE-DIENE (EPDM)	R																				
ETHYLENE-PROPYLENE TERPOLYMER (EPT)	R	—	—	—	—																
FKM (Viton A)	R	—	—	—	—	—	—	—													
HARD RUBBER	R	—	—	—	—	—	—	—													
SOFT RUBBER																					
ISOPRENE (IR)	R	—	—																		
KALREZ (FPM)	R	—	—	—	—	—	—	—													
KOROSEAL	R	—																			
NATURAL RUBBER (GRS)	R	—	—																		
NEOPRENE GR-M (CR)	R	—	—	—	—	—	—	—													
NITRILE BUNA-N (NBR)	U																				
NORDEL (EPDM)																					
POLYBUTADIENE (BR)																					
POLYESTER (PE)	U																				
POLYETHER-URETHANE (EU)	U																				
POLYISOPRENE (IR)																					
POLYSULFIDES (T)	U																				
POLYURETHANE (AU)																					
SBR STYRENE (BUNA-S)	U																				
SILICONE RUBBERS	U																				

°F 60 80 100 120 140 160 180 200 220 240 260 280 300 320 340 360 380 400 420 440 460
°C 15 26 38 49 60 71 82 93 104 116 127 138 149 160 171 182 193 204 216 227 238

*See also Nylon 11 under PLASTICS

HYDROFLUORIC ACID 70%

METALS	°F 60 / °C 15	80 / 26	100 / 38	120 / 49	140 / 60	160 / 71	180 / 82	200 / 93	220 / 104	240 / 116	260 / 127	280 / 138	300 / 149	320 / 160	340 / 171	360 / 182	380 / 193	400 / 204	420 / 216	440 / 227	460 / 238	480 / 249	500 / 260	520 / 271	540 / 282	560 / 293
ADMIRALTY BRASS	U																									
ALUMINUM	U																									
ALUMINUM BRONZE	U																									
BRASS, RED	U																									
BRONZE	U																									
CARBON STEEL	U																									
COLUMBIUM (NIOBIUM)	U																									
COPPER	U																									
HASTELLOY B/B-2	G																									
HASTELLOY C/C-276	G																									
HASTELLOY D																										
HASTELLOY G/G-3																										
HIGH SILICON IRON	U																									
INCONEL	U																									
INCOLLOY	U																									
LEAD	U																									
MONEL 2	E	G																								
NAVAL BRONZE	U																									
NICKEL 2	G																									
NI-RESIST																										
STAINLESS STEELS																										
Type 304/347	U																									
Type 316	U																									
Type 410	U																									
Type 904-L																										
17-4 PH																										
20 Cb 3	U																									
E-Bright 26-1																										
SILICON BRONZE	U																									
SILICON COPPER	U																									
STELLITE																										
TANTALUM	U																									
TITANIUM	U																									
ZIRCONIUM	U																									

FOR METALS

E = < 2 Mils Penetration/Year; G = < 20 Mils Penetration /year

S = < 50 Mils Penetration/Year; U = > 50 Mils Penetration/year

FOR NONMETALLICS

R = Resistant

U = Unsatisfactory

HYDROFLUORIC ACID 70%

PLASTICS	15	26	38	49	60	71	82	93	104	116	127	138	149	160	171	182	193	204	216	227	238
°F	60	80	100	120	140	160	180	200	220	240	260	280	300	320	340	360	380	400	420	440	460
ABS	U																				
ACETAL	U																				
ACRYLICS																					
ASBESTOS REINFORCED EPOXY																					
ASBESTOS REINFORCED PHENOLIC																					
CHLORINATED POLYETHER (Penton)	U																				
CPVC	R																				
E-CTFE (Halar)	R																				
EPOXY	U																				
ETFE (Tefzel)	R																				
FEP	R																				
FURAN (FURFURAL ALCOHOL)																					
NORYL	U																				
NYLON 6	U																				
NYLON 11	U																				
NYLON 66	U																				
PHENOLIC	U																				
PFA (Teflon)	R																				
POLYAMIDE-IMIDE																					
POLYESTERS																					
Bisphenol A–fumarate																					
Halogenated																					
Hydrogenated Bisphenol A –Bisphenol A	U																				
Isophthalic	U																				
Terephthalate (PET)	U																				
POLYETHERIMIDE (ULTEM)																					
POLYETHERETHERKETONE (PEEK)	U																				
POLYETHERSULFONE (PES)																					
POLYETHYLENE HMW	U																				
POLYETHYLENE UHMW	U																				
POLYMETHYLPENTENE																					
POLYPROPYLENE (PP)	R																				
POLYSTYRENE																					
POLYSULFONE																					
POLYVINYLIDENE CHLORIDE																					
PVC TYPE 1	R																				
PVC TYPE 2																					
PVDF (Kynar)	R																				
RYTON																					
TFE (Teflon)	R																				
VINYL ESTER	U																				
°F	60	80	100	120	140	160	180	200	220	240	260	280	300	320	340	360	380	400	420	440	460
°C	15	26	38	49	60	71	82	93	104	116	127	138	149	160	171	182	193	204	216	227	238

HYDROFLUORIC ACID 70%

Temperature scale: °F 60, 80, 100, 120, 140, 160, 180, 200, 220, 240, 260, 280, 300, 320, 340, 360, 380, 400, 420, 440, 460 / °C 15, 26, 38, 49, 60, 71, 82, 93, 104, 116, 127, 138, 149, 160, 171, 182, 193, 204, 216, 227, 238

ELASTOMERS* AND LININGS	Rating	Bar extends to (°F)	(°C)
ACRYLATE-BUTADIENE (ABR)			
BUTYL GR-1 (IIR)	R	140	60
CARBOXYLIC-ACRYLONITRILE-BUTADIENE (XNBR)			
CHEMRAZ (FPM)	R	300	149
CHLORO-ISOBUTENE-ISOPRENE (CIIR)	U		
HYPALON (CHLORO-SULFONYL-POLYETHYLENE (CSM)	R	between 140/160	between 60/71
ETHYLENE-ACRYLIC (EA)			
ETHYLENE-PROPYLENE (EPM)			
ETHYLENE-PROPYLENE-DIENE (EPDM)	U		
ETHYLENE-PROPYLENE TERPOLYMER (EPT)	U		
FKM (Viton A)	R	340	171
HARD RUBBER	U		
SOFT RUBBER	U		
ISOPRENE (IR)	U		
KALREZ (FPM)	R	200	93
KOROSEAL	R	80	26
NATURAL RUBBER (GRS)	U		
NEOPRENE GR-M (CR)	R	between 180/200	between 82/93
NITRILE BUNA-N (NBR)	U		
NORDEL (EPDM)	U		
POLYBUTADIENE (BR)			
POLYESTER (PE)	U		
POLYETHER-URETHANE (EU)	U		
POLYISOPRENE (IR)			
POLYSULFIDES (T)	U		
POLYURETHANE (AU)			
SBR STYRENE (BUNA-S)	U		
SILICONE RUBBERS	U		

*See also Nylon 11 under PLASTICS

HYDROFLUORIC ACID 100%

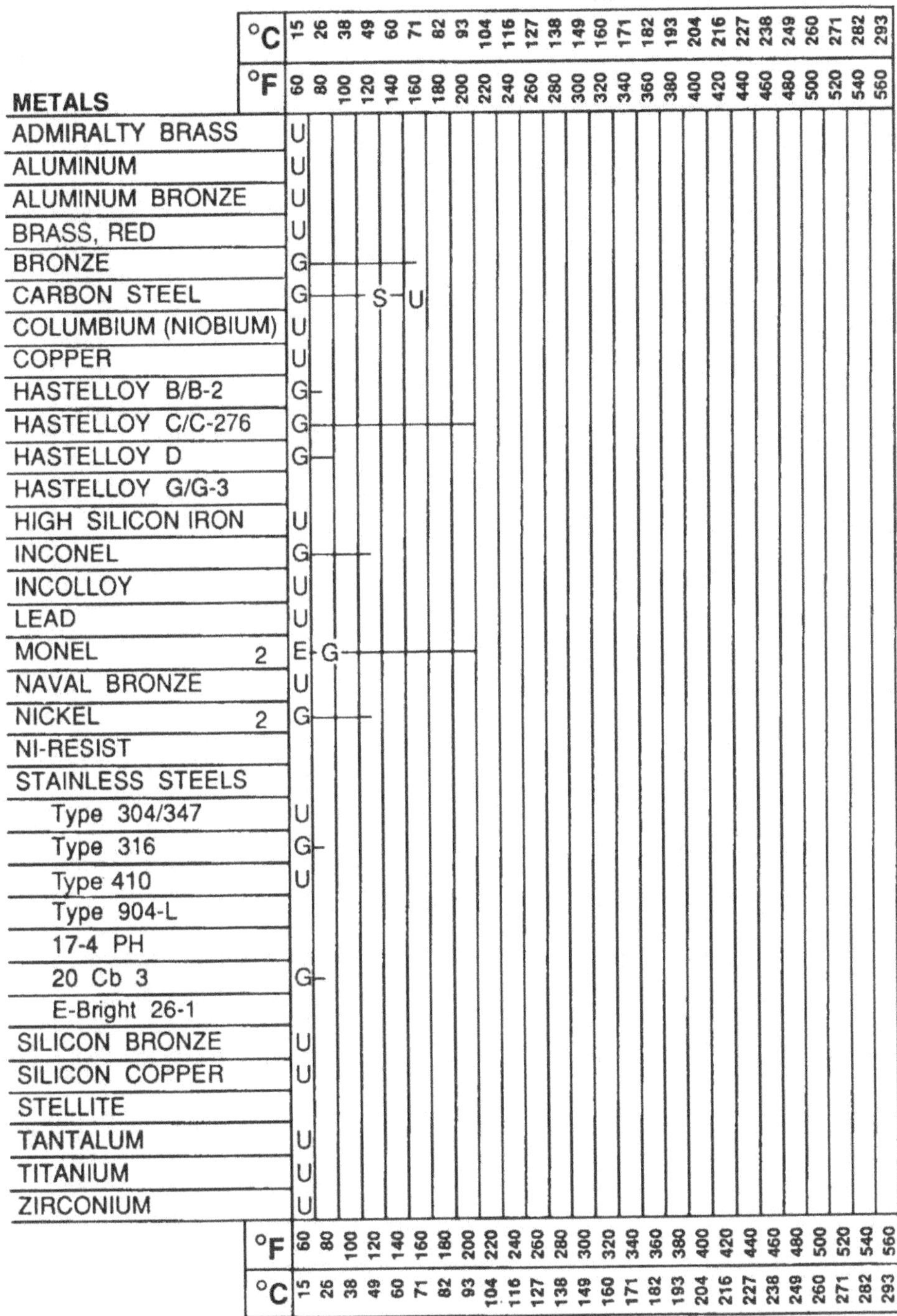

METALS	°F 60	80	100	120	140	160	180	200	220	240	260	280	300	320	340	360	380	400	420	440	460	480	500	520	540	560
°C	15	26	38	49	60	71	82	93	104	116	127	138	149	160	171	182	193	204	216	227	238	249	260	271	282	293
ADMIRALTY BRASS	U																									
ALUMINUM	U																									
ALUMINUM BRONZE	U																									
BRASS, RED	U																									
BRONZE	G																									
CARBON STEEL	G				S	U																				
COLUMBIUM (NIOBIUM)	U																									
COPPER	U																									
HASTELLOY B/B-2	G																									
HASTELLOY C/C-276	G																									
HASTELLOY D	G																									
HASTELLOY G/G-3																										
HIGH SILICON IRON	U																									
INCONEL	G																									
INCOLLOY	U																									
LEAD	U																									
MONEL 2	E	G																								
NAVAL BRONZE	U																									
NICKEL 2	G																									
NI-RESIST																										
STAINLESS STEELS																										
Type 304/347	U																									
Type 316	G																									
Type 410	U																									
Type 904-L																										
17-4 PH																										
20 Cb 3	G																									
E-Bright 26-1																										
SILICON BRONZE	U																									
SILICON COPPER	U																									
STELLITE																										
TANTALUM	U																									
TITANIUM	U																									
ZIRCONIUM	U																									
°F	60	80	100	120	140	160	180	200	220	240	260	280	300	320	340	360	380	400	420	440	460	480	500	520	540	560
°C	15	26	38	49	60	71	82	93	104	116	127	138	149	160	171	182	193	204	216	227	238	249	260	271	282	293

FOR METALS

E = < 2 Mils Penetration/Year; G = < 20 Mils Penetration /year

S = < 50 Mils Penetration/Year; U = > 50 Mils Penetration/year

FOR NONMETALLICS

R = Resistant

U = Unsatisfactory

HYDROFLUORIC ACID 100%

PLASTICS	Rating	Max. °F	Max. °C
ABS	U		
ACETAL	U		
ACRYLICS	R	80	26
ASBESTOS REINFORCED EPOXY	R	80	26
ASBESTOS REINFORCED PHENOLIC	U		
CHLORINATED POLYETHER (Penton)	U		
CPVC	U		
E-CTFE (Halar)	R	240	116
EPOXY	U		
ETFE (Tefzel)	R	220	104
FEP	R	400	204
FURAN (FURFURAL ALCOHOL)	R	280	138
NORYL	U		
NYLON 6	U		
NYLON 11	U		
NYLON 66	U		
PHENOLIC	U		
PFA (Teflon)	R	200	93
POLYAMIDE-IMIDE			
POLYESTERS			
Bisphenol A–fumarate			
Halogenated			
Hydrogenated Bisphenol A –Bisphenol A	U		
Isophthalic	U		
Terephthalate (PET)	U		
POLYETHERIMIDE (ULTEM)			
POLYETHERETHERKETONE (PEEK)	U		
POLYETHERSULFONE (PES)			
POLYETHYLENE HMW			
POLYETHYLENE UHMW			
POLYMETHYLPENTENE			
POLYPROPYLENE (PP)	R	200	93
POLYSTYRENE	R	80	26
POLYSULFONE			
POLYVINYLIDENE CHLORIDE	U		
PVC TYPE 1			
PVC TYPE 2			
PVDF (Kynar)	R	200	93
RYTON			
TFE (Teflon)	R	460	238
VINYL ESTER	U		

HYDROFLUORIC ACID 100%

ELASTOMERS* AND LININGS	60	80	100	120	140	160	180	200	220	240	260	280	300	320	340	360	380	400	420	440	460
°C	15	26	38	49	60	71	82	93	104	116	127	138	149	160	171	182	193	204	216	227	238
ACRYLATE-BUTADIENE (ABR)																					
BUTYL GR-1 (IIR)	U																				
CARBOXYLIC-ACRYLONITRILE-BUTADIENE (XNBR)																					
CHEMRAZ (FPM)	R	—	—	—	—	—	—	—	—	—	—	—	—	—	—	—	—	—			
CHLORO-ISOBUTENE-ISOPRENE (CIIR)																					
HYPALON (CHLORO-SULFONYL-POLYETHYLENE (CSM)	R	—	—	—	—	—															
ETHYLENE-ACRYLIC (EA)																					
ETHYLENE-PROPYLENE (EPM)																					
ETHYLENE-PROPYLENE-DIENE (EPDM)	U																				
ETHYLENE-PROPYLENE TERPOLYMER (EPT)	U																				
FKM (Viton A)	R	U																			
HARD RUBBER	U																				
SOFT RUBBER																					
ISOPRENE (IR)	U																				
KALREZ (FPM)	R	—	—	—	—	—	—	—													
KOROSEAL	R	—																			
NATURAL RUBBER (GRS)	U																				
NEOPRENE GR-M (CR)	U																				
NITRILE BUNA-N (NBR)	U																				
NORDEL (EPDM)	U																				
POLYBUTADIENE (BR)																					
POLYESTER (PE)	U																				
POLYETHER-URETHANE (EU)	U																				
POLYISOPRENE (IR)																					
POLYSULFIDES (T)	U																				
POLYURETHANE (AU)																					
SBR STYRENE (BUNA-S)	U																				
SILICONE RUBBERS	U																				

°F 60 80 100 120 140 160 180 200 220 240 260 280 300 320 340 360 380 400 420 440 460
°C 15 26 38 49 60 71 82 93 104 116 127 138 149 160 171 182 193 204 216 227 238

*See also Nylon 11 under PLASTICS

MAGNESIUM CHLORIDE

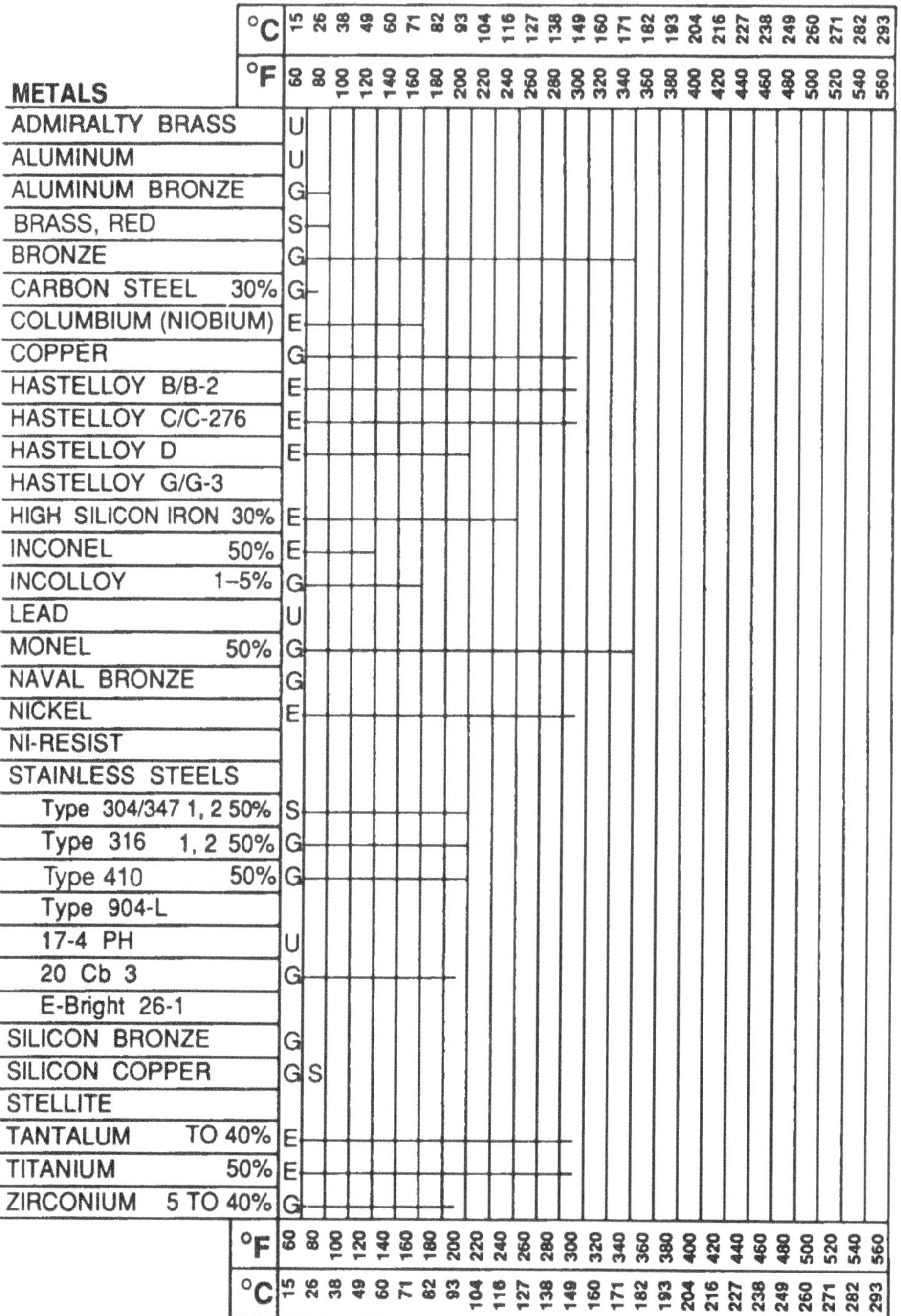

METALS	°C 15 / °F 60	26 / 80	38 / 100	49 / 120	60 / 140	71 / 160	82 / 180	93 / 200	104 / 220	116 / 240	127 / 260	138 / 280	149 / 300	160 / 320	171 / 340	182 / 360	193 / 380	204 / 400	216 / 420	227 / 440	238 / 460	249 / 480	260 / 500	271 / 520	282 / 540	293 / 560
ADMIRALTY BRASS	U																									
ALUMINUM	U																									
ALUMINUM BRONZE	G																									
BRASS, RED	S																									
BRONZE	G																									
CARBON STEEL 30%	G																									
COLUMBIUM (NIOBIUM)	E																									
COPPER	G																									
HASTELLOY B/B-2	E																									
HASTELLOY C/C-276	E																									
HASTELLOY D	E																									
HASTELLOY G/G-3																										
HIGH SILICON IRON 30%	E																									
INCONEL 50%	E																									
INCOLLOY 1–5%	G																									
LEAD	U																									
MONEL 50%	G																									
NAVAL BRONZE	G																									
NICKEL	E																									
NI-RESIST																										
STAINLESS STEELS																										
Type 304/347 1, 2 50%	S																									
Type 316 1, 2 50%	G																									
Type 410 50%	G																									
Type 904-L																										
17-4 PH	U																									
20 Cb 3	G																									
E-Bright 26-1																										
SILICON BRONZE	G																									
SILICON COPPER	G	S																								
STELLITE																										
TANTALUM TO 40%	E																									
TITANIUM 50%	E																									
ZIRCONIUM 5 TO 40%	G																									

FOR METALS

E = < 2 Mils Penetration/Year; G = < 20 Mils Penetration /year

S = < 50 Mils Penetration/Year; U = > 50 Mils Penetration/year

FOR NONMETALLICS

R = Resistant

U = Unsatisfactory

MAGNESIUM CHLORIDE

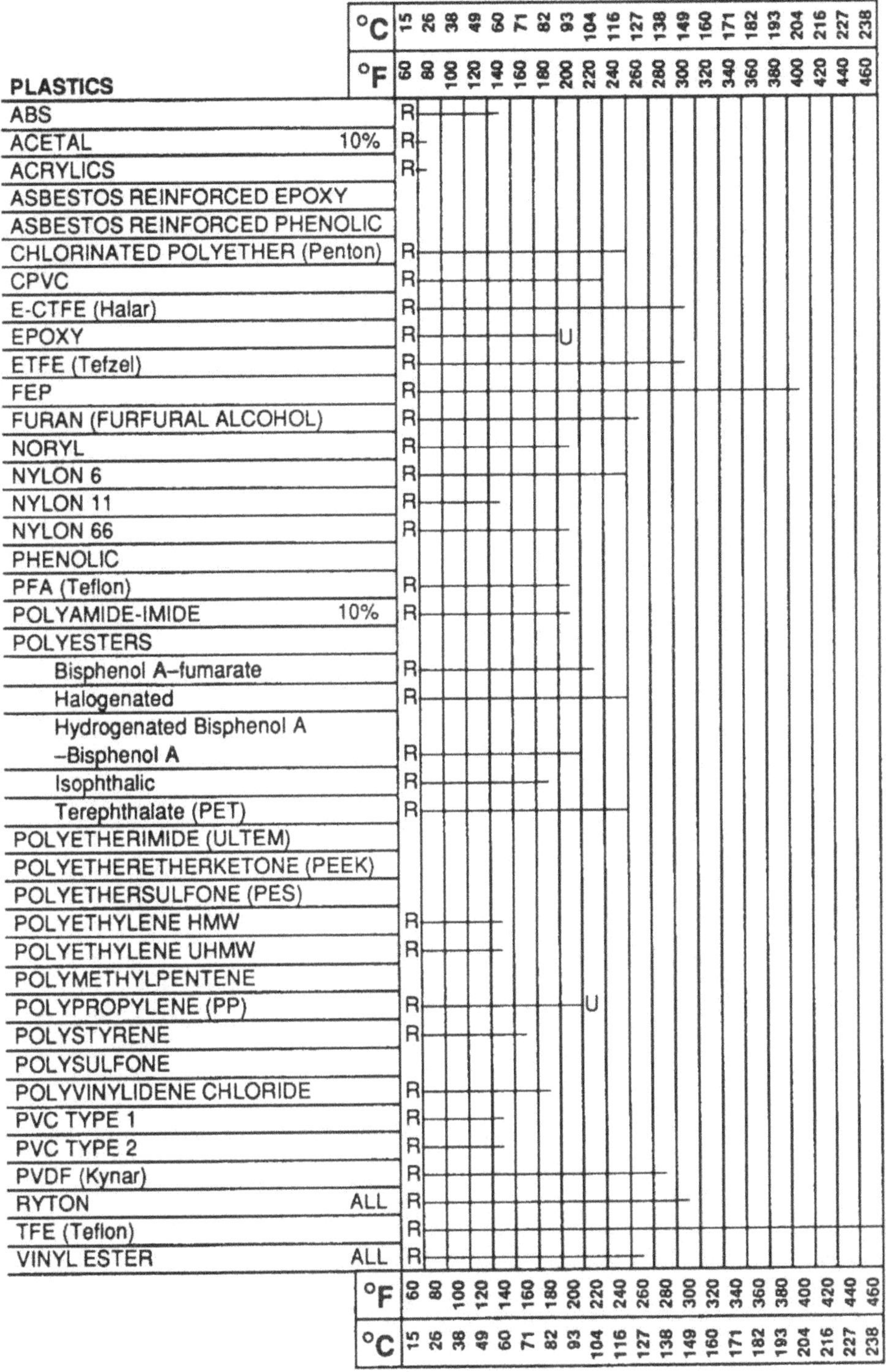

°C: 15 26 38 49 60 71 82 93 104 116 127 138 149 160 171 182 193 204 216 227 238

°F: 60 80 100 120 140 160 180 200 220 240 260 280 300 320 340 360 380 400 420 440 460

ELASTOMERS* AND LININGS	
ACRYLATE-BUTADIENE (ABR)	
BUTYL GR-1 (IIR)	R
CARBOXYLIC-ACRYLONITRILE-BUTADIENE (XNBR)	R
CHEMRAZ (FPM)	R
CHLORO-ISOBUTENE-ISOPRENE (CIIR)	R
HYPALON (CHLORO-SULFONYL-POLYETHYLENE (CSM)	R
ETHYLENE-ACRYLIC (EA)	
ETHYLENE-PROPYLENE (EPM)	R
ETHYLENE-PROPYLENE-DIENE (EPDM)	R
ETHYLENE-PROPYLENE TERPOLYMER (EPT)	R
FKM (Viton A)	R
HARD RUBBER	R
SOFT RUBBER	R
ISOPRENE (IR)	R
KALREZ (FPM)	R
KOROSEAL	R
NATURAL RUBBER (GRS)	R
NEOPRENE GR-M (CR)	R
NITRILE BUNA-N (NBR)	R
NORDEL (EPDM)	R
POLYBUTADIENE (BR)	
POLYESTER (PE)	R
POLYETHER-URETHANE (EU)	R
POLYISOPRENE (IR)	
POLYSULFIDES (T)	
POLYURETHANE (AU)	R
SBR STYRENE (BUNA-S)	
SILICONE RUBBERS	R

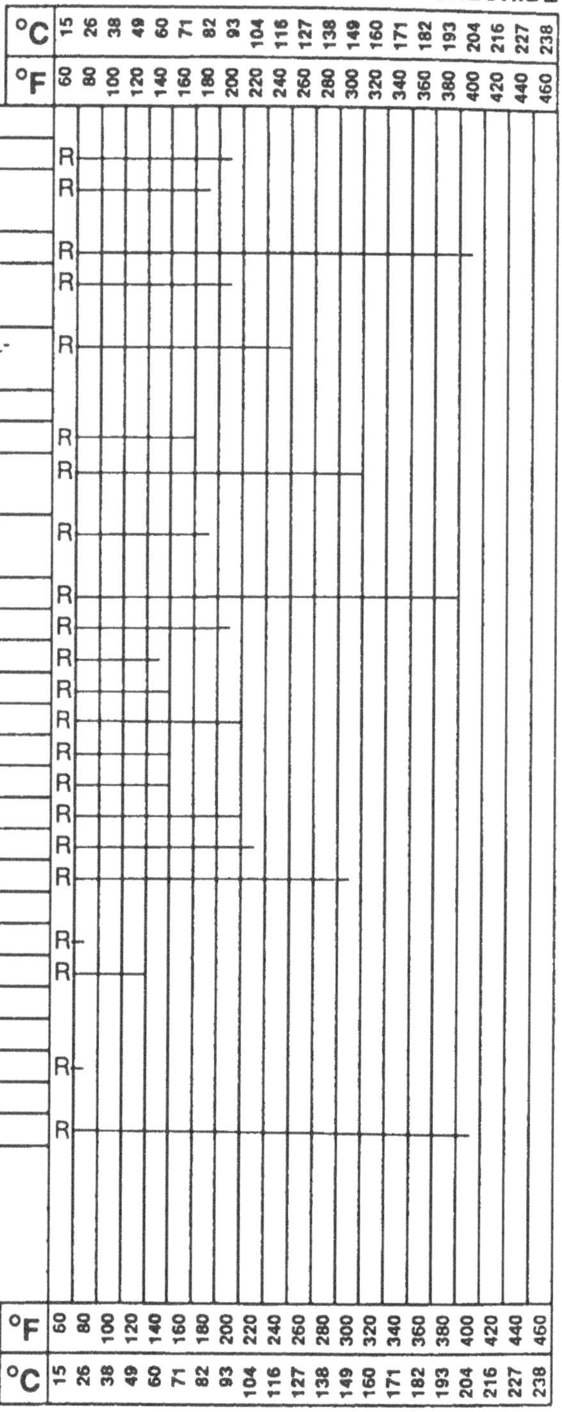

*See also Nylon 11 under PLASTICS

°F: 60 80 100 120 140 160 180 200 220 240 260 280 300 320 340 360 380 400 420 440 460

°C: 15 26 38 49 60 71 82 93 104 116 127 138 149 160 171 182 193 204 216 227 238

NITRIC ACID 5%

METALS		60	80	100	120	140	160	180	200	220	240	260	280	300	320	340	360	380	400	420	440	460	480	500	520	540	560
°C		15	26	38	49	60	71	82	93	104	116	127	138	149	160	171	182	193	204	216	227	238	249	260	271	282	293
ADMIRALTY BRASS		U																									
ALUMINUM		U																									
ALUMINUM BRONZE		U																									
BRASS, RED		U																									
BRONZE		U																									
CARBON STEEL		U																									
COLUMBIUM (NIOBIUM)		G	—	—	—	—	—	—	—																		
COPPER		U																									
HASTELLOY B/B-2		U																									
HASTELLOY C/C-276		E	—	—	—	—	—	G	—																		
HASTELLOY D		U																									
HASTELLOY G/G-3																											
HIGH SILICON IRON		E	—	G	—	—	S	—	—																		
INCONEL		G	—	U																							
INCOLLOY		E	—																								
LEAD		U																									
MONEL		U																									
NAVAL BRONZE		U																									
NICKEL		U																									
NI-RESIST																											
STAINLESS STEELS																											
Type 304/347		E	—	—	—	—	—	—	—																		
Type 316	4	E	—	—	—	—	—	—	—																		
Type 410		E	—																								
Type 904-L																											
17-4 PH																											
20 Cb 3		E	—	—	—	—	—	—	—																		
E-Bright 26-1		G	—	—	—	—	—	—	—	—	—	—	—	—	—												
SILICON BRONZE		U																									
SILICON COPPER		U																									
STELLITE																											
TANTALUM		E	—	—	—	—	—	—	—	—	—	—	—	—	—	—	—										
TITANIUM		E	—	—	—	—	—	—	—	—	—	—	—	G	—	—	—	—									
ZIRCONIUM		E	—	—	—	—	—	—	—	—	—	—	—	—	—	—	—	—	—	—	—	—	—	—			
°F		60	80	100	120	140	160	180	200	220	240	260	280	300	320	340	360	380	400	420	440	460	480	500	520	540	560
°C		15	26	38	49	60	71	82	93	104	116	127	138	149	160	171	182	193	204	216	227	238	249	260	271	282	293

FOR METALS

E = < 2 Mils Penetration/Year; G = < 20 Mils Penetration /year

S = < 50 Mils Penetration/Year; U = > 50 Mils Penetration/year

FOR NONMETALLICS

R = Resistant

U = Unsatisfactory

NITRIC ACID 5%

°C: 15, 26, 38, 49, 60, 71, 82, 93, 104, 116, 127, 138, 149, 160, 171, 182, 193, 204, 216, 227, 238

°F: 60, 80, 100, 120, 140, 160, 180, 200, 220, 240, 260, 280, 300, 320, 340, 360, 380, 400, 420, 440, 460

PLASTICS	Rating
ABS	R
ACETAL	U
ACRYLICS	R
ASBESTOS REINFORCED EPOXY	R … U
ASBESTOS REINFORCED PHENOLIC	R … U
CHLORINATED POLYETHER (Penton)	R
CPVC	R
E-CTFE (Halar)	R
EPOXY	R
ETFE (Tefzel)	R
FEP	R
FURAN (FURFURAL ALCOHOL)	R … U
NORYL	R
NYLON 6	U
NYLON 11	U
NYLON 66	U
PHENOLIC	U
PFA (Teflon)	R
POLYAMIDE-IMIDE	
POLYESTERS	
Bisphenol A–fumarate	R
Halogenated	R
Hydrogenated Bisphenol A –Bisphenol A	R
Isophthalic	R
Terephthalate (PET)	R
POLYETHERIMIDE (ULTEM)	R
POLYETHERETHERKETONE (PEEK)	R
POLYETHERSULFONE (PES)	R
POLYETHYLENE HMW	R
POLYETHYLENE UHMW	R
POLYMETHYLPENTENE	R
POLYPROPYLENE (PP)	R
POLYSTYRENE	R
POLYSULFONE	U
POLYVINYLIDENE CHLORIDE	R
PVC TYPE 1	R
PVC TYPE 2	R
PVDF (Kynar)	R
RYTON	R
TFE (Teflon)	R
VINYL ESTER	R

°F: 60, 80, 100, 120, 140, 160, 180, 200, 220, 240, 260, 280, 300, 320, 340, 360, 380, 400, 420, 440, 460

°C: 15, 26, 38, 49, 60, 71, 82, 93, 104, 116, 127, 138, 149, 160, 171, 182, 193, 204, 216, 227, 238

NITRIC ACID 5%

°C	15	26	38	49	60	71	82	93	104	116	127	138	149	160	171	182	193	204	216	227	238
°F	60	80	100	120	140	160	180	200	220	240	260	280	300	320	340	360	380	400	420	440	460

ELASTOMERS* AND LININGS	Rating	Line extends to (°F)
ACRYLATE-BUTADIENE (ABR)		
BUTYL GR-1 (IIR)	R	160
CARBOXYLIC-ACRYLONITRILE-BUTADIENE (XNBR)		
CHEMRAZ (FPM)	R	400
CHLORO-ISOBUTENE-ISOPRENE (CIIR)	R	200
HYPALON (CHLORO-SULFONYL-POLYETHYLENE (CSM)	R	100
ETHYLENE-ACRYLIC (EA)		
ETHYLENE-PROPYLENE (EPM)	R	140
ETHYLENE-PROPYLENE-DIENE (EPDM)	R	
ETHYLENE-PROPYLENE TERPOLYMER (EPT)	U	
FKM (Viton A)	R	400
HARD RUBBER	R	140
SOFT RUBBER	U	
ISOPRENE (IR)	U	
KALREZ (FPM)	R	200
KOROSEAL	R	140
NATURAL RUBBER (GRS)	U	
NEOPRENE GR-M (CR)	U	
NITRILE BUNA-N (NBR)	U	
NORDEL (EPDM)	R	300
POLYBUTADIENE (BR)	R	60
POLYESTER (PE)	U	
POLYETHER-URETHANE (EU)		
POLYISOPRENE (IR)		
POLYSULFIDES (T)	U	
POLYURETHANE (AU)	U	
SBR STYRENE (BUNA-S)	U	
SILICONE RUBBERS	R	60

*See also Nylon 11 under PLASTICS

°F	60	80	100	120	140	160	180	200	220	240	260	280	300	320	340	360	380	400	420	440	460
°C	15	26	38	49	60	71	82	93	104	116	127	138	149	160	171	182	193	204	216	227	238

NITRIC ACID 20%

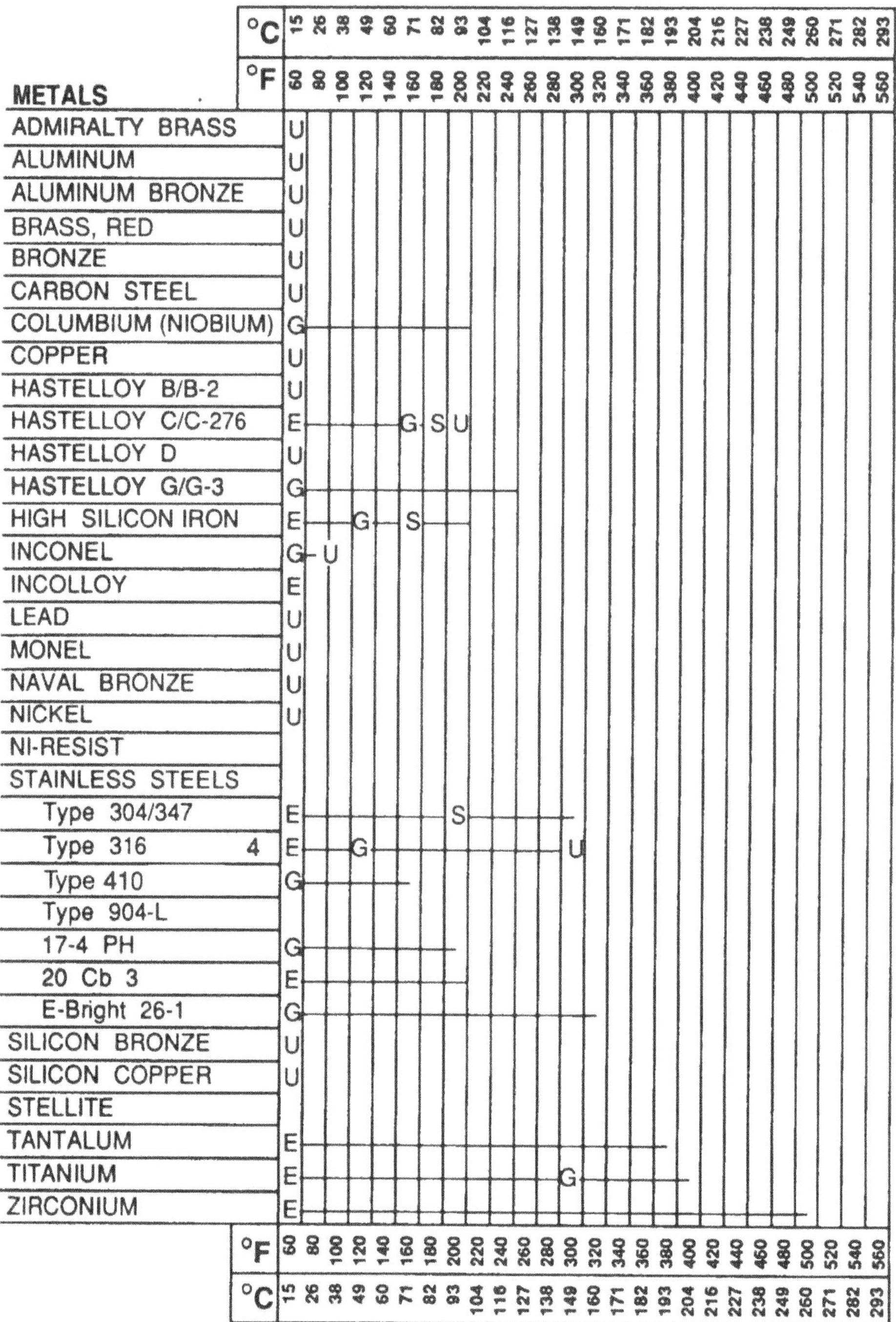

METALS	60	80	100	120	140	160	180	200	220	240	260	280	300	320	340	360	380	400	420	440	460	480	500	520	540	560
°C	15	26	38	49	60	71	82	93	104	116	127	138	149	160	171	182	193	204	216	227	238	249	260	271	282	293
ADMIRALTY BRASS	U																									
ALUMINUM	U																									
ALUMINUM BRONZE	U																									
BRASS, RED	U																									
BRONZE	U																									
CARBON STEEL	U																									
COLUMBIUM (NIOBIUM)	G	—	—	—	—	—	—	—																		
COPPER	U																									
HASTELLOY B/B-2	U																									
HASTELLOY C/C-276	E	—	—	—	—	G	S	U																		
HASTELLOY D	U																									
HASTELLOY G/G-3	G	—	—	—	—	—	—	—	—	—																
HIGH SILICON IRON	E	—	—	G	—	S	—	—																		
INCONEL	G	U																								
INCOLLOY	E																									
LEAD	U																									
MONEL	U																									
NAVAL BRONZE	U																									
NICKEL	U																									
NI-RESIST																										
STAINLESS STEELS																										
Type 304/347	E	—	—	—	—	—	—	S	—	—	—	—	—													
Type 316 4	E	—	—	G	—	—	—	—	—	—	—	—	U													
Type 410	G	—	—	—	—	—																				
Type 904-L																										
17-4 PH	G	—	—	—	—	—	—	—																		
20 Cb 3	E	—	—	—	—	—	—	—																		
E-Bright 26-1	G	—	—	—	—	—	—	—	—	—	—	—	—	—												
SILICON BRONZE	U																									
SILICON COPPER	U																									
STELLITE																										
TANTALUM	E	—	—	—	—	—	—	—	—	—	—	—	—	—	—	—	—									
TITANIUM	E	—	—	—	—	—	—	—	—	—	—	—	G	—	—	—	—	—								
ZIRCONIUM	E	—	—	—	—	—	—	—	—	—	—	—	—	—	—	—	—	—	—	—	—	—	—			

FOR METALS

E = < 2 Mils Penetration/Year; G = < 20 Mils Penetration /year

S = < 50 Mils Penetration/Year; U = > 50 Mils Penetration/year

FOR NONMETALLICS

R = Resistant

U = Unsatisfactory

NITRIC ACID 20%

PLASTICS	60 °F / 15 °C	80 / 26	100 / 38	120 / 49	140 / 60	160 / 71	180 / 82	200 / 93	220 / 104	240 / 116	260 / 127	280 / 138	300 / 149	320 / 160	340 / 171	360 / 182	380 / 193	400 / 204	420 / 216	440 / 227	460 / 238
ABS	R	—	—	—	U																
ACETAL	U																				
ACRYLICS	U																				
ASBESTOS REINFORCED EPOXY																					
ASBESTOS REINFORCED PHENOLIC																					
CHLORINATED POLYETHER (Penton)	R																				
CPVC	R	—	—	—	—	—															
E-CTFE (Halar)	R	—	—	—	—	—	—	—	—	—											
EPOXY	R	—	—																		
ETFE (Tefzel)	R	—	—	—	—																
FEP	R	—	—	—	—	—	—	—	—	—	—	—	—	—	—	—	—	—			
FURAN (FURFURAL ALCOHOL)	U																				
NORYL	R	—																			
NYLON 6	U																				
NYLON 11	U																				
NYLON 66	U																				
PHENOLIC																					
PFA (Teflon)	R	—	—	—	—	—	—	—													
POLYAMIDE-IMIDE																					
POLYESTERS																					
Bisphenol A–fumarate	R	—	—																		
Halogenated	R																				
Hydrogenated Bisphenol A –Bisphenol A																					
Isophthalic	U																				
Terephthalate (PET)	R																				
POLYETHERIMIDE (ULTEM)	R																				
POLYETHERETHERKETONE (PEEK)	R	—	—	—	—	—	—	—													
POLYETHERSULFONE (PES)	U																				
POLYETHYLENE HMW	R	—	—	—	—																
POLYETHYLENE UHMW	R	—	—	—	—																
POLYMETHYLPENTENE	R	—	—	—	—																
POLYPROPYLENE (PP)	R	—	—	—	—																
POLYSTYRENE	R	—																			
POLYSULFONE	U																				
POLYVINYLIDENE CHLORIDE	R	—	—	—	—																
PVC TYPE 1	R	—	—	—	—																
PVC TYPE 2	R	—	—	—	—																
PVDF (Kynar)	R	—	—	—	—	—	—														
RYTON	R	—	—																		
TFE (Teflon)	R	—	—	—	—	—	—	—	—	—	—	—	—	—	—	—	—	—	—	—	—
VINYL ESTER	R	—	—	—	—																

NITRIC ACID 20%

°C: 15 | 26 | 38 | 49 | 60 | 71 | 82 | 93 | 104 | 116 | 127 | 138 | 149 | 160 | 171 | 182 | 193 | 204 | 216 | 227 | 238

°F: 60 | 80 | 100 | 120 | 140 | 160 | 180 | 200 | 220 | 240 | 260 | 280 | 300 | 320 | 340 | 360 | 380 | 400 | 420 | 440 | 460

ELASTOMERS* AND LININGS	
ACRYLATE-BUTADIENE (ABR)	
BUTYL GR-1 (IIR)	R
CARBOXYLIC-ACRYLONITRILE-BUTADIENE (XNBR)	
CHEMRAZ (FPM)	R
CHLORO-ISOBUTENE-ISOPRENE (CIIR)	R
HYPALON (CHLORO-SULFONYL-POLYETHYLENE (CSM)	R
ETHYLENE-ACRYLIC (EA)	
ETHYLENE-PROPYLENE (EPM)	R
ETHYLENE-PROPYLENE-DIENE (EPDM)	R
ETHYLENE-PROPYLENE TERPOLYMER (EPT)	U
FKM (Viton A)	R
HARD RUBBER	U
SOFT RUBBER	U
ISOPRENE (IR)	U
KALREZ (FPM)	R
KOROSEAL	R
NATURAL RUBBER (GRS)	U
NEOPRENE GR-M (CR)	U
NITRILE BUNA-N (NBR)	U
NORDEL (EPDM)	R
POLYBUTADIENE (BR)	R
POLYESTER (PE)	U
POLYETHER-URETHANE (EU)	
POLYISOPRENE (IR)	
POLYSULFIDES (T)	U
POLYURETHANE (AU)	U
SBR STYRENE (BUNA-S)	U
SILICONE RUBBERS	U

*See also Nylon 11 under PLASTICS

°F: 60 | 80 | 100 | 120 | 140 | 160 | 180 | 200 | 220 | 240 | 260 | 280 | 300 | 320 | 340 | 360 | 380 | 400 | 420 | 440 | 460

°C: 15 | 26 | 38 | 49 | 60 | 71 | 82 | 93 | 104 | 116 | 127 | 138 | 149 | 160 | 171 | 182 | 193 | 204 | 216 | 227 | 238

NITRIC ACID 70%

METALS	60	80	100	120	140	160	180	200	220	240	260	280	300	320	340	360	380	400	420	440	460	480	500	520	540	560
°C	15	26	38	49	60	71	82	93	104	116	127	138	149	160	171	182	193	204	216	227	238	249	260	271	282	293
ADMIRALTY BRASS	U																									
ALUMINUM	U																									
ALUMINUM BRONZE	U																									
BRASS, RED	U																									
BRONZE	U																									
CARBON STEEL	U																									
COLUMBIUM (NIOBIUM)	G																									
COPPER	U																									
HASTELLOY B/B-2	U																									
HASTELLOY C/C-276	E				G				U																	
HASTELLOY D																										
HASTELLOY G/G-3	E																									
HIGH SILICON IRON	E																									
INCONEL	U																									
INCOLLOY																										
LEAD	U																									
MONEL	U																									
NAVAL BRONZE	U																									
NICKEL	U																									
NI-RESIST																										
STAINLESS STEELS																										
Type 304/347	E		G				S																			
Type 316 4	E		G																							
Type 410	G																									
Type 904-L																										
17-4 PH	U																									
20 Cb 3	E	G																								
E-Bright 26-1	G																									
SILICON BRONZE	U																									
SILICON COPPER	U																									
STELLITE																										
TANTALUM	E																									
TITANIUM	E												G					S					U			
ZIRCONIUM	E																									
°F	60	80	100	120	140	160	180	200	220	240	260	280	300	320	340	360	380	400	420	440	460	480	500	520	540	560
°C	15	26	38	49	60	71	82	93	104	116	127	138	149	160	171	182	193	204	216	227	238	249	260	271	282	293

FOR METALS

E = < 2 Mils Penetration/Year; G = < 20 Mils Penetration /year

S = < 50 Mils Penetration/Year; U = > 50 Mils Penetration/year

FOR NONMETALLICS

R = Resistant

U = Unsatisfactory

NITRIC ACID 70%

PLASTICS	15	26	38	49	60	71	82	93	104	116	127	138	149	160	171	182	193	204	216	227	238
°F	60	80	100	120	140	160	180	200	220	240	260	280	300	320	340	360	380	400	420	440	460
ABS	U																				
ACETAL	U																				
ACRYLICS	U																				
ASBESTOS REINFORCED EPOXY																					
ASBESTOS REINFORCED PHENOLIC																					
CHLORINATED POLYETHER (Penton)	R	U																			
CPVC	R	—	—	—	—	—	—														
E-CTFE (Halar)	R	—	—	—	—	U															
EPOXY	U																				
ETFE (Tefzel)	R																				
FEP	R	—	—	—	—	—	—	—	—	—	—	—	—	—	—	—	—	—			
FURAN (FURFURAL ALCOHOL)	U																				
NORYL	U																				
NYLON 6	U																				
NYLON 11	U																				
NYLON 66	U																				
PHENOLIC																					
PFA (Teflon)	R	—	—	—	—	—	—	—	—	—											
POLYAMIDE-IMIDE																					
POLYESTERS																					
Bisphenol A–fumarate																					
Halogenated	R																				
Hydrogenated Bisphenol A –Bisphenol A																					
Isophthalic	U																				
Terephthalate (PET)	R																				
POLYETHERIMIDE (ULTEM)																					
POLYETHERETHERKETONE (PEEK)																					
POLYETHERSULFONE (PES)																					
POLYETHYLENE HMW	U																				
POLYETHYLENE UHMW	U																				
POLYMETHYLPENTENE																					
POLYPROPYLENE (PP)	U																				
POLYSTYRENE	U																				
POLYSULFONE	U																				
POLYVINYLIDENE CHLORIDE	U																				
PVC TYPE 1	R	—	—	—	—																
PVC TYPE 2	R	—	—	—	—																
PVDF (Kynar)	R	—	—	—																	
RYTON																					
TFE (Teflon)	R	—	—	—	—	—	—	—	—	—	—	—	—	—	—	—	—	—	—	—	—
VINYL ESTER	U																				

°F 60 80 100 120 140 160 180 200 220 240 260 280 300 320 340 360 380 400 420 440 460
°C 15 26 38 49 60 71 82 93 104 116 127 138 149 160 171 182 193 204 216 227 238

NITRIC ACID 70%

ELASTOMERS* AND LININGS	60	80	100	120	140	160	180	200	220	240	260	280	300	320	340	360	380	400	420	440	460
°C	15	26	38	49	60	71	82	93	104	116	127	138	149	160	171	182	193	204	216	227	238
ACRYLATE-BUTADIENE (ABR)																					
BUTYL GR-1 (IIR)	R																				
CARBOXYLIC-ACRYLONITRILE-BUTADIENE (XNBR)																					
CHEMRAZ (FPM)	U																				
CHLORO-ISOBUTENE-ISOPRENE (CIIR)	U																				
HYPALON (CHLORO-SULFONYL-POLYETHYLENE (CSM)	U																				
ETHYLENE-ACRYLIC (EA)																					
ETHYLENE-PROPYLENE (EPM)	U																				
ETHYLENE-PROPYLENE-DIENE (EPDM)	U																				
ETHYLENE-PROPYLENE TERPOLYMER (EPT)	U																				
FKM (Viton A)	R																				
HARD RUBBER	U																				
SOFT RUBBER	U																				
ISOPRENE (IR)	U																				
KALREZ (FPM)	R																				
KOROSEAL	U																				
NATURAL RUBBER (GRS)	U																				
NEOPRENE GR-M (CR)	U																				
NITRILE BUNA-N (NBR)	U																				
NORDEL (EPDM)	U																				
POLYBUTADIENE (BR)	U																				
POLYESTER (PE)	U																				
POLYETHER-URETHANE (EU)																					
POLYISOPRENE (IR)																					
POLYSULFIDES (T)	U																				
POLYURETHANE (AU)	U																				
SBR STYRENE (BUNA-S)	U																				
SILICONE RUBBERS	U																				
°F	60	80	100	120	140	160	180	200	220	240	260	280	300	320	340	360	380	400	420	440	460
°C	15	26	38	49	60	71	82	93	104	116	127	138	149	160	171	182	193	204	216	227	238

*See also Nylon 11 under PLASTICS

NITRIC ACID 100% (ANHYDROUS)

METALS	°F 60	80	100	120	140	160	180	200	220	240	260	280	300	320	340	360	380	400	420	440	460	480	500	520	540	560
°C	15	26	38	49	60	71	82	93	104	116	127	138	149	160	171	182	193	204	216	227	238	249	260	271	282	293
ADMIRALTY BRASS	U																									
ALUMINUM	E	—																								
ALUMINUM BRONZE	U																									
BRASS, RED	U																									
BRONZE	U																									
CARBON STEEL	U																									
COLUMBIUM (NIOBIUM)																										
COPPER	U																									
HASTELLOY B/B-2	U																									
HASTELLOY C/C-276	G	—																								
HASTELLOY D																										
HASTELLOY G/G-3																										
HIGH SILICON IRON	E	—	—	—	—	—	—	—	—	—																
INCONEL	U																									
INCOLLOY	G	—	—	—	—	—	—	—																		
LEAD	U																									
MONEL	U																									
NAVAL BRONZE	U																									
NICKEL	U																									
NI-RESIST																										
STAINLESS STEELS																										
Type 304/347	E	U																								
Type 316 4	E	—	—	U																						
Type 410	S																									
Type 904-L																										
17-4 PH																										
20 Cb 3	E	U																								
E-Bright 26-1																										
SILICON BRONZE	U																									
SILICON COPPER	U																									
STELLITE																										
TANTALUM	E	—	—	—	—	—	—	—	—	—	—	—	—													
TITANIUM	G	—	—	—	—	—	—	—																		
ZIRCONIUM	E	—																								
°F	60	80	100	120	140	160	180	200	220	240	260	280	300	320	340	360	380	400	420	440	460	480	500	520	540	560
°C	15	26	38	49	60	71	82	93	104	116	127	138	149	160	171	182	193	204	216	227	238	249	260	271	282	293

FOR METALS

E = < 2 Mils Penetration/Year; G = < 20 Mils Penetration /year

S = < 50 Mils Penetration/Year; U = > 50 Mils Penetration/year

FOR NONMETALLICS

R = Resistant

U = Unsatisfactory

NITRIC ACID 100% (ANHYDROUS)

PLASTICS																					
°C	15	26	38	49	60	71	82	93	104	116	127	138	149	160	171	182	193	204	216	227	238
°F	60	80	100	120	140	160	180	200	220	240	260	280	300	320	340	360	380	400	420	440	460
ABS	U																				
ACETAL	U																				
ACRYLICS	U																				
ASBESTOS REINFORCED EPOXY	U																				
ASBESTOS REINFORCED PHENOLIC	R																				
CHLORINATED POLYETHER (Penton)	U																				
CPVC	U																				
E-CTFE (Halar)	R					U															
EPOXY	U																				
ETFE (Tefzel)	U																				
FEP	R																				
FURAN (FURFURAL ALCOHOL)	U																				
NORYL	U																				
NYLON 6	U																				
NYLON 11	U																				
NYLON 66	U																				
PHENOLIC	R																				
PFA (Teflon) 90%	R																				
POLYAMIDE-IMIDE																					
POLYESTERS																					
Bisphenol A–fumarate																					
Halogenated																					
Hydrogenated Bisphenol A –Bisphenol A																					
Isophthalic	U																				
Terephthalate (PET)																					
POLYETHERIMIDE (ULTEM)																					
POLYETHERETHERKETONE (PEEK)																					
POLYETHERSULFONE (PES)																					
POLYETHYLENE HMW	U																				
POLYETHYLENE UHMW	U																				
POLYMETHYLPENTENE																					
POLYPROPYLENE (PP)	U																				
POLYSTYRENE	U																				
POLYSULFONE	U																				
POLYVINYLIDENE CHLORIDE	U																				
PVC TYPE 1	U																				
PVC TYPE 2	U																				
PVDF (Kynar)	R																				
RYTON																					
TFE (Teflon)	R																				
VINYL ESTER	U																				
°F	60	80	100	120	140	160	180	200	220	240	260	280	300	320	340	360	380	400	420	440	460
°C	15	26	38	49	60	71	82	93	104	116	127	138	149	160	171	182	193	204	216	227	238

NITRIC ACID 100% (ANHYDROUS)

ELASTOMERS* AND LININGS	15	26	38	49	60	71	82	93	104	116	127	138	149	160	171	182	193	204	216	227	238
	60	80	100	120	140	160	180	200	220	240	260	280	300	320	340	360	380	400	420	440	460
ACRYLATE-BUTADIENE (ABR)																					
BUTYL GR-1 (IIR)	U																				
CARBOXYLIC-ACRYLONITRILE-BUTADIENE (XNBR)																					
CHEMRAZ (FPM)	U																				
CHLORO-ISOBUTENE-ISOPRENE (CIIR)	U																				
HYPALON (CHLORO-SULFONYL-POLYETHYLENE (CSM)	U																				
ETHYLENE-ACRYLIC (EA)																					
ETHYLENE-PROPYLENE (EPM)	U																				
ETHYLENE-PROPYLENE-DIENE (EPDM)	U																				
ETHYLENE-PROPYLENE TERPOLYMER (EPT)	U																				
FKM (Viton A)	R																				
HARD RUBBER	U																				
SOFT RUBBER	U																				
ISOPRENE (IR)	U																				
KALREZ (FPM)	U																				
KOROSEAL	U																				
NATURAL RUBBER (GRS)	U																				
NEOPRENE GR-M (CR)	U																				
NITRILE BUNA-N (NBR)	U																				
NORDEL (EPDM)																					
POLYBUTADIENE (BR)	U																				
POLYESTER (PE)	U																				
POLYETHER-URETHANE (EU)																					
POLYISOPRENE (IR)																					
POLYSULFIDES (T)	U																				
POLYURETHANE (AU)	U																				
SBR STYRENE (BUNA-S)	U																				
SILICONE RUBBERS	U																				
°F	60	80	100	120	140	160	180	200	220	240	260	280	300	320	340	360	380	400	420	440	460
°C	15	26	38	49	60	71	82	93	104	116	127	138	149	160	171	182	193	204	216	227	238

*See also Nylon 11 under PLASTICS

OLEUM

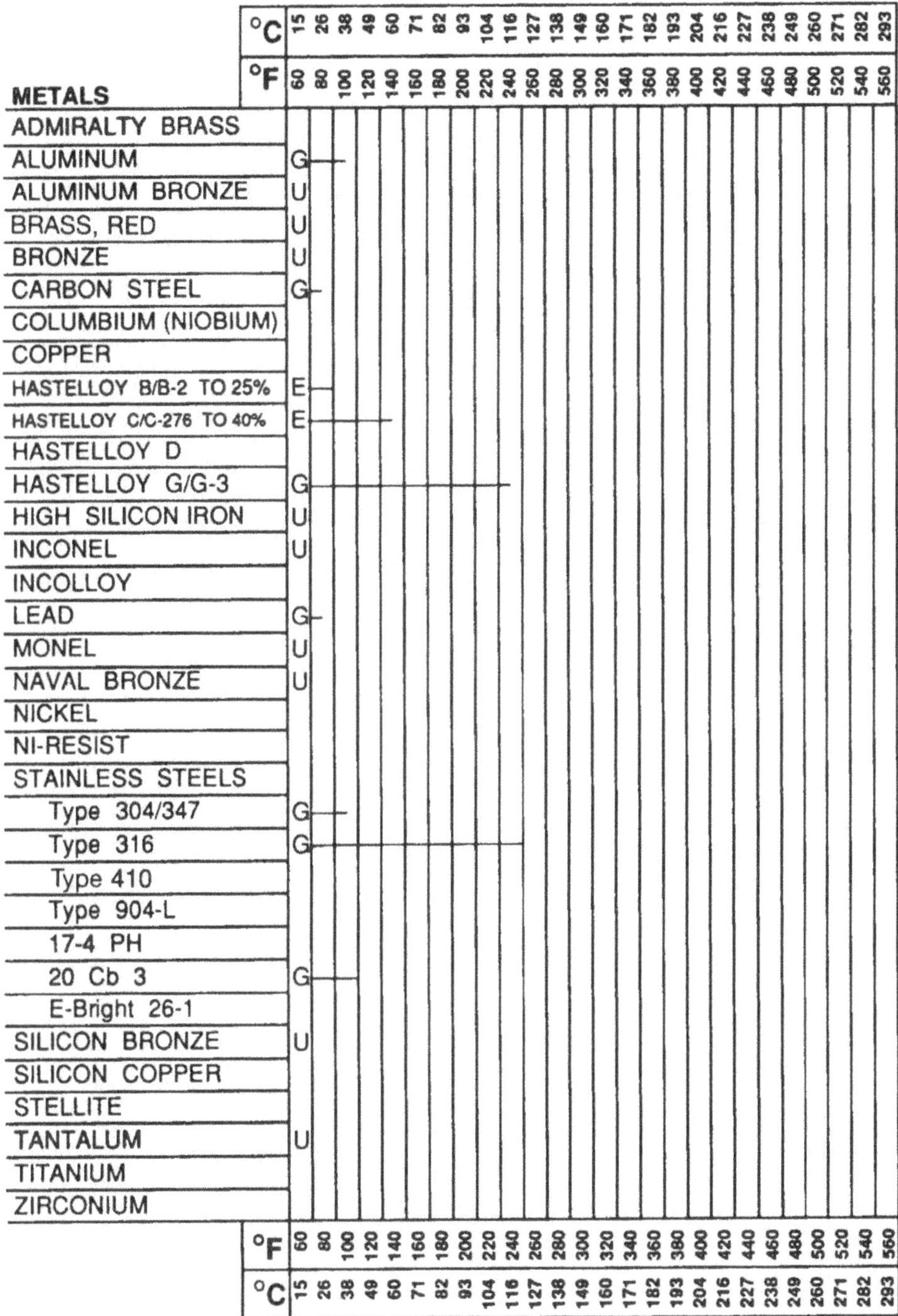

METALS	Rating
ADMIRALTY BRASS	
ALUMINUM	G
ALUMINUM BRONZE	U
BRASS, RED	U
BRONZE	U
CARBON STEEL	G
COLUMBIUM (NIOBIUM)	
COPPER	
HASTELLOY B/B-2 TO 25%	E
HASTELLOY C/C-276 TO 40%	E
HASTELLOY D	
HASTELLOY G/G-3	G
HIGH SILICON IRON	U
INCONEL	U
INCOLLOY	
LEAD	G
MONEL	U
NAVAL BRONZE	U
NICKEL	
NI-RESIST	
STAINLESS STEELS	
Type 304/347	G
Type 316	G
Type 410	
Type 904-L	
17-4 PH	
20 Cb 3	G
E-Bright 26-1	
SILICON BRONZE	U
SILICON COPPER	
STELLITE	
TANTALUM	U
TITANIUM	
ZIRCONIUM	

FOR METALS
E = < 2 Mils Penetration/Year; G = < 20 Mils Penetration /year
S = < 50 Mils Penetration/Year; U = > 50 Mils Penetration/year

FOR NONMETALLICS
R = Resistant
U = Unsatisfactory

PLASTICS	60	80	100	120	140	160	180	200	220	240	260	280	300	320	340	360	380	400	420	440	460
°C	15	26	38	49	60	71	82	93	104	116	127	138	149	160	171	182	193	204	216	227	238
ABS	U																				
ACETAL	U																				
ACRYLICS	U																				
ASBESTOS REINFORCED EPOXY																					
ASBESTOS REINFORCED PHENOLIC																					
CHLORINATED POLYETHER (Penton)																					
CPVC	U																				
E-CTFE (Halar)	R	U																			
EPOXY	U																				
ETFE (Tefzel)	R	—	—	—	—																
FEP	R	—	—	—	—	—	—	—	—	—	—	—	—	—	—	—	—				
FURAN (FURFURAL ALCOHOL)	R	—	—	—	—	—	—	U													
NORYL																					
NYLON 6																					
NYLON 11																					
NYLON 66																					
PHENOLIC																					
PFA (Teflon)	R																				
POLYAMIDE-IMIDE	R	—	—																		
POLYESTERS																					
Bisphenol A–fumarate	U																				
Halogenated	U																				
Hydrogenated Bisphenol A –Bisphenol A	U																				
Isophthalic	U																				
Terephthalate (PET)																					
POLYETHERIMIDE (ULTEM)																					
POLYETHERETHERKETONE (PEEK)																					
POLYETHERSULFONE (PES)																					
POLYETHYLENE HMW																					
POLYETHYLENE UHMW																					
POLYMETHYLPENTENE																					
POLYPROPYLENE (PP)	U																				
POLYSTYRENE																					
POLYSULFONE																					
POLYVINYLIDENE CHLORIDE	U																				
PVC TYPE 1	U																				
PVC TYPE 2	U																				
PVDF (Kynar)	U																				
RYTON	R																				
TFE (Teflon)	R	—	—	—	—	—	—	—	—	—	—	—	—	—	—	—	—	—	—	—	—
VINYL ESTER	U																				
°F	60	80	100	120	140	160	180	200	220	240	260	280	300	320	340	360	380	400	420	440	460
°C	15	26	38	49	60	71	82	93	104	116	127	138	149	160	171	182	193	204	216	227	238

	15	26	38	49	60	71	82	93	104	116	127	138	149	160	171	182	193	204	216	227	238
ELASTOMERS* AND LININGS °F	60	80	100	120	140	160	180	200	220	240	260	280	300	320	340	360	380	400	420	440	460
ACRYLATE-BUTADIENE (ABR)																					
BUTYL GR-1 (IIR)	U																				
CARBOXYLIC-ACRYLONITRILE-BUTADIENE (XNBR)																					
CHEMRAZ (FPM)	R																				
CHLORO-ISOBUTENE-ISOPRENE (CIIR)																					
HYPALON (CHLORO-SULFONYL-POLYETHYLENE (CSM)	U																				
ETHYLENE-ACRYLIC (EA)																					
ETHYLENE-PROPYLENE (EPM)	U																				
ETHYLENE-PROPYLENE-DIENE (EPDM)	U																				
ETHYLENE-PROPYLENE TERPOLYMER (EPT)	U																				
FKM (Viton A)	R																				
HARD RUBBER	U																				
SOFT RUBBER																					
ISOPRENE (IR)	U																				
KALREZ (FPM)	R																				
KOROSEAL																					
NATURAL RUBBER (GRS)	U																				
NEOPRENE GR-M (CR)	U																				
NITRILE BUNA-N (NBR)	U																				
NORDEL (EPDM)	U																				
POLYBUTADIENE (BR)																					
POLYESTER (PE)	U																				
POLYETHER-URETHANE (EU)																					
POLYISOPRENE (IR)																					
POLYSULFIDES (T)																					
POLYURETHANE (AU)	U																				
SBR STYRENE (BUNA-S)	U																				
SILICONE RUBBERS	U																				
°F	60	80	100	120	140	160	180	200	220	240	260	280	300	320	340	360	380	400	420	440	460
°C	15	26	38	49	60	71	82	93	104	116	127	138	149	160	171	182	193	204	216	227	238

*See also Nylon 11 under PLASTICS

PHOSPHORIC ACID 50–85%

METALS	60°F (15°C)	80°F (26°C)	100°F (38°C)	120°F (49°C)	140°F (60°C)	160°F (71°C)	180°F (82°C)	200°F (93°C)	220°F (104°C)	240°F (116°C)	260°F (127°C)	280°F (138°C)	300°F (149°C)	320°F (160°C)	340°F (171°C)	360°F (182°C)
ADMIRALTY BRASS	U															
ALUMINUM	U															
ALUMINUM BRONZE	U															
BRASS, RED	U															
BRONZE	U															
CARBON STEEL	U															
COLUMBIUM (NIOBIUM)	E															
COPPER	U															
HASTELLOY B/B-2	E					G										
HASTELLOY C/C-276	E															
HASTELLOY D	E					G										
HASTELLOY G/G-3	G															
HIGH SILICON IRON	G															
INCONEL	G							U								
INCOLLOY																
LEAD	G															
MONEL	S															
NAVAL BRONZE	U															
NICKEL	U															
NI-RESIST	U															
STAINLESS STEELS																
Type 304/347 4	E															
Type 316 4	G															
Type 410	S															
Type 904-L	G															
17-4 PH 70%	U															
20 Cb 3	E			G												
E-Bright 26-1																
SILICON BRONZE	U															
SILICON COPPER	G															
STELLITE																
TANTALUM	E															U
TITANIUM	S															
ZIRCONIUM	G															

FOR METALS

E = < 2 Mils Penetration/Year; G = < 20 Mils Penetration /year

S = < 50 Mils Penetration/Year; U = > 50 Mils Penetration/year

FOR NONMETALLICS

R = Resistant

U = Unsatisfactory

PHOSPHORIC ACID 50–85%

PLASTICS	60	80	100	120	140	160	180	200	220	240	260	280	300	320	340	360	380	400	420	440	460
°C	15	26	38	49	60	71	82	93	104	116	127	138	149	160	171	182	193	204	216	227	238
ABS	R	—	—	—	U																
ACETAL	U																				
ACRYLICS	R	—																			
ASBESTOS REINFORCED EPOXY	R	—	—	—	—	—	—	—													
ASBESTOS REINFORCED PHENOLIC	R	—	—	—	—	—	—	—	—	—	—	—	—								
CHLORINATED POLYETHER (Penton)	R	—	—	—	—	—	—	—	—	—											
CPVC	R	—	—	—	—	—	—														
E-CTFE (Halar)	R	—	—	—	—	—	—	—	—	—											
EPOXY	R	—	—	U																	
ETFE (Tefzel)	R	—	—	—	—	—	—	—	—	—	—										
FEP	R	—	—	—	—	—	—	—	—	—	—	—	—	—	—	—	—	—			
FURAN (FURFURAL ALCOHOL)	R	—	—	—	—	—	—	—	—	—	—										
NORYL	R	—	—	—	—																
NYLON 6	U																				
NYLON 11	U																				
NYLON 66	U																				
PHENOLIC	U																				
PFA (Teflon)	R	—	—	—	—	—	—	—													
POLYAMIDE-IMIDE																					
POLYESTERS																					
Bisphenol A–fumarate	R	—	—	—	—	—	—	—	—												
Halogenated	R	—	—	—	—	—	—	—	—	—											
Hydrogenated Bisphenol A –Bisphenol A	R	—	—	—	—	—	—	—													
Isophthalic	R	—	—	—	—	—	—														
Terephthalate (PET)	R	—	—	—	—	—	—	—	—	—											
POLYETHERIMIDE (ULTEM)	R	—																			
POLYETHERETHERKETONE (PEEK)	R	—	—	—	—	—	—	—													
POLYETHERSULFONE (PES)	R	—	—	—	—	—	—	—													
POLYETHYLENE HMW	R	—	—																		
POLYETHYLENE UHMW	R	—	—																		
POLYMETHYLPENTENE	R	—	—	—	—																
POLYPROPYLENE (PP)	R	—	—	—	—	—	—	—													
POLYSTYRENE	R	—	—	—	—	—															
POLYSULFONE	R	—																			
POLYVINYLIDENE CHLORIDE	R	—	—	—																	
PVC TYPE 1	R	—	—	—	—																
PVC TYPE 2	R	—	—	—	—																
PVDF (Kynar)	R	—	—	—	—	—	—	—	—	—											
RYTON	R	—	—	—	—	—	—	—	—												
TFE (Teflon)	R	—	—	—	—	—	—	—	—	—	—	—	—	—	—	—	—	—	—	—	—
VINYL ESTER	R	—	—	—	—	—	—	—													
°F	60	80	100	120	140	160	180	200	220	240	260	280	300	320	340	360	380	400	420	440	460
°C	15	26	38	49	60	71	82	93	104	116	127	138	149	160	171	182	193	204	216	227	238

PHOSPHORIC ACID 50–85%

ELASTOMERS* AND LININGS		Rating	Max. temp. °F (°C)
ACRYLATE-BUTADIENE (ABR)			
BUTYL GR-1 (IIR)		R	150 (66)
CARBOXYLIC-ACRYLONITRILE-BUTADIENE (XNBR)			
CHEMRAZ (FPM)		R	400 (204)
CHLORO-ISOBUTENE-ISOPRENE (CIIR)		R	150 (66)
HYPALON (CHLORO-SULFONYL-POLYETHYLENE (CSM)		R	200 (93)
ETHYLENE-ACRYLIC (EA)			
ETHYLENE-PROPYLENE (EPM)		R	170 (77)
ETHYLENE-PROPYLENE-DIENE (EPDM)		R	140 (60)
ETHYLENE-PROPYLENE TERPOLYMER (EPT)		R	180 (82)
FKM (Viton A)		R	300 (149)
HARD RUBBER		R	200 (93)
SOFT RUBBER		R	150 (66)
ISOPRENE (IR)		R; U	R to 100 (38); U at 120 (49)
KALREZ (FPM)		R	210 (99)
KOROSEAL		R	150 (66)
NATURAL RUBBER (GRS)		R; U	R to 100 (38); U at 120 (49)
NEOPRENE GR-M (CR)	50%	R	140 (60)
NITRILE BUNA-N (NBR)		U	
NORDEL (EPDM)		R	300 (149)
POLYBUTADIENE (BR)			
POLYESTER (PE)			
POLYETHER-URETHANE (EU)			
POLYISOPRENE (IR)			
POLYSULFIDES (T)		U	
POLYURETHANE (AU)			
SBR STYRENE (BUNA-S)			
SILICONE RUBBERS		U	

Temperature scale: °F 60, 80, 100, 120, 140, 160, 180, 200, 220, 240, 260, 280, 300, 320, 340, 360, 380, 400, 420, 440, 460; °C 15, 26, 38, 49, 60, 71, 82, 93, 104, 116, 127, 138, 149, 160, 171, 182, 193, 204, 216, 227, 238

*See also Nylon 11 under PLASTICS

SODIUM CHLORIDE

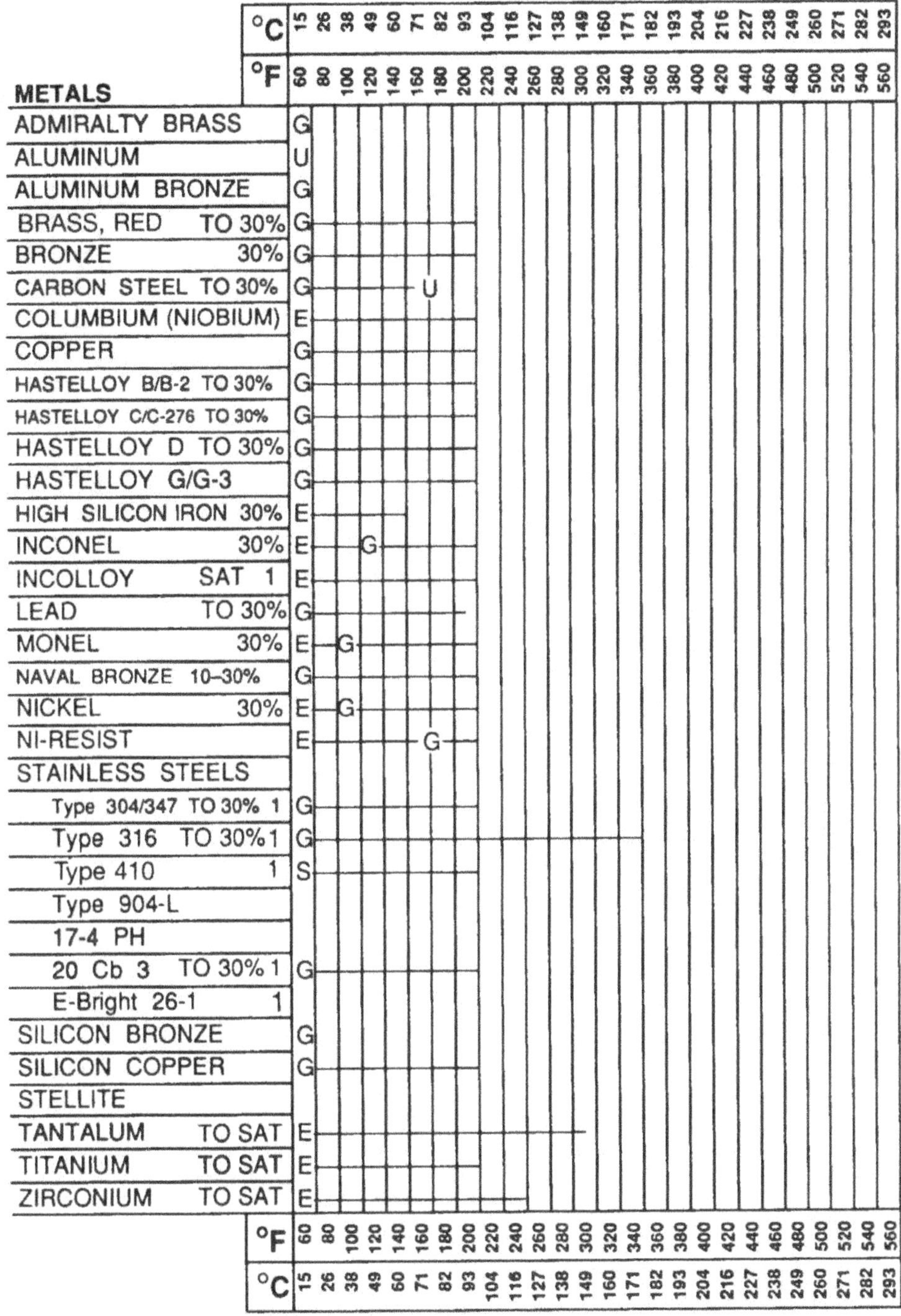

°C: 15, 26, 38, 49, 60, 71, 82, 93, 104, 116, 127, 138, 149, 160, 171, 182, 193, 204, 216, 227, 238, 249, 260, 271, 282, 293

°F: 60, 80, 100, 120, 140, 160, 180, 200, 220, 240, 260, 280, 300, 320, 340, 360, 380, 400, 420, 440, 460, 480, 500, 520, 540, 560

METALS	Concentration	Note	Rating
ADMIRALTY BRASS			G
ALUMINUM			U
ALUMINUM BRONZE			G
BRASS, RED	TO 30%		G
BRONZE	30%		G
CARBON STEEL	TO 30%		G – U
COLUMBIUM (NIOBIUM)			E
COPPER			G
HASTELLOY B/B-2	TO 30%		G
HASTELLOY C/C-276	TO 30%		G
HASTELLOY D	TO 30%		G
HASTELLOY G/G-3			G
HIGH SILICON IRON	30%		E
INCONEL	30%		E – G
INCOLLOY	SAT	1	E
LEAD	TO 30%		G
MONEL	30%		E – G
NAVAL BRONZE	10–30%		G
NICKEL	30%		E – G
NI-RESIST			E – G
STAINLESS STEELS			
Type 304/347	TO 30%	1	G
Type 316	TO 30%	1	G
Type 410		1	S
Type 904-L			
17-4 PH			
20 Cb 3	TO 30%	1	G
E-Bright 26-1		1	
SILICON BRONZE			G
SILICON COPPER			G
STELLITE			
TANTALUM	TO SAT		E
TITANIUM	TO SAT		E
ZIRCONIUM	TO SAT		E

FOR METALS

E = < 2 Mils Penetration/Year; G = < 20 Mils Penetration /year

S = < 50 Mils Penetration/Year; U = > 50 Mils Penetration/year

FOR NONMETALLICS

R = Resistant

U = Unsatisfactory

SODIUM CHLORIDE

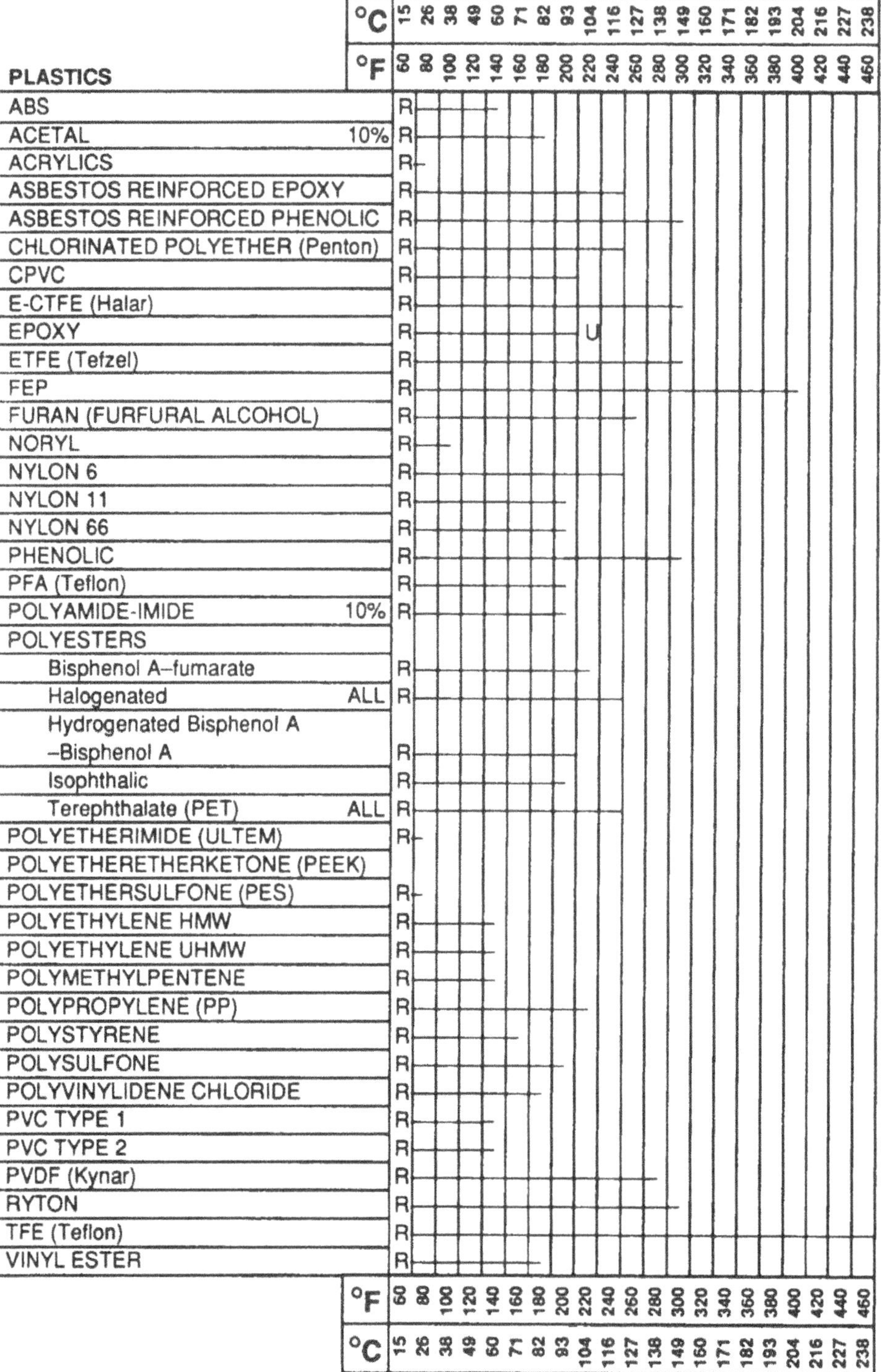

PLASTICS		
ABS		R
ACETAL	10%	R
ACRYLICS		R
ASBESTOS REINFORCED EPOXY		R
ASBESTOS REINFORCED PHENOLIC		R
CHLORINATED POLYETHER (Penton)		R
CPVC		R
E-CTFE (Halar)		R
EPOXY		R U
ETFE (Tefzel)		R
FEP		R
FURAN (FURFURAL ALCOHOL)		R
NORYL		R
NYLON 6		R
NYLON 11		R
NYLON 66		R
PHENOLIC		R
PFA (Teflon)		R
POLYAMIDE-IMIDE	10%	R
POLYESTERS		
Bisphenol A–fumarate		R
Halogenated	ALL	R
Hydrogenated Bisphenol A –Bisphenol A		R
Isophthalic		R
Terephthalate (PET)	ALL	R
POLYETHERIMIDE (ULTEM)		R
POLYETHERETHERKETONE (PEEK)		
POLYETHERSULFONE (PES)		R
POLYETHYLENE HMW		R
POLYETHYLENE UHMW		R
POLYMETHYLPENTENE		R
POLYPROPYLENE (PP)		R
POLYSTYRENE		R
POLYSULFONE		R
POLYVINYLIDENE CHLORIDE		R
PVC TYPE 1		R
PVC TYPE 2		R
PVDF (Kynar)		R
RYTON		R
TFE (Teflon)		R
VINYL ESTER		R

SODIUM CHLORIDE

ELASTOMERS* AND LININGS		60	80	100	120	140	160	180	200	220	240	260	280	300	320	340	360	380	400	420	440	460
°C		15	26	38	49	60	71	82	93	104	116	127	138	149	160	171	182	193	204	216	227	238
ACRYLATE-BUTADIENE (ABR)																						
BUTYL GR-1 (IIR)		R	—	—	—	—	—	—														
CARBOXYLIC-ACRYLONITRILE-BUTADIENE (XNBR)		R	—	—	—	—	—	—														
CHEMRAZ (FPM)		R	—	—	—	—	—	—	—	—	—	—	—	—	—	—	—	—	—			
CHLORO-ISOBUTENE-ISOPRENE (CIIR)		R	—	—	—	—	—	—														
HYPALON (CHLORO-SULFONYL-POLYETHYLENE (CSM)		R	—	—	—	—	—	—	—	—	—											
ETHYLENE-ACRYLIC (EA)																						
ETHYLENE-PROPYLENE (EPM)		R	—	—	—	—	—															
ETHYLENE-PROPYLENE-DIENE (EPDM)		R	—	—	—	—	—	—	—	—	—	—	—	—								
ETHYLENE-PROPYLENE TERPOLYMER (EPT)		R	—	—	—	—	—	—														
FKM (Viton A)		R	—	—	—	—	—	—	—	—	—	—	—	—	—	—	—	—	—			
HARD RUBBER		R	—	—	—	—	—	—	—													
SOFT RUBBER		R	—	—	—	—																
ISOPRENE (IR)		R	—	—	—																	
KALREZ (FPM)		R	—	—	—	—	—	—	—													
KOROSEAL		R	—	—	—	—																
NATURAL RUBBER (GRS)		R	—	—	—																	
NEOPRENE GR-M (CR)		R	—	—	—	—	—	—	—													
NITRILE BUNA-N (NBR)		R	—	—	—	—	—	—	—	—												
NORDEL (EPDM)		R	—	—	—	—	—	—	—	—	—	—	—	—								
POLYBUTADIENE (BR)																						
POLYESTER (PE)		R																				
POLYETHER-URETHANE (EU)		R	—	—	—																	
POLYISOPRENE (IR)																						
POLYSULFIDES (T)		R																				
POLYURETHANE (AU)		R																				
SBR STYRENE (BUNA-S)		R	—	—	—	—	—	—	—													
SILICONE RUBBERS	10%	R	—	—	—	—	—	—	—	—	—	—	—	—	—	—	—	—	—			

*See also Nylon 11 under PLASTICS

SODIUM HYDROXIDE 10%

METALS	Note	Rating (°F range as charted)
ADMIRALTY BRASS		G (60)
ALUMINUM		U (60)
ALUMINUM BRONZE		G (60)
BRASS, RED		G (60–200)
BRONZE		E (60); G (80–180)
CARBON STEEL		G (60–200)
COLUMBIUM (NIOBIUM)		G (60–80)
COPPER		E (60); G (80–200)
HASTELLOY B/B-2	2	E (60–220); G (240)
HASTELLOY C/C-276		G (60–220)
HASTELLOY D		G (60–80)
HASTELLOY G/G-3		
HIGH SILICON IRON		G (60–100); S (120–160); U (180)
INCONEL		E (60–200); G (220–300)
INCOLLOY		E (60–80)
LEAD		U (60)
MONEL	2	E (60–200); G (220–340)
NAVAL BRONZE	2	G (60–200)
NICKEL	2	E (60–200)
NI-RESIST		G (60–160)
STAINLESS STEELS		
Type 304/347		E (60–200)
Type 316		E (60–340)
Type 410		G (60–200)
Type 904-L		
17-4 PH		E (60–180)
20 Cb 3		E (60–180); G (200–300)
E-Bright 26-1		E (60–200)
SILICON BRONZE		G (60–80)
SILICON COPPER		G (60)
STELLITE		
TANTALUM		U (60)
TITANIUM		E (60–200)
ZIRCONIUM		E (60–200)

Temperature scale: °F 60, 80, 100, 120, 140, 160, 180, 200, 220, 240, 260, 280, 300, 320, 340, 360, 380, 400, 420, 440, 460, 480, 500, 520, 540, 560; °C 15, 26, 38, 49, 60, 71, 82, 93, 104, 116, 127, 138, 149, 160, 171, 182, 193, 204, 216, 227, 238, 249, 260, 271, 282, 293

FOR METALS
E = < 2 Mils Penetration/Year; G = < 20 Mils Penetration /year
S = < 50 Mils Penetration/Year; U = > 50 Mils Penetration/year

FOR NONMETALLICS
R = Resistant
U = Unsatisfactory

SODIUM HYDROXIDE 10%

PLASTICS		60°F (15°C)	80°F (26°C)	100°F (38°C)	120°F (49°C)	140°F (60°C)	160°F (71°C)	180°F (82°C)	200°F (93°C)	220°F (104°C)	240°F (116°C)	260°F (127°C)	280°F (138°C)	300°F (149°C)	320°F (160°C)	340°F (171°C)	360°F (182°C)	380°F (193°C)	400°F (204°C)	420°F (216°C)	440°F (227°C)	460°F (238°C)
ABS		R																				
ACETAL		R																				
ACRYLICS		R																				
ASBESTOS REINFORCED EPOXY		R																				
ASBESTOS REINFORCED PHENOLIC		U																				
CHLORINATED POLYETHER (Penton)		R																				
CPVC		R																				
E-CTFE (Halar)		R																				
EPOXY		R							U													
ETFE (Tefzel)		R																				
FEP		R																				
FURAN (FURFURAL ALCOHOL)		U																				
NORYL		R																				
NYLON 6		R																				
NYLON 11		R																				
NYLON 66		R																				
PHENOLIC		U																				
PFA (Teflon)		R																				
POLYAMIDE-IMIDE		U																				
POLYESTERS																						
Bisphenol A–fumarate		R				U																
Halogenated		R			U																	
Hydrogenated Bisphenol A –Bisphenol A		R																				
Isophthalic		U																				
Terephthalate (PET)		R																				
POLYETHERIMIDE (ULTEM)		R																				
POLYETHERETHERKETONE (PEEK)		R																				
POLYETHERSULFONE (PES)		R																				
POLYETHYLENE HMW		R																				
POLYETHYLENE UHMW		R																				
POLYMETHYLPENTENE		R																				
POLYPROPYLENE (PP)		R																				
POLYSTYRENE		R																				
POLYSULFONE		R																				
POLYVINYLIDENE CHLORIDE		R		U																		
PVC TYPE 1		R																				
PVC TYPE 2		R																				
PVDF (Kynar)	2	R																				
RYTON		R																				
TFE (Teflon)		R																				
VINYL ESTER		R						U														

SODIUM HYDROXIDE 10%

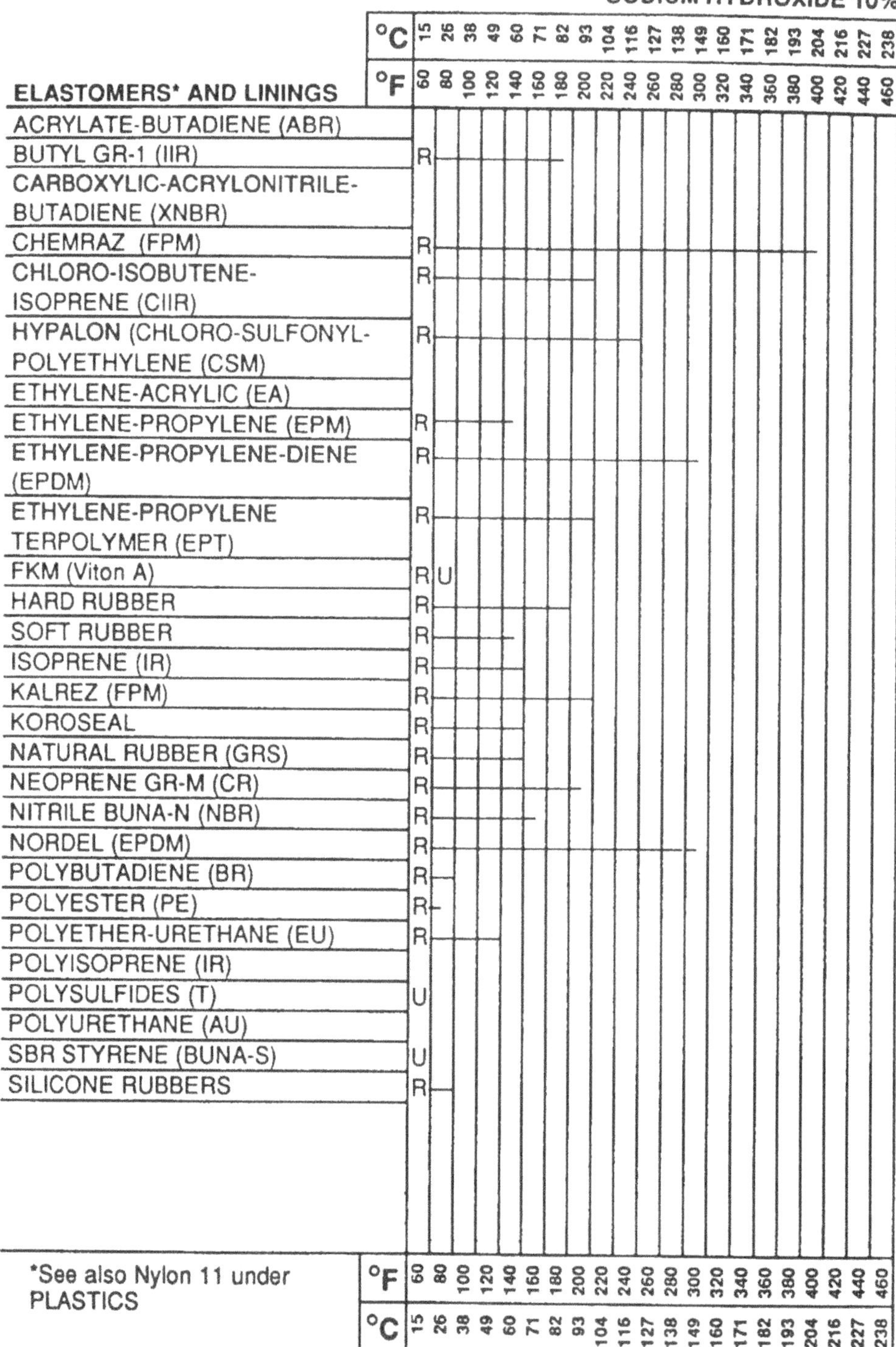

°C	15	26	38	49	60	71	82	93	104	116	127	138	149	160	171	182	193	204	216	227	238
ELASTOMERS* AND LININGS °F	60	80	100	120	140	160	180	200	220	240	260	280	300	320	340	360	380	400	420	440	460
ACRYLATE-BUTADIENE (ABR)																					
BUTYL GR-1 (IIR)	R																				
CARBOXYLIC-ACRYLONITRILE-BUTADIENE (XNBR)																					
CHEMRAZ (FPM)	R																				
CHLORO-ISOBUTENE-ISOPRENE (CIIR)	R																				
HYPALON (CHLORO-SULFONYL-POLYETHYLENE (CSM)	R																				
ETHYLENE-ACRYLIC (EA)																					
ETHYLENE-PROPYLENE (EPM)	R																				
ETHYLENE-PROPYLENE-DIENE (EPDM)	R																				
ETHYLENE-PROPYLENE TERPOLYMER (EPT)	R																				
FKM (Viton A)	R	U																			
HARD RUBBER	R																				
SOFT RUBBER	R																				
ISOPRENE (IR)	R																				
KALREZ (FPM)	R																				
KOROSEAL	R																				
NATURAL RUBBER (GRS)	R																				
NEOPRENE GR-M (CR)	R																				
NITRILE BUNA-N (NBR)	R																				
NORDEL (EPDM)	R																				
POLYBUTADIENE (BR)	R																				
POLYESTER (PE)	R																				
POLYETHER-URETHANE (EU)	R																				
POLYISOPRENE (IR)																					
POLYSULFIDES (T)	U																				
POLYURETHANE (AU)																					
SBR STYRENE (BUNA-S)	U																				
SILICONE RUBBERS	R																				
°F	60	80	100	120	140	160	180	200	220	240	260	280	300	320	340	360	380	400	420	440	460
°C	15	26	38	49	60	71	82	93	104	116	127	138	149	160	171	182	193	204	216	227	238

*See also Nylon 11 under PLASTICS

METALS		Rating
ADMIRALTY BRASS		G
ALUMINUM		U
ALUMINUM BRONZE		U
BRASS, RED		U
BRONZE		S — U
CARBON STEEL		S — U
COLUMBIUM (NIOBIUM)		
COPPER		S — U
HASTELLOY B/B-2	2	E —
HASTELLOY C/C-276		E — G —
HASTELLOY D		E —
HASTELLOY G/G-3		
HIGH SILICON IRON		S — U
INCONEL	2	E — G —
INCOLLOY		
LEAD		U
MONEL		E — G —
NAVAL BRONZE		U
NICKEL	2	E — G —
NI-RESIST		G —
STAINLESS STEELS		
Type 304/347		E — G —
Type 316	2	E — G —
Type 410		G
Type 904-L		
17-4 PH		U
20 Cb 3	2	G —
E-Bright 26-1		E —
SILICON BRONZE		U
SILICON COPPER		U
STELLITE		
TANTALUM		U
TITANIUM		E — G —
ZIRCONIUM		G —

°F: 60 80 100 120 140 160 180 200 220 240 260 280 300 320 340 360 380 400 420 440 460 480 500 520 540 560

°C: 15 26 38 49 60 71 82 93 104 116 127 138 149 160 171 182 193 204 216 227 238 249 260 271 282 293

FOR METALS

E = < 2 Mils Penetration/Year; G = < 20 Mils Penetration /year

S = < 50 Mils Penetration/Year; U = > 50 Mils Penetration/year

FOR NONMETALLICS

R = Resistant

U = Unsatisfactory

SODIUM HYDROXIDE 50%

°F: 60, 80, 100, 120, 140, 160, 180, 200, 220, 240, 260, 280, 300, 320, 340, 360, 380, 400, 420, 440, 460

°C: 15, 26, 38, 49, 60, 71, 82, 93, 104, 116, 127, 138, 149, 160, 171, 182, 193, 204, 216, 227, 238

PLASTICS		Rating	Line extends to (°F)
ABS		R	140
ACETAL		U	
ACRYLICS		R	80
ASBESTOS REINFORCED EPOXY		R	190; U at 200
ASBESTOS REINFORCED PHENOLIC		U	
CHLORINATED POLYETHER (Penton)		R	250
CPVC		R	180
E-CTFE (Halar)		R	250
EPOXY		R	200
ETFE (Tefzel)		R	230
FEP		R	400
FURAN (FURFURAL ALCOHOL)		U	
NORYL		R	140
NYLON 6		R	250
NYLON 11		R	200
NYLON 66		R	200
PHENOLIC		U	
PFA (Teflon)		R	250
POLYAMIDE-IMIDE		U	
POLYESTERS			
Bisphenol A–fumarate		R	220
Halogenated		U	
Hydrogenated Bisphenol A –Bisphenol A		U	
Isophthalic		U	
Terephthalate (PET)		U	
POLYETHERIMIDE (ULTEM)			
POLYETHERETHERKETONE (PEEK)		R	180
POLYETHERSULFONE (PES)		R	80
POLYETHYLENE HMW		R	160
POLYETHYLENE UHMW		R	180
POLYMETHYLPENTENE		R	140
POLYPROPYLENE (PP)		R	220
POLYSTYRENE		R	180
POLYSULFONE		R	200
POLYVINYLIDENE CHLORIDE		R	160
PVC TYPE 1		R	140
PVC TYPE 2		R	140
PVDF (Kynar)	2	R	220
RYTON		R	210
TFE (Teflon)		R	460
VINYL ESTER		R	220; U at 230

SODIUM HYDROXIDE 50%

ELASTOMERS* AND LININGS	Rating (60°F/15°C)	Line extends to °F	Line extends to °C
ACRYLATE-BUTADIENE (ABR)			
BUTYL GR-1 (IIR)	R	180	82
CARBOXYLIC-ACRYLONITRILE-BUTADIENE (XNBR)			
CHEMRAZ (FPM)	R	400	204
CHLORO-ISOBUTENE-ISOPRENE (CIIR)	R	180	82
HYPALON (CHLORO-SULFONYL-POLYETHYLENE (CSM)	R	240	116
ETHYLENE-ACRYLIC (EA)			
ETHYLENE-PROPYLENE (EPM)	R	140	60
ETHYLENE-PROPYLENE-DIENE (EPDM)	R	300	149
ETHYLENE-PROPYLENE TERPOLYMER (EPT)	R	200	93
FKM (Viton A)	R (U at 80°F/26°C)		
HARD RUBBER	R	200	93
SOFT RUBBER	U		
ISOPRENE (IR)	R	140	60
KALREZ (FPM)	R	200	93
KOROSEAL	R	140	60
NATURAL RUBBER (GRS)	R	140	60
NEOPRENE GR-M (CR)	R	200	93
NITRILE BUNA-N (NBR)	R	140	60
NORDEL (EPDM)	R	300	149
POLYBUTADIENE (BR)	R	80	26
POLYESTER (PE)			
POLYETHER-URETHANE (EU)			
POLYISOPRENE (IR)			
POLYSULFIDES (T)	U		
POLYURETHANE (AU)	R	80	26
SBR STYRENE (BUNA-S)	U		
SILICONE RUBBERS	R	80	26

Temperature scale: °F 60, 80, 100, 120, 140, 160, 180, 200, 220, 240, 260, 280, 300, 320, 340, 360, 380, 400, 420, 440, 460; °C 15, 26, 38, 49, 60, 71, 82, 93, 104, 116, 127, 138, 149, 160, 171, 182, 193, 204, 216, 227, 238

*See also Nylon 11 under PLASTICS

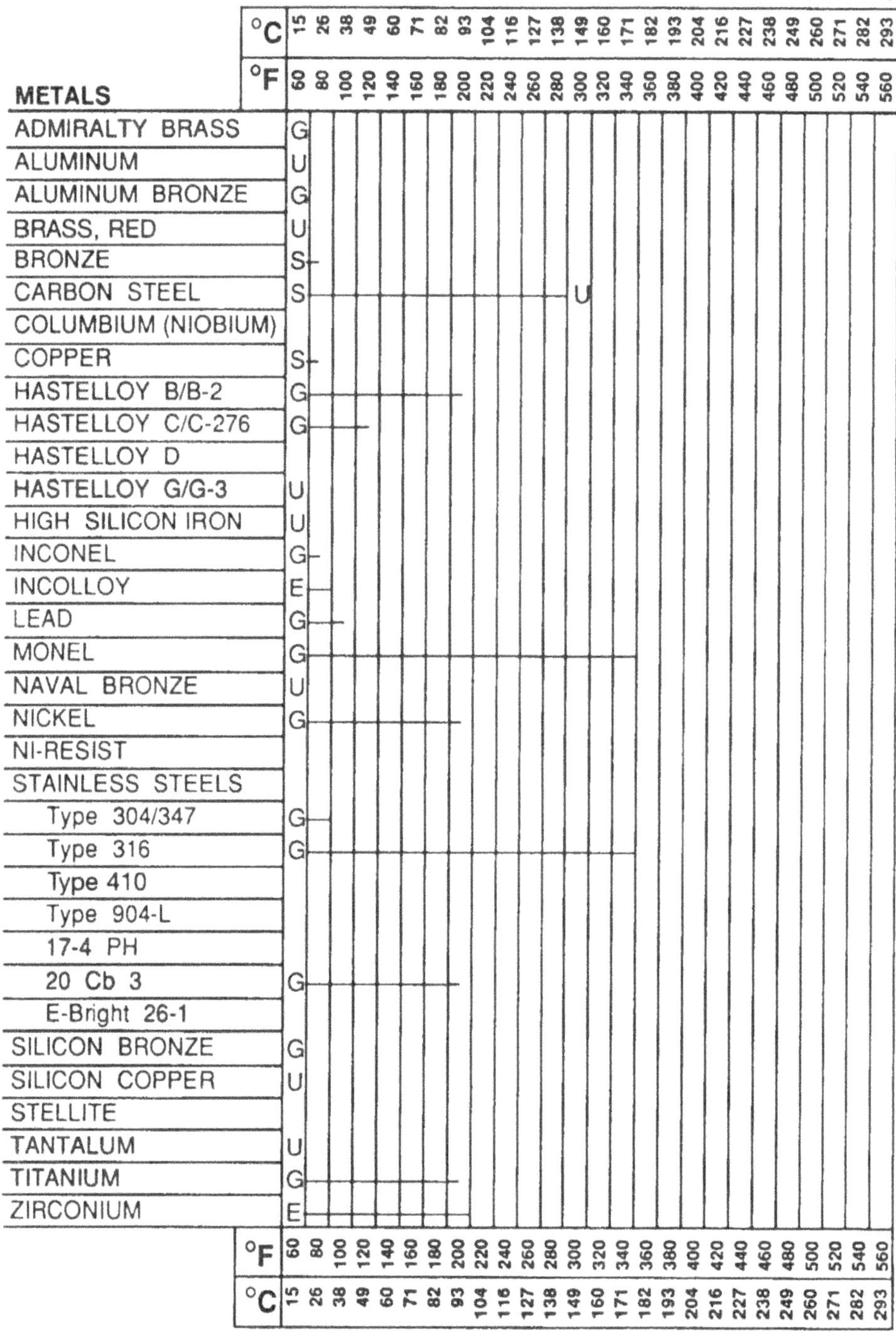

METALS	Rating at 60°F (15°C)	Rating line extends to (°F)
ADMIRALTY BRASS	G	
ALUMINUM	U	
ALUMINUM BRONZE	G	
BRASS, RED	U	
BRONZE	S	80
CARBON STEEL	S	280; U at 300
COLUMBIUM (NIOBIUM)		
COPPER	S	80
HASTELLOY B/B-2	G	200
HASTELLOY C/C-276	G	120
HASTELLOY D		
HASTELLOY G/G-3	U	
HIGH SILICON IRON	U	
INCONEL	G	80
INCOLLOY	E	80
LEAD	G	100
MONEL	G	340
NAVAL BRONZE	U	
NICKEL	G	200
NI-RESIST		
STAINLESS STEELS		
Type 304/347	G	80
Type 316	G	340
Type 410		
Type 904-L		
17-4 PH		
20 Cb 3	G	200
E-Bright 26-1		
SILICON BRONZE	G	
SILICON COPPER	U	
STELLITE		
TANTALUM	U	
TITANIUM	G	200
ZIRCONIUM	E	200

°F: 60 80 100 120 140 160 180 200 220 240 260 280 300 320 340 360 380 400 420 440 460 480 500 520 540 560

°C: 15 26 38 49 60 71 82 93 104 116 127 138 149 160 171 182 193 204 216 227 238 249 260 271 282 293

FOR METALS

E = < 2 Mils Penetration/Year; G = < 20 Mils Penetration /year

S = < 50 Mils Penetration/Year; U = > 50 Mils Penetration/year

FOR NONMETALLICS

R = Resistant

U = Unsatisfactory

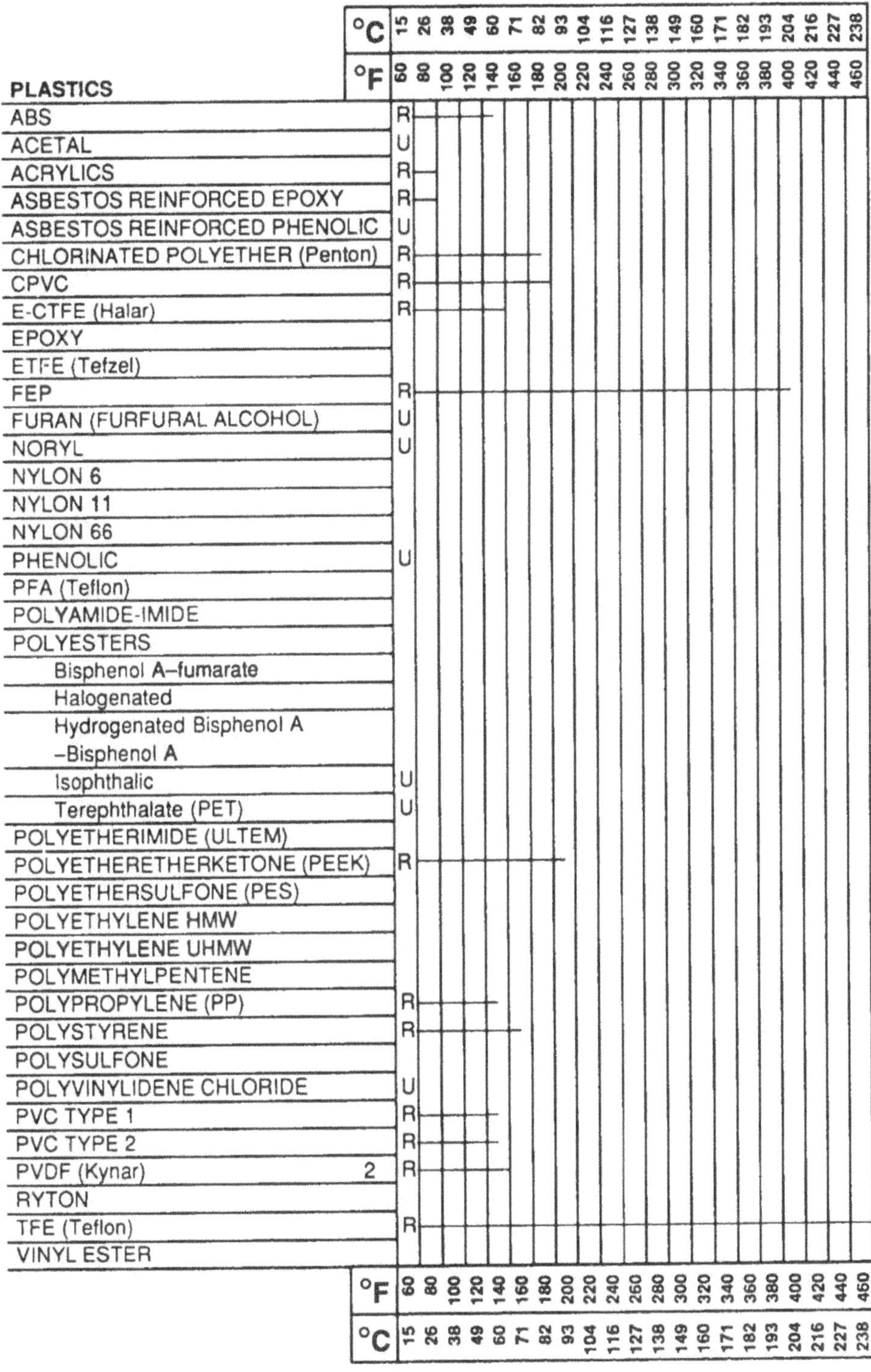

SODIUM HYDROXIDE SOLUTION (CONC.)

°C: 15, 26, 38, 49, 60, 71, 82, 93, 104, 116, 127, 138, 149, 160, 171, 182, 193, 204, 216, 227, 238

°F: 60, 80, 100, 120, 140, 160, 180, 200, 220, 240, 260, 280, 300, 320, 340, 360, 380, 400, 420, 440, 460

PLASTICS	
ABS	R
ACETAL	U
ACRYLICS	R
ASBESTOS REINFORCED EPOXY	R
ASBESTOS REINFORCED PHENOLIC	U
CHLORINATED POLYETHER (Penton)	R
CPVC	R
E-CTFE (Halar)	R
EPOXY	
ETFE (Tefzel)	
FEP	R
FURAN (FURFURAL ALCOHOL)	U
NORYL	U
NYLON 6	
NYLON 11	
NYLON 66	
PHENOLIC	U
PFA (Teflon)	
POLYAMIDE-IMIDE	
POLYESTERS	
Bisphenol A–fumarate	
Halogenated	
Hydrogenated Bisphenol A –Bisphenol A	
Isophthalic	U
Terephthalate (PET)	U
POLYETHERIMIDE (ULTEM)	
POLYETHERETHERKETONE (PEEK)	R
POLYETHERSULFONE (PES)	
POLYETHYLENE HMW	
POLYETHYLENE UHMW	
POLYMETHYLPENTENE	
POLYPROPYLENE (PP)	R
POLYSTYRENE	R
POLYSULFONE	
POLYVINYLIDENE CHLORIDE	U
PVC TYPE 1	R
PVC TYPE 2	R
PVDF (Kynar) 2	R
RYTON	
TFE (Teflon)	R
VINYL ESTER	

SODIUM HYDROXIDE SOLUTION (CONC.)

ELASTOMERS* AND LININGS	60	80	100	120	140	160	180	200	220	240	260	280	300	320	340	360	380	400	420	440	460
°C	15	26	38	49	60	71	82	93	104	116	127	138	149	160	171	182	193	204	216	227	238
ACRYLATE-BUTADIENE (ABR)																					
BUTYL GR-1 (IIR)	R	—	—	—	—	—	—														
CARBOXYLIC-ACRYLONITRILE-BUTADIENE (XNBR)																					
CHEMRAZ (FPM)	R	—	—	—	—	—	—	—	—	—	—	—	—	—	—	—	—	—			
CHLORO-ISOBUTENE-ISOPRENE (CIIR)	R	—	—	—	—	—	—														
HYPALON (CHLORO-SULFONYL-POLYETHYLENE (CSM)	R	—	—	—	—	—	—	—	—	—											
ETHYLENE-ACRYLIC (EA)																					
ETHYLENE-PROPYLENE (EPM)																					
ETHYLENE-PROPYLENE-DIENE (EPDM)	R	—	—	—	—	—	—	—	—	—	—	—	—								
ETHYLENE-PROPYLENE TERPOLYMER (EPT)	R	–																			
FKM (Viton A)	R	U																			
HARD RUBBER	R	—	—	—	—	—	—														
SOFT RUBBER	U																				
ISOPRENE (IR)	R	—	—	—	—																
KALREZ (FPM)	R	—	—	—	—	—	—	—													
KOROSEAL	R	—	—	—	—																
NATURAL RUBBER (GRS)	R	—	—	—	—																
NEOPRENE GR-M (CR)	R	—	—	—	—	—	—	–													
NITRILE BUNA-N (NBR)	R	—	—	—	—																
NORDEL (EPDM)	R	—	—	—	—	—	—	—	—	—	—	—	—								
POLYBUTADIENE (BR)	R	—																			
POLYESTER (PE)																					
POLYETHER-URETHANE (EU)																					
POLYISOPRENE (IR)																					
POLYSULFIDES (T)	U																				
POLYURETHANE (AU)																					
SBR STYRENE (BUNA-S)	U																				
SILICONE RUBBERS	R	—																			
°F	60	80	100	120	140	160	180	200	220	240	260	280	300	320	340	360	380	400	420	440	460
°C	15	26	38	49	60	71	82	93	104	116	127	138	149	160	171	182	193	204	216	227	238

*See also Nylon 11 under PLASTICS

SODIUM HYPOCHLORITE 20%

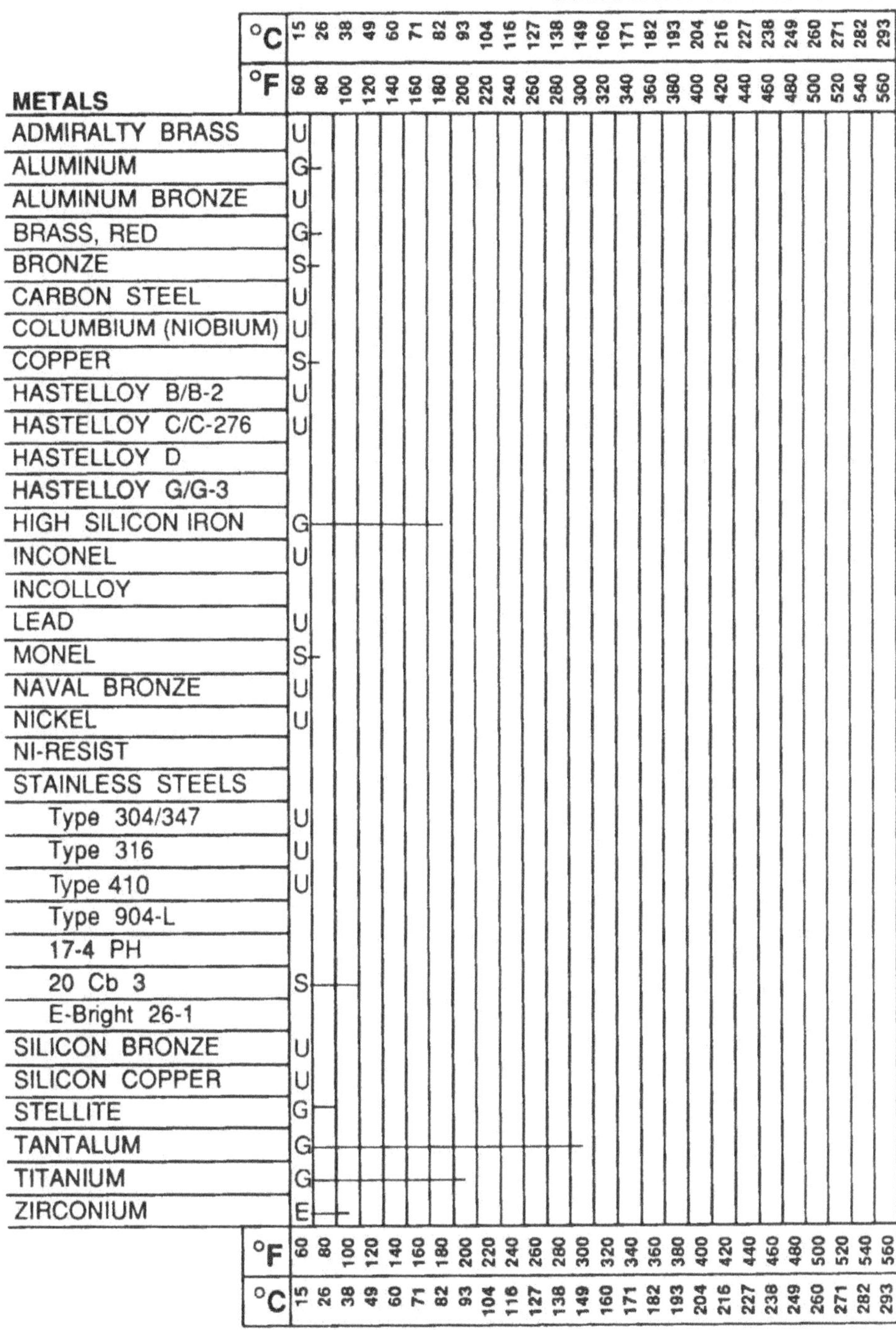

METALS	60 °F / 15 °C	80 / 26	100 / 38	120 / 49	140 / 60	160 / 71	180 / 82	200 / 93	220 / 104	240 / 116	260 / 127	280 / 138	300 / 149	320 / 160	340 / 171	360 / 182	380 / 193	400 / 204	420 / 216	440 / 227	460 / 238	480 / 249	500 / 260	520 / 271	540 / 282	560 / 293
ADMIRALTY BRASS	U																									
ALUMINUM	G																									
ALUMINUM BRONZE	U																									
BRASS, RED	G																									
BRONZE	S																									
CARBON STEEL	U																									
COLUMBIUM (NIOBIUM)	U																									
COPPER	S																									
HASTELLOY B/B-2	U																									
HASTELLOY C/C-276	U																									
HASTELLOY D																										
HASTELLOY G/G-3																										
HIGH SILICON IRON	G																									
INCONEL	U																									
INCOLLOY																										
LEAD	U																									
MONEL	S																									
NAVAL BRONZE	U																									
NICKEL	U																									
NI-RESIST																										
STAINLESS STEELS																										
Type 304/347	U																									
Type 316	U																									
Type 410	U																									
Type 904-L																										
17-4 PH																										
20 Cb 3	S																									
E-Bright 26-1																										
SILICON BRONZE	U																									
SILICON COPPER	U																									
STELLITE	G																									
TANTALUM	G																									
TITANIUM	G																									
ZIRCONIUM	E																									

FOR METALS
E = < 2 Mils Penetration/Year; G = < 20 Mils Penetration /year
S = < 50 Mils Penetration/Year; U = > 50 Mils Penetration/year

FOR NONMETALLICS
R = Resistant
U = Unsatisfactory

SODIUM HYPOCHLORITE 20%

°C: 15, 26, 38, 49, 60, 71, 82, 93, 104, 116, 127, 138, 149, 160, 171, 182, 193, 204, 216, 227, 238

°F: 60, 80, 100, 120, 140, 160, 180, 200, 220, 240, 260, 280, 300, 320, 340, 360, 380, 400, 420, 440, 460

PLASTICS		
ABS		R
ACETAL		U
ACRYLICS	15%	R
ASBESTOS REINFORCED EPOXY		
ASBESTOS REINFORCED PHENOLIC		
CHLORINATED POLYETHER (Penton)		R
CPVC		R
E-CTFE (Halar)		R
EPOXY	5%	R – U
ETFE (Tefzel)		R
FEP		R
FURAN (FURFURAL ALCOHOL)	5%	U
NORYL	20%	R
NYLON 6		U
NYLON 11		U
NYLON 66		U
PHENOLIC		U
PFA (Teflon)	20%	R
POLYAMIDE-IMIDE	10%	R
POLYESTERS		
Bisphenol A–fumarate	15%	U
Halogenated		U
Hydrogenated Bisphenol A –Bisphenol A	10%	R
Isophthalic		U
Terephthalate (PET)		R
POLYETHERIMIDE (ULTEM)		R
POLYETHERETHERKETONE (PEEK)		R
POLYETHERSULFONE (PES)		R
POLYETHYLENE HMW		R
POLYETHYLENE UHMW		R
POLYMETHYLPENTENE		R
POLYPROPYLENE (PP)		R
POLYSTYRENE		R
POLYSULFONE		R
POLYVINYLIDENE CHLORIDE	10%	R
PVC TYPE 1		R
PVC TYPE 2		R
PVDF (Kynar)		R
RYTON	5%	R
TFE (Teflon)		R
VINYL ESTER	15%	R

°F: 60, 80, 100, 120, 140, 160, 180, 200, 220, 240, 260, 280, 300, 320, 340, 360, 380, 400, 420, 440, 460

°C: 15, 26, 38, 49, 60, 71, 82, 93, 104, 116, 127, 138, 149, 160, 171, 182, 193, 204, 216, 227, 238

SODIUM HYPOCHLORITE 20%

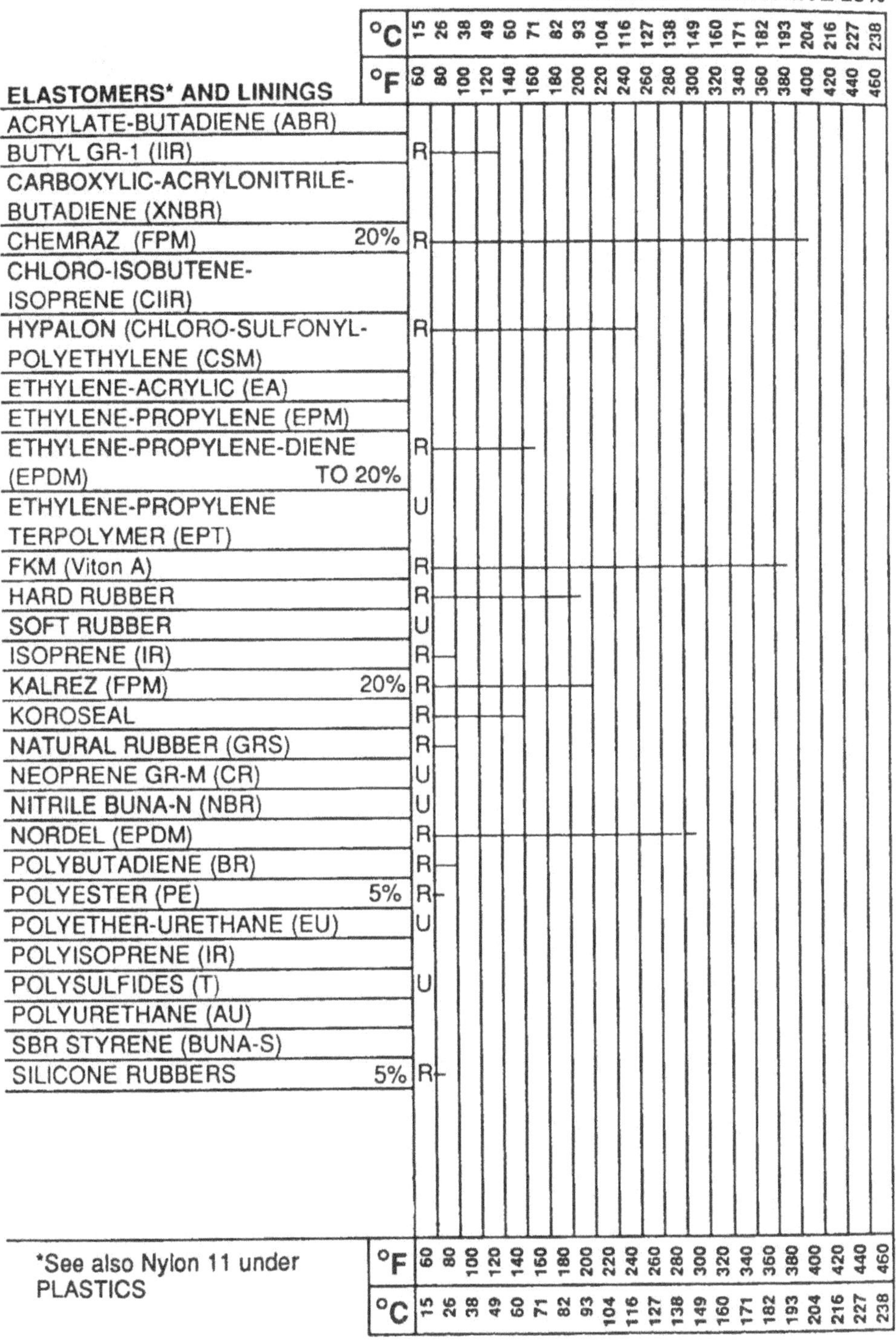

°C: 15, 26, 38, 49, 60, 71, 82, 93, 104, 116, 127, 138, 149, 160, 171, 182, 193, 204, 216, 227, 238

°F: 60, 80, 100, 120, 140, 160, 180, 200, 220, 240, 260, 280, 300, 320, 340, 360, 380, 400, 420, 440, 460

ELASTOMERS* AND LININGS		
ACRYLATE-BUTADIENE (ABR)		
BUTYL GR-1 (IIR)		R
CARBOXYLIC-ACRYLONITRILE-BUTADIENE (XNBR)		
CHEMRAZ (FPM)	20%	R
CHLORO-ISOBUTENE-ISOPRENE (CIIR)		
HYPALON (CHLORO-SULFONYL-POLYETHYLENE (CSM)		R
ETHYLENE-ACRYLIC (EA)		
ETHYLENE-PROPYLENE (EPM)		
ETHYLENE-PROPYLENE-DIENE (EPDM)	TO 20%	R
ETHYLENE-PROPYLENE TERPOLYMER (EPT)		U
FKM (Viton A)		R
HARD RUBBER		R
SOFT RUBBER		U
ISOPRENE (IR)		R
KALREZ (FPM)	20%	R
KOROSEAL		R
NATURAL RUBBER (GRS)		R
NEOPRENE GR-M (CR)		U
NITRILE BUNA-N (NBR)		U
NORDEL (EPDM)		R
POLYBUTADIENE (BR)		R
POLYESTER (PE)	5%	R
POLYETHER-URETHANE (EU)		U
POLYISOPRENE (IR)		
POLYSULFIDES (T)		U
POLYURETHANE (AU)		
SBR STYRENE (BUNA-S)		
SILICONE RUBBERS	5%	R

*See also Nylon 11 under PLASTICS

METALS	°F 60	80	100	120	140	160	180	200	220	240	260	280	300	320	340	360	380	400	420	440	460	480	500	520	540	560
°C	15	26	38	49	60	71	82	93	104	116	127	138	149	160	171	182	193	204	216	227	238	249	260	271	282	293
ADMIRALTY BRASS	U																									
ALUMINUM	U																									
ALUMINUM BRONZE	U																									
BRASS, RED	U																									
BRONZE	U																									
CARBON STEEL	U																									
COLUMBIUM (NIOBIUM)	U																									
COPPER	U																									
HASTELLOY B/B-2	U																									
HASTELLOY C/C-276 10%	E		G																							
HASTELLOY D	U																									
HASTELLOY G/G-3	G																									
HIGH SILICON IRON																										
INCONEL	U																									
INCOLLOY 5% 1	G																									
LEAD	U																									
MONEL	U																									
NAVAL BRONZE	U																									
NICKEL	U																									
NI-RESIST																										
STAINLESS STEELS																										
Type 304/347 10%	S																									
Type 316 10%	S																									
Type 410	U																									
Type 904-L																										
17-4 PH																										
20 Cb 3	S																									
E-Bright 26-1 30%	G																									
SILICON BRONZE	U																									
SILICON COPPER	S																									
STELLITE																										
TANTALUM	G																									
TITANIUM 6%	E																									
ZIRCONIUM 6%	G																									
°F	60	80	100	120	140	160	180	200	220	240	260	280	300	320	340	360	380	400	420	440	460	480	500	520	540	560
°C	15	26	38	49	60	71	82	93	104	116	127	138	149	160	171	182	193	204	216	227	238	249	260	271	282	293

FOR METALS

E = < 2 Mils Penetration/Year; G = < 20 Mils Penetration /year

S = < 50 Mils Penetration/Year; U = > 50 Mils Penetration/year

FOR NONMETALLICS

R = Resistant

U = Unsatisfactory

SODIUM HYPOCHLORITE (CONC.)

PLASTICS	60	80	100	120	140	160	180	200	220	240	260	280	300	320	340	360	380	400	420	440	460
°C	15	26	38	49	60	71	82	93	104	116	127	138	149	160	171	182	193	204	216	227	238
ABS	R	—	—	—	—																
ACETAL	U																				
ACRYLICS																					
ASBESTOS REINFORCED EPOXY	U																				
ASBESTOS REINFORCED PHENOLIC	U																				
CHLORINATED POLYETHER (Penton)	R	—	—	—	—	—	—	—													
CPVC	R	—	—	—	—	—	—														
E-CTFE (Halar)	R	—	—	—	—	—	—	—	—	—	—	—	—								
EPOXY																					
ETFE (Tefzel)	R	—	—	—	—	—	—	—	—	—	—	—	—								
FEP	R	—	—	—	—	—	—	—	—	—	—	—	—	—	—	—	—	—			
FURAN (FURFURAL ALCOHOL)	U																				
NORYL	R																				
NYLON 6	U																				
NYLON 11	U																				
NYLON 66	U																				
PHENOLIC	U																				
PFA (Teflon)																					
POLYAMIDE-IMIDE	U																				
POLYESTERS																					
Bisphenol A–fumarate	R	—	—	—	—																
Halogenated	U																				
Hydrogenated Bisphenol A –Bisphenol A																					
Isophthalic	U																				
Terephthalate (PET)	R																				
POLYETHERIMIDE (ULTEM)	R																				
POLYETHERETHERKETONE (PEEK)	R																				
POLYETHERSULFONE (PES)	R																				
POLYETHYLENE HMW	R	—	—	—	—																
POLYETHYLENE UHMW	R	—	—	—	—																
POLYMETHYLPENTENE	R	—	—	—	—																
POLYPROPYLENE (PP)	R	—	—	U																	
POLYSTYRENE	R																				
POLYSULFONE	R	—	—	—	—	—	—	—	—	—	—	—	—								
POLYVINYLIDENE CHLORIDE	R	—	—	—																	
PVC TYPE 1	R	—	—	—	—																
PVC TYPE 2	R	—	—	—	—																
PVDF (Kynar)	R	—	—	—	—	—	—	—	—	—	—	—									
RYTON																					
TFE (Teflon)	R	—	—	—	—	—	—	—	—	—	—	—	—	—	—	—	—	—	—	—	—
VINYL ESTER	R	—	—																		

°F 60 80 100 120 140 160 180 200 220 240 260 280 300 320 340 360 380 400 420 440 460
°C 15 26 38 49 60 71 82 93 104 116 127 138 149 160 171 182 193 204 216 227 238

SODIUM HYPOCHLORITE (CONC.)

ELASTOMERS* AND LININGS	60	80	100	120	140	160	180	200	220	240	260	280	300	320	340	360	380	400	420	440	460
°C	15	26	38	49	60	71	82	93	104	116	127	138	149	160	171	182	193	204	216	227	238
ACRYLATE-BUTADIENE (ABR)																					
BUTYL GR-1 (IIR)	R	—																			
CARBOXYLIC-ACRYLONITRILE-BUTADIENE (XNBR)																					
CHEMRAZ (FPM)																					
CHLORO-ISOBUTENE-ISOPRENE (CIIR)																					
HYPALON (CHLORO-SULFONYL-POLYETHYLENE (CSM)	R																				
ETHYLENE-ACRYLIC (EA)																					
ETHYLENE-PROPYLENE (EPM)																					
ETHYLENE-PROPYLENE-DIENE (EPDM)	R	—	—	—	—																
ETHYLENE-PROPYLENE TERPOLYMER (EPT)	U																				
FKM (Viton A)	R	—	—	—	—	—	—	—	—	—	—	—	—	—	—	—	—	—			
HARD RUBBER	R	—	—																		
SOFT RUBBER	U																				
ISOPRENE (IR)	R	—																			
KALREZ (FPM)	R	—	—	—	—	—	—	—													
KOROSEAL	R	—	—	—	—																
NATURAL RUBBER (GRS)	R	—																			
NEOPRENE GR-M (CR)	U																				
NITRILE BUNA-N (NBR)	U																				
NORDEL (EPDM)	R	—	—	—	—	—	—	—	—	—	—	—	—								
POLYBUTADIENE (BR)	R																				
POLYESTER (PE)																					
POLYETHER-URETHANE (EU)																					
POLYISOPRENE (IR)																					
POLYSULFIDES (T)	U																				
POLYURETHANE (AU)	U																				
SBR STYRENE (BUNA-S)																					
SILICONE RUBBERS																					
°F	60	80	100	120	140	160	180	200	220	240	260	280	300	320	340	360	380	400	420	440	460
°C	15	26	38	49	60	71	82	93	104	116	127	138	149	160	171	182	193	204	216	227	238

*See also Nylon 11 under PLASTICS

SULFURIC ACID 10%

METALS	60 °F / 15 °C	80 / 26	100 / 38	120 / 49	140 / 60	160 / 71	180 / 82	200 / 93	220 / 104	240 / 116	260 / 127	280 / 138	300 / 149	320 / 160	340 / 171	360 / 182	380 / 193	400 / 204	420 / 216	440 / 227	460 / 238	480 / 249	500 / 260	520 / 271	540 / 282	560 / 293
ADMIRALTY BRASS	U																									
ALUMINUM	U																									
ALUMINUM BRONZE	U																									
BRASS, RED	G																									
BRONZE	U																									
CARBON STEEL	U																									
COLUMBIUM (NIOBIUM)	G																									
COPPER	U																									
HASTELLOY B/B-2	E				G																					
HASTELLOY C/C-276	E							S																		
HASTELLOY D	E							G																		
HASTELLOY G/G-3	G																									
HIGH SILICON IRON	E																									
INCONEL	U																									
INCOLLOY 825	G									U																
LEAD	G																									
MONEL	S		U																							
NAVAL BRONZE	G																									
NICKEL	S		U																							
NI-RESIST																										
STAINLESS STEELS																										
Type 304/347	U																									
Type 316	U																									
Type 410	U																									
Type 904-L	S																									
17-4 PH																										
20 Cb 3	E			G					U																	
E-Bright 26-1	U																									
SILICON BRONZE	U																									
SILICON COPPER	U																									
STELLITE	U																									
TANTALUM	G																									
TITANIUM	U																									
ZIRCONIUM	E																									

FOR METALS

E = < 2 Mils Penetration/Year; G = < 20 Mils Penetration /year

S = < 50 Mils Penetration/Year; U = > 50 Mils Penetration/year

FOR NONMETALLICS

R = Resistant

U = Unsatisfactory

PLASTICS	°F 60–460 / °C 15–238
ABS	R
ACETAL	U
ACRYLICS	R
ASBESTOS REINFORCED EPOXY	R
ASBESTOS REINFORCED PHENOLIC	R
CHLORINATED POLYETHER (Penton)	R
CPVC	R
E-CTFE (Halar)	R
EPOXY	R ... U
ETFE (Tefzel)	R
FEP	R
FURAN (FURFURAL ALCOHOL)	R
NORYL	R
NYLON 6	U
NYLON 11	R
NYLON 66	U
PHENOLIC	R
PFA (Teflon)	R
POLYAMIDE-IMIDE	R
POLYESTERS	
Bisphenol A–fumarate	R
Halogenated	R
Hydrogenated Bisphenol A –Bisphenol A	R
Isophthalic	R
Terephthalate (PET)	R
POLYETHERIMIDE (ULTEM)	R
POLYETHERETHERKETONE (PEEK)	R
POLYETHERSULFONE (PES)	R
POLYETHYLENE HMW	R
POLYETHYLENE UHMW	R
POLYMETHYLPENTENE	R
POLYPROPYLENE (PP)	R
POLYSTYRENE	R
POLYSULFONE	R
POLYVINYLIDENE CHLORIDE	R
PVC TYPE 1	R
PVC TYPE 2	R
PVDF (Kynar)	R
RYTON	R
TFE (Teflon)	R
VINYL ESTER	R

°F: 60, 80, 100, 120, 140, 160, 180, 200, 220, 240, 260, 280, 300, 320, 340, 360, 380, 400, 420, 440, 460

°C: 15, 26, 38, 49, 60, 71, 82, 93, 104, 116, 127, 138, 149, 160, 171, 182, 193, 204, 216, 227, 238

SULFURIC ACID 10%

ELASTOMERS* AND LININGS	Rating	Shown to °F (°C)
ACRYLATE-BUTADIENE (ABR)		
BUTYL GR-1 (IIR)	R	180 (82)
CARBOXYLIC-ACRYLONITRILE-BUTADIENE (XNBR)		
CHEMRAZ (FPM)	R	400 (204)
CHLORO-ISOBUTENE-ISOPRENE (CIIR)	R	200 (93)
HYPALON (CHLORO-SULFONYL-POLYETHYLENE (CSM)	R	240 (116)
ETHYLENE-ACRYLIC (EA)		
ETHYLENE-PROPYLENE (EPM)	R	160 (71)
ETHYLENE-PROPYLENE-DIENE (EPDM)	R	140 (60)
ETHYLENE-PROPYLENE TERPOLYMER (EPT)	R	200 (93)
FKM (Viton A)	R	340 (171)
HARD RUBBER	R	200 (93)
SOFT RUBBER	R	140 (60)
ISOPRENE (IR)	R	140 (60)
KALREZ (FPM)	R	240 (116)
KOROSEAL	R	140 (60)
NATURAL RUBBER (GRS)	R	140 (60)
NEOPRENE GR-M (CR)	R	200 (93)
NITRILE BUNA-N (NBR)	R	140 (60)
NORDEL (EPDM)	R	300 (149)
POLYBUTADIENE (BR)	R	60 (15)
POLYESTER (PE)	R	60 (15)
POLYETHER-URETHANE (EU)		
POLYISOPRENE (IR)		
POLYSULFIDES (T)	U	
POLYURETHANE (AU)	U	
SBR STYRENE (BUNA-S)	U	
SILICONE RUBBERS	U	

Temperature scale — °F: 60, 80, 100, 120, 140, 160, 180, 200, 220, 240, 260, 280, 300, 320, 340, 360, 380, 400, 420, 440, 460; °C: 15, 26, 38, 49, 60, 71, 82, 93, 104, 116, 127, 138, 149, 160, 171, 182, 193, 204, 216, 227, 238

*See also Nylon 11 under PLASTICS

SULFURIC ACID 50%

METALS	60	80	100	120	140	160	180	200	220	240	260	280	300	320	340	360	380	400	420	440	460	480	500	520	540	560
°C	15	26	38	49	60	71	82	93	104	116	127	138	149	160	171	182	193	204	216	227	238	249	260	271	282	293
ADMIRALTY BRASS	U																									
ALUMINUM	U																									
ALUMINUM BRONZE	U																									
BRASS, RED	G																									
BRONZE	U																									
CARBON STEEL	U																									
COLUMBIUM (NIOBIUM)	G	U																								
COPPER	U																									
HASTELLOY B/B-2	E									G																
HASTELLOY C/C-276	E				G					U																
HASTELLOY D	E									G																
HASTELLOY G/G-3																										
HIGH SILICON IRON	E					G																				
INCONEL	U																									
INCOLLOY 825	G								U																	
LEAD	G																									
MONEL	G		S																							
NAVAL BRONZE	G																									
NICKEL	S																									
NI-RESIST																										
STAINLESS STEELS																										
Type 304/347	U																									
Type 316	U																									
Type 410	U																									
Type 904-L																										
17-4 PH	U																									
20 Cb 3	E			S		U																				
E-Bright 26-1	U																									
SILICON BRONZE	U																									
SILICON COPPER	U																									
STELLITE	U																									
TANTALUM	G																									
TITANIUM	U																									
ZIRCONIUM	E									G																
°F	60	80	100	120	140	160	180	200	220	240	260	280	300	320	340	360	380	400	420	440	460	480	500	520	540	560
°C	15	26	38	49	60	71	82	93	104	116	127	138	149	160	171	182	193	204	216	227	238	249	260	271	282	293

FOR METALS
E = < 2 Mils Penetration/Year; G = < 20 Mils Penetration /year
S = < 50 Mils Penetration/Year; U = > 50 Mils Penetration/year

FOR NONMETALLICS
R = Resistant
U = Unsatisfactory

SULFURIC ACID 50%

PLASTICS	60 °F / 15 °C	80 °F / 26 °C	100 °F / 38 °C	120 °F / 49 °C	140 °F / 60 °C	160 °F / 71 °C	180 °F / 82 °C	200 °F / 93 °C	220 °F / 104 °C	240 °F / 116 °C	260 °F / 127 °C	280 °F / 138 °C	300 °F / 149 °C	320 °F / 160 °C	340 °F / 171 °C	360 °F / 182 °C	380 °F / 193 °C	400 °F / 204 °C	420 °F / 216 °C	440 °F / 227 °C	460 °F / 238 °C
ABS	R				U																
ACETAL	U																				
ACRYLICS	R																				
ASBESTOS REINFORCED EPOXY	U																				
ASBESTOS REINFORCED PHENOLIC	R																				
CHLORINATED POLYETHER (Penton)	R																				
CPVC	R																				
E-CTFE (Halar)	R																				
EPOXY	R			U																	
ETFE (Tefzel)	R																				
FEP	R																				
FURAN (FURFURAL ALCOHOL)	R																				
NORYL	R																				
NYLON 6	U																				
NYLON 11	U																				
NYLON 66	U																				
PHENOLIC	R																				
PFA (Teflon)	R																				
POLYAMIDE-IMIDE																					
POLYESTERS																					
Bisphenol A–fumarate	R																				
Halogenated	R																				
Hydrogenated Bisphenol A –Bisphenol A	R																				
Isophthalic	R																				
Terephthalate (PET)	R																				
POLYETHERIMIDE (ULTEM)	R																				
POLYETHERETHERKETONE (PEEK)	R																				
POLYETHERSULFONE (PES)	U																				
POLYETHYLENE HMW	R																				
POLYETHYLENE UHMW	R																				
POLYMETHYLPENTENE	R																				
POLYPROPYLENE (PP)	R																				
POLYSTYRENE	R																				
POLYSULFONE	R																				
POLYVINYLIDENE CHLORIDE	U																				
PVC TYPE 1	R																				
PVC TYPE 2	R																				
PVDF (Kynar)	R																				
RYTON	R																				
TFE (Teflon)	R																				
VINYL ESTER	R																				

°F 60 80 100 120 140 160 180 200 220 240 260 280 300 320 340 360 380 400 420 440 460
°C 15 26 38 49 60 71 82 93 104 116 127 138 149 160 171 182 193 204 216 227 238

SULFURIC ACID 50%

ELASTOMERS* AND LININGS	Rating	Temperature limit, °F (°C)
ACRYLATE-BUTADIENE (ABR)		
BUTYL GR-1 (IIR)	R	150 (66)
CARBOXYLIC-ACRYLONITRILE-BUTADIENE (XNBR)		
CHEMRAZ (FPM)	R	400 (204)
CHLORO-ISOBUTENE-ISOPRENE (CIIR)		
HYPALON (CHLORO-SULFONYL-POLYETHYLENE (CSM)	R	250 (121)
ETHYLENE-ACRYLIC (EA)		
ETHYLENE-PROPYLENE (EPM)	R	140 (60)
ETHYLENE-PROPYLENE-DIENE (EPDM)	R	140 (60)
ETHYLENE-PROPYLENE TERPOLYMER (EPT)	R	210 (99)
FKM (Viton A)	R	350 (177)
HARD RUBBER	R	200 (93)
SOFT RUBBER	U	
ISOPRENE (IR)	R	100 (38)
KALREZ (FPM)	R	250 (121)
KOROSEAL	R	150 (66)
NATURAL RUBBER (GRS)	R	100 (38)
NEOPRENE GR-M (CR)	R	200 (93)
NITRILE BUNA-N (NBR)	R	200 (93)
NORDEL (EPDM)	R	300 (149)
POLYBUTADIENE (BR)	R	80 (26)
POLYESTER (PE)	U	
POLYETHER-URETHANE (EU)		
POLYISOPRENE (IR)		
POLYSULFIDES (T)	U	
POLYURETHANE (AU)	U	
SBR STYRENE (BUNA-S)	U	
SILICONE RUBBERS	U	

Temperature scale: °F 60, 80, 100, 120, 140, 160, 180, 200, 220, 240, 260, 280, 300, 320, 340, 360, 380, 400, 420, 440, 460; °C 15, 26, 38, 49, 60, 71, 82, 93, 104, 116, 127, 138, 149, 160, 171, 182, 193, 204, 216, 227, 238

*See also Nylon 11 under PLASTICS

SULFURIC ACID 70%

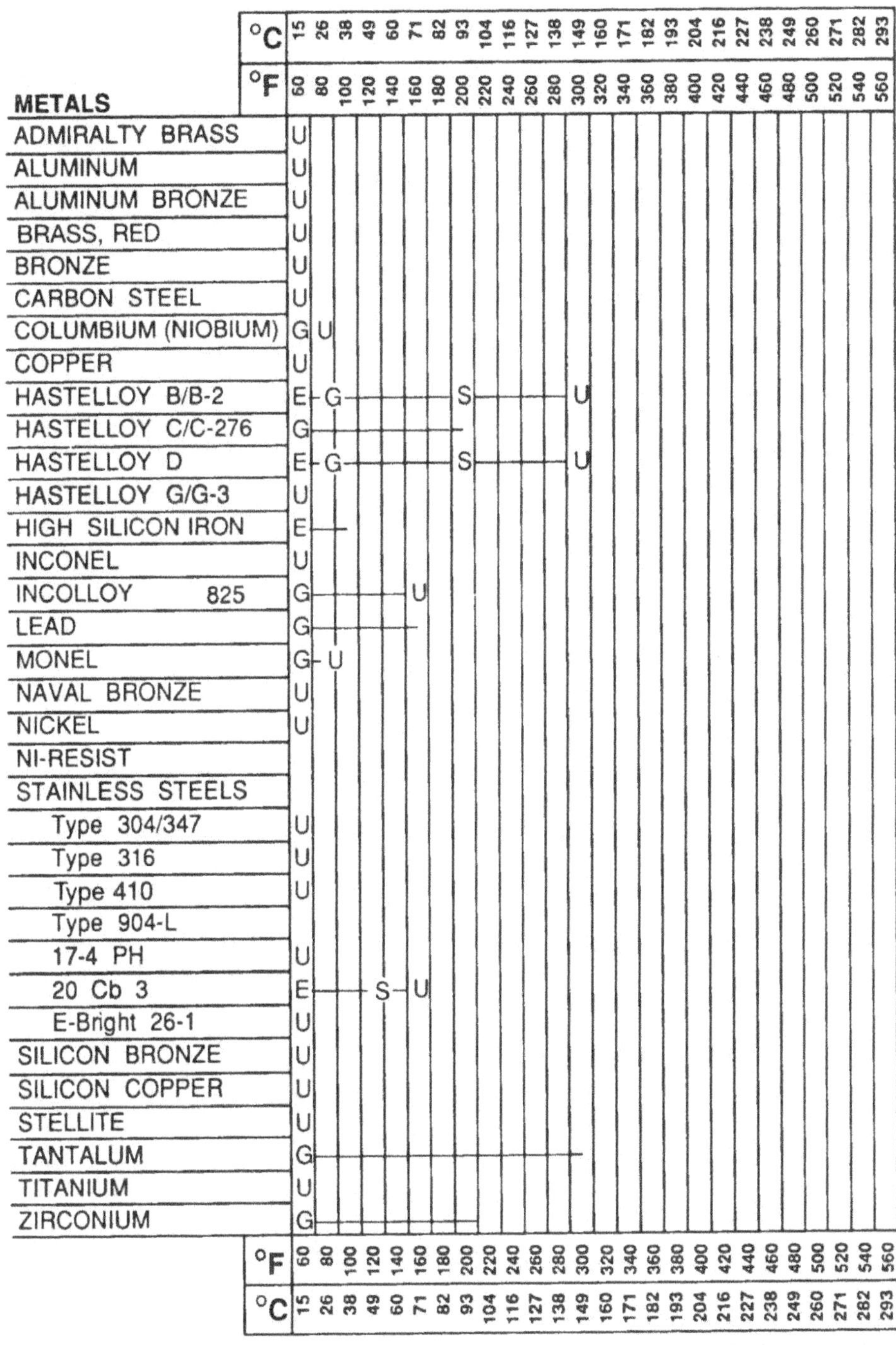

METALS	60	80	100	120	140	160	180	200	220	240	260	280	300	320	340	360	380	400	420	440	460	480	500	520	540	560
°C	15	26	38	49	60	71	82	93	104	116	127	138	149	160	171	182	193	204	216	227	238	249	260	271	282	293
ADMIRALTY BRASS	U																									
ALUMINUM	U																									
ALUMINUM BRONZE	U																									
BRASS, RED	U																									
BRONZE	U																									
CARBON STEEL	U																									
COLUMBIUM (NIOBIUM)	G	U																								
COPPER	U																									
HASTELLOY B/B-2	E	G						S					U													
HASTELLOY C/C-276	G																									
HASTELLOY D	E	G						S					U													
HASTELLOY G/G-3	U																									
HIGH SILICON IRON	E																									
INCONEL	U																									
INCOLLOY 825	G					U																				
LEAD	G																									
MONEL	G	U																								
NAVAL BRONZE	U																									
NICKEL	U																									
NI-RESIST																										
STAINLESS STEELS																										
Type 304/347	U																									
Type 316	U																									
Type 410	U																									
Type 904-L																										
17-4 PH	U																									
20 Cb 3	E				S	U																				
E-Bright 26-1	U																									
SILICON BRONZE	U																									
SILICON COPPER	U																									
STELLITE	U																									
TANTALUM	G																									
TITANIUM	U																									
ZIRCONIUM	G																									
°F	60	80	100	120	140	160	180	200	220	240	260	280	300	320	340	360	380	400	420	440	460	480	500	520	540	560
°C	15	26	38	49	60	71	82	93	104	116	127	138	149	160	171	182	193	204	216	227	238	249	260	271	282	293

FOR METALS
E = < 2 Mils Penetration/Year; G = < 20 Mils Penetration /year
S = < 50 Mils Penetration/Year; U = > 50 Mils Penetration/year

FOR NONMETALLICS
R = Resistant
U = Unsatisfactory

SULFURIC ACID 70%

Temperature scale (°F / °C): 60/15, 80/26, 100/38, 120/49, 140/60, 160/71, 180/82, 200/93, 220/104, 240/116, 260/127, 280/138, 300/149, 320/160, 340/171, 360/182, 380/193, 400/204, 420/216, 440/227, 460/238

PLASTICS	Rating	Line extends to (°F)
ABS	U	
ACETAL		
ACRYLICS	R	80
ASBESTOS REINFORCED EPOXY	U	
ASBESTOS REINFORCED PHENOLIC	R … U	110 (U at 120)
CHLORINATED POLYETHER (Penton)	R	220
CPVC	R	200
E-CTFE (Halar)	R	250
EPOXY	R … U	110 (U at 120)
ETFE (Tefzel)	R	300
FEP	R	400
FURAN (FURFURAL ALCOHOL)	R	260
NORYL	R	90
NYLON 6	U	
NYLON 11	U	
NYLON 66	U	
PHENOLIC	R	200
PFA (Teflon)	R	250
POLYAMIDE-IMIDE		
POLYESTERS		
Bisphenol A–fumarate	R	160
Halogenated	R … U	190 (U at 200)
Hydrogenated Bisphenol A –Bisphenol A	R	90
Isophthalic	U	
Terephthalate (PET)	U	
POLYETHERIMIDE (ULTEM)		
POLYETHERETHERKETONE (PEEK)	U	
POLYETHERSULFONE (PES)		
POLYETHYLENE HMW	R	80
POLYETHYLENE UHMW	R	80
POLYMETHYLPENTENE		
POLYPROPYLENE (PP)	R	180
POLYSTYRENE	R	90
POLYSULFONE	U	
POLYVINYLIDENE CHLORIDE	U	
PVC TYPE 1	R	140
PVC TYPE 2	R	140
PVDF (Kynar)	R	220
RYTON	R	250
TFE (Teflon)	R	460
VINYL ESTER	R	180

SULFURIC ACID 70%

ELASTOMERS* AND LININGS	Rating	Bar extends to (°F / °C)
ACRYLATE-BUTADIENE (ABR)		
BUTYL GR-1 (IIR)	R	100 / 38
CARBOXYLIC-ACRYLONITRILE-BUTADIENE (XNBR)		
CHEMRAZ (FPM)	R	400 / 204
CHLORO-ISOBUTENE-ISOPRENE (CIIR)	U	
HYPALON (CHLORO-SULFONYL-POLYETHYLENE (CSM)	R	160 / 71
ETHYLENE-ACRYLIC (EA)		
ETHYLENE-PROPYLENE (EPM)	R	140 / 60
ETHYLENE-PROPYLENE-DIENE (EPDM)	R	140 / 60
ETHYLENE-PROPYLENE TERPOLYMER (EPT)	R	210 / 99
FKM (Viton A)	R	350 / 177
HARD RUBBER	U	
SOFT RUBBER	U	
ISOPRENE (IR)	U	
KALREZ (FPM)	R	150 / 66
KOROSEAL	U	
NATURAL RUBBER (GRS)	U	
NEOPRENE GR-M (CR)	R	200 / 93
NITRILE BUNA-N (NBR)	U	
NORDEL (EPDM)	R	300 / 149
POLYBUTADIENE (BR)	R	90 / 32
POLYESTER (PE)	U	
POLYETHER-URETHANE (EU)		
POLYISOPRENE (IR)		
POLYSULFIDES (T)	U	
POLYURETHANE (AU)	U	
SBR STYRENE (BUNA-S)	U	
SILICONE RUBBERS	U	

Temperature scale: °C 15, 26, 38, 49, 60, 71, 82, 93, 104, 116, 127, 138, 149, 160, 171, 182, 193, 204, 216, 227, 238; °F 60, 80, 100, 120, 140, 160, 180, 200, 220, 240, 260, 280, 300, 320, 340, 360, 380, 400, 420, 440, 460

*See also Nylon 11 under PLASTICS

SULFURIC ACID 90%

METALS	60	80	100	120	140	160	180	200	220	240	260	280	300	320	340	360	380	400	420	440	460	480	500	520	540	560
°C	15	26	38	49	60	71	82	93	104	116	127	138	149	160	171	182	193	204	216	227	238	249	260	271	282	293
ADMIRALTY BRASS	U																									
ALUMINUM	U																									
ALUMINUM BRONZE	U																									
BRASS, RED	U																									
BRONZE	U																									
CARBON STEEL	U																									
COLUMBIUM (NIOBIUM)	G	U																								
COPPER	U																									
HASTELLOY B/B-2	E	—	—	—	—	G	—	U																		
HASTELLOY C/C-276	E	—	—	—	—	G	—	U																		
HASTELLOY D	E	—	G	—	—	—	—	U																		
HASTELLOY G/G-3																										
HIGH SILICON IRON	E	—	—	—	—	—	—	—	—	—																
INCONEL	U																									
INCOLLOY 825	G	—	—	—	—	—	—	U																		
LEAD	E	—	—																							
MONEL	U																									
NAVAL BRONZE	U																									
NICKEL	U																									
NI-RESIST																										
STAINLESS STEELS																										
Type 304/347 4	G	—	U																							
Type 316 4	G	—	U																							
Type 410	U																									
Type 904-L																										
17-4 PH	U																									
20 Cb 3	E	—	—	S	—	—	U																			
E-Bright 26-1	U																									
SILICON BRONZE	U																									
SILICON COPPER	U																									
STELLITE	U																									
TANTALUM	G	—	—	—	—	—	—	—	—	—	—	—	—													
TITANIUM																										
ZIRCONIUM																										

FOR METALS

E = < 2 Mils Penetration/Year; G = < 20 Mils Penetration /year

S = < 50 Mils Penetration/Year; U = > 50 Mils Penetration/year

FOR NONMETALLICS

R = Resistant

U = Unsatisfactory

SULFURIC ACID 90%

PLASTICS	60	80	100	120	140	160	180	200	220	240	260	280	300	320	340	360	380	400	420	440	460
°C	15	26	38	49	60	71	82	93	104	116	127	138	149	160	171	182	193	204	216	227	238
°F	60	80	100	120	140	160	180	200	220	240	260	280	300	320	340	360	380	400	420	440	460
ABS	U																				
ACETAL	U																				
ACRYLICS	U																				
ASBESTOS REINFORCED EPOXY	U																				
ASBESTOS REINFORCED PHENOLIC	R	—	—	U																	
CHLORINATED POLYETHER (Penton)	R	—	—	—	—																
CPVC	U																				
E-CTFE (Halar)	R	—	—	—	—																
EPOXY	U																				
ETFE (Tefzel)	R	—	—	—	—	—	—	—	—	—	—	—	—								
FEP	R	—	—	—	—	—	—	—	—	—	—	—	—	—	—	—	—	—			
FURAN (FURFURAL ALCOHOL)	U																				
NORYL	R																				
NYLON 6	U																				
NYLON 11	U																				
NYLON 66	U																				
PHENOLIC	R																				
PFA (Teflon)	R	—	—	—	—	—	—	—	—	—											
POLYAMIDE-IMIDE																					
POLYESTERS																					
Bisphenol A–fumarate	U																				
Halogenated	U																				
Hydrogenated Bisphenol A –Bisphenol A	U																				
Isophthalic	U																				
Terephthalate (PET)	U																				
POLYETHERIMIDE (ULTEM)																					
POLYETHERETHERKETONE (PEEK)	U																				
POLYETHERSULFONE (PES)																					
POLYETHYLENE HMW	U																				
POLYETHYLENE UHMW	U																				
POLYMETHYLPENTENE																					
POLYPROPYLENE (PP)	R	—	—	—	—	—	—														
POLYSTYRENE																					
POLYSULFONE	U																				
POLYVINYLIDENE CHLORIDE	U																				
PVC TYPE 1	R	—	—	—	—																
PVC TYPE 2	U																				
PVDF (Kynar)	R	—	—	—	—	—	—	—													
RYTON	R	—	—	—	—	—	—	—	—												
TFE (Teflon)	R	—	—	—	—	—	—	—	—	—	—	—	—	—	—	—	—	—	—	—	—
VINYL ESTER	U																				
°F	60	80	100	120	140	160	180	200	220	240	260	280	300	320	340	360	380	400	420	440	460
°C	15	26	38	49	60	71	82	93	104	116	127	138	149	160	171	182	193	204	216	227	238

SULFURIC ACID 90%

ELASTOMERS* AND LININGS	60	80	100	120	140	160	180	200	220	240	260	280	300	320	340	360	380	400	420	440	460
°C	15	26	38	49	60	71	82	93	104	116	127	138	149	160	171	182	193	204	216	227	238
ACRYLATE-BUTADIENE (ABR)																					
BUTYL GR-1 (IIR)	U																				
CARBOXYLIC-ACRYLONITRILE-BUTADIENE (XNBR)																					
CHEMRAZ (FPM)	R																				
CHLORO-ISOBUTENE-ISOPRENE (CIIR)	U																				
HYPALON (CHLORO-SULFONYL-POLYETHYLENE (CSM)	U																				
ETHYLENE-ACRYLIC (EA)																					
ETHYLENE-PROPYLENE (EPM)	R																				
ETHYLENE-PROPYLENE-DIENE (EPDM)	U																				
ETHYLENE-PROPYLENE TERPOLYMER (EPT)	R																				
FKM (Viton A)	R																				
HARD RUBBER	U																				
SOFT RUBBER	U																				
ISOPRENE (IR)	U																				
KALREZ (FPM)	R																				
KOROSEAL	U																				
NATURAL RUBBER (GRS)	U																				
NEOPRENE GR-M (CR)	U																				
NITRILE BUNA-N (NBR)	U																				
NORDEL (EPDM)																					
POLYBUTADIENE (BR)																					
POLYESTER (PE)	U																				
POLYETHER-URETHANE (EU)																					
POLYISOPRENE (IR)																					
POLYSULFIDES (T)	U																				
POLYURETHANE (AU)	U																				
SBR STYRENE (BUNA-S)	U																				
SILICONE RUBBERS	U																				

°F 60 80 100 120 140 160 180 200 220 240 260 280 300 320 340 360 380 400 420 440 460
°C 15 26 38 49 60 71 82 93 104 116 127 138 149 160 171 182 193 204 216 227 238

*See also Nylon 11 under PLASTICS

SULFURIC ACID 98%

METALS	°F 60	80	100	120	140	160	180	200	220	240	260	280	300	320	340	360	380	400	420	440	460	480	500	520	540	560
	°C 15	26	38	49	60	71	82	93	104	116	127	138	149	160	171	182	193	204	216	227	238	249	260	271	282	293
ADMIRALTY BRASS	U																									
ALUMINUM	U																									
ALUMINUM BRONZE	U																									
BRASS, RED	U																									
BRONZE	U																									
CARBON STEEL	S																									
COLUMBIUM (NIOBIUM)	E																									
COPPER	U																									
HASTELLOY B/B-2	G												U													
HASTELLOY C/C-276	E			G																						
HASTELLOY D	G												U													
HASTELLOY G/G-3	E																									
HIGH SILICON IRON	E																									
INCONEL	U																									
INCOLLOY 825	G								U																	
LEAD	E																									
MONEL	U																									
NAVAL BRONZE	U																									
NICKEL	U																									
NI-RESIST																										
STAINLESS STEELS																										
Type 304/347 4	G																									
Type 316 4	G																									
Type 410	U																									
Type 904-L																										
17-4 PH	E		G																							
20 Cb 3	E			G																						
E-Bright 26-1	E																									
SILICON BRONZE	U																									
SILICON COPPER	U																									
STELLITE	U																									
TANTALUM	G												U													
TITANIUM	U																									
ZIRCONIUM	U																									

FOR METALS
E = < 2 Mils Penetration/Year; G = < 20 Mils Penetration /year
S = < 50 Mils Penetration/Year; U = > 50 Mils Penetration/year

FOR NONMETALLICS
R = Resistant
U = Unsatisfactory

SULFURIC ACID 98%

PLASTICS	60	80	100	120	140	160	180	200	220	240	260	280	300	320	340	360	380	400	420	440	460
°C	15	26	38	49	60	71	82	93	104	116	127	138	149	160	171	182	193	204	216	227	238
°F	60	80	100	120	140	160	180	200	220	240	260	280	300	320	340	360	380	400	420	440	460
ABS	U																				
ACETAL	U																				
ACRYLICS	U																				
ASBESTOS REINFORCED EPOXY	U																				
ASBESTOS REINFORCED PHENOLIC	U																				
CHLORINATED POLYETHER (Penton)	U																				
CPVC	U																				
E-CTFE (Halar)	R	—	—	—	—																
EPOXY	U																				
ETFE (Tefzel)	R	—	—	—	—	—	—	—	—	—	—	—	—								
FEP	R	—	—	—	—	—	—	—	—	—	—	—	—	—	—	—	—	—			
FURAN (FURFURAL ALCOHOL)	U																				
NORYL	R	—	—	—	—	—	—														
NYLON 6	U																				
NYLON 11	U																				
NYLON 66	U																				
PHENOLIC	U																				
PFA (Teflon)	R	—	—	—	—	—	—	—													
POLYAMIDE-IMIDE																					
POLYESTERS																					
Bisphenol A–fumarate	U																				
Halogenated	U																				
Hydrogenated Bisphenol A –Bisphenol A	U																				
Isophthalic	U																				
Terephthalate (PET)	U																				
POLYETHERIMIDE (ULTEM)																					
POLYETHERETHERKETONE (PEEK)	U																				
POLYETHERSULFONE (PES)																					
POLYETHYLENE HMW	U																				
POLYETHYLENE UHMW	U																				
POLYMETHYLPENTENE																					
POLYPROPYLENE (PP)	R	—	—	—																	
POLYSTYRENE	U																				
POLYSULFONE	U																				
POLYVINYLIDENE CHLORIDE	U																				
PVC TYPE 1	U																				
PVC TYPE 2	U																				
PVDF (Kynar)	R	—	—	—	—																
RYTON																					
TFE (Teflon)	R	—	—	—	—	—	—	—	—	—	—	—	—	—	—	—	—	—	—	—	—
VINYL ESTER	U																				

°F 60 80 100 120 140 160 180 200 220 240 260 280 300 320 340 360 380 400 420 440 460
°C 15 26 38 49 60 71 82 93 104 116 127 138 149 160 171 182 193 204 216 227 238

SULFURIC ACID 98%

ELASTOMERS* AND LININGS	°C 15	26	38	49	60	71	82	93	104	116	127	138	149	160	171	182	193	204	216	227	238
	°F 60	80	100	120	140	160	180	200	220	240	260	280	300	320	340	360	380	400	420	440	460
ACRYLATE-BUTADIENE (ABR)																					
BUTYL GR-1 (IIR)	U																				
CARBOXYLIC-ACRYLONITRILE-BUTADIENE (XNBR)																					
CHEMRAZ (FPM)	R	—	—	—	—	—	—	—	—	—	—	—	—	—	—	—	—	—	—		
CHLORO-ISOBUTENE-ISOPRENE (CIIR)																					
HYPALON (CHLORO-SULFONYL-POLYETHYLENE (CSM)	R	—	—	U																	
ETHYLENE-ACRYLIC (EA)																					
ETHYLENE-PROPYLENE (EPM)	R	—																			
ETHYLENE-PROPYLENE-DIENE (EPDM)	U																				
ETHYLENE-PROPYLENE TERPOLYMER (EPT)	U																				
FKM (Viton A)	R	—	—	—	—	—	—	—	—	—	—	—	—	—	—	—	—				
HARD RUBBER	U																				
SOFT RUBBER	U																				
ISOPRENE (IR)	U																				
KALREZ (FPM)	R	—	—	—	—	—	—	—													
KOROSEAL	U																				
NATURAL RUBBER (GRS)	U																				
NEOPRENE GR-M (CR)	U																				
NITRILE BUNA-N (NBR)	U																				
NORDEL (EPDM)																					
POLYBUTADIENE (BR)																					
POLYESTER (PE)	U																				
POLYETHER-URETHANE (EU)																					
POLYISOPRENE (IR)																					
POLYSULFIDES (T)	U																				
POLYURETHANE (AU)	U																				
SBR STYRENE (BUNA-S)	U																				
SILICONE RUBBERS	U																				
	°F 60	80	100	120	140	160	180	200	220	240	260	280	300	320	340	360	380	400	420	440	460
	°C 15	26	38	49	60	71	82	93	104	116	127	138	149	160	171	182	193	204	216	227	238

*See also Nylon 11 under PLASTICS

SULFURIC ACID 100%

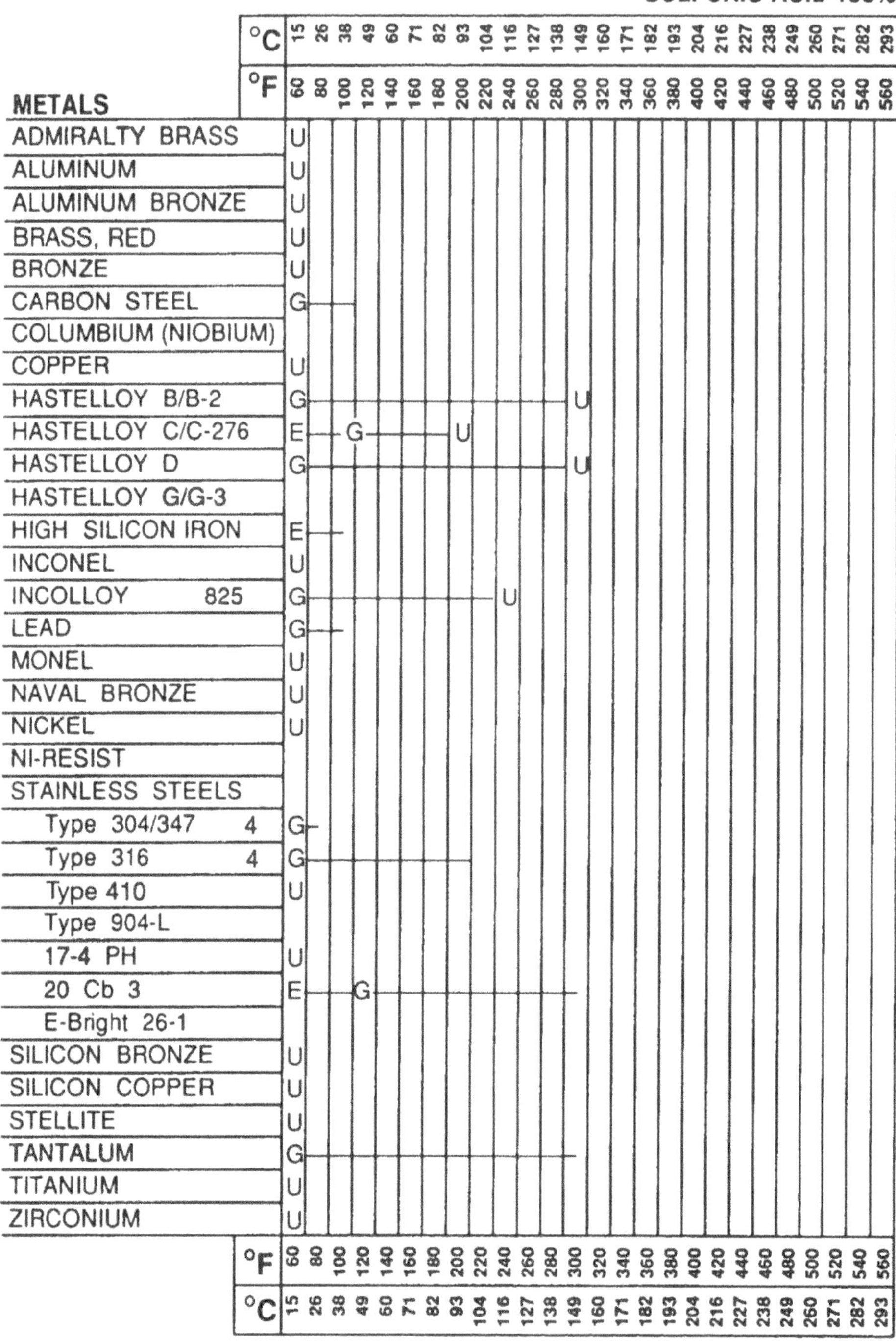

METALS °C	15	26	38	49	60	71	82	93	104	116	127	138	149	160	171	182	193	204	216	227	238	249	260	271	282	293
°F	60	80	100	120	140	160	180	200	220	240	260	280	300	320	340	360	380	400	420	440	460	480	500	520	540	560
ADMIRALTY BRASS	U																									
ALUMINUM	U																									
ALUMINUM BRONZE	U																									
BRASS, RED	U																									
BRONZE	U																									
CARBON STEEL	G																									
COLUMBIUM (NIOBIUM)																										
COPPER	U																									
HASTELLOY B/B-2	G												U													
HASTELLOY C/C-276	E			G				U																		
HASTELLOY D	G												U													
HASTELLOY G/G-3																										
HIGH SILICON IRON	E																									
INCONEL	U																									
INCOLLOY 825	G									U																
LEAD	G																									
MONEL	U																									
NAVAL BRONZE	U																									
NICKEL	U																									
NI-RESIST																										
STAINLESS STEELS																										
Type 304/347 4	G																									
Type 316 4	G																									
Type 410	U																									
Type 904-L																										
17-4 PH	U																									
20 Cb 3	E			G																						
E-Bright 26-1																										
SILICON BRONZE	U																									
SILICON COPPER	U																									
STELLITE	U																									
TANTALUM	G																									
TITANIUM	U																									
ZIRCONIUM	U																									
°F	60	80	100	120	140	160	180	200	220	240	260	280	300	320	340	360	380	400	420	440	460	480	500	520	540	560
°C	15	26	38	49	60	71	82	93	104	116	127	138	149	160	171	182	193	204	216	227	238	249	260	271	282	293

FOR METALS

E = < 2 Mils Penetration/Year; G = < 20 Mils Penetration /year

S = < 50 Mils Penetration/Year; U = > 50 Mils Penetration/year

FOR NONMETALLICS

R = Resistant

U = Unsatisfactory

SULFURIC ACID 100%

PLASTICS																					
°C	15	26	38	49	60	71	82	93	104	116	127	138	149	160	171	182	193	204	216	227	238
°F	60	80	100	120	140	160	180	200	220	240	260	280	300	320	340	360	380	400	420	440	460
ABS	U																				
ACETAL	U																				
ACRYLICS																					
ASBESTOS REINFORCED EPOXY	U																				
ASBESTOS REINFORCED PHENOLIC																					
CHLORINATED POLYETHER (Penton)	U																				
CPVC	U																				
E-CTFE (Halar)	R																				
EPOXY	U																				
ETFE (Tefzel)	R																				
FEP	R																				
FURAN (FURFURAL ALCOHOL)	U																				
NORYL																					
NYLON 6	U																				
NYLON 11	U																				
NYLON 66	U																				
PHENOLIC	U																				
PFA (Teflon)																					
POLYAMIDE-IMIDE																					
POLYESTERS																					
Bisphenol A–fumarate																					
Halogenated																					
Hydrogenated Bisphenol A –Bisphenol A	U																				
Isophthalic	U																				
Terephthalate (PET)	U																				
POLYETHERIMIDE (ULTEM)																					
POLYETHERETHERKETONE (PEEK)	U																				
POLYETHERSULFONE (PES)																					
POLYETHYLENE HMW	U																				
POLYETHYLENE UHMW	U																				
POLYMETHYLPENTENE																					
POLYPROPYLENE (PP)	U																				
POLYSTYRENE	U																				
POLYSULFONE	U																				
POLYVINYLIDENE CHLORIDE	U																				
PVC TYPE 1	U																				
PVC TYPE 2	U																				
PVDF (Kynar)	U																				
RYTON																					
TFE (Teflon)	R																				
VINYL ESTER	U																				
°F	60	80	100	120	140	160	180	200	220	240	260	280	300	320	340	360	380	400	420	440	460
°C	15	26	38	49	60	71	82	93	104	116	127	138	149	160	171	182	193	204	216	227	238

SULFURIC ACID 100%

ELASTOMERS* AND LININGS	°C	15	26	38	49	60	71	82	93	104	116	127	138	149	160	171	182	193	204	216	227	238
	°F	60	80	100	120	140	160	180	200	220	240	260	280	300	320	340	360	380	400	420	440	460
ACRYLATE-BUTADIENE (ABR)																						
BUTYL GR-1 (IIR)		U																				
CARBOXYLIC-ACRYLONITRILE-BUTADIENE (XNBR)																						
CHEMRAZ (FPM)		R	—	—	—	—	—	—	—	—	—	—	—	—	—	—	—	—	—	—		
CHLORO-ISOBUTENE-ISOPRENE (CIIR)		U																				
HYPALON (CHLORO-SULFONYL-POLYETHYLENE (CSM)		U																				
ETHYLENE-ACRYLIC (EA)																						
ETHYLENE-PROPYLENE (EPM)		R–																				
ETHYLENE-PROPYLENE-DIENE (EPDM)		U																				
ETHYLENE-PROPYLENE TERPOLYMER (EPT)		U																				
FKM (Viton A)		R	—	—	—	—	—	—														
HARD RUBBER		U																				
SOFT RUBBER		U																				
ISOPRENE (IR)		U																				
KALREZ (FPM)																						
KOROSEAL		U																				
NATURAL RUBBER (GRS)		U																				
NEOPRENE GR-M (CR)		U																				
NITRILE BUNA-N (NBR)		U																				
NORDEL (EPDM)																						
POLYBUTADIENE (BR)																						
POLYESTER (PE)		U																				
POLYETHER-URETHANE (EU)																						
POLYISOPRENE (IR)																						
POLYSULFIDES (T)		U																				
POLYURETHANE (AU)		U																				
SBR STYRENE (BUNA-S)		U																				
SILICONE RUBBERS		U																				
	°F	60	80	100	120	140	160	180	200	220	240	260	280	300	320	340	360	380	400	420	440	460
	°C	15	26	38	49	60	71	82	93	104	116	127	138	149	160	171	182	193	204	216	227	238

*See also Nylon 11 under PLASTICS

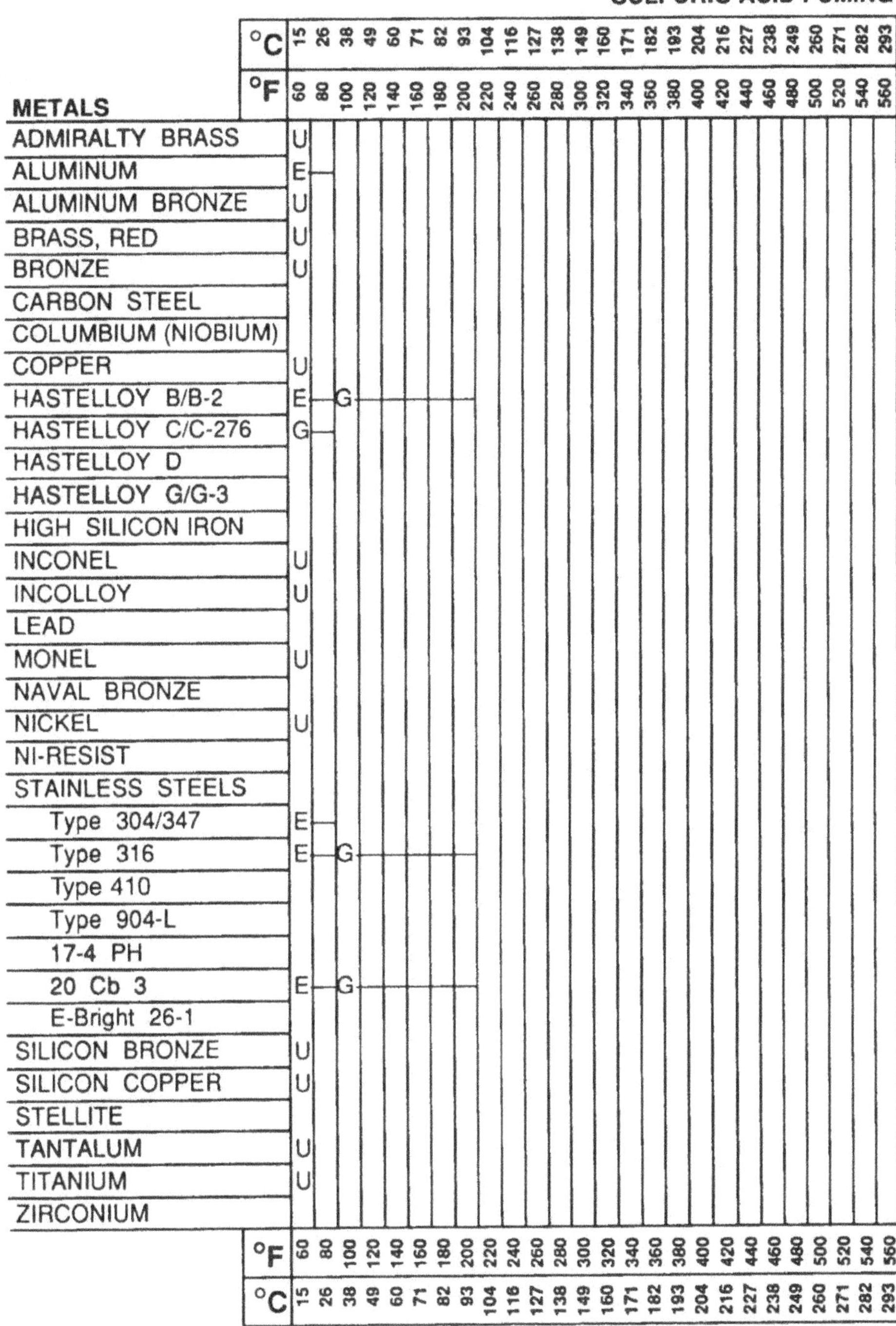

SULFURIC ACID FUMING

°C	15	26	38	49	60	71	82	93	104	116	127	138	149	160	171	182	193	204	216	227	238	249	260	271	282	293
METALS °F	60	80	100	120	140	160	180	200	220	240	260	280	300	320	340	360	380	400	420	440	460	480	500	520	540	560
ADMIRALTY BRASS	U																									
ALUMINUM	E																									
ALUMINUM BRONZE	U																									
BRASS, RED	U																									
BRONZE	U																									
CARBON STEEL																										
COLUMBIUM (NIOBIUM)																										
COPPER	U																									
HASTELLOY B/B-2	E		G																							
HASTELLOY C/C-276	G																									
HASTELLOY D																										
HASTELLOY G/G-3																										
HIGH SILICON IRON																										
INCONEL	U																									
INCOLLOY	U																									
LEAD																										
MONEL	U																									
NAVAL BRONZE																										
NICKEL	U																									
NI-RESIST																										
STAINLESS STEELS																										
Type 304/347	E																									
Type 316	E		G																							
Type 410																										
Type 904-L																										
17-4 PH																										
20 Cb 3	E		G																							
E-Bright 26-1																										
SILICON BRONZE	U																									
SILICON COPPER	U																									
STELLITE																										
TANTALUM	U																									
TITANIUM	U																									
ZIRCONIUM																										
°F	60	80	100	120	140	160	180	200	220	240	260	280	300	320	340	360	380	400	420	440	460	480	500	520	540	560
°C	15	26	38	49	60	71	82	93	104	116	127	138	149	160	171	182	193	204	216	227	238	249	260	271	282	293

FOR METALS

E = < 2 Mils Penetration/Year; G = < 20 Mils Penetration /year

S = < 50 Mils Penetration/Year; U = > 50 Mils Penetration/year

FOR NONMETALLICS

R = Resistant

U = Unsatisfactory

PLASTICS	°C 15	26	38	49	60	71	82	93	104	116	127	138	149	160	171	182	193	204	216	227	238
	°F 60	80	100	120	140	160	180	200	220	240	260	280	300	320	340	360	380	400	420	440	460
ABS	U																				
ACETAL	U																				
ACRYLICS	U																				
ASBESTOS REINFORCED EPOXY																					
ASBESTOS REINFORCED PHENOLIC																					
CHLORINATED POLYETHER (Penton)																					
CPVC	R	U																			
E-CTFE (Halar)	R																				
EPOXY	U																				
ETFE (Tefzel)	R																				
FEP	R																				
FURAN (FURFURAL ALCOHOL)	U																				
NORYL	R																				
NYLON 6	U																				
NYLON 11	U																				
NYLON 66	U																				
PHENOLIC																					
PFA (Teflon)	R																				
POLYAMIDE-IMIDE	R																				
POLYESTERS																					
Bisphenol A–fumarate																					
Halogenated																					
Hydrogenated Bisphenol A –Bisphenol A																					
Isophthalic	U																				
Terephthalate (PET)																					
POLYETHERIMIDE (ULTEM)	U																				
POLYETHERETHERKETONE (PEEK)	U																				
POLYETHERSULFONE (PES)																					
POLYETHYLENE HMW	U																				
POLYETHYLENE UHMW	U																				
POLYMETHYLPENTENE																					
POLYPROPYLENE (PP)	U																				
POLYSTYRENE	U																				
POLYSULFONE	U																				
POLYVINYLIDENE CHLORIDE	U																				
PVC TYPE 1	U																				
PVC TYPE 2	U																				
PVDF (Kynar)	U																				
RYTON	R																				
TFE (Teflon)	R																				
VINYL ESTER	U																				
	°F 60	80	100	120	140	160	180	200	220	240	260	280	300	320	340	360	380	400	420	440	460
	°C 15	26	38	49	60	71	82	93	104	116	127	138	149	160	171	182	193	204	216	227	238

ELASTOMERS* AND LININGS	60	80	100	120	140	160	180	200	220	240	260	280	300	320	340	360	380	400	420	440	460
°C	15	26	38	49	60	71	82	93	104	116	127	138	149	160	171	182	193	204	216	227	238
ACRYLATE-BUTADIENE (ABR)																					
BUTYL GR-1 (IIR)																					
CARBOXYLIC-ACRYLONITRILE-BUTADIENE (XNBR)																					
CHEMRAZ (FPM)	R	—	—	—	—	—	—	—	—	—	—	—	—	—	—	—	—	—			
CHLORO-ISOBUTENE-ISOPRENE (CIIR)																					
HYPALON (CHLORO-SULFONYL-POLYETHYLENE (CSM)	U																				
ETHYLENE-ACRYLIC (EA)																					
ETHYLENE-PROPYLENE (EPM)																					
ETHYLENE-PROPYLENE-DIENE (EPDM)	U																				
ETHYLENE-PROPYLENE TERPOLYMER (EPT)																					
FKM (Viton A)	R	—	—	—	—	—	—	—													
HARD RUBBER																					
SOFT RUBBER																					
ISOPRENE (IR)																					
KALREZ (FPM)	R	—	—	—	—	—	—	—													
KOROSEAL	R	—	—	—	—																
NATURAL RUBBER (GRS)																					
NEOPRENE GR-M (CR)	U																				
NITRILE BUNA-N (NBR)	U																				
NORDEL (EPDM)																					
POLYBUTADIENE (BR)																					
POLYESTER (PE)	U																				
POLYETHER-URETHANE (EU)																					
POLYISOPRENE (IR)																					
POLYSULFIDES (T)	U																				
POLYURETHANE (AU)																					
SBR STYRENE (BUNA-S)	U																				
SILICONE RUBBERS	U																				
°F	60	80	100	120	140	160	180	200	220	240	260	280	300	320	340	360	380	400	420	440	460
°C	15	26	38	49	60	71	82	93	104	116	127	138	149	160	171	182	193	204	216	227	238

*See also Nylon 11 under PLASTICS

REFERENCE

Schweitzer, Philip A., *Corrosion Resistance Tables, Metals, Nonmetals, Coatings, Mortars, Plastics, Elastomers and Linings, and Fabrics*, Fourth Edition, Parts A–C, Marcel Dekker, Inc., New York, (1995).

Appendix B
Conversion Units

Length

1 m = 3.28 ft
1 cm = 2.54 in.
1 mm = 0.0394 in.
1 km = 0.622 mile

1 Å = 10^{-8} cm
1 μm = 10^{-6}m
1 nm = 10^{-9} m
1 mil = 0.0254 mm

Area

1 m^2 = 10.76 ft^2
1 cm^2 = 0.155 ft^2

1 yd^2 = 0.837 m^2
1 ft^2 = 0.093 m^2

Volume

1 cm^3 = 0.0610 $in.^3$
1 m^3 = 35.32 $ft.^3$

1 $in.^3$ = 16.39 cm^3
1 $ft.^3$ = 0.0283 m^3

Mass

1 kg = 2.205 lb
1 g = 2.205 × 10^{-3} lb
1 lb = 453.6 g

1 oz = 28.34 g
1 troy oz = 31.1 g

Density

1 kg/cm^3 = 0.0624 lb_m/ft^3
1 g/cm^3 = 0.0361 $lb_m/in.^3$

1 lb_m/ft^3 = 16.02 kg/m^3
1 $lb_m/in.^3$ = 27.7 g/m^3

Force

1 N = 0.2248 lb_f
1 N = 10^5 dyn

1 lb_f = 4.44 N
1 dyn = 10^{-5} N

Current

1 A = 1 C/s

1 C = 3 × 10^9 esu of charge

Stress

1 N/m^2 = 1 Pa
1 kg/m^2 = 9.8 Pa
1 g/mm^2 = 1.422 psi
1 dyn/cm^2 = 0.10 Pa

1 Pa = 0.145 × 10^{-3} psi
1 ksi = 6.89 MPa
10^6 psi = 6.89 GPa
1 psi = 0.703 g/mm^2

Viscosity

1 poise = 1 g/cm · s

Impact Energy

1 ft-lb$_f$ = 1.355 J; divide net value of energy by specimen width to obtain J/m or ft-lb/in.
1 ft-lb/in. = 1.355 J/m

Fracture Toughness

1 MPa $\sqrt{m}$ = 0.91 ksi $\sqrt{in.}$

1 ksi $\sqrt{in.}$ = 1.1 MPa $\sqrt{m}$

Energy

1 J = 1 N · m
1 cal = 4.18 J
1 cal = 3.97 × 10^{-3} Btu
1 J = 0.738 ft-lb$_f$

1 eV = 1.6 × 10^{-19} J
1 Btu = 1054 J
1 erg = 10^{-7} J
1 Btu = 252 cal

Power

1 W = 1 J/s

1 cal/s = 14.29 Btu/hr

Thermal Conductivity

1 W/m · K = 2.39 × 10^{-3} cal/g · K
1 Btu/lb$_m$ · °F = 1.0 cal/g · K
1 cal/cm · s · K = 24.8 Btu/ft · hr · °F

Specific Heat

1 cal/g · K = 1.0 Btu/lb$_m$ · °F

1 cal/g · K = 4.184 J/g · K

Appendix C
Material Costs

C.1.1 Composites

Material	Form	Price (Jan. 1992) ($/lb)*
Boron filament–epoxy	Impregnated tape	~400
Boron filament–Al matrix	Impregnated tape	400–500
SiC filament–Al 6061	Impregnated tape	Not quoted > boron–Al
SiC particulate 20% vol.–Al 6061	Ingot	3.30
Al_2O_3 particulate 10% vol.–Al 6061	Ingot	3.15
Kelvar-reinforced epoxy	Molding type	50.00
Glass-fiber-reinforced epoxy	Varies	5.00

*Depends on form and quantity; prices above are for small quantities.

C.1.2 Polymers

Material	Form	Price (Jan. 1992) ($/lb)*
Polyethylene HD	0.5-in.-thick sheet	1.73†
Polycarbonate	0.5-in.-diameter rod	15.00†
PVC	0.5-in. sheet	3.60†
Polycarbonate + 30% glass fibers	0.5 in. rod	37.00†
Acrylics	0.125 in. sheet	3.00†
Polyethylene HD resin	Raw material	0.40‡
Polystyrene resin	Raw material	0.30‡
PET resin	Raw material	0.60‡
Polyamide resin	Raw material	1.75‡
PEEK	Raw material	25.00‡

*Depends on form and quantity.
†Price for small quantity.
‡Price for tonnage quantity.

Relative Cost of Polymers Based on Density (Polyethylene = unity)

Polyethylene HD	1.0	Nylon 66 30% glass filled	7.2
Polyproplyene	1.1	Nylon 66	6.6
Polystyrene	1.7	Nylon 6	4.9
Polystyrene 30% glass filled	1.8	Polycarbonate	5.6
PVC	1.6	PET 30% glass filled	6.8
ABS	2.4	Polysulfone 30% glass filled	15.6
ABS 30% glass	3.4	Polyamide-imide	59.2
Acrylic	3.6	PEEK	88.0

C.1.3 Metals

Material	Form	Price (Jan. 1992) ($/lb)*
Low-alloy steel	Plate	0.28
Plain carbon steel	Hot-rolled sheet	0.28
Al alloy 384	Ingot	0.59
Al alloy 319	Ingot	0.62
2024 Al T6	Plate	1.75
7075 Al T6	Plate	1.75
304 Stainless steel	Coiled sheet	1.47
316 Stainless steel	Coiled sheet	1.99
310 Stainless steel	Coiled sheet	3.10
304 Stainless steel	Plate	1.32
316 Stainless steel	Plate	1.73
310 Stainless steel	Plate	3.03
410 Martensitic stainless steel	Bar	1.50
17-4 PH	Bar	1.80–2.00
Copper	Wire-bar	1.10
Cu-Be alloy	Strip	8.40
Cu--Zn (brass)	Rod	2.29
Nickel 200 (high purity)	Rod	14.00
Nickel	Melt stock	4.00
Chromium	Electrolytic	4.62
Titanium	Sponge	4.40
Zinc	As refined	0.29
Vanadium	As refined	9.50
Zirconium	—	11.00–17.00
Aluminum	Electrolytic	0.65
Ti-6% Al-4% V	Plate	8.00
Monel 400	—	5.00
Monel K500	Rod	10.00
Tin	Melt stock	2.43
Magnesium	—	1.43
Lead	—	0.35
Cobalt	Cathode	12.50

*100- to 1000-lb quantities.

Index

Milton Keynes UK
Ingram Content Group UK Ltd.
UKHW052029071024
449327UK00027B/2491

9 780367 400927